The Geological Society of America, Inc.
Memoir 157

Tectonic and Stratigraphic Studies in the Eastern Great Basin

Edited by

David M. Miller
Victoria R. Todd
Keith A. Howard

1983

Published by The Geological Society of America, Inc.
3300 Penrose Place, P.O. Box 9140, Boulder, Colorado 80301

Printed in U.S.A.

Library of Congress Cataloging in Publication Data
Main entry under title:

Tectonic and stratigraphic studies in the eastern Great
    Basin.

    (Memoir / Geological Society of America ; 157)
    "Compiled as a tribute to Max D. Crittenden, Jr."—
Introd.
    Includes bibliographies.
    1. Geology, Structural—Addresses, essays, lectures.
2. Geology—Great Basin—Addresses, essays, lectures.
3. Crittenden, Max D. I. Miller, David M. (David
MacArthur), 1951-   . II. Todd, Victoria R.
(Victoria Roy) III. Howard, Keith A. IV. Crittenden,
Max D. V. Geological Society of America. VI. Series:
Memoir (Geological Society of America) ; 157.
QE601.T38   1983   551.8'0979   83-5632
ISBN 0-8137-1157-6

Contents

To Max D. Crittenden, Jr.

This volume is dedicated to Max D. Crittenden, Jr., in recognition of his long and continuing influence on geologic understanding of the eastern Great Basin and Wasatch Mountain regions. His encouragement to others, dedication to his own work, and wide-ranging contributions have all led directly to the present geologic framework of the region.

Max began his studies in the eastern Great Basin in 1942 as a member of the U.S. Geological Survey, first in the Drum Mountains under D. Foster Hewett and later in the Wasatch Mountains under the guidance of Frank Calkins and A. A. Baker. He earned his doctorate at the University of California, Berkeley, in 1949, after which he continued his studies in the Wasatch Range with the Survey.

At the time Max began his Wasatch studies, low-angle faults were poorly understood; their presence was even denied by many geologists working in the region despite the earlier documentation of such faults farther to the north. Max's recognition of thrust faults in the northern Wasatch Mountains led to his collaboration with Baker and began a lifelong interest in thrust-fault tectonics. Max and his coworkers demonstrated not only that thrust faults are major structural features of the region but also that the thrust sheets have characteristic geometries and juxtapose contrasting stratigraphic sequences derived from widely separated parts of the Cordillera. Max's detailed studies of structure and of Precambrian and Paleozoic stratigraphy helped erect the foundation for our present knowledge of the complex thrust relations in the Idaho-Wyoming-Utah region, and proved essential to the recognition of the Mesozoic foreland thrust belt as an essentially continuous feature of the Cordillera.

As the geometry and timing of thrust-belt evolution was clarified, the genesis and source regions for these great thrust sheets continued to be enigmatic. Following the lead of Peter Misch and his students in their tectonic studies in Nevada west of the foreland thrust belt, Max began to focus attention on hinterland structures, leading him to collaborate with Ralph Roberts on a model of gravity-propelled thrust sheets for the origin of the thrust belt. Max began promoting studies of the tectonics of northwestern Utah, including the Raft River–Grouse Creek–Albion met-

amorphic core complex. A Penrose Conference on Cordilleran metamorphic core complexes that he helped convene in 1977 successfully assembled workers from diverse segments of the Cordillera to appraise the possible links between hinterland and foreland structures. As a consequence, the overprinting of the Nevada-Utah hinterland by Tertiary plutonism, metamorphism, and low-angle faulting was brought into sharper focus. Since then, Max has continued his effort to unravel the Mesozoic and Cenozoic tectonics of the hinterland through mapping in the Promontory Range, Utah, and through his informal coordination of studies in northwestern Utah by U.S. Geological Survey, state, and university geologists.

Max has always been inquisitive and versatile. He has embraced topics ranging from isostatic rebound, paleomagnetism, and geochronology, to the gamut of field geologic studies: Quaternary studies, Precambrian and Phanerozoic stratigraphy, Precambrian glaciogenic deposits, tectonics, igneous and metamorphic petrology, mineralization, geothermal and hydrocarbon resources, and geologic hazards evaluation. His studies of Precambrian glaciogenic deposits gained him international recognition and improved the understanding of Precambrian stratigraphy throughout western North America. In another landmark study, he used isostatic rebound of Pleistocene Lake Bonneville shorelines to interpret mantle and crustal rheology. Also, Max participated in the U.S. Geological Survey's strategic minerals program during and after World War II under D. Foster Hewett, and in time he became one of the nation's leading authorities on the geology of manganese deposits, serving for many years as the manganese commodity geologist for the U.S. Government.

As chief of the Southwestern Branch of Regional Geology from 1965 to 1969, Max helped shape the U.S. Geological Survey's regional mapping program. Moreover, his encouragement and support of collaborative studies with state and university geologists greatly enhanced the scope as well as professional capability of many Survey programs. Max also served as chairman of the U.S. Geothermal Environmental Advisory Panel from 1974 to 1978. In recognition of his outstanding achievements, he was presented in 1977 the Meritorious Service Award by the U.S. Department of the Interior. He is an Associate Editor for the *Bulletin* of the Geological Society of America and has furthered the objectives of this and other professional organizations by serving on diverse committees and working groups. He maintains a strong interest in scientific education and has taught courses at the University of California, Santa Cruz.

Max Crittenden's influence results not only from his incisive geologic concepts but also from his talents in communication and as a stimulating teacher. His reports are models of brevity, clarity, and organization; his oral presentations are articulate; his broadminded pursuit of science is laudatory; and his open exchange of ideas and data is enviable. The diverse papers in this volume sample Max's breadth of interest and demonstrate his impact on geologic thinking. He is a model to those who have had the honor of working with him and who are trying to emulate his wealth of knowledge and to apply his methods and thought processes.

This volume is a token of appreciation on the part of the contributors and many others who have been and continue to be inspired by Max D. Crittenden, Jr.

Geological Society of America
Memoir 157
1983

Introduction

David M. Miller
Keith A. Howard
U.S. Geological Survey
345 Middlefield Road
Menlo Park, California 94025

Tectonic and Stratigraphic Studies in the Eastern Great Basin is compiled as a tribute to Max D. Crittenden, Jr., whose geologic leadership and exemplary studies in the eastern Great Basin region have stimulated significant scientific advances. Many of the authors in this volume conducted their studies with assistance from Max, and much of the foundation for these studies is his work on the foreland thrust belt in the Wasatch Mountains of Utah.

The part of the eastern Great Basin and adjoining Rocky Mountains that is the subject of this volume includes central and northern Utah, eastern Nevada, and southern Idaho. The region is geologically diverse; it contains rocks from Archean, Proterozoic, and every Phanerozoic system, and complex structures that include a major Mesozoic overthrust belt, low- and high-angle extensional faults, and metamorphic core complexes (Armstrong, 1968, 1972; Hintze, 1973; Stewart, 1980). With the exception of studies concerning metamorphic core complexes (*in* Crittenden and others, 1980), few published studies in this region have been compiled in an easily accessible form. This volume is a collection of many recently completed studies and makes a state-of-the-art perspective available to the interested student of Great Basin geology.

The papers in this volume range widely in both geologic specialty and geographic location. The chief subject is tectonics; the authors address tectonic problems by means of a variety of tools, such as structural analysis, stratigraphy, geochronology, and geophysics. To emphasize continuity of stratigraphy and structure, the papers are ordered geographically, beginning with the Wasatch front and progressing westward.

The Wasatch front, a discontinuous physiographic zone that follows the western side of the Wasatch Mountains and continues southward along the western margin of the Colorado Plateau to southern Nevada, has marked fundamental tectonic boundaries in every era since the Proterozoic. Near the Wasatch front, thick miogeoclinal strata of Proterozoic and Paleozoic age are thrust over thin cratonal sections along Mesozoic foreland thrusts of the Sevier orogenic belt. This same zone is the approximate eastern limit of pronounced Cenozoic crustal extension that is manifested as the Basin and Range province. The hinterland of the Sevier belt lies westward and northwestward of the Wasatch front, well within the Basin and Range province. In central Nevada, the hinterland is superposed Paleozoic and Mesozoic orogenic belts. In parts of the Basin and Range province, Cenozoic extension—particularly where associated with metamorphic core complexes—was pronounced; it was accomplished by complex low- and high-angle faults.

The modern Wasatch front is a dramatic physiographic break between the Rocky Mountains and Colorado Plateaus provinces and the Basin and Range province, but its geophysical signature is one of gradual transition. The origin and nature of the deep crustal transition between the provinces are problematic. To address these problems, Zoback examines the mechanisms of uplift by means of detailed seismological and other geophysical studies, and Naeser and others investigate the uplift history by means of fission-track dating.

The foreland thrust belt in the Wasatch front area has been studied in some detail, but irregularities in the thrusts and uncertainties in the nature of the transition from foreland structures (which place older rocks over younger) to hinterland structures (which commonly place younger over older) remain controversial. Some or all of the younger-over-older faults may be Cenozoic. A long-standing question has been whether foreland thrusts root immediately west of the foreland thrust belt, or sole into miogeoclinal sections in the hinterland, or surface westward as the rear part of gravity slide sheets. Stratigraphic and structural assemblages defining thrust sheets of the foreland belt, and the geometry of these thrusts, are described by Tooker, by Morris, by Bohannon, by Christie-Blick, and by Allmendinger and Platt. The last three papers also deal specifically with the geometry and age of younger-over-older (extensional) faults in the transition between the hinterland and foreland, providing new perspectives on this important zone. Witkind describes complexities in the foreland thrusts that he attributes to salt diapirism.

Knowledge of the sedimentology, stratigraphy, and geochronology of Precambrian and Paleozoic rocks is necessary for interpreting the tectonic environments during deposition and the subsequent tectonic displacements. Hedge and others examine the age of crystalline basement in the Wasatch Range. Papers by Link and by Miller describe early deposits of the Cordilleran miogeocline that contrain models of its tectonic development. A major problem in northern Utah is to define the shifting basins and source areas of the Pennsylvanian and Permian. Detailed stratigraphic studies, such as those by Stevens and Armin and by Mytton and others, are required to identify the extent and nature of these basins; these studies provide the paleogeographic data necessary for an estimate of structural duplications, of the sort described by Allmendinger and Platt.

Structural events in the hinterland, particularly low-angle faulting, are much harder to date than those in the foreland because synorogenic deposits are commonly lacking. The only absolute dating is by geochronology of igneous and metamorphic rocks. Moore and McKee provide much new geochronologic data from plutons and volcanic rocks in central and western Utah that are useful for constraining tectonic events.

Metamorphic core complexes in the hinterland present a host of unresolved problems. In the Raft River–Grouse Creek–Albion Mountains metamorphic terrane, major questions include the ages and affinities of metamorphosed strata and the relations of the unmetamorphosed rocks surrounding the terrane to the metamorphic rocks within it. Miller addresses correlations of metasedimentary rocks and explores the tectonic consequences of these correlations for the terrane. Jordan describes a section of Paleozoic strata at the margin of the terrane and discusses its relations to the metamorphic strata. Papers by Todd, by Compton, and by Covington consider relations among and tectonic implications of low-angle fault slices of upper Paleozoic, Mesozoic, and Cenozoic strata faulted against the margins of the metamorphic terrane.

Northeastern Nevada contains multiple generations of Mesozoic thrusts that utilized and displaced Paleozoic thrusts, resulting in a complicated array of disparate thrusts and stratigraphic sequences. Coats and Riva describe these complicated relationships and discuss orogenic models to account for the distribution of thrust faults.

This volume includes papers containing abundant new data and interpretations, as well as summary papers capping topics of long-continued study. It is a suitable tribute to a man who helped shape the course of study in the region; we believe that papers included here will serve as a cornerstone for future grand syntheses of the region's geology.

REFERENCES CITED

Armstrong, R. L., 1968, Sevier orogenic belt in Nevada and Utah: Geological Society of America Bulletin, v. 79, p. 429–458.

Armstrong, R. L., 1972, Low-angle (denudation) faults, hinterland of the Sevier orogenic belt, eastern Nevada and western Utah: Geological Society of America Bulletin, v. 83, p. 1729–1754.

Crittenden, M. D., Jr., Coney, P. J., and Davis, G. H., 1980, eds., Cordilleran metamorphic core complexes: Geological Society of America Memoir 153, 490 p.

Hintze, L. F., 1973, Geologic history of Utah: Brigham Young University Geology Studies, v. 20, pt. 3, Studies for Students No. 8, 181 p.

Stewart, J. H., 1980, Geology of Nevada: Nevada Bureau of Mines and Geology Special Publication 4, 136 p.

Manuscript Accepted by the Society August 20, 1982

Geological Society of America
Memoir 157
1983

Structure and Cenozoic tectonism
along the Wasatch fault zone, Utah

Mary Lou Zoback
U.S. Geological Survey
345 Middlefield Road
Menlo Park, California 94025

ABSTRACT

Geology, gravity, and earthquake data are examined in a 200-km-wide zone along the Wasatch fault zone, the eastern boundary of the Basin and Range province in northern Utah. Gravity data define the geometry of crustal segmentation responsible for the major basin-and-range blocks. Locally, basin fill is more than 4 km thick, and minimum vertical offsets across the major normal fault zones vary between 3 and more than 5 km. East-west-trending transverse zones truncate individual basin-and-range blocks. Some zones are bounded by faults; along other zones the margins are gently downwarped. It has been suggested that the transverse zones delimit distinct, independent fault segments; however, at least one example of a major continuous zone of Quaternary scarps cross-cutting one of these zones has been identified. Preexisting structure apparently accounts for some of the transverse zones.

Major west-dipping normal fault zones and gently east-tilted range blocks coupled with widespread major Mesozoic thrust faults dipping gently to the west throughout the region favor a listric-fault or tilted-block model to account for the tectonic extension characterizing the area. However, simple models of extension are complicated by major east-dipping fault zones identified from the gravity data and by earthquake focal mechanisms that appear to rule out seismic slip on low-angle faults (dips $< 30°$). In addition, earthquake focal depths suggest that at least some faulting penetrates to depths within the Precambrian crystalline basement (depths greater than ~10 km).

Correlations between earthquakes, surface faults, and even major fault zones identified by the gravity data are poor. Well-studied earthquake sequences indicate that the major seismically active fault planes at depth trend obliquely to the surface trend of basin-range blocks and the faults that bound them.

Current extensional deformation across the region can be accounted for by an east-west to west-southwest–east-northeast least principal stress/strain axis. Analyses of slickenside data along major fault zones with young activity show a similar pattern of deformation extending back possibly 0.5 to 1.0 m.y. The east-west to west-southwest–east-northeast least principal stress/strain axis in northern Utah represents an approximately 30° counterclockwise rotation of deformational axes relative to most of the rest of the Basin and Range province.

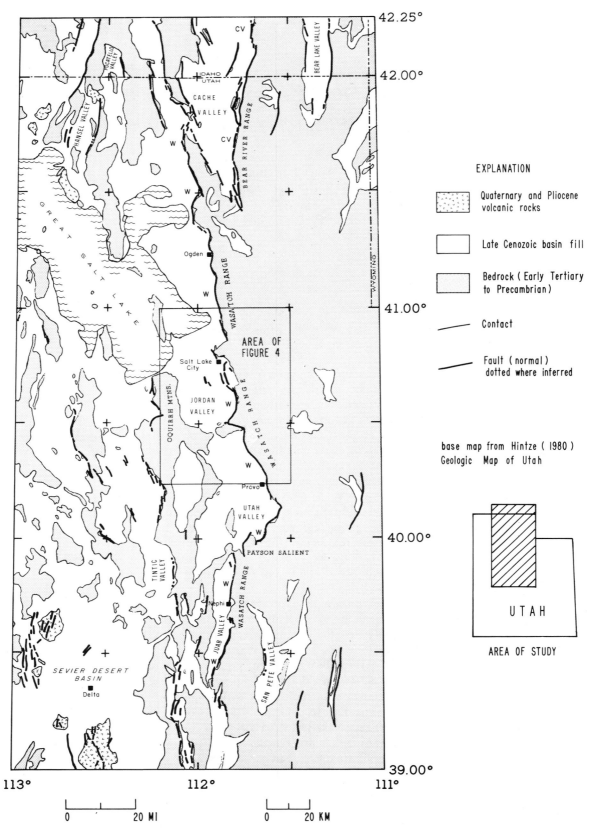

Figure 1. Generalized geologic map of the Wasatch front region, north-central Utah. Wasatch (W) and Cache Valley (CV) fault zones are indicated.

INTRODUCTION

The upper crust of the Basin and Range province in Nevada and Utah has been segmented and extended by late Cenozoic normal faulting. While the timing of initiation of basin-range structure in northern Utah remains obscure, the style of extensional tectonism has formed relative uplifted blocks separated by downdropped basin blocks filled with as much as 4.0 km of late Cenozoic sediments. Individual basin-and-range blocks are typically 10 to 15 km wide, and total vertical relief across the major fault zones locally exceeds 5 km.

The eastern structural boundary of the Basin and Range province in central and southern Utah generally coincides with the boundary between the intensely deformed eastern edge of the Paleozoic miogeocline in the Basin and Range and the relatively stable cratonic block of the Colorado Plateau. In northern Utah the physiographic boundary between the actively extending Basin and Range and the Rocky Mountain province is generally considered to be the north-south-trending Wasatch and Cache Valley fault zones, together referred to as the Wasatch front (Fig. 1). The 200-km-wide region straddling the Wasatch front from lat. 39°N to 42.25°N and from long. 111°W to 113°W constitutes the area of this report. Geophysical evidence (including heat flow, earthquake, and magnetic data) suggests that this boundary is not sharp but rather a broad zone of transition extending 50 to 100 km into the Colorado Plateau/Rocky Mountain provinces proper (Shuey and others, 1973; Thompson and Zoback, 1979). The eastern boundary of the Basin and Range province in Utah lies within the intermountain seismic belt defined by Smith and Sbar (1974) and is one of two active seismic zones within the province. Many Quaternary fault scarps along the Wasatch fault zone suggest frequent seismic activity during the past 2 m.y. (Cluff and others, 1975; Swan and others, 1980).

In this paper, kinematic data, including studies of earthquake focal mechanisms, fault slip (slickenside) measurements, and geodetic strain measurements, are examined and together with gravity data help to clarify the pattern and style of the current deformation in the region. The pattern of slip on normal faults and interaction of the major fault blocks are discussed and compared with seismic constraints on the subsurface geometry of faulting. The influence of preexisting structural grain on the late Cenozoic extensional segmentation of the crust is also examined.

DELINEATION OF BASIN AND RANGE BLOCKS USING GRAVITY DATA

The complete Bouguer gravity map for the study area shown in Figure 2 is based on U.S. Geological Survey data files (T. G. Hildenbrand, 1981, written commun.) compiled as part of Department of Energy Radioactive Waste Isolation, Task I program contracted to the USGS. Terrain corrections were computed for a radial zone 0.895 to 167 km using digital terrain tapes (30-s digitization level) produced by the Defense Mapping Agency. Shown superimposed on the complete Bouguer map are the bedrock-alluvium contacts and the major mapped normal faults after Hintze (1980).

In general, the gravity values decrease eastward toward the regionally high mountainous areas forming the western margin of the Colorado Plateau/Rocky Mountains provinces. The trend of decreasing Bouguer anomalies (decrease of 70 to 90 mgal) corresponds to an increase in mean elevation (800 to 900 m, Diment and Urban, 1981) across the region and suggests that the bedrock highlands east of the Wasatch fault zone are isostatically compensated. The exact nature of this compensation is complex because crustal, upper mantle, and lithospheric-level structural and density changes are occurring along the Great Basin–Colorado Plateau/Rocky Mountains transition in this region (Thompson and Zoback, 1979). Crustal thickness alone varies from about 25 km beneath the Basin and Range province in the western part of the study area to about 40 km beneath the Colorado Plateau province in the eastern part of the study area (Smith, 1978).

The late Cenozoic basins generally are well defined by gravity lows, indicating low-density alluvial fill. In fact, most of the linear steep gravity gradients shown on the map indicate both mapped and inferred major basin-bounding fault zones. However, some of the variation in gravity values is due to variations in bedrock density, such as thick sequences of low-density Tertiary volcanic rocks.

Figure 3 delineates the major basin-and-range blocks in the region on the basis of the geologic and gravity data. With the exception of the San Pete Valley in the south and Bear Lake Valley to the north (Fig. 1), all major basins lie west of the Wasatch and Cache Valley fault zones. Although normal faulting and small basins containing late Cenozoic sediments have been recognized east of these fault zones, the lack of major gravity lows (> 10 mgal) associated with these basins demonstrates that these so-called back basins do not contain a significant amount of fill, which suggests that the magnitude of the basin-related fault offsets in this region is small. The one exception to this general pattern is Ogden Valley, located due east of Ogden (Fig. 1). Gravity modeling by Stewart (1958) indicated up to 1.8 km of fill in this valley (see Fig. 6). Located just south of Cache Valley, near the junction of the Cache Valley fault zone and the main Wasatch fault zone (Fig. 1), Ogden Valley may thus be considered to lie within a broad fault zone rather than to be a true back basin.

Despite the fact that most mapped range-bounding faults and zones of Quaternary to Holocene faulting in the region occur on the west sides of the ranges and dip westward, the gravity data indicate that many of the basins west

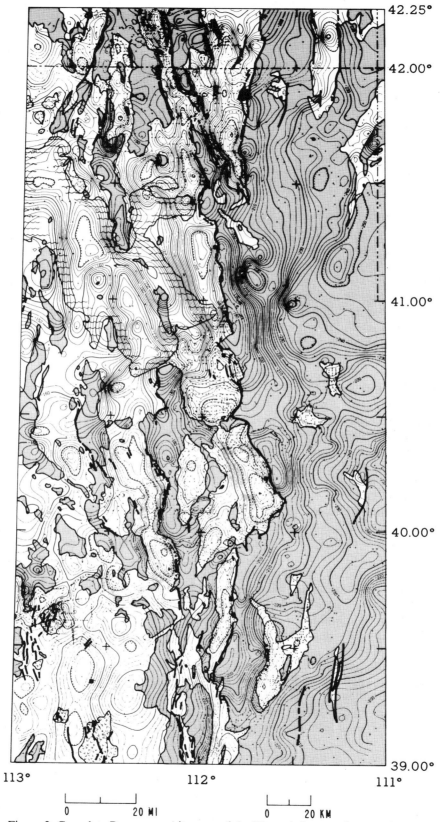

Figure 2. Complete Bouguer gravity map of the Wasatch front region, north-central Utah. Gravity stations are indicated by crosses. Contour interval 2.5 mgal. See Figure 1 for explanation of map symbols.

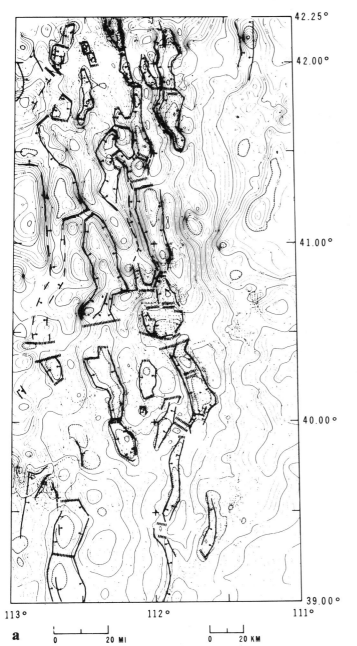

a

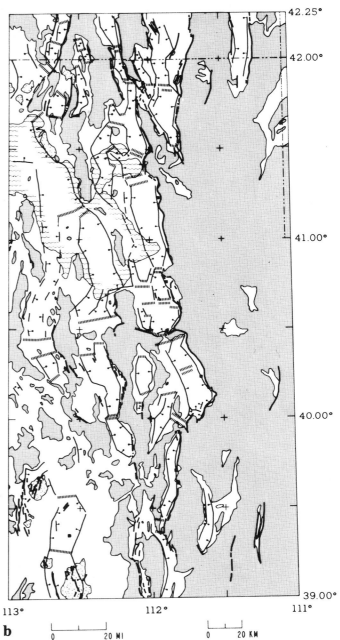

b

Figure 3. Basin-and-range blocks in the Wasatch front region delineated by gravity shown superimposed on (a) the complete Bouguer gravity map (Fig. 2) and (b) the generalized geologic map (Fig. 1). Bounding normal fault zones defined by the gravity data

are indicated by heavy lines with a ball on the downthrown side. Transverse zones defined both geologically and geophysically are marked by hachured lines. Double hachured lines represent transverse structural ridges or saddles.

of the Wasatch and Cache Valley fault zones have normal faults on both sides. In some areas of dense gravity data coverage asymmetric basin structure is suggested by a much steeper gravity gradient occurring on one side of the basin than on the other; in other areas the data suggest symmetrical basins. The details of basin subsurface structure are discussed in a later section; however, surface data indicate that most of the major bedrock blocks defining the ranges

are tilted gently eastward (between < 5° and 20°) (Stewart, 1978). Along the western edge of the study area (west of about long. 112.75°), however, the major range blocks show a reversal in direction, displaying tilt to the west (Stewart, 1978).

As shown in Figures 3a and 3b, most basin-range blocks in the region trend north to north-northwest. The elongate basins formed by the main downdropped and, in

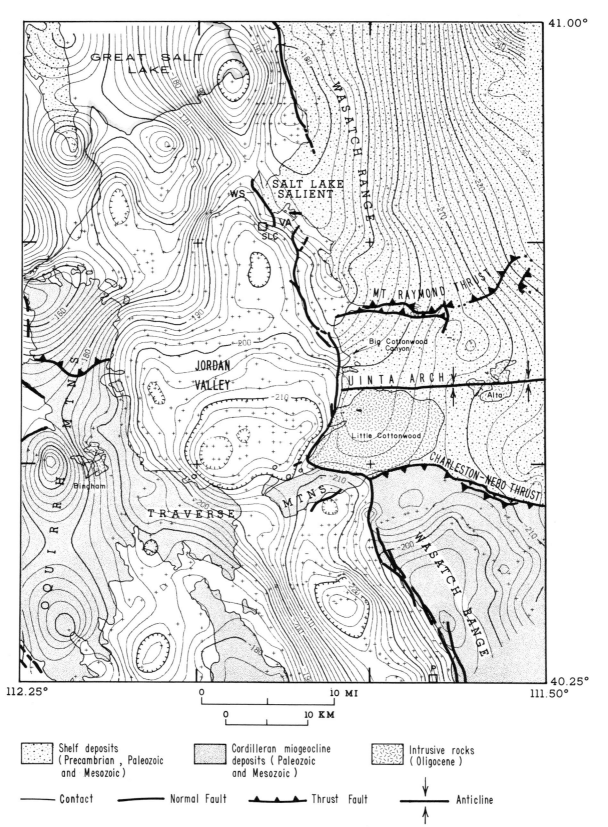

Figure 4. Generalized geologic map of the Jordan (Salt Lake) Valley area with the complete Bouguer gravity map superimposed. Contour interval 2.0 mgal. Generalized geology and structure after Crittenden (1964) and Hintze (1980).

some cases, tilted blocks vary in width from 6 to 20 km, most being 10 to 15 km. In some areas the main basin-bounding fault zone defined by the gravity data appears to coincide with a range-bounding fault mapped at the surface. Elsewhere, offsets between steep gravity gradients and range-front scarps suggest that pediments up to 5 km wide or downdropped, step-faulted blocks occur along the range front.

The elongate basins are terminated or interrupted by transverse zones trending approximately east-west (Figs. 3a, 3b). In some areas these transverse zones appear as east-west-trending structural spurs or bedrock salients. Several excellent examples can be seen between lat. 40.4°N and 41.0°N (Fig. 4). The Salt Lake salient, bounding Jordan (Salt Lake) Valley on the north, seems to be an unfaulted extension of the main Wasatch range (Eardley, 1944). It is bounded on its western edge by the Warm Springs fault, a major normal fault. A gravity saddle west of the Warm Springs fault indicates that a downfaulted extension of the salient exists in the subsurface as a structural ridge beneath the basin (Cook and Berg, 1961). Displacement along the main Wasatch fault zone appears to die out where it crosses the salient. The north and south boundaries of the salient seem to be the result of a gentle downwarping, no major east-west-trending fault zones have been identified (Eardley, 1944). A minor east-west-trending fault, with probably Quaternary displacement, has been mapped along the southern margin of the salient (Kaliser, 1976). Slickensides on this fault indicate a combination of dip-slip and right-lateral strike-slip movement (Pavlis and Smith, 1980).

The Traverse Mountains block, forming the southern boundary of Jordan Valley (Fig. 4), is characterized by east-west-trending anticlinal folds in Permian-Pennsylvanian age rocks. The Traverse Mountains themselves are a continuation of the Oquirrh Mountains that was separated from the main Wasatch Range to the east by a fault zone with an offset of 1.5 to 3.0 km (Crittenden, 1964). East-west-trending fault zones with probably Quaternary displacement have been identified along the northeast margin of the Traverse Mountains (Cluff and others, 1970). A steep gravity gradient as well as drill-hole data (Mattick, 1970) also suggest a buried east-west-trending fault zone in this region.

Another example of the east-west-trending structural ridges is the Payson salient at the south end of Utah Valley (Fig. 1). Here the main fault zone makes a broad arcuate bend to the southwest, and a major 10-km westward shift in the fault zone occurs through a zone of distributed north-south-trending, down-to-the-west normal faults. East-west-trending segments of the main zone of normal faulting help to define the northern margin of the Payson salient.

Many of the transverse zones identified by the gravity data are buried beneath basins and mark major longitudi-

nal changes in basin subsurface structure. More of these buried transverse zones probably remain undetected because of poor gravity coverage, particularly between lat. 40.7°N and lat. 41°N. Effimoff and Pinezich (1981) reported similar east-west-trending structural axes (or highs) beneath basins indicated by gravity and seismic reflection data in northeastern Nevada. In northern Utah these transverse zones or buried ridges often intersect major normal faults at localities where the main fault zone changes trend by 20° to almost 90° (Fig. 3b).

Individual basin lengths vary from less than 10 km to 150 km, the length of the main continuous basin west of the Wasatch fault zone that extends from approximately lat. 40.75°N to 42.25°N. In general, basins in the northern part of the area (north of approximately 40°N) appear longer and more continuous than those in the south; some of this apparent structural simplicity may be due to poor gravity coverage, particularly in the area of the Great Salt Lake.

The Sevier Desert Basin area, south of approximately 40°N in the southwestern part of the region (Fig. 1), has an apparent style of crustal extension and segmentation that contrasts with the surrounding region. Bedrock ranges in the Sevier Desert Basin region are irregular in shape rather than elongate and are surrounded by broad pediments, as indicated by both the gravity data and small bedrock outcrops in the basins. Results from seismic reflection profiling (discussed in a following section) reveal shallow subsurface structures in the Sevier Desert area—structures that may be responsible for the contrasting style of segmentation in that region.

INFLUENCE OF PREEXISTING STRUCTURE ON BASIN–RANGE FAULTING

Two major tectonic elements of the western United States intersect in north-central Utah: (1) a great east-west-trending upwarp—the Uinta arch, and (2) a north-south-trending arcuate belt of Mesozoic folds and thrust faults that extends from southern Nevada into Idaho and Montana. Several of the structures comprising these tectonic elements are shown superimposed on the generalized geologic map and the basin-range block map in Figures 5a and 5b. The thrust structure and the inferred connections across alluvial valleys are after Crittenden (1976) and Morris (1977). As has been noted previously by numerous workers, many of the major jogs or changes in trend of the main Wasatch fault zone seem to occur near its intersection with these earlier established structural features. Some segments of the Wasatch fault zone actually appear to coincide with preexisting features. A good example of such coincidence occurs at the south end of Jordan Valley where the main Wasatch fault zone bounding the range follows the western margin of the Little Cottonwood stock of Oligocene age (Fig. 4) (M. D. Crittenden, 1982, oral commun.). Farther

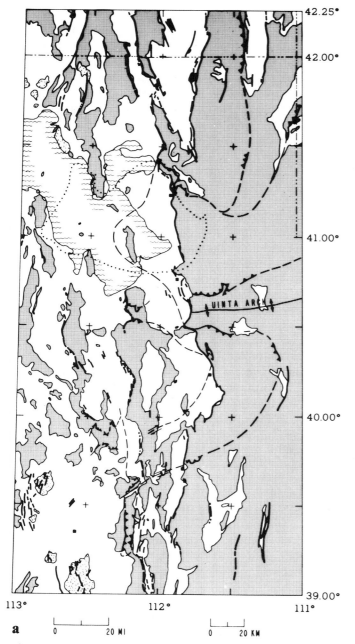 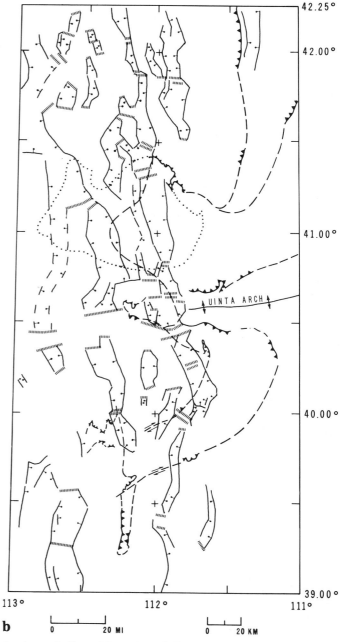

a   0   20 MI   0   20 KM

b   0   20 MI   0   20 KM

Figure 5. Major preexisting structural features influencing the region superimposed on (a) the generalized geologic map showing major normal fault zones (Fig. 1) and (b) a map of basin-and-range blocks defined by gravity (Fig. 3). Major Mesozoic thrust sheets are indicated by sawtooth lines. Heavy dashed lines mark the inferred connection of these sheets. Heavy dotted line outlines the extent of the northern Utah uplift (NUU) after Eardley (1969).

south the fault zone makes a right-angle bend for 7 km to the east at the eastern edge of the Traverse Mountains before returning to its north-south trend near Alpine, Utah. This eastward offset in the Wasatch fault zone is located at the intersection of the Wasatch with the east-trending edge of the Charlestone-Nebo thrust fault and the east-trending Deer Creek fault (Fig. 4). The Deek Creek fault is a normal fault probably of middle Tertiary age (Crittenden and others, 1952). Thus, basin-range faulting along the Wasatch fault zone apparently followed for a short distance the preexisting zones of weakness with nearly east-west trends before bending back to the more representative north-south trend.

One of the major features influencing the present structure of the region is the Uinta arch. In Precambrian time the east-west-trending axis of the modern Uinta arch probably coincided with an aulacogen filled with sediments, most of which are more than 1 b.y. old (Stokes,

1976). By late Precambrian time the basin emerged along the present Uinta axis, and during Early Ordovician time, uplift along this axis extended westward into central Nevada (Cortez-Uinta axis of Roberts and others, 1965). In Laramide time (latest Mesozoic–earliest Tertiary) a broad anticlinal structure developed, forming the modern arch and the Uinta Range to the east of the study area. As can be seen in Figure 5b, there are many east-west-trending transverse "ridges" along and west of the central Wasatch fault zone near the latitude of the intersection with the Uinta arch. In fact, a zone of nearly continuous bedrock can be traced westward from the Wasatch fault zone across the area at approximately lat. 40.5°N (Fig. 5a). Total east-west extension across this zone is limited to the bedrock fault offsets and appears to be much less than in regions to the north and south that contain major intervening basins.

The localization of major east-west-trending transverse zones along the westerly projection of the Uinta arch suggests a genetic relationship between the arch and those features. As has been pointed out by numerous investigators, this intersection is also the site of a major east-west zone of Oligocene igneous intrusions including the Bingham, Little Cottonwood, and Alta stocks (Fig. 4). Aeromagnetic data suggest a westerly continuation of this east-west-trending zone across the map area. The amplitude of the magnetic anomalies requires a greater abundance of intrusive rock in the subsurface than exposed at the surface (Mabey and others, 1978). Intrusions of middle Tertiary age along the east-west trends suggest the presence of deep-seated structures, possibly related to the westward extension of the Uinta axis (also called the Cortez-Uinta axis). These same structures may have influenced the formation of the east-west-trending transverse structural zones during basin-range extension.

Segmentation of Jordan (Salt Lake) Valley by east-west transverse features is indicated on Figure 4. Details of the subsurface structure of Jordan Valley have been discussed by Cook and Berg (1961), Arnow and Mattick (1968), and Mattick (1970) and are consequently only summarized here. As can be seen on Figure 4, Jordan Valley is truncated on the north and south ends by east-west-trending structural highs—the Salt Lake salient to the north and the Tranverse Mountains salient to the south. A major east-west-trending gravity gradient extending westward from approximately the mouth of Big Cottonwood Canyon near lat. 40.65°N separates the basin into a northern and a southern segment. The main graben in the northern part of the basin is narrow, being flanked on the west by a major bedrock pediment extending out from the north end of the Oquirrh Mountains. South of the east-west zone defined by the gravity gradient, the basin is broad and the data suggest a possible small intrabasin graben along the east flank of the Oquirrh Mountains.

The source of this gravity gradient crossing the valley near the latitude of Big Cottonwood Canyon (Fig. 4) is enigmatic. Gravity control in both the Oquirrh and Wasatch Mountains is not good enough to indicate if the gradient extends into the ranges or if it is restricted to the valley. The northward-dipping thrust fault mapped within the Oquirrh Mountains near the latitude of the gradient (Fig. 4) is an intraformation thrust with probably little or no density contrast across it. The Mount Raymond thrust in the Wasatch Range north of the latitude of the gradient also dips to the north; downfaulting of that feature along the Wasatch fault zone would presumably shift its trace southward. However, as this thrust juxtaposes upper Paleozoic and lower Mesozoic units (limestones and quartzites) on a lower plate containing the identical units as well as Precambrian quartzite and argillite, there is probably little or no density contrast across it either. The sense of the gravity gradient indicates a deficiency of mass to the south and probably largely reflects a much thicker section of valley fill in the southern part of the valley. Throughout most of the valley the gradient is gentle. If the anomaly represents a thickening section of valley fill to the south, then the underlying bedrock surface could be either gently downwarped or broken by a broad zone of east-west-trending step faults.

Adjacent to the range front near Big Cottonwood Canyon, the east-west gravity contours are closely spaced, and two-dimensional modeling of the gradient suggests a shallow source (< 5-km depth). If this gradient is modeled by an east-west-trending fault down to the south, juxtaposing basin fill against bedrock (assuming a density contrast of -0.4 g/cm^3), a steeply dipping fault ($\sim 60°$S) with a throw of about 1 km is suggested. However, no east-west-trending faults or lines of weakness have been identified in the bedrock in the area. Although an east-west-trending normal fault downdropped on the south is consistent with the gravity data, no independent corroborating evidence supports that interpretation.

Another possible interpretation for the gravity anomaly is that it represents the buried northern edge of the Little Cottonwood stock and indicates a continuation of the stock beneath the basin. There are several difficulties with this interpretation: (1) The midpoint of the gradient (approximate edge of the causative subsurface body) is 5 km north of the northern edge of the main Little Cottonwood stock. (2) Foliation within the stock suggests that its mapped northern edge may closely coincide with the true margin of the pluton (M. D. Crittenden, Jr., 1982, oral commun.); two small outcrops of intrusive rock in the range south of Big Cottonwood Canyon probably represent minor satellitic bodies. (3) The density contrast is probably very small between the quartz monzonite of Little Cottonwood stock and the Precambrian Big Cottonwood Formation, a quartzite and argillite unit, into which it is intruded.

Regardless of the source of the main east-west-

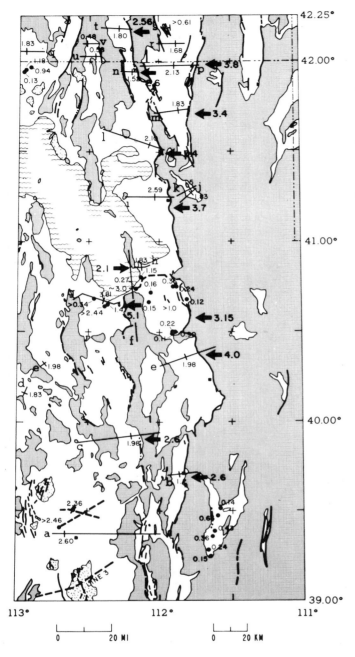

Figure 6. Available data on basin subsurface and fault offsets. Locations of wells (dots) that have penetrated the valley fill section (fill thickness in kilometers), modeled gravity profiles (solid lines), as well as the location of seismic data (heavy dashed lines) are shown. Numbers along gravity profiles indicate maximum depth of valley fill in kilometers based on two-dimensional modeling. Heavy arrows and numbers indicate minimum vertical offsets in kilometers across fault zones. Seismic reflection data in the Sevier Desert Basin from McDonald (1976). Seismic refraction data in Jordan Valley from Arnow and Mattick (1968). Letters refer to modeled gravity profiles shown on Figure 7. See Figure 1 for explanation of map symbols.

trending gravity gradient, it is significant that the young fault scarps along the range front extend continuously across this transverse structure and are apparently uninfluenced by the buried, shallow source of the anomaly. Also noteworthy is that no east-west-trending structures have been mapped in the adjacent bedrock of the Wasatch Mountains with which the gravity anomaly can be correlated.

Another preexisting structure possibly influencing current deformation of the region is the northern Utah uplift (Fig. 5) (Eardley, 1939, 1969)—an area of extensive exposure of Precambrian rocks. This uplift coincides with the major reentrant in the Mesozoic frontal thrust system. The region of the northern Utah uplift has also been mapped as a Pennsylvanian platform and in part coincides with an uplift in late Precambrian time (Eardley, 1969). This uplift now represents the lowest structural level exposed in the region and is characterized by low seismicity; however, its possible influence on present seismicity will be discussed later.

SUBSURFACE STRUCTURE OF BASINS

Seismic reflection and refraction data, gravity data, and data from deep wells all provided information on the depth and subsurface form of basins in the study area. This information is summarized on Figure 6, together with the location of the detailed gravity profiles that have been modeled by numerous workers. The individual gravity profiles and models are shown in Figure 7. Depths to bedrock beneath the basins, as obtained from well data and from gravity modeling, are given on Figure 6. Fill thickness in the major basins (as computed from the two-dimensional profiles shown) ranges from about 1.25 km to about 3.8 km, with a mean depth of about 2.0 km. Also shown on Figure 6 are the estimated minimum vertical offsets across some of the major fault zones, computed from the thickness of the valley fill and the elevation of adjacent high points in the ranges. Minimum estimates of vertical offsets along the

Figure 7. Residual gravity profiles and two-dimensional model of basin structure for profiles located on Figure 6. Basin fill designated by stippling. Solid curve gives residual profile; dots indicate calculated gravity values for the model shown. Profiles are grouped with common length scales, and there is no vertical exaggeration on models. The sources for the gravity models together with their constant assumed density contrasts (in g/cm^3) between bedrock and basin fill are as follows: (a) Isherwood (1967), −0.5; (b) and (e′) this study, −0.5; (c) Mabey and Morris (1967), −0.45 for basin fill and −0.22 for underlying volcanics (shown by cross hatching); (d) and (e) Johnson and Cook (1957), −0.4; (f) Mabey and others (1963), a quantitative model was not calculated for this profile, so no density contrast is given; (g) and (h) Cook and others (1966), −0.5; (i) Cook and others (1967), −0.5; (j) and (k) Stewart (1958); (l), (n), and (t) Peterson (1974), −0.5; (m), (p), (r), and (s) Peterson and Oriel (1970), −0.5; (q) Cook and others (1964), −0.5; and (u) and (v) Harr and Mabey (1976), −0.55.

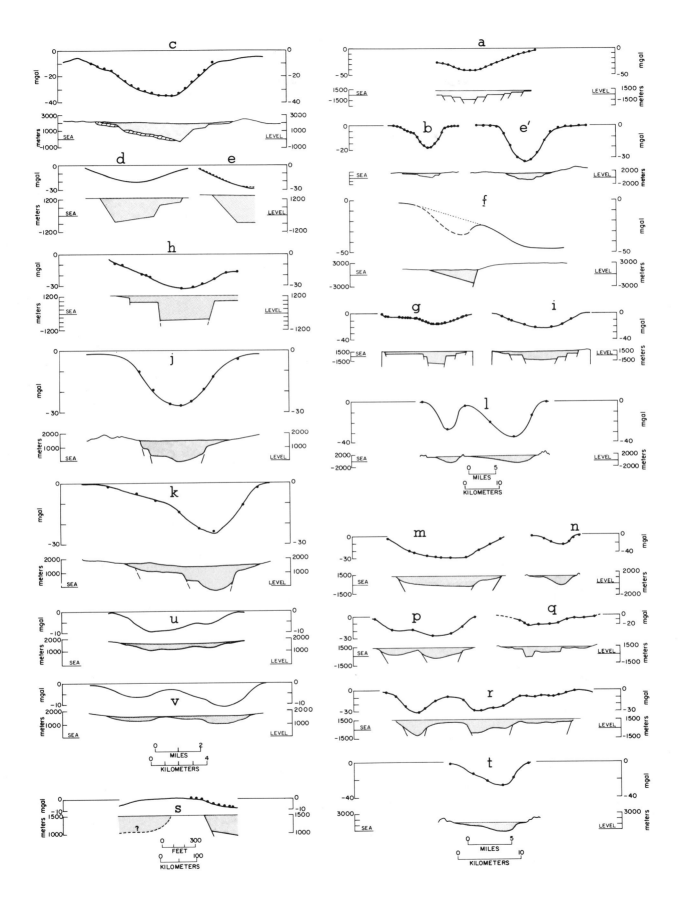

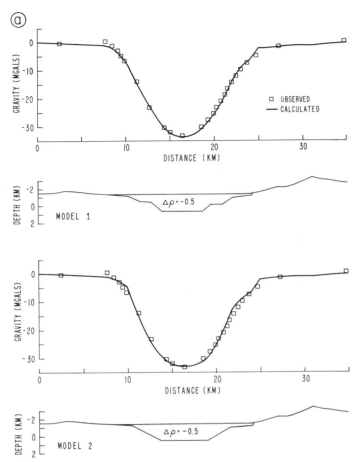

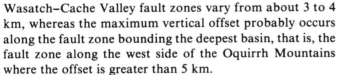

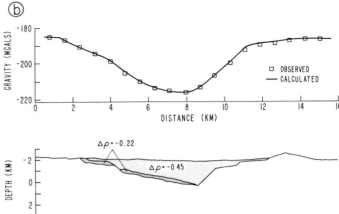

Figure 8. (a) Observed gravity profile across the northern end of Utah Valley (profile e′ on Fig. 6) demonstrating the graben-type structure of the valley. Two models satisfying the observed data are given. Basin fill indicated by stippling, density contrast, $\Delta\rho$ (in g/cm^3), used for modeling is shown. The dip of the boundaries of the main graben probably represents an effective dip of a bounding zone of step faults, not a single fault. (b) Observed gravity profile and model across the Tintic Valley (profile c on Fig. 6) demonstrating the gravity character of an asymmetric, tilted block basin (from Mabey and Morris, 1967). Basin fill indicated by stippling, modeled with a density contrast of $\Delta\rho = -0.45$ g/cm^3. Middle Tertiary volcanic rocks (welded tuffs and pyroclastic and flow rocks ranging from composition from rhyolite to basalt) are inferred to underlie the basin modeled with a density contrast, $\Delta\rho = -0.22$ g/cm^3.

Wasatch–Cache Valley fault zones vary from about 3 to 4 km, whereas the maximum vertical offset probably occurs along the fault zone bounding the deepest basin, that is, the fault zone along the west side of the Oquirrh Mountains where the offset is greater than 5 km.

A detailed discussion of the subsurface structure of each basin is beyond the scope of this study. The modeled gravity profiles shown with similar length scales on Figure 7 give some indication of the variation of two-dimensional subsurface structure that can be inferred from gravity data. The profiles were modeled by various workers using constant density contrasts between the basin fill and bedrock that ranged between –0.40 and –0.55 g/cm^3. Most profiles were modeled with a density contrast of –0.50 g/cm^3 (the actual values used in each case are given in the caption for Fig. 7). The constant density contrasts used to model the basins are an obvious oversimplification; density probably varies both laterally (due to facies variations) and vertically (largely due to compaction) within the basin fill section. However, without additional data such as seismic reflection or refraction studies for an individual basin, the use of a constant average density contrast to constrain basin geometry is assumed justified, particularly when comparing the overall shape and depth of a variety of basins.

As mentioned in the previous section, many of the basins have a symmetric graben-type structure with a major fault zone on both sides. Utah Valley (Fig. 1) is a dramatic example of such a basin; Figure 8a shows the gravity profile (profile e′) across the north end of this valley (see Fig. 6 for profile location) together with two possible models that match the observed data. (Model 1 is also shown in Fig. 7, profile e′). In other basins the gravity data indicate an asymmetric subsurface structure, one margin of the basin being defined by a steep gravity gradient (probably reflecting a major fault zone) and the other margin being marked by a very gentle gradient. The gentle gravity gradient probably reflects structure that resulted from tilting, downwarping, and/or step faulting. Profile c, together with its gravity model (Fig. 8b and also Fig. 7), demonstrates the typical gravity expression of an asymmetric basin. As shown in Figure 7, profiles b, c, d, f, l, p, r, and t were modeled with an asymmetric basin subsurface structure. Profiles b, c, l, and t show eastward tilts of the basin floors, profiles p, d, and r show westward tilts, and profile f suggests a southward tilt. Throughout much of the area, however, the data coverage is too sparse to define the gradients accurately, and hence the detailed subsurface structure of the basins is not known.

SUBSURFACE GEOMETRY OF FAULTING

Dips measured on bedrock scarps range between 44° and 70° along segments of the Wasatch fault zone and adjacent faults that show evidence of Quaternary activity (see Table 1). Gilluly (1928) reported 14 dip measurements made along the frontal fault zone along the west side of the Oquirrh Range. The measured dips ranged from 40° to 80°, averaging 57° along this fault zone whose most recent movement was probably in late Pleistocene time (Anderson and Miller, 1979).

TABLE 1.   MEAN ATTITUDE AND SLIP DIRECTIONS FROM BEDROCK
SCARPS IN NORTHERN UTAH

Site	Locality	Strike	Dip	Rake*
a	Bear Lake I	N3°W	53°W	81°
b	Bear Lake II	N5°E	57°W	93°
c	Utah Hot Springs	N35°W	70°W	79°
d	Warm Springs	N45°W	70°W	85° Y[†]
				56° O
e	Virginia Street	N80°E	68°S	143°
				172°
f	Mercur mine	N10°W	46.5°W	82°
g	Rock Canyon	N3°W	44°W	98.5°
h	Springville	N20°W	51°W	93°
i	Little Birch Canyon	N40°E	68°W	100°

*Rake is defined as the angle, in the fault plane, between horizontal and the down-dip slip direction. The values here are all referenced to horizontal on the north (or, for site e, the east-northeast) end of the fault plane.

[†]Two sets of slickensides were observed; superposition of the two sets indicated relative age; Y-younger, O-older.

Gilbert (1928) in his classic study of normal faulting along the Wasatch front described bedrock fault scarps with dips ranging from 29° to 45°, with an average dip of 33°. Although Gilbert observed a few much steeper slickensided surfaces (~70° dip), he considered that these surfaces represented subsidiary fault planes within the footwall. Gilbert was no doubt influenced by his view that the triangular facets (rising from the base of the range with slopes of ~30°) represented the true footwall of the range-front fault little modified by erosion. Most subsequent observers agree that the facets are not remnants of the fault surface but rather represent an erosional surface along the fault zone. Pack (1926) recognized this relation nearly 60 years ago and pointed out that many of the facets terminate downward into unfaulted bedrock. At about the same time, Gilluly (1928) noted numerous examples of steep bedrock fault planes at the base of much more shallowly dipping triangu-

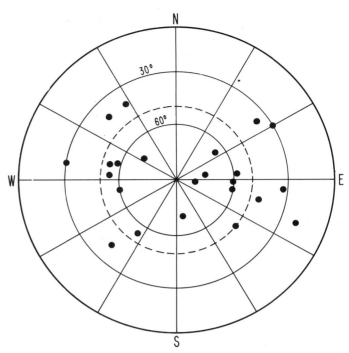

Figure 9. Diagram showing dips and dip azimuths of all nodal planes of northern Utah earthquake focal mechanisms. Data from Smith and Lindh (1978), Arabasz and others (1981), and Bache and others (1980). Dashed circle shows mean dip of 49°.

lar facets along ranges throughout the northern Great Basin. Gilbert did not observe many of the fault planes given in Table 1, because most of them have only recently been exposed in gravel pits and road cuts. Some of the fault surfaces described by Gilbert have been examined by the author, many of these are probably older faults within the frontal fault zone and may represent subsidiary faulting within that zone.

Information on the subsurface attitude of seismically active fault planes can be inferred from nodal plane data from earthquake focal mechanisms and, in some cases, the distribution of hypocenters with depth. Each focal mechanism provides two possible orthogonal fault planes: one plane represents the true fault plane, and the other defines the slip on the fault plane. In general, it is not always possible to define the proper fault plane. However, consideration of all possible nodal planes defines a range of possible subsurface dips of active faults. The dip data from all nodal planes in the northern Utah area (with no attempt to discriminate the actual fault plane) are shown in Figure 9 and suggest possible subsurface dips that range between 20° and 80°. The mean dip is 49°, and the median dip is 49° to 54°. These seismically inferred dips agree fairly well with the range of fault dips measured at the surface (40° to 80°, Table 1, and Gilluly, 1928).

The regular eastward tilt of the major range blocks in the region (Stewart, 1978, 1980) and the predominantly westward dip of the major normal fault zones are probably genetically related. This pattern of opposing fault dip and block tilt can result from either a primarily listric style of normal faulting or a more planar style of normal faulting in which both the faults and the blocks they bound tilt uniformly. The predominance of major west-dipping normal fault zones in the near surface, coupled with the eastward tilt of the range blocks and the presence of major Mesozoic thrust faults that dip gently to the west, is commonly cited as evidence of a listric character for the major fault zones. The focal-mechanism dip data, however, do not support major subhorizontal seismic slip alng the "soles" of listric faults and rather suggest subsurface slip on fault planes with a wide range of subsurface dips. In the next section the focal mechanisms are examined in detail, and by using independent evidence where available, the most likely fault planes are identified (Table 2). For eight events the fault dips ranged between 30° and 80°, with a mean value of 54° and a median value between 55° and 58°. Although it appears likely that many fault zones decrease in dip with depth, others (as indicated by the focal mechanisms) remain relatively steep.

TABLE 2. FOCAL MECHANISMS WHERE A PREFERRED FAULT PLANE CAN BE SELECTED

Site	Nodal Plane A		Nodal Plane B*		Strike	Dip	Rake[†]
	Az.	Dip	Az.	Dip			
1	319°	49°NE	15°	55°W	N15°E	55°W	130°
2	120°	20°S	7°	80°E	N7°E	80°E	112°
3	1°	60°E	45°	39°W	N45°E	39°W	127°
4	5°	60°E	10°	30°W	N10°E	30°W	95°
5	5°	32°E	17°	58°W	N17°E	58°W	96°
6	351°	60°SW	330°	30°NE	N30°W	30°NE	71°
7	290°	30°S	323°	64°N	N37°W	64°NE	74°
8	58°	40°NW	170°	74°NE	N10°W	74°E	56°

Note: Data from Smith and Lindh (1978), Arabasz and others (1981), and Bache and others (1980).

*Nodal plane B is the selected fault plane.

[†]Rake is defined as the angle, in the fault plane, between horizontal and the down-dip slip direction. The values here are all referenced to horizontal on the north end of the fault plane.

Additional information on the subsurface configuration of major normal fault zones comes from detailed seismicity studies. One of the best documented earthquake sequences in the eastern Basin and Range province is the 1975 (M = 6.0) Pocatello Valley earthquake and its aftershocks. The main shock occurred at lat. 42.06°N, long. 112.53°W, 6 km north of the Idaho-Utah border, at a depth of 8.7 km (Bache and others, 1980; Arabasz and others, 1981). The preferred orientation for the fault plane has a N45°E strike and a northwest dip. This orientation is in agreement with (1) source modeling of the focal-mechanism data (Bache and others, 1980), (2) the spatial distribution of

aftershock foci in the hypocentral area (Arabasz and others, 1981), (3) the trend of a fault within the basin inferred from gravity data (Harr and Mabey, 1976), and (4) modeling of the geodetically observed subsidence (Arabasz and others, 1981). The dip of the nodal plane for the main shock, as obtained by Bache and others (1980) from source modeling, is 39°NW. This dip compares with an average dip of 50°NW obtained from aftershock hypocentral cross-sections and a dip of 60°NW from dislocation modeling (Arabasz and others, 1981). As pointed out by Arabasz and others (1981), the focal-mechanism dip of 39° at 8.7-km depth is not incompatible with steeper dips shallower in the crust if the fault is listric. The seismically inferred fault plane (with its N45°E strike and northwest dip) obliquely intersects the north-trending bounding faults and fault blocks in the Pocatello Valley, complicating the deformation pattern in this region. In addition, the overall epicentral trend of all the earthquakes in the sequence is generally northwesterly (Arabasz and others, 1981).

Seismic reflection data bearing on the subsurface configuration of faults are limited. A seismic reflection profile that crosses the Wasatch fault zone near Nephi has recently been purchased by the author, and a detailed discussion and interpretation of this seismic profile and additional new gravity data are the subject of a separate study in progress. In general, this seismic profile indicates an asymmetric graben structure for Juab Valley (Fig. 1). Numerous small step faults form an east-dipping flank along the western margin of the valley, whereas the eastern margin of the basin appears to be truncated by a major west-dipping fault zone. The overall structure of the basin floor seems to be an asymmetric sag or downwarp, with basinward-dipping flanks on both sides. This type of basin structure has been reconized previously on seismic reflection profiles in other parts of the Basin and Range province (Anderson and Zoback, 1981a, 1981b; Anderson and others, 1982).

Seismic reflection data are also available for the topographically and structurally low area forming the Sevier Desert Basin (Fig. 1). Published seismic profiles for this basin (McDonald, 1976) show a conspicuous thin zone of continuous, gently west-sloping (4° to 8° dip) reflectors at a depth of less than 1 to 5 km. Above this zone the seismic profiles show other conspicuous reflectors which, on the basis of scanty drill-hole data, probably represent Oligocene and younger strata (Lindsey and others, 1981). The gently west-sloping zone of reflectors is unbroken by the numerous overlying normal faults related to a major intrabasin graben defined by the gravity data (Fig. 3). Thus, the lower zone of reflectors appears to represent a region of structural accommodation for fault-controlled extension forming the overlying basin and is accordingly considered a planar detachment fault. The graben-bounding faults above the detachment surface, as best they can be interpreted from the seismic reflection data, appear to be both planar and listric in form. The actual angle of intersection of the fault plane with the detachment surface is difficult to determine. The apparent fault dips on the seismic section vary between 35° and 60°. However, this range actually represents minimum dips determined on the published two-way travel time sections. Because of the general increase in velocity with depth, determining true dips will require detailed knowledge of the subsurface velocity structure in order to convert the time sections to true depth sections.

The west dip of the detachment surface observed in the seismic data would allow for its projection beneath the ranges bounding the Sevier Desert Basin on the west. Unfortunately, only one profile of the available seismic reflection data extends to the west margin of the basin (line 3, Fig. 6), and this profile does not extend into the range. Thus, it is not known whether the detachment surface is offset at the bounding range-front faults (a prominent zone of Holocene scarps does extend along the eastern base of the next range front to the north near lat. 39.5° N) or whether the detachment extends continuously westward beneath the range.

Interpretation of the seismic data and speculation as to the nature of the detachment surface have been presented elsewhere (Anderson and others, 1982) and will be discussed only briefly here. On the basis of stratigraphic evidence from three deep drill holes in the basin, Lindsey and others (1981) have interpreted the Sevier Desert Basin as the site of a broad Oligocene evaporite basin probably formed by broad warping, minor block faulting, or possibly damming of drainages by eruptive products from nearby volcanic centers. They concluded, however, that development of the present structural basins of the Sevier Desert probably began during late Miocene time when thick conglomeratic units were deposited. MacDonald (1976) suggested that the major west-dipping detachment surface beneath the basin may be a Mesozoic thrust reactivated in a reversed sense. The structural complexity of thrusts exposed in the adjacent ranges, however, is difficult to reconcile with the extensive, nearly continuous surface (at least 3,700 km^2) recorded on the seismic data. An alternate interpretation of the detachment surface is that it represents a major low-angle normal fault formed by extensional tectonism during late Cenozoic time (Wernicke, 1981).

It is difficult to assess how far the thin-skin-style extensional tectonics seen in the Sevier Desert Basin may extend into north-central Utah. As mentioned previously, the character of the gravity anomalies in the Sevier Desert Basin is distinctive. Well-defined basin-and-range blocks are not observed either in the gravity data or at the surface. Broad pediments and shallow intervening basins between the bedrock making up the ranges are common. Thus, the thin-skin style of extensional tectonism may be limited to areas characterized by the particular structure of the Sevier Desert Basin.

CORRELATION OF SEISMICITY WITH BASIN-RANGE BLOCKS

A preliminary map of instrumentally located epicenters for earthquakes recorded by the University of Utah seismograph network from 1962 to September 1980 (R. B. Smith, 1981, written commun.) is superimposed on a map of bedrock-alluvium contacts with normal faults in Figure 10a and on the map of major bains-range blocks defined by sdurface geology and gravity data in Figure 10b. Significantly, a large number of the recorded events in this time period lie between 15 and 30 km east of the Wasatch–Cache Valley fault zones, largely in regions where surface and gravity evidence does not indicate major fault zones. The accuracy of epicentral location in the study area varies between ±1 to 2 km for the 1975 to 1978 events and ±5 km for the 1962 to 1974 events (Arabasz and others, 1980). The seismicity data show that few if any earthquakes during the past 20 years have occurred along any of the fault zones showing high levels of Quaternary activity. Furthermore, few earthquakes occur along the major fault zones mapped at the surface or those delineated by the gravity data. However, epicenters displaced from the surface trace of a dipping fault may correspond to the hypocenters of earthquakes occurring on fault planes at depth. Hypocentral depths for northern Utah earthquakes range from about 0 to 20 km, with about 80% of the events occurring at depths less than 10 km. Although, in general, the frequency of events decreases with increasing depth in the region of the Wasatch front (long. 111.25° to 112.25° W), there appears to be a bimodal distribution of events, with peaks centered at depths of 2 to 3 km and 7 to 8 km (Arabasz and others, 1980).

Because 80% of the earthquakes in the region occur at depths of less than 10 km, the minimum lateral offset between the surface trace of a fault and activity along that same fault plane at depth is about 6 km for a 60° dip, which is typical of the surface dip of many of the faults (as discussed previously). However, decreasing fault dips at depth (typical of listric normal faults) could produce significantly greater lateral offsets.

As numerous workers have noted, very little of the seismic activity in the time period of seismic monitoring could be attributed to slip along either the main trace of the Wasatch or Cache Valley fault zones, although a few events near 41.7° N as well as some events near the southern end of Juab Valley (~39.5° N) may indicate minor slip along the Wasatch fault zone. As can be seen in Figures 10a and 10b, some events along the Wasatch front appear to more closely coincide with the transverse structural features or "ridges" defined by surface geology and gravity data. A significant amount of seismicity has been recorded directly east of the Cache Valley fault zone where events appear to define an east-dipping fault zone beneath the Bear River

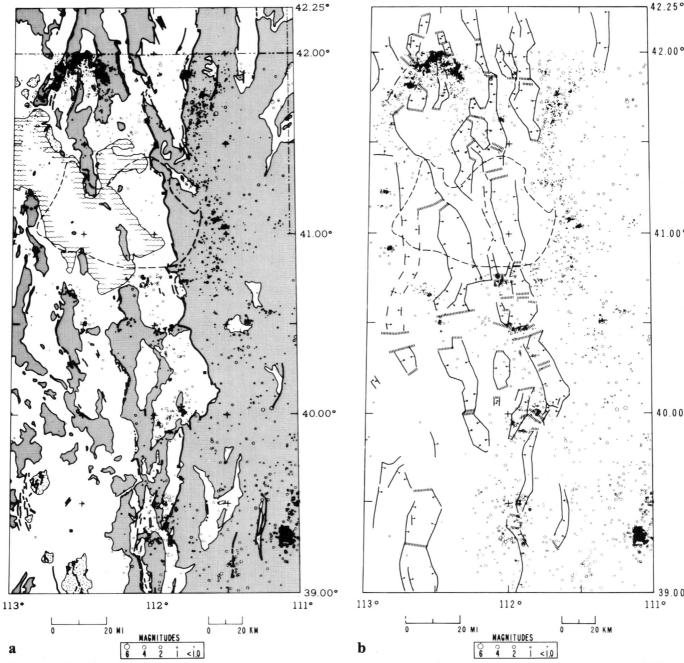

a

b

Figure 10. Preliminary epicenter (circles) map of north-central Utah seismicity (1962–September 1980) from R. B. Smith (1981, written commun.) superimposed on (a) the generalized geologic map showing bedrock outcrops and normal fault zones (Fig. 1) and (b) a map of basin-and-range blocks delineated by the gravity data (Fig. 3). Dashed line outlines the northern Utah uplift.

Range. These earthquakes are thus probably unrelated to the major west-dipping Cache Valley fault zone bounding the range (Westphal and Lange, 1966).

Seismicity in the Basin and Range province west of the Wasatch–Cache Valley fault zones is dominated by earthquakes in the northwestern part of the study area near the Idaho-Utah border. The high level of activity here is largely related to aftershocks of the Pocatello Valley 1975

(M = 6.0) earthquakes. Immediately to the south is a broad region characterized by a low level of seismicity coinciding in part with the Great Salt Lake and the northern Utah uplift defined by Eardley (1939, 1969) (Fig. 5, 10) and described previously. If the pattern revealed in the short period of seismic monitoring (1962 to 1980) is meaningful, this region lacks the background low level of seismicity seen in surrounding regions. Possibly the shallow Precambrian basement rocks in this area act more competently and are characterized by major events with little intervening activity, whereas surrounding areas of thick Phanerozoic sections cut by many preexisting faults may respond with a continuous low level of seismicity.

In other areas west of the Wasatch front the hypocenters are approximately evenly divided between basin blocks and range blocks. The seismicity centered beneath range blocks is difficult to reconcile with a graben-type model for basin-range extension that predicts extensional deformation localized beneath the basins. This seismicity, however, is consistent with seismic slip on low-angle detachment faults extending beneath ranges, although, as discussed previously, the available data on the dip of seismically active fault planes (from earthquake focal mechanisms) appear to rule out low-angle slip planes (dips <30°) in the focal depth region of <10 to 12 km. The possibility exists that slip on low-angle faults is aseismic and that the events located at depth beneath the range may represent minor adjustments above zones of aseismic slip. It is puzzling that adjacent major fault zones bounding these range blocks show no activity.

The lowest level of seismicity over the 1962 to 1980 period of monitoring is in the Sevier Desert Basin area. It is difficult to say whether this low level of activity indicates that local thin-skin deformation occurs aseismically or simply reflects the absence of recent deformation in this area. In any case, it is noteworthy that a 35-km-long fault zone of probable Holocene age (Drum Mountains scarp) and a 20-km-long zone of probable late Quaternary age faulting (Clear Lake scarp) have been mapped in this area (Bucknam and Anderson, 1980). These scarps are similar in appearance to recent scarps throughout the Basin and Range province and are generally interpreted as indicating slip associated with major prehistoric earthquakes.

KINEMATICS OF FAULTING

Earthquake focal mechanisms and fault slickenside measurements can provide information on the pattern and sense of slip in an actively deforming region. The range of attitudes of slipping fault planes and the direction of slip on each can be used to constrain a regional stress tensor responsible for the deformation. While focal mechanisms reflect the recent deformation, slickensides on bedrock fault surfaces observed at the surface reflect older deformation, although the age of slickensides is difficult to establish. In some places the observed slickensides record the last fault movement, hence the faulting may be late Pliocene, Quaternary, or Holocene in age as determined from geologic evidence found elsewhere along the fault trace. In other places, the conditions required to form slickensides may have occurred at deeper levels, and subsequent slip of fanglomerate gravels against the bedrock scarp may not have

altered slickensides formed earlier. In any case, the slickensides represent a much longer period of deformation than do the earthquake focal mechanisms. The extent to which these two indicators agree indicates the uniformity through time of the regional stress field.

Earthquake Focal-Mechanism Data

Seismicity data along the Wasatch fault zone and surrounding regions in northern Utah have been summarized by Arabasz and others (1981). The available focal mechanisms as well as the trends of the tension axes are shown in Figure 11. Notably, none of the mechanisms (with the possible exception of event 8 near Nephi) can be attributed directly to slip along the Wasatch fault zone. The focal-mechanism data indicate generally normal faulting along northeast- to northwest-trending fault planes and suggest a broad zone of east-west extension (Smith and Sbar, 1974; Arabasz and others, 1981). The pressure (P) and tension (T) axes derived from the focal mechanisms (Fig. 12d) suggest a nearly vertical mean P axis (greatest principal stress) and an approximately east-west orientation for the mean T axis (least principal stress).

Two possible fault planes (nodal planes) are suggested by earthquake focal mechanisms. Surface faulting, seismicity and aftershock alignments, as well as gravity-inferred fault patterns in the epicentral region, allow, in some cases, selection of the most likely fault plane. The pole of the remaining nodal plane then defines the slip vector on the selected fault plane. The preferred fault planes and slip directions from the well-constrained focal mechanisms listed in Table 2 are shown on Figure 12b. The horizontal components of the inferred slip vectors vary between N21°E and S31°E. As already noted, the fault dips range from 30° to 80°.

A moment tensor analysis by Doser and Smith (1982) of these same earthquakes (and a few small aftershocks in the Pocatello Valley area) yields the net regional strain produced by these events, weighted by magnitude. The result indicates a near-vertical greatest principal compressional strain and a horizontal extensional least principal strain trending N78°E.

Fault Slip Data

As part of this study detailed measurements were made of slickensides on numerous bedrock scarps in north-central Utah. The data were collected from scarps along fault zones where evidence for Quaternary movement was present somewhere along their trace. The site localities are shown on Figure 11 together with the earthquake focal mechanisms. Descriptions of the sites, data collection techniques, and the actual measurements taken are part of a manuscript currently being prepared by the author. An at-

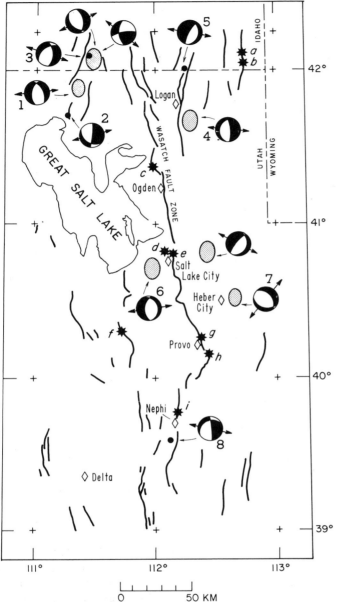

Figure 11. Summary of earthquake focal mechanisms (lower hemisphere projections) and slip data localities (stars) in the Wasatch front area. Numbered focal mechanisms refer to entries in Table 2 and lettered fault-slip sites refer to Table 1. Focal mechanisms after Arabasz and others (1980). Compressional quadrants are shaded, and the trends of T axes are shown by heavy arrows. Large dots show location of single-event solutions; lined zones show sample area for composite solutions.

tempt was made to determine a mean orientation for the fault plane as well as the slip vector by using 10 to 30 measurements at each site. These data have been supplemented with a detailed study of two fault zones north of Salt Lake City (sites d and e on Figure 11) made by Pavlis and Smith (1980) and by a measurement reported by Everitt and Kaliser (1980) along the west side of the Oquirrh Mountains (site f on Figure 11). The data are presented in

Table 1 and shown on a stereographic projection in Figure 12a. As can be seen in the table, in some localities more than one sense of slip was recorded.

Comparison of Fault Slip and Focal Mechanism Data

Since both the attitude of the fault plane and the slip vector are measured from the slickensided bedrock scarps, pseudofocal mechanisms can be constructed for these data. The resulting P and T axes from these pseudofocal mechanisms (Fig. 12c) show a pattern similar to that of the earthquake focal mechanisms (Fig. 12d). The inherent uncertainties in defining stress orientations from P and T axes have been discussed by numerous authors (see McKenzie, 1969; Raleigh and others, 1972). In general, for a region thought to be deforming under a relatively uniform regional stress field the mean P and T axes can be used to infer stress orientations with about a ± 20° uncertainty (Zoback and Zoback, 1980b). Thus, both the earthquake and fault-slip data suggest an approximately east-west least principal stress across the region.

Because fault-slip data lack the ambiguities of focal mechanisms with regard to true fault plane and slip direction, it is possible to invert these slip observations to obtain orientations of the mean regional principal stresses if one assumes that the direction of slip on the fault plane is in the direction of maximum resolved shear stress on that plane. Various methods of inverting the fault-slip data have been proposed (Angelier, 1977, 1979; Carey, 1979). All assume that (1) the faults slip independently, that is, the direction of slip is in the direction of maximum resolved shear stress on the fault plane, and (2) the deformation relates to a single tectonic event governed by a single regional stress tensor. Justification for these seemingly constricting assumptions has been discussed elsewhere (Angelier, 1979). Perhaps the best justification of these assumptions is that the method seems to work; using a well-constrained data set, the observed and predicted slip directions agree well.

The Utah slip data have been inverted using a modification of a method described by Angelier (1979) that minimizes the component of shear stress acting on the fault plane in a direction perpendicular to the observed slip vector and constrains the resulting principal stress axes to lie in vertical and horizontal planes. The results are shown on Figure 12a. The method and numerical procedure used, as well as application to the complete set of data from Utah and some from Nevada, are described elsewhere (Zoback, 1981, and a manuscript in preparation). The same analysis was applied to the well-constrained focal-mechanism data for which a fault plane could be selected (Table 2); the results are shown in Figure 12b. Both data sets indicate deformation resulting from a regional principal stress field with the greatest compressive stress being vertical and the least principal stress oriented approximately east-west.

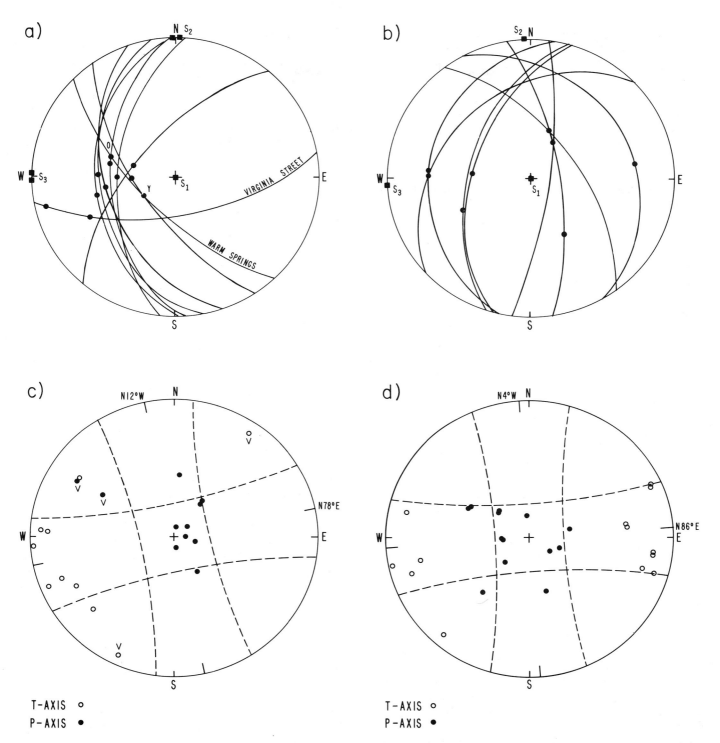

Figure 12. Lower hemisphere equal-area projections of fault-slip and focal-mechanism data in the Wasatch front area. (a) Mean fault attitudes and slip directions (heavy dots) measured on bedrock scarps (site localities a to i shown in Fig. 11 and listed in Table 1). Boxes show orientations of principal stress directions (S_1 = greatest principal stress, S_2 = intermediate principal stress, S_3 = least principal stress) obtained from an inversion of the slip data described in the text. (b) Fault attitudes and slip directions from focal mechanisms in which a preferred fault plane could be selected (Table 2). Boxes give principal stress directions as described in (a). (c) Orientation of P and T axes inferred from pseudo-focal mechanisms constructed from the slip data shown in (a). Mean horizontal stress orientations (with a ± 20° uncertainty indicated with dashed lines) inferred from the P and T axes are shown. Apparently anomalous P and T axes for the Virginia Street fault are indicated by V. (d) Orientation of P and T axes for all northern Utah focal mechanisms shown in Figure 11. Mean horizontal stress orientations (with a ± 20° uncertainty indicated with dashed lines) inferred from the P and T axes are shown.

Analysis of the fault-slip data given in Figure 12a indicates two possible azimuths for the horizontal principal stresses that differ by 2°. These two results represent solutions obtained using the two different slip directions observed for the Warm Springs fault (Fig. 4) that were reported by Pavlis and Smith (1980). Measurements of slickensides on the nearby Virginia Street fault (also reported by Pavlis and Smith), which lies along the southern margin of the Salt Lake salient (Fig. 4), were not included in the inversion to determine the stress tensor. The largely strike-slip offset observed on the Virginia Street fault and the close correspondence in horizontal slip vectors between this fault and the nearby Warm Springs fault indicate that the observed slip on the Virginia Street fault is an excellent example of "second-order" displacement imposed by "first-order" motion of a major discrete crustal block, in this case the Jordan Valley basin block (Pavlis and Smith, 1980). Thus, slip on the Virginia Street fault does not appear independent and thus does not meet the criteria outlined above for the inversion of the slip data.

The anomalous P and T axes derived from the Virginia Street fault indicated on Figure 12c are in contrast to the generally east-west-trending T axes determined from the earthquake focal mechanisms (Fig. 12d) and the other pseudofocal mechanisms from slip data shown on (Fig. 12c). As pointed out by Pavlis and Smith (1980), the apparent anomalous stress field for this fault is probably the result of the displacement imposed by motion of major discrete crustal blocks. Such block motion may also explain the wide range in azimuth of slip vectors inferred from focal mechanisms of moderate size earthquakes (Fig. 12b) compared to the much smaller range in azimuth of the measured slip vectors from the main fault zones along the Wasatch front (Fig. 12a). This latter set of data is probably more representative of the faults and slip controlling major block motion.

REGIONAL STRESS–STRAIN ANALYSIS

Data from an in situ stress study made 20 km east of the Wasatch fault zone near Provo (Zoback and others, 1981) and from geodetic strain data collected near Ogden and Salt Lake City (Prescott and others, 1979; Soler and others, 1981) provide the additional constraints on recent deformation. The stress investigation involved hydraulic fracturing in a well 0.6 km deep and indicated a normal-faulting stress regime with a least principal stress orientation of N73°E ± 15°.

The geodetic strain data present a picture of complex deformation. A 6-year resurvey of a strain net that crosses the Wasatch fault zone near Ogden (Fig. 1) revealed significant compressional deformation in a direction N73°E ± 10°—a direction nearly perpendicular to the Wasatch fault (Prescott and others, 1979). Triangulation-trilateration

data from north of Salt Lake City, obtained by the National Ocean Survey/National Geodetic Survey (NOS/NGS) between 1962 and 1974 (Soler and others, 1981), yielded similar results. Their data indicated a maximum horizontal compressional strain oriented N55°E ± 10°. However, NOS/NGS strain data for the area around and south of Salt Lake City led to an estimate of N79°E + 14° for the horizontal extensional strain direction, consistent with the estimates from the earthquake focal mechanisms, the fault-slip data, and the in-situ stress measurements.

The unexpected orientations of maximum horizontal compression nearly perpendicular to the Wasatch fault zone found north of Salt Lake City and near Ogden probably represent local, short-term variations in the strain field. One proposed explanation of the anomalous compressional strain direction near Ogden involves westward motion of a portion of the adjacent Wasatch Mountains block, along a low-angle detachment surface within the mountain range east of the Wasatch fault zone (Zandt and Richins, 1981). Thus, the apparent contradictory strain orientations measured within the strain nets may be the localized consequence of deformation in surrounding areas. The anomalous compressional strains perpendicular to the Wasatch fault may thus reflect the direction of nearby extensional deformation.

Considered together, all the available data on deformation along the Wasatch–Cache Valley fault zones indicate an east-west to west-southwest–east-northeast (~N75°) least principal stress/strain direction. This direction is nearly perpendicular to the average trend of the basin-and-range blocks; thus, the region appears to be expanding in simple extension. Local complexities, however, do exist. Slip on obliquely oriented faults can be determined either by the resolution of the regional stress field on the fault plane or by an imposed requirement for block motion (Pavlis and Smith, 1980). This requirement for block motion probably results in a component of transcurrent motion along the margins of the transverse structural ridges.

DISCUSSION

This study has revealed several perplexing relations between basin-range structural style and the deformation responsible for the extensional tectonism. One indication of the complexity of the structural style of recent deformation in north-central Utah comes from seismic reflection data from the Sevier Desert Basin in the southwestern part of the region. Here listric and relatively steep (~50°) planar normal faults appear to intersect a gently dipping detachment surface at very shallow depths (3 to 4 km). These data corroborate geologic mapping (see Chamberlain, 1978; Wernicke and Burchfiel, 1982) that demonstrates that tilted blocks are not necessarily a result of listric normal faults, that is, relatively steep, nearly planar faults and a domino-style extension can also produce the tilted blocks (Morton and Black, 1975).

The detachment surface in the Sevier Desert Basin accommodates fault-controlled extension above an apparent stable block. Barring any anomalously shallow ductile flow, the nature of extension beneath the detachment surface, if in fact it occurs, remains unclear. Wernicke (1981) proposed that the shallow detachment seen on the seismic data is part of a major low-angle fault that extends down-dip westerly through the crust and possibly the lithosphere. This model allows for deep-seated extension far removed laterally from its surface expression. However, this hypothesis for modern basin-range extension fails to explain the rather regular spacing and size of the major basin-and-range blocks throughout the province. In addition, the Sevier Desert Basin is the site of the most voluminous recent volcanism in the study area. Basalt, basaltic andesite, and rhyolite ranging in age from 6.9 to 0.88 m.y. crop out in the basin (Mehnert and others, 1978; Peterson and others, 1978; Galyardt and Rush, 1979). This volcanism suggests active extensional tectonism within the crust directly beneath the Sevier Desert Basin.

Seismicity data provide little help in understanding the present style of deformation in the Sevier Desert Basin because of the low level of earthquake activity in this area. A major zone of Holocene fault scarps 35 km long was mapped near the base of one of the ranges along the western margin of the basin (Bucknam and Anderson, 1979). Unfortunately, available seismic reflection data do not cover this area, so the relations of this faulting to the detachment surface remains unknown. Bucknam and Anderson (1980) mapped a second scarp of probable Quaternary age (the Clear Lake scarp) in the Sevier Desert Basin. This scarp was crossed by a seismic reflection line (line 3, Fig. 6) and appears to correlate with a major listric fault that intersects the detachment at a depth of 2.0 to 2.3 km. Both of these scarps are similar in appearance to scarps in alluvium elsewhere in the Basin and Range province. The distinction of faults that terminate at such shallow depths from those that extend to 10-km depth or more poses a major problem for earthquake hazard assessment in the Basin and Range province; nevertheless, the potential earthquake hazard for faults that are so limited vertically will be far less than that for alluvial scarps over major, deeply penetrating normal fault zones.

The lateral extent of the thin-skin style extension seen in the Sevier Desert Basin is not known. Seismic reflection data from the Raft River region near the Nevada-Utah-Idaho border area indicate a similar style of thin-skin extension over a shallow detachment fault (Covington, 1980). The gravity data and surface mapping indicate a rather unusual pattern of segmentation of the crust in the Sevier

Desert region in contrast to surrounding areas, which are cut by deep intervening basins. In addition, the earthquake focal-depth distribution provides strong evidence that if a similar style of deformation is occurring elsewhere the detachment faults must be much deeper (>10 to 20 km) than the 2- to 4-km depth in the Sevier Desert Basin.

Another aspect of the problem associated with the subsurface configuration of the major normal fault zones is the relation between the mapped surface pattern of faults and the seismically active fault planes at depth. In the best-studied earthquake sequence in the area, the Pocatello Valley (M = 6.0) 1976 earthquake, the main shock and aftershock distributions as well as the focal-mechanism data appear to define seismically active fault planes that trend obliquely to the trend of the main range block and the mapped faults in the area (Arabasz and others, 1981; Bache and others, 1980). Similar, but not as complete, data for the Cache Valley (M = 5.7) 1962 earthquake and for subsequent activity in the region show that the events occurred beneath the range (east of the major west-dipping Cache Valley fault zone) and seem to indicate a major east-dipping fault zone beneath the range (Westphal and Lange, 1966).

Similarly, for the Fairview Peak, Nevada (M = 7.25) earthquake of 1954 the surface ground breakage along a preexisting range front fault does not correlate with the fault plane determined from an earthquake focal-mechanism study or the aftershock distribution. There the nodal plane selected as the fault plane trends N11°W (Romney, 1957; Stauder and Ryall, 1967), whereas the range-bounding fault and surface rupture resulting from the earthquake trends N12°E (Slemmons, 1957). Support for the focal mechanism comes from an excellent correspondence between the horizontal component of the slip vector determined from the focal mechanism (Romney, 1957) and geodetically determined horizontal components of slip in the epicentral region (Whitten, 1957).

Another excellent example of the discordance between surface fault patterns and earthquake-related fault slip was noted by Wallace (1979, 1980) for the 1915 (M = 7.8) Pleasant Valley, Nevada, earthquake. Faults that were activated during the earthquake all lay within a band 6 km wide and 60 km long trending N10° to 20°E which crossed the trends of four separate fault-bounded range blocks. Individual preexisting faults, even though they extended far beyond the band, broke only along segments that were restricted to this relatively narrow band (Wallace, 1980).

These observations emphasize that a possible mismatch between surficial structural trends and subsurface deformation may help explain the general lack of correlation between the seismicity in north-central Utah and the major fault zones bounding the basin-and-range blocks. The seismicity beneath the range blocks could be related to low-angle fault planes that extend beneath the ranges. However, as noted above, the nodal-plane data for earthquake focal mechanisms in northern Utah appear to rule out seismic slip on major low-angle surfaces (<30° dip).

CONCLUSIONS

The region in northern Utah west of the Wasatch–Cache Valley faults exhibits classic normal fault-bounded basin-range structure with many active fault scarps in contrast to the region to the east, which shows only minor evidence for normal faulting and probably represents a transitional zone to the tectonically more stable Colorado Plateau and Rocky Mountains provinces. The thermal and crustal structure in this transition zone is similar to the Basin and Range province (Shuey and others, 1973; Thompson and Zoback, 1979), although the lack of major deformation indicates much lower strain rates.

Extension in the area has resulted in segmentation and minor tilting of the upper crust into the modern basin-and-range blocks. The gravity data indicate that the subsurface structure of basins includes both symmetric graben and asymmetric tilted blocks. Maximum thicknesses of basin fill vary between 1.2 and 3.8 km. Minimum vertical offsets across the major fault zones are from about 3 to 5 km. The averaged long-term uplift rate of 0.4 mm/yr for the Wasatch fault zone proposed by Naeser and others (1980) based on fission-track dating implies a minimum age of 12 m.y. for the onset of basin-range faulting in this region.

Major east-west-trending transverse zones (in some areas structural ridges and in others possible fault zones with components of vertical and horizontal offset across them) commonly truncate the individual basins. These east-west transverse features may be controlled by preexisting structures. The effect of these transverse ridges on fault behavior, such as possibly limiting the fault length in a single major seismic event, requires further investigation. At least one example exists along the Wasatch fault zone south of Salt Lake City where Quaternary fault scarps extend continuously across a major east-west zone (defined by the gravity data), indicating a major change in basin subsurface structure.

The subsurface configuration of the major fault zones remains a mystery. Bedrock scarps and Quaternary scarps in alluvium generally dip to the west and occur on the west sides of ranges. These west-dipping normal fault zones, coupled with the general eastward tilt of the range blocks in the area favor a listrict-fault or tilted-block model to account for extension and suggest the presence of detachment or decoupling zones at some depth in the crust. Stratigraphically controlled, gently westward-dipping Mesozoic thrust faults within the region have been suggested as likely candidates for these detachment surfaces. Complicating this simple configuration of extensional tectonism are major east-dipping normal fault zones recognized in the gravity data. These data suggest that a graben-type model

for extension must apply locally. Perhaps the most significant contradiction to the simple west-dipping master listric fault model comes from available earthquake focal mechanisms that rule out seismic slip on low-angle surfaces (dips <30°) at focal depths less than 10 to 12 km. This apparent contradiction to the listric fault model based on nodal-plane dips can be countered with the suggestion that slip on gently dipping detachment surfaces or the low-dip segments of listric normal faults occurs aseismically and that only the high-angle portions of the normal faults slip seismically (Smith, 1981). A reported bimodal distribution of earthquake focal depths centered at 2 to 3 km and 7 to 8 km lends support to this hypothesis; however, 20% of the events are at depths greater than 10 km, suggesting some deep fault penetration by relatively steep fault planes. Depths greater than 10 km are probably within the Precambrian crystalline basement in most of the region, hence these faults would have cut any low-angle thrusts or detachment zones within the Phanerozoic section.

Additional difficulties arise in trying to apply seismicity data to the mapped pattern of surface faulting. As has been noted previously by numerous workers, the correlation in this region between seismicity and both the surface faults and the major fault zones identified with the gravity data is poor. One possible explanation for this lack of correlation is that major slip events at depth may be more representative of the lower crustal geometry of extension with the surface faults accommodating the strain by slip on preexisting fault surfaces (Wallace, 1980). In any case, the few well-studied earthquake-aftershock sequences within the Basin and Range province (both in north-central Utah and in Nevada) all indicate that the major seismically active fault plane at depth trends obliquely to the surface trend of the basin-range blocks and the faults that bound them. Examples exist in both Nevada and Utah in which the seismic fault plane apparently passes beneath several range blocks.

Recent extensional deformation across the region results from an east-west to west-southwest–east-northeast oriented least principal stress/strain axis. Analysis of slickenside data along the major fault zones with Quaternary activity extends a similar pattern of deformation back possibly 0.5 to 1.0 m.y. This least principal stress/strain axis represents an approximately 30° counterclockwise rotation of deformational axes relative to most of the rest of the northern Basin and Range province. The counterclockwise rotation may be due to a superposition of approximately north-south compression related to the clockwise rotation of the Colorado Plateau kinematically required by the extensional opening in the Rio Grande rift (G. A. Thompson, 1982, oral commun.). Alternately, the local east-west to west-southwest–east-northeast least principal stress/strain in north-central Utah may be a result of the prominent north-south structural grain in the region which was probably inherited from the late Precambrian rifted margin.

ACKNOWLEDGMENTS

Special thanks go to T. G. Hildenbrand for providing a tape of the gravity data and to R. B. Smith for providing the seismicity map. Conversations with M. D. Crittenden, Jr., regarding structure in the Jordan Valley area were extremely helpful. Reviews by R. E. Anderson, C. Campbell, D. P. Hill, R. B. Smith, and R. E. Wallace greatly improved the manuscript, and their prompt attention to it was truly appreciated.

REFERENCES CITED

Anderson, L. W., and Miller, D. G., 1979, Quaternary fault map of Utah, Long Beach, California, Fugro, Inc., scale 1:500,000.

Anderson, R. E., and Zoback, M. L., 1981a, The subsurface geometry of faults and the evolution of basins in the northern Basin and Range province [abs.], *in* Papers Presented to the Conference on Processes of Planetary Rifting: Houston, Lunar and Planetary Institute, Contribution no. 457, p. 93–95.

——1981b, Subsurface geometry of normal faults in the northern Basin and Range province inferred from seismic reflection data: EOS (American Geophysical Union Transactions), v. 62, p. 960.

Anderson, R. E., Zoback, M. L., and Thompson, G. A., 1982, Implications of selected subsurface data on the structural form and evolution of some basins in the northern Basin and Range province, Nevada and Utah: Geological Society of America Bulletin (in press).

Angelier, J., 1977, La reconstitution dynamique et geometrique de la tectonique defailles a partir de mesures locales (pland de faille, stries, sens de jeu, rejets): quelques precisions: Paris, Comptes Rendus Academie des Sciences, Series D, v. 285. p. 637–640.

——1979, Determination of the mean principal directions of stresses for a given fault population: Tectonophysics, v. 56, p. T17–26.

Arnow, T., and Mattick, R. E., 1968, Thickness of valley fill in the Jordan Valley east of the Great Salt Lake, Utah: U.S. Geological Survey Professional Paper 600-B, p. B79–82.

Arabasz, W. J., Smith, R. B., and Richins, W. D., 1980, Earthquake studies along the Wasatch front, Utah: Network monitoring, seismicity, and seismic hazards: Seismological Society of America Bulletin, v. 70 p. 1479–1500.

Arabasz, W. J., Richins, W. D., and Langer, C. J., 1981, The Pocatello Valley (Idaho-Utah border) earthquake sequence of March to April 1975: Seismological Society of America Bulletin, v. 71, p. 803–826.

Bache, T. C., Lambert, D. G., and Barker, T. G., 1980, A source model for the March 18, 1975, Pocatello Valley earthquake from time domain modeling of teleseismic P waves: Seismological Society of America Bulletin, v. 70, p. 405–418.

Bucknam, R. C., and Anderson, R. E., 1979, Estimation of fault scarp ages from a scarp height-slope angle relationship: Geology, v. 7, p. 11–14.

——1980, Map of fault scarps on unconsolidated sediments, Delta 1° x 2° quadrangle, Utah: U.S. Geological Survey Open-File Report 79-366, scale 1:250,000.

Carey, E., 1979, Recherche des directions principales de contraintes associees au jeu d'un population de failles: Revue de Geologie Dynamique et de Geographie Physique, v. 21, p. 57–67.

Chamberlain, R. M., 1978, Structural development of the Lemitar Mountains, an intrarift tilted fault-block uplift, central New Mexico, *in*

International Symposium on the Rio Grande Rift, Programs and Abstracts: Los Alamos, New Mexico, Los Alamos Scientific Laboratory, p. 22–24.

Cluff, L. S., Hintze, L. F., Brogan, G. E., and Glass, C. E., 1975, Recent activity of the Wasatch fault, northwestern Utah, U.S.A.: Tectonophysics, v. 29, p. 161–168.

Cook, K. L., and Berg, J. W., Jr., 1961, Regional gravity survey along the central and southern Wasatch front, Utah: U.S. Geological Survey Professional Paper 316-E, p. 75–89.

Cook, K. L., Halverson, M. D., Stepp, J. C., and Berg, J. W., Jr., 1964, Regional gravity survey of the northern Great Salt Lake Desert and adjacent areas in Utah, Nevada, and Idaho: Geological Society of America Bulletin, v. 75, p. 715–740.

Cook, K. L., and Berg, J. W., Jr., Johnson, W. W., and Novotny, R. T., 1966, Some Cenozoic structural basins in the Great Salt Lake area, Utah, indicated by regional gravity surveys: Guidebook to Geology of Utah, Utah Geological Society, v. 20, p. 57–75.

Cook, K. L., and Berg, J. W., Jr., and Lum, D., 1967, Seismic and gravity profile across the northern Wasatch trench, Utah, *in* Musgrove, A. W., ed., Seismic refraction prospecting: Tulsa, Oklahoma, Society of Exploration Geophysicists, p. 539–549.

Covington, H. R., 1980, Subsurface geology of the Raft River geothermal area, Idaho: Geothermal Resources Council, Transactions, v. 4, p. 113–115.

Crittenden, M. D., Jr., 1964, General geology of Salt Lake County: Utah Geological and Mineralogical Survey Bulletin 69, p. 11–48.

——1976, Stratigraphic and structural setting of the Cottonwood area, Utah, *in* Hill, J. G., ed., Geology of the Cordilleran hingeline: Denver, Rocky Mountain Association of Geologists, p. 281–317.

Crittenden, M. D., Jr., Granger, A. E., Sharp, B. J., and Calkins, F. C., 1952, Geology of the Wasatch Mountains east of Salt Lake City: Utah Geological Society Guidebook, v. 8, p. 1–37.

Diment, W. H., and Urban, T. C., 1981, Average elevation map of the conterminous United States (Gilluly Averaging Method): U.S. Geological Survey Geophysical Investigations Map GP-933, scale 1:2,500,000.

Doser, D. I., and Smith, R. B., 1982, Seismic moment rates in the Utah region: Seismological Society of America Bulletin, v. 72, p. 525–551.

Eardley, A. J., 1939, Structure of the Wasatch–Great Basin region: Geological Society of America Bulletin, v. 50, p. 1277–1310.

——1944, Geology of the north-central Wasatch Mountains, Utah: Geological Society of America Bulletin, v. 55, p. 819–894.

——1969, Willard thrust and the Cache uplift: Geological Society of America Bulletin, v. 86, p. 669–680.

Effimoff, I., and Pinezich, A. R., 1981, Tertiary structural development of selected valleys based on seismic data: Basin and Range province, northeastern Nevada: Philosophical Transactions of the Royal Society of London A, v. 300, p. 435–442.

Everitt, B. L., and Kaliser, B. N., 1980, Geology for assessment of seismic risk in the Tooele and Rush Valleys, Tooele County, Utah: Utah Geological and Mineralogical Survey Special Studies, no. 51, 33 p.

Galyardt, G. L., and Rush, F. E., 1979, Geologic map of the Crater Springs known geothermal resources area and vicinity, Juab and Millard Counties, Utah: U.S. Geological Survey Open-File Report, 79-1158, scale 1:24,000.

Gilbert, G. K., 1928, Studies of basin-range structure: U.S. Geological Survey Professional Paper 153, 89 p.

Gilluly, J., 1928, Basin-range faulting along the Oquirrh Range, Utah: Geological Society of America Bulletin, v. 39, p. 1103–1130.

Harr, C. J., and Mabey, D. R., 1976, Gravity survey of Pocatello Valley, Idaho and Utah: U.S. Geological Survey Open-File Report 76-766, 12 p.

Hintze, L. F., 1980, Geologic map of Utah: Salt Lake City, Utah, Utah Geological and Mineralogical Survey Map, scale 1:500,000.

Isherwood, W. F., 1967, Regional gravity survey of parts of Millard, Juab, and Sevier Counties, Utah [M.S. Thesis]: Salt Lake City, University of Utah, 32 p.

Johnson, J. B., Jr., and Cook, K. L., 1957, Regional gravity survey of parts of Tooele, Juab, and Millard Counties, Utah: Geophysics, v. 22, p. 48–61.

Kaliser, B. B., 1976, Earthquake fault map of a portion of Salt Lake County, Utah: Utah Geological and Mineral Survey, Map 42, scale 1:150,000.

Lindsey, D. A., Glanzman, R. K., Naeser, C. W., and Nichols, D. J., 1981, Upper Oligocene evaporites in basin fill of Sevier Desert region, western Utah: American Association of Petroleum Geologists Bulletin, v. 65-2, p. 251–260.

Mabey, D. R., and Morris, H. T., 1967, Geologic interpretation of gravity and aeromagnetic maps of Tintic Valley and adjacent areas, Tooele and Juab counties, Utah: U.S. Geological Survey Professional Paper 516-D, p. 1–10.

Mabey, D. R., Tooker, E. W., and Roberts, R. J., 1963, Gravity and magnetic anomalies in the northern Oquirrh Mountains, Utah: U.S. Geological Survey Professional Paper 450-E, p. E28–31.

Mabey, D. R., Zeitz, I., Eaton, G. P., and Kleinkopf, M. D., 1978, Regional magnetic patterns in part of the Cordillera in the western United States: Geological Society of America Memoir 152, p. 93–106.

Mattick, R. E., 1970, Thickness of unconsolidated to semiconsolidated sediments in Jordan Valley, Utah: U.S. Geological Survey Professional Paper 700-C, p. C119–124.

McDonald, R. E., 1976, Tertiary tectonics and sedimentary rocks along the transition: Basin and Range province to Plateau and thrust belt province, Utah, in Hill, J. G., ed., Geology of the Cordilleran hingeline: Denver, Rocky Mountain Association of Geologists, p. 281–317.

McKenzie, D. P., 1969, The relationship between fault plane solutions for earthquakes and the directions of the principal stresses: Seismological Society of America Bulletin, v. 59, p. 591–601.

Mehnert, H. H., Rowley, P. D., and Lipman, P. W., 1978, K-Ar ages and geothermal implications of young rhyolites in west-central Utah: Isochron/West, no. 21, p. 3–7.

Morris, H. T., 1977, Geologic map and sections of the Furner Ridge quadrangle, Juab County, Utah: U.S. Geological Survey Miscellaneous Investigations Map, I-1045.

Morton, W. H., and Black, R., 1975, Crustal attenuation in Afar, in Pilger, A., and Rosler, A., eds., Afar depression of Ethiopia, Inter-Union Commission on Geodynamics: International Symposium on the Afar Region and Related Rift Problems, Stuttgart, Germany, E. Schweizerbart'sche Verlagsbuchhandlung, Proceedings, Scientific Report No. 14, p. 55–65.

Naeser, C. W., Bryant, B. R., Crittenden, M. D., Jr., and Sorenson, M. L., 1980, Fission-track dating in the Wasatch Mts., Utah: U.S. Geological Survey Open-File Report 80-801, p. 634–646.

Pack, F. J., 1926, New discoveries relating to the Wasatch fault: American Journal of Science, v. 11, p. 399–410.

Pavlis, T. L., and Smith, R. B., 1980, Slip vectors on faults near Salt Lake City from Quaternary displacements and seismicity: Seismological Society of America Bulletin, v. 70, p. 1521–1526.

Peterson, D. L., 1974, Bouguer gravity map of part of the northern Lake Bonneville basin, Utah and Idaho: U.S. Geological Survey Miscellaneous Field Studies Map MF-627, scale 1:250,000.

Peterson, D. L., and Oriel, S. S., 1970, Gravity anomalies in Cache Valley, Cache and Box Elder Counties, Utah, and Bannock and Franklin Counties, Idaho: U.S. Geological Survey Professional Paper 700-C, p. C114–118.

Peterson, J., Turley, C., Nash, W. P., and Brown, F. H., 1978, Late Cenozoic basalt-rhyolite volcanism in west-central Utah: Geological Society of America Abstracts with Programs, v. 10, p. 236.

Prescott, W. H., Savage, J. C., and Kinoshita, W. T., 1979, Strain accumulation rates in the western United States between 1970 and 1978:

Journal of Geophysical Research, v. 84, p. 5423–5435.

Raleigh, C. B., Healy, J. H., and Bredehoeft, J. D., 1972, Faulting and crustal stress at Rangely, Colorado: American Geophysical Union Monograph, v. 16, p. 275–284.

Roberts, R. J., Crittenden, M. D., Jr., Tooker, E. W., Morris, H. T., Hose, R. K., and Cheney, T. M., 1965, Pennsylvanian and Permian basins in northwestern Utah, northeastern Nevada and south-central Idaho: American Association of Petroleum Geologists Bulletin, v. 49, p. 1926–1956.

Romney, C., 1957, Seismic waves from the Dixie Valley–Fairview Peak (Nevada) earthquakes: Seismological Society of America Bulletin, v. 47, p. 301–319.

Shuey, R. T., Shellinger, D. K., Johnson, E. G., and Alley, L. B., 1973, Aeromagnetics and the transition between the Colorado Plateau and Basin and Range provinces: Geology, v. 1, p. 107–110.

Slemmons, D. B., 1957, The Dixie Valley–Fairview Peak, Nevada, earthquakes of December 16, 1954: Geological effects: Seismological Society of America Bulletin, v. 47, p. 353–376.

Smith, R. B., 1978, Seismicity, crustal structure, and intraplate tectonics of the interior of the western Cordillera: Geological Society of America Memoir 152, p. 111–144.

Smith, R. B., 1981, Listric faults and earthquakes, what evidence?: EOS (American Geophysical Union Transactions), v. 62, p. 961.

Smith, R. B., and Lindh, A. G., 1978, Fault-plane solutions of the western United States: A compilation: Geological Society of America Memoir 152, p. 107–110.

Smith, R. B., and Sbar, M. L., 1974, Contemporary tectonics and seismicity of the western United States with emphasis on the intermountain seismic belt: Geological Society of America, v. 85, p. 1205–1218.

Soler, T., Snay, R. A., and Smith, R. B., 1981, Geodetically derived strain near Salt Lake City, Utah: EOS (American Geophysical Union Transactions), v. 62, p. 394.

Stauder, W., and Ryall, A., 1967, Spatial distribution and source mechanism of microearthquakes in central Nevada: Seismological Society of America Bulletin, v. 57, p. 1317–1345.

Stewart, J. H., 1978, Basin and range structure in western North America: A review: Geological Society of America Memoir 152, p. 1–31.

—— 1980, Regional tilt patterns of late Cenozoic basin-range fault blocks, western United States: Geological Society of America Bulletin, v. 91, p. 460–464.

Stewart, S. W., 1958, Gravity survey of Ogden Valley in the Wasatch Mountains, north-central Utah: EOS (American Geophysical Union Transactions), v. 39, p. 1151–1157.

Stokes, W. L., 1976, What is the Wasatch Line? in Hill, J. G., ed., Geology of the Cordilleran hingeline: Denver, Rocky Mountain Association of Geologists, p. 1–25.

Swan, F. H., III, Schwartz, D. P., and Cluff, L. S., 1980, Recurrence of moderate to large magnitude earthquakes produced by surface faulting on the Wasatch fault zone, Utah: Seismological Society of America Bulletin, v. 70, p. 1431–1462.

Thompson, G. A., and Zoback, M. L., 1979, Regional geophysics of the Colorado Plateau: Tectonophysics, v. 61, p. 149–181.

Wallace, R. E., 1979, Strain pattern represented by scarps formed during the earthquakes of October 2, 1915, Pleasant Valley, Nevada: Tectonophysics, v. 52, p. 599.

—— 1980, Listric-faulting and seismicity, northern Nevada [abs.]: Earthquake Notes, v. 50, no. 4, p. 67–68.

Wernicke, B., 1981, Low-angle normal faults in the Basin and Range province: Nappe tectonics in an extending orogen: Nature, v. 291, p. 645–648.

Wernicke, B., and Burchfiel, B. C., 1982, Modes of extensional tectonics: Journal of Structural Geology, v. 4, p. 105–115.

Westphal, W. H., and Lange, A. L., 1966, The distribution of earthquake aftershock foci, Cache Valley, Utah, September 1962: EOS (American Geophysical Union Transactions), v. 47, p. 428.

Whitten, C. A., 1957, The Dixie Valley–Fairview Peak, Nevada, earthquakes of December 16, 1954: Geodetic measurements: Seismological Society of America Bulletin, v. 47, p. 321–328.

Zandt, G., and Richins, W. D., 1981, Interaction of high- and low-angle normal faults along the eastern Basin and Range, northern Utah: EOS (American Geophysical Union Transactions, v. 62, p. 960.

Zoback, M. D., Zoback, M. L., Svitek, J., and Liechti, R., 1981, Hydraulic-fracturing stress measurements near the Wasatch fault, central Utah: EOS (American Geophysical Union Transactions), v. 62, p. 394.

Zoback, M. L., 1981, State of stress inferred from fault slip data in the northern Basin and Range: EOS (American Geophysical Union Transactions), v. 62, p. 394.

Zoback, M. L., and Zoback, M. D., 1980a, Faulting patterns in north-central Nevada and strength of the crust: Journal of Geophysical Research, v. 85, p. 275–284.

——1980b, State of stress in the conterminous United States: Journal of Geophysical Research, v. 85, p. 6113–6156.

MANUSCRIPT ACCEPTED BY THE SOCIETY AUGUST 20, 1982

Printed in U.S.A.

Geological Society of America
Memoir 157
1983

Fission-track ages of apatite in the Wasatch Mountains, Utah: An uplift study

C. W. Naeser
Bruce Bryant
U.S. Geological Survey
Box 25046, Denver Federal Center
Denver, Colorado 80225

M. D. Crittenden, Jr.*
M. L. Sorensen
U.S. Geological Survey
345 Middlefield Road
Menlo Park, California 94025

ABSTRACT

Apatite fission-track ages from basement rocks in the Wasatch Mountains between Ogden and Bountiful, Utah, range from 5 m.y. near the Wasatch fault along the west margin of the mountains to 94 m.y. on the crest of the range and show a correlation with altitude within individual fault blocks. Analysis of these ages yields an uplift rate of 0.4 mm/yr for the last 10 m.y. for the fault blocks having the most sustained rapid uplift. A variety of intermediate ages of apatite from the crest of the range and at lower altitudes is interpreted as representing fault blocks with rocks having mixed ages due to various degrees of tilting and varying but lesser amounts of uplift.

INTRODUCTION

Several studies over the past decade have applied fission-track dating of apatite to tectonic and uplift problems (Naeser and Faul, 1969; Wagner and Reimer, 1972; Wagner and others, 1977; Naeser, 1979a, 1979b; Bryant and Naeser, 1980). Naeser and Faul (1969) showed that fission tracks in apatite would be annealed at temperatures as low as about 150°C if the sample were held at that temperature for at least 10^6 yr. This conclusion came from extrapolating laboratory annealing data to times and temperatures of geologic significance. Naeser and Forbes (1976), Brookins and others (1977), and Naeser (1981) have studied the change of apparent age of apatite at increasing temperatures with depth in drill holes. Both the laboratory and drill-hole data show that the duration of the thermal event determines the temperature at which the apatite gives a "zero" apparent age. A rapid cooling rate (equivalent to a

short thermal event) requires a higher temperature to achieve a zero apparent age than does a slower cooling rate. The drill-hole data suggest somewhat lower annealing temperatures than those indicated by the laboratory experiments.

The apparent age of apatite records the time when the rock containing the dated apatite passed through the appropriate annealing temperature. It is therefore possible to calculate an uplift rate from the apatite age and a postulated geothermal gradient. Uplift rates can also be determined when more than one apatite age is available in vertical sequence; such a calculation has the advantage that it does not require knowledge of either the geothermal gradient or annealing temperature, if the cooling rate was constant over that interval.

This study resulted from an apatite age reported by Crittenden and others (1973) of 8.5 ± 1.0 m.y. from their westernmost sample of the Little Cottonwood stock 60 km south of this study area. Because the Little Cottonwood stock is about 26 m.y. old (Crittenden and others, 1973),

*Deceased.

C. W. Naeser and Others

the young apatite age near the Wasatch fault was interpreted as reflecting recent cooling (uplift). To estimate the rate of recent uplift of the Wasatch Mountains this study was undertaken in an area of Precambrian basement rock types that contained apatite in quantities suitable for dating.

We present here apatite fission-track ages from a 30-km-long segment of the Wasatch Mountains between Ogden and Bountiful, Utah (Fig. 1). The Wasatch Mountains form the eastern boundary of the Basin and Range physiographic province, although geophysical parameters of the crust underlying the Basin and Range province extend 50 km farther east (Shuey and others, 1973). The segment of the Wasatch Mountains in this study consists of a rather even crested block 5 to 10 km wide and 2,600 to 2,950 m in altitude rising above the Great Salt Lake (1,280 m) on the west side and the Morgan valley (1,500 m) on the east side. The range is underlain by Precambrian crystalline rocks of the Archean and Proterozoic Farmington Canyon Complex. No Tertiary intrusive igneous rocks have been found in this part of the Wasatch Mountains; therefore, heat introduced by magma is not a problem in the interpretation of these apatite fission-track ages.

GEOLOGIC SETTING

Farmington Canyon Complex is composed of a variety of high-grade metamorphic rocks derived from sedimentary and igneous rocks metamorphosed about 2,600 m.y. ago and intruded and remetamorphosed about 1,800 m.y. ago (Hedge and others, 1983). It was subjected to retrogressive metamorphism in the late Proterozoic and/or Cretaceous (Bryant, 1980). During Paleozoic and most of Mesozoic time the basement rocks were buried under an increasing thickness of sedimentary rock. Extrapolation across what is possibly a major thrust fault south of the Farmington Canyon Complex to the section in the Parleys Canyon syncline suggests that 7 or 8 km of sedimentary rock overlay the basement rocks by Coniacian time in the Late Cretaceous (Granger, 1953; Crittenden and Van Horn, *in* U.S. Geological Survey, 1978, p. 71; Bryant, unpub. data).

Episodic uplift accompanying movement on thrust sheets during the Sevier orogeny progressed from west to east throughout most of Cretaceous time. The youngest sedimentary rocks close to the Farmington Canyon Complex that are older than the main Seiver orogenic movements are Coniacian or about 82 to 88 m.y. old. The oldest nearby dated rocks younger than the main thrust movements are near East Canyon Reservoir (about 18 km to the east) and are Maestrichtian and, perhaps, late Campanian (about 68 m.y. old; Nichols and Warner, 1978).

The Farmington Canyon Complex was uplifted during

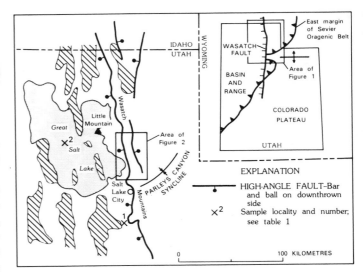

Figure 1. Geographic location and setting of study area in Wasatch Mountains, Utah, from King and Beikman (1974), and location of dated tuffs in Salt Lake Formation.

the Sevier orogeny 88 m.y. ago or later. We have no firm evidence of when the basement rocks were first exposed, except that exposure occurred before the deposition of the Wasatch Formation began about 55 m.y. ago. Approximately 500 m of Wasatch Formation was deposited on the Farmington Canyon Complex, and on the east side of the Wasatch block about 1,500 m of tuff and tuffaceous sediment were deposited above the Wasatch Formation during the interval 35 to 40 m.y. ago. Uplift of the modern Wasatch Mountains is reflected in the appearance of basement clasts in conglomerates exposed just southwest of the Farmington Canyon Complex. We do not know the precise age of those conglomerates except that they are younger than 35- to 37-m.y.-old volcanic rocks in the Salt Lake salient (Van Horn, 1981).

The Farmington Canyon Complex is exposed in a horst bounded by the Wasatch fault on the west and a complex system of faults on the east (Fig. 1). The fault zone on the east side dies out south of Morgan Valley. Many fault and fracture systems also occur within the range south of the Weber River and north of Farmington Canyon (Bell, 1952; Bryant, 1979). Minimum Neogene structural relief between basement rocks under Morgan Valley and the crest of the range is 3.3 km. Gravity data in the Farmington graben west of the range suggest that there is at least 5 km of relief between the crest of the range and the base of the graben fill (McDonald, 1976). The time of formation of the modern basins and ranges in the Salt Lake region has not been precisely determined, but fission-track ages from tuffs in the Salt Lake Formaton in the Salt Lake Valley indicate that deposition of that formation, which forms much of the basin fill, began before 10 m.y. ago (Table 1).

SAMPLES AND PROCEDURES

Fifty-eight apatite samples were dated in this study. Fifty-six of these were from forty-six localities in the Farmington Canyon Complex of the Wasatch Mountains (Fig. 2) and two were from upper Proterozoic diamictite-exposed Little Mountain, Utah, a horst (Peterson, 1974; McDonald, 1976) 30 km west of the Wasatch Mountains in the Basin and Range province (Fig. 1). Most of the Wasatch Mountain samples are from the range crest or west of the crest; only in Weber Canyon do we have samples from the eastern margin. We also lack samples from the numerous fault blocks east of the crest. The Wasatch Formation overlies the Farmington Canyon Complex in most of these blocks, and we expect that the apatite in them would yield ages greater than 55 m.y.

The procedures used to date the apatite are described by Naeser (1979b). All but two of the apatite concentrates were dated using the population method. The splits used for the determination of induced track density were annealed (500° for 12 h) before being irradiated. The other two samples, 41 and 42, were dated with the external detector method. The apatites were all etched in 7% HNO3 for 30 s at 25°C. The neutron fluence was determined with National Bureau of Standards Standard Reference Materials 962 and 963. The primary copper activation calibration as determined by Carpenter and Reimer (1974) was used in calculating the doses used for calculating the ages reported in Tables 1 and 2. Errors on population method ages calculated by combining standard error of mean of induced and fossil counts.

RESULTS

While the apatites studied range in age from 4.9 m.y. to 94.4 m.y. (nos. 3, 38, Table 2), there is a general increase in age with altitude within the mountain block (Fig. 3). Samples collected at lower elevations on the west side near the Wasatch fault have apatite fission-track ages as young as 5

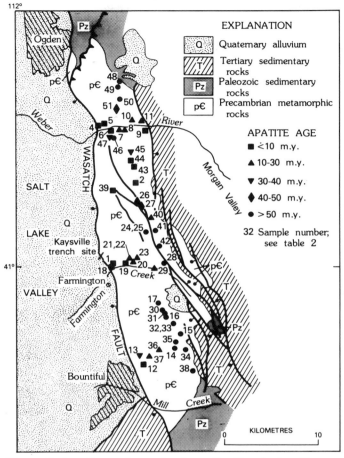

Figure 2. Locations of samples from the Farmington Canyon Complex in the Wasatch Mountains. Geology simplified from Bryant (1979).

m.y., wheras most samples along the crest of the range have ages greater than 50 m.y. The oldest sample (no. 38) has an age of 94.4 m.y. and was collected directly beneath the Wasatch Formation at 2,615-m altitude on the crest of the range. Four samples from the crest of the range (nos. 2, 40, 43, and 44) have "low ages" (Fig. 2), but they are all from

TABLE 1. FISSION-TRACK DATA FOR ZIRCON FROM SALT LAKE FORMATION IN THE SALT LAKE VALLEY

No.	Sample	Altitude (m)	Number of grains counted	s,r*	Fossil tracks/cm² x 10⁶	Induced tracks/cm² x 10⁵	Neutrons/cm² x 10⁵	T ± 2σ (m.y.)
1.	DF-2897 J-2-e	1422	6	0.98(r)	1.10 (230)	1040 (1083)	1.03	6.5 ± 0.5
2.	DF-3224 Am-2	-1451	6	0.81(r)	0.881 (204)	5.26 (609)	1.03	10.3 ± 1.0

Note: Sample localities and descriptions:
1. Jordan Narrows unit of Salt Lake Group of Slentz (1955); 294 m S60°W of dam at Jordan Narrows, 40°28'28"N 111°55'30"W. Very light gray airfall tuff containing glass and fragments of plagioclase crystals.
2. Salt Lake Formation from 2,721 m depth in Amoco Production Co., Bridge State of Utah well in Great Salt Lake; 41°26'28"N, 112°33'30"W. Green, water-laid, crystal-vitric tuff with some of the glass altered to analcite and clay. Some euhedral crystals and anhedral crystal fragments of plagioclase as much as 1.2 mm long and with a composition of An 45 to 52. Minor amounts of chlorite, a zeolite, and zircon.
*Altitude relative to sea level.

TABLE 2. FISSION-TRACK DATA FOR APATITE FROM THE PRECAMBRIAN OF THE WASATCH MOUNTAINS AND LITTLE MOUNTAIN, UTAH

Map No.	Sample	Elevation (m)	Number of grains counted	s,r*	Fossil tracks/cm^2 X10^6	Induced tracks/cm^2 X10^6	Neutrons/cm^2 X10^1	T±2σ (m.y.)
Wasatch Mts.								
1	72MC113g DF-220	1454	50[+]	0.07(s)	0.122(254)[§]	1.09(2270)[§]	1.19	8.0 ± 1.1
2	72MC-116 DF-221	2958	50/40	0.06(s)	0.194(405)	1.44(2408)	1.19	9.6 ± 1.0
3	72MS-339 DF-22	1890	50	0.09(s)	0.030(28)	0.441(408)	1.19	4.9 ± 1.9
4	73MS27a DF-606	1448	50	0.06(s)	0.164(152)	1.43(1323)	1.13	7.8 ± 1.3
4	73MS27b DF-607	1448	50	0.08(s)	0.131(121)	1.67(1546)	1.13	5.3 ± 1.0
5	73MS28a DF-608	1433	50	0.09(s)	0.238(220)	2.04(1887)	1.13	7.9 ± 1.1
5	73MS28b DF-609	1433	50	0.07(s)	0.136(126)	1.61(1487)	1.13	5.7 ± 1.1
6	73MS29a DF-610	1463	50	0.12(s)	0.104(96)	0.646(598)	1.13	10.8 ± 2.4
6	73MS29b DF-611	1463	50	0.10(s)	0.194(180)	1.75(1617)	1.13	7.5 ± 1.2
7	73MS30 DF-612	1433	50	0.08(s)	0.248(230)	1.16(1076)	1.13	14.4 ± 2.1
8	73MS31a DF-613	1463	50	0.08(s)	0.202(187)	1.15(1065)	1.13	11.9 ± 1.9
8	73MS31b DF-614	1463	50	0.14(s)	0.132(122)	0.717(664)	1.13	12.4 ± 2.4
9	73MS32 DF-615	1509	100/50	0.07(s)	0.076(141)	0.571(529)	1.13	9.0 ± 1.7
10	73MS33a DF-616	1463	50	0.06(s)	0.377(349)	1.72(1595)	1.13	14.7 ± 1.7
10	73MS33b DF-617	1463	50	0.11(s)	0.397(368)	1.43(1326)	1.13	18.7 ± 2.2
11	73MS34 DF-618	1463	50	0.08(s)	0.269(248)	0.909(842)	1.13	20.0 ± 2.9
12	73MS36 DF-619	1768	50	0.05(s)	0.181(168)	1.23(1138)	1.13	10.0 ± 1.6
13	73MS37 DF-620	1920	50	0.12(s)	0.149(138)	0.347(321)	1.22	31.3 ± 6.3
14	73MS39 DF-621	2377	50	0.07(s)	0.613(568)	0.733(679)	1.13	56.3 ± 6.4
15	73MS40 DF-622	2824	50	0.11(s)	0.172(159)	0.154(143)	1.13	74.8 ±17.2
16	73MS42 DF-623	2652	50	0.08(s)	1.10(1015)	1.06(985)	1.13	69.3 ± 6.2
17	73MS43b DF-624	2774	50	0.07(s)	0.953(882)	0.972(900)	1.13	65.9 ± 6.2

TABLE 2. (Continued)

Map No.	Sample	Elevation (m)	Number of grains counted	s,r*	Fossil tracks/cm² X10⁶	Induced tracks/cm² X10⁶	Neutrons/cm² X10¹	T±2σ (m.y.)
	Wasatch Mts. (Cont'd)							
18	75MC7 DF-1027	1456	50	0.06(s)	0.252(525)	2.92(6078)	1.45	7.5 ± 0.7
19	75MC8 DF-1028	1646	50	0.09(s)	0.242(505)	2.18(4535)	1.45	9.7 ± 0.9
20	75MC9 DF-1029	1750	50	0.06(s)	0.708(656)	3.92(3630)	1.45	15.7 ± 1.3
21	75MC11 DF-1030	1839	50	0.07(s)	0.302(280)	1.50(1387)	1.45	15.7 ± 2.3
22	75MC12 DF-1031	1839	50	0.09(s)	0.231(2656)	1.28(2656)	1.45	14.7 ± 1.5
23	75MC13 DF-1032	1890	50	0.09(s)	0.293(271)	1.49(1379)	1.45	17.0 ± 1.3
24	75MC14 DF-1033	2908	50	0.10(s)	0.113(236)	8.160(333)	1.45	61.2 ± 10.4
25	75MC15 DF-1034	2908	50	0.10(s)	0.348(726)	0.460(958)	1.45	65.4 ± 3.2
26	75MC16 DF-1035	2893	50	0.07(s)	0.423(882)	0.777(1619)	1.45	47.1 ± 4.0
27	75MC18 DF-1036	2774	50	0.09(s)	0.728(674)	1.36(1264)	1.45	46.1 ± 4.4
28	74MC20 DF-1037	2512	50	0.08(s)	0.903(1881)	1.05(2186)	1.45	74.3 ± 4.6
29	75MC21 DF-1038	2012	50	0.13(s)	0.231(482)	0.505(1052)	1.45	39.6 ± 4.4
30	75M2C23 DF-1039	2707	50	0.08(s)	0.333(693)	0.339(707)	1.45	84.5 ± 9.0
31	75MC24 DF-1040	2707	50	0.05(s)	0.724(1508)	1.04(2175)	1.45	59.9 ± 4.0
32	75MC25 DF-1041	2627	50	0.07(s)	0.391(815)	0.392(815)	1.45	67.3 ± 3.6
33	75MC26 DF-1042	2633	50	0.14(s)	0.883(818)	0.869(805)	1.45	87.6 ± 8.6
34	75MC28 DF-1043	2548	50	0.06(s)	1.47(1359)	1.77(1642)	1.45	31.4 ± 5.2
35	75MC29 DF-1044	2426	50	0.06(s)	0.154(321)	0.204(424)	1.45	65.4 ± 9.6
36	75MC31 DF-1045	2179	50	0.10(s)	0.091(190)	0.406(846)	1.45	19.5 ± 3.1
37	75MC32 DF-1046	2000	50	0.06(s)	0.107(222)	0.671(1397)	1.45	13.8 ± 2.0
38	B-112 DF-1873	2615	50	0.08(s)	0.839(777)	0.586(543)	1.11	94.4 ± 10.5
39	K-12 DF-1874	2234	50	0.05(s)	0.084(176)	0.710(1479)	1.11	7.9 ± 1.3

TABLE 2. (Continued)

Map No.	Sample	Elevation (m)	Number of grains counted	s,r*	Fossil tracks/cm² X106	Induced tracks/cm² X10⁶	Neutrons/cm² X10¹	T±2σ (m.y.).
Wasatch Mts. (Cont'd)								
40	P-77 DF-2164	2816	50	0.21(s)	0.059(74)	0.192(240)	0.961	17.7 ± 4.7
41	P-78 DF-2165	2816	6	0.89(r)	1.11(335)	5.81(853)	4.72	55.2 ± 4.0
42	P-79 DF-2166	2713	6	0.77(r)	1.52(400)	7.50(965)	4.69	56.7 ± 4.4
43	P-87 DF-2952	2901	50	0.10(s)	0.076	0.559	1.07	8.6 ± 3.3
44	P-88 DF-2953	2774	50	0.10(s)	0.031	0.267	1.06	7.3 ± 3.5
45	P-89 DF=2954	2731	50	0.08(5) (461)	0.22 (746)	0.36	1.06	39.1 ± 11.5
46	S-56 DF-2951	2206	50	0.06(s) (1032)	1.11 (1222)	1.32	1.07	53.9 ± 11.2
47	0-66 DF-2946	1974	50	0.08(s) (407)	0.44 (719)	0.78	1.10	37.2 ± 10.6
48	S-54 DF-2949	2791	50	0.07(s) (721)	0.78 (860)	0.93	1.08	54.0 ± 12.5
49	S-53 DF-2948	2740	50	0.11(s) (288)	1.39 (1678)	1.81	1.09	49.9 ± 15.5
50	S-52 DF-2947	2608	50	0.07(s) (871)	0.94 (857)	0.93	1.09	66.0 ± 13.6
51	S-55 DF-2950	2273	50	0.11(s) (198)	0.285 (287)	0.413	1.08	44.4 ± 14.2
Little Mt.								
52	72MC140a DF-223	1292	50	0.07(s)	0.605(1262)	0.571(1189)	1.19	75.2 ± 6.1
53	72MC140 DF-224	1292	50	0.10(s)	0.179(166)	0.190(176)	1.19	66.9 ± 14.5

Note: $\lambda_F = 7.03 \times 10^{-17} yr^{-1}$.

*(s) = relative standard error of the mean of the induced count. r = correlation coefficient.

[†]If one number present, same number of grains counted for fossil and induced counts; if two numbers present, first is number of grains counted for fossil, and second is number of grains counted for the induced count.

[§]Number of tracks counted is determining fossil or induced track density.

an area of rock separated by faults from areas of more "normal" ages for that altitude.

A series of samples collected along the Weber River, in Weber Canyon (4 to 11) at roughly the same elevation, show a slight increase in age from west to east over the first 5 km. The apatite from the easternmost sample (9), near the fault bounding the eastern margin of the mountain block, has a younger fission-track age.

DISCUSSION OF RESULTS

These apatite ages reflect the cooling history of the Wasatch Mountain block, and since no Tertiary igneous rocks occur in this part of the block, cooling must have been due only to uplift and erosion. The cooling rate indicated by the altitude distribution of the youngest samples plotted on Figure 3 is sufficiently high that a temperature of 120°C is assumed for the apparent closing temperature. Given this assumption, these ages record the passage of the rocks through approximately the 120°C isotherm, which is at a depth of about 3.5 km below the land surface. This depth is determined by the intersection of the line drawn through the youngest ages at the different altitudes with the abscissa shown on Figure 3. The figures for depths and temperatures are compatible with bottom-hole temperature measurements of 124°C at 3.3 km and 154°C at 3.2 km made by Amoco Production Co. in oil exploration wells in the deep graben in the central part of the Great Salt Lake. Assuming a geothermal gradient of 30°/km, there may have been at least 3 to 4 km of uplift during the last 5 m.y. in order to have old apatite with apparent ages of 5 m.y. exposed at the surface near the base of the mountains. Rate of uplift over that time interval would be about 0.8 mm/yr. However, in view of the analytical uncertainties of the ages, which range from 5% to 30%, it may be more accurate to evaluate the rate of uplift from the slope of the best-fit line through the younger ages over a range of altitude as shown on Figure 3. Based on the line the rate of uplift for the past 10 m.y. must have been about 0.4 mm/yr. As mentioned above, the available geologic evidence suggests that the initiation uplift of the Wasatch block predates 10 m.y. ago.

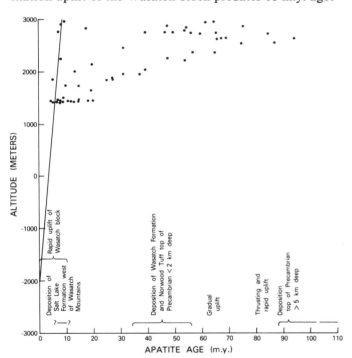

Figure 3. Plot of apatite fission-track ages from the Farmington Canyon Complex in the Wasatch Mountains against altitude. Solid line indicates the rate of uplift of the fault blocks having the greatest uplift rate sustained over 10 m.y. Inferred geologic history of the Farmington Canyon Complex in Late Cretaceous and younger time shown.

Two trenching studies have determined displacement across the Wasatch fault during Holocene time. Swan and others (1980) reported 11 m of offset over the past 6,000 years at Kaysville in the area of our study (Fig. 2). This indicates a displacement rate of 1.8 + 1.0 − 0.6 mm/yr. At Hobble Creek 90 km to the south they found 11.5 to 13.5 m of displacement during the past 12,000 to 13,000 years, or

an average rate of 1.0 ± 0.1 mm/yr.

The uplift rates of 0.8 mm/yr to 0.4 mm/yr for the past 5 and 10 m.y., respectively, are in reasonable agreement with the rates of 1.0 and 1.8 mm/yr reported by Swan and others (1980) and for the past 12,000 and 6,000 years, respectively.

An uplift rate of 0.4 mm/yr over a period of 10 m.y. would result in a total of 4 km of displacement between the range and the adjacent basin. This figure is compatible with estimates of total structural relief between the Wasatch Mountains and the adjacent basins and therefore approximates the time of the formation of these basin-and-range structures.

Dated apatite from much of the Farmington Canyon Complex now at the surface in the Wasatch Mountains delineates a zone of mixed ages between those of the initial Phanerozoic uplift about 85 m.y. ago and renewed rapid uplift 10 m.y. ago. Mixed ages result from the partial loss of tracks due to incomplete annealing. The rocks of this zone are now found at a variety of altitudes because of complex faulting and tilting within the mountain block.

If we can assume the previously extrapolated 7- to 8-km burial of the Precambrian crystalline rocks at the end of sedimentation 85 m.y. ago, the uplift of discrete blocks during the Sevier orogeny must have been very rapid, because a few samples have ages older than 85 m.y. Subsequent uplift was slower and apparently interrupted by an episode of downwarping and burial under as much as 2 km of sediment 55 to 35 m.y. ago. Some apatite was partially annealed during that burial, and some of the mixed ages may be from rock which was 1 to 4 km below the base of the Wasatch 35 m.y. ago and partially annealed.

The range from 7.3 to 94.4 m.y. in age of samples from the crest of the mountains indicates that the evenness of the crest does not represent a cyclic erosional surface as suggested by Eardley (1944). Likewise, the range in age of 4.9 to 94.4 m.y. on the present land surface between the foot and the crest of the mountains makes it unlikely that the slopes west of the crest represent a tilted Pliocene erosional surface as suggested by Eardley (1944) and Bell (1952). Perhaps the concept of dynamic equilibrium (Hack, 1960) can better be applied to the landforms in this part of the Wasatch Mountains. These landforms have been and continue to be sculpted across the differential uplift movement of the Wasatch Mountains and all their internal fault blocks.

The two samples from an altitude of 1,300 m on Little Mountain have an average age of about 70 m.y. Samples with a similar age occur at altitudes of 2,500 to 2,950 m in the Wasatch Mountains (Fig. 3). This suggests a minimum relative displacement between the Little Mountain block and the Wasatch block of 1.5 km sometime during the past 70 m.y., although the maximum possible displacement is about 5 km.

CONCLUSION

Although the information obtained in this study is much more complex than that obtained in many similar studies of this kind, it can nevertheless be interpreted in a reasonable fashion. The apatite fission-track ages show that the Wasatch Mountain block has been uplifted at a rate of about 0.4 mm/yr over the past 10 m.y. and that the rocks have been cooling at a rate of 12°C/m.y. The fission-track-derived uplift rate appears to be compatible with the assumptions concerning depth and temperature with other data available from the region. The scatter of older ages can be explained by exposure at the present surface of the zone of mixed ages produced by uplift concomitant with and after the Sevier orogeny, Eocene and early Oligocene downwarping and various amounts of Neogene uplift, and the relatively large analytical uncertainty associated with these ages. The distribution of the apatite ages argues against the presence of cyclical Pliocene and older erosional surfaces on the Wasatch Mountain block.

ACKNOWLEDGMENTS

We thank Amoco Production Co. for samples of core and for furnishing us with bottom-hole temperatures from two deep wells drilled during exploration beneath the Great Salt Lake. We also thank R. A. Zimmermann, W. E. Scott, and V. A. Frizzell for their comments on the manuscript.

REFERENCES CITED

Bell, G. L., 1952, Geology of the northern Farmington Mountains; *in* Marsell, R. E., ed., Geology of the Central Wasatch Mountains: Guidebook to the Geology of Utah No. 8, Utah Geological Society, p. 38–51.

Brookins, D. G., Forbes, R. B., Turner, D. L. Laughlin, A. W., and Naeser, C. W., 1977, Rb-Sr, K-Ar, and fission-track geochronological studies from LASL drill holes GT-1, GT-2, and EE-1: Los Alamos Scientific Laboratories Informal Report LA-6829-MS, 27 p.

Bryant, Bruce, 1979, Reconnaissance geologic map of the Precambrian Farmington Canyon Complex and surrounding rocks in the Wasatch Mountains between Ogden and Bountiful, Utah: U.S. Geological Survey Open-File Report 79–709, scale 1:50,000.

——1980, Metamorphic and structural history of the Farmington Canyon Complex, Wasatch Mountains, Utah [abs.]: Geological Society of America Abstracts with Programs, v. 12, no. 6, p. 269.

Bryant, Bruce, and Naeser, C. W., 1980, The significance of fission-track ages of apatite in relation to the tectonic history of the Front and Sawatch Ranges, Colorado: Geological Society of America Bulletin, Part 1, v. 91, p. 156–164.

Carpenter, B. S., and Reimer, G. M., 1974, Calibrated glass standards for fission-track use: National Bureau of Standards Special Publication 160-49, p. 1–16.

Crittenden, M. D., Jr., Stuckless, J. S., Kistler, R. W., and Stern, T. W., 1973, Radiometric dating of intrusive rocks in the Cottonwood area, Utah: U.S. Geological Survey Journal of Research, v. 1, no. 2, p. 173–178.

Eardley, A. J., 1944, Geology of the north-central Wasatch Mountains, Utah: Geological Society of America Bulletin, v. 55, p. 819–894.

Granger, A. E., 1953, Stratigraphy of the Wasatch Range near Salt Lake City, Utah: U.S. Geological Survey Circular 296, 14 p.

Hack, J. T., 1960, Interpretation of erosional topography in humid temperate regions: American Journal of Science, v. 258-A, p. 80–97.

Hedge, C. E., Stacey, J. S., and Bryant, Bruce, 1983, Geochronology of the Farmington Canyon Complex, *in* Miller, D. M., Todd, V. R., and Howard, K. A., eds., Tectonic and stratigraphic studies in the eastern Great Basin: Geological Society of America Memoir 157 (this volume).

King, P. B., and Beikman, H. M., 1974, Geologic map of the United States: U.S. Geological Survey Map, scale 1:2,500,000.

McDonald, R. E., 1976, Tertiary tectonics and sedimentary rocks along the transiton—Basin and Range province to Plateau and Thrust Belt province, Utah; *in* Hill, J. G., ed., Geology of the Cordilleran hinge-line: Rocky Mountain Assoication of Geologists, p. 281–317.

Naeser, C. W., 1979a, Thermal history of sedimentary basins: Fission-track dating of subsurface rocks: SEPM Special Publication No. 26, p. 109–112.

——1979b, Fission-track dating and geologic annealing of fission tracks; *in* Jager, E., and Hunziker, J. C., eds., Lectures in isotopes geology: Heidelberg, Springer-Verlag, p. 154–169.

——1981, The fading of fission tracks in the geologic environment-data from deep drill holes: Nuclear Tracks, v. 5, p. 248–250.

Naeser, C. W., and Faul, H., 1969, Fission track annealing in apatite and sphene: Journal of Geophysical Research, v. 74, p. 705–710.

Naeser, C. W., and Forbes, R. B., 1976, Variation of fission-track ages with depth in two deep drill holes: EOS (American Geophysical Union Transactions) v. 57, p. 353.

Nichols, D. J., and Warner, N. A., 1978, Palynology, age, and correlation of the Wanship Formation and their implications for the tectonic history of northeastern Utah: Geology, v. 6, p. 430–433.

Peterson, D. L., 1974, Bouger gravity map of part of the northern Lake Bonneville Basin, Utah and Idaho: U.S. Geological Survey Miscellaneous Field Studies Map MF-627, scale 1:250,000.

Shuey, R. T., Schellinger, D. K., Johnson, E. H., and Alley, L. B., 1973, Aeromagnetics and the transition between the Colorado Plateau and Basin and Range provinces: Geology, v. 1, p. 107–110.

Slentz, L. W., 1955, Salt Lake Group in lower Jordan Valley, Utah: Guidebook to Geology of Utah, no. 10, p. 23–36.

Swan, F. H., III, Schwartz, D. P., and Cluff, L. S., 1980, Recurrance of moderate to large magnitude earthquakes produced by surface faulting on the Wasatch fault zone, Utah: Bulletin of the Seismological Society of America, v. 70, no. 5, p. 1431–1432.

U.S. Geological Survey, 1978, Geological Survey Research 1978: U.S. Geological Survey Professional Paper 1100, 464 p.

Van Horn, Richard, 1981, Geologic map of pre-Quaternary rocks of the Salt Lake North Quadrangle, Davis and Salt Lake Counties, Utah: U.S. Geological Survey Miscellaneous Geologic Investigations Map I-1330, scale 1:24,000.

Wagner, G. A., and Reimer, G. M., 1972, Fission-track tectonics: The tectonic interpretation of fission track apatite ages: Earth and Planetary Science Letters, v. 14, p. 263–268.

Wagner, G. A., Reimer, G. M., and Jager, E., 1977, Cooling ages derived by apatite fission-track, mica Rb-Sr and K-Ar dating: The uplift and cooling history of the central Alps: Memorie degli Instituti geologia a Mineralogia dell Universite di Padova, v. 30, p. 1–29.

MANUSCRIPT ACCEPTED BY THE SOCIETY AUGUST 20, 1982

Geological Society of America
Memoir 157
1983

Geochronology of the Farmington Canyon Complex, Wasatch Mountains, Utah

Carl E. Hedge
John S. Stacey
Bruce Bryant
U.S. Geological Survey
Box 25046, Denver Federal Center
Denver, Colorado 80225

ABSTRACT

High-grade metamorphic rocks and migmatites in the Farmington Canyon Complex were derived from igneous and sedimentary rocks possibly as old as 3000 m.y. They were probably metamorphosed about 2600 m.y. ago, and they were severely metamorphosed, migmatized, and intruded by quartz monzonite 1790 m.y. ago. A very high initial Sr^{87}/Sr^{86} ratio of 0.769 indicates that the quartz monzonite magma was derived by melting of the more leucocratic parts of the layered gneisses. On and just west of Antelope Island, plutons of granite were emplaced 2020 m.y. ago.

INTRODUCTION

Crystalline basement rocks exposed in north-central Utah in the eastern part of the Sevier orogenic belt (Fig. 1) are known as the Farmington Canyon Complex (Eardley and Hatch, 1940). These rocks are exposed in the Wasatch Mountains between Bountiful and Ogden and on Antelope Island in the Great Salt Lake. They were brought to the surface in uplifted blocks of late Tertiary and Quaternary age in the eastern part of the Basin and Range province. The crystalline rocks occur in a terrane that is para-autocthonous in relation to the far-traveled Willard thrust sheet which overlies them north of Ogden Canyon (Crittenden, 1972). At least one thrust of Late Cretaceous age, the Ogden thrust, passes into the Farmington Canyon Complex south of Ogden Canyon, and branches of it apparently pass above rocks of the complex on the east side of the Wasatch Mountains. The whole complex may be allochthonous, for a sole thrust is necessary to explain the configuration of the numerous eastward-directed thrusts east of the Wasatch Mountains (Royce and others, 1975), unless there has been major transport along the upper split of the Ogden thrust.

Bryant (1980) has divided the rocks of the Farmington Canyon Complex into four major units: (1) quartz monzo-nite gneiss; (2) migmatite, gneiss, and schist; (3) gneiss and schist; and (4) quartzite gneiss and schist.

The quartz monzonite gneiss occurs in the northern part of the complex (Fig. 2). It is a fairly uniform rock, of medium grain size, and has a well-developed foliation. It locally has a streaky and spotty distribution of mafic minerals and grades into faintly layered rock, especially near its contact with migmatite. The unit contains numerous pegmatites, lenses and pods of amphibolite, and scattered inclusions of quartzite. The quartz monzonite gneiss is interpreted as being a catazonal, syntectonic pluton.

South of the quartz monzonite gneiss is a large area of migmatite, gneiss, and schist. This is a heterogenous unit containing biotite-hornblende-quartz-feldspar gneiss, garnet-quartz-feldspar gneiss, biotite-garnet-quartz-feldspar gneiss, garnet-biotite schist, and sillimanite-garnet-biotite schist with or without microcline. The migmatite unit contains amphibolite layers, lenses, and pods, as well as both concordant and discordant pegmatites. Contacts between the migmatite and quartz monzonite gneiss are gradational.

Farther to the south, migmatite becomes much less prevalent and the rocks are mapped as gneiss and schist. The lithologies of this unit are biotite-feldspar-quartz

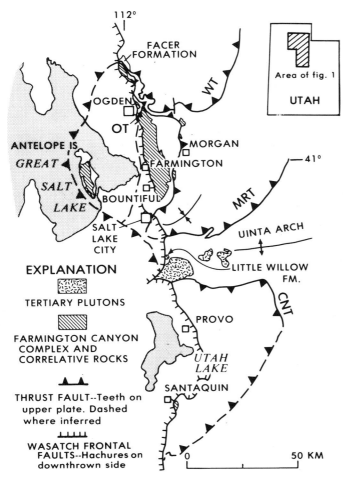

Figure 1. Location and tectonic setting of the Farmington Canyon Complex. OT, Ogden thrust; WT Willard thrust; MRT, Mount Raymond thrust; CNT, Charleston-Nebo thrust. Major inferred faults not named. Modified from Stokes (1963) and Crittenden (1972).

gneiss, garnet-biotite-feldspar-quartz gneiss, and sillimanite-biotite schist. Amphibolite pods, layers, and lenses are scattered throughout this unit also. Large, sharply bounded pegmatites are more numerous than in the migmatite, gneiss, and schist unit.

In the southern part of the Farmington Canyon Complex, significant quartzite is interlayered with the gneisses and schists. The quartzite is white to greenish gray and contains some light-green muscovite.

With the exception of the quartz monzonite gneiss, the lithologies of the Farmington Canyon Complex suggest that rocks were derived from a dominantly sedimentary sequence, although some of the quartz-feldspar gneiss may originally have been felsic volcanic rock, and the chemical compositions of the amphibolites suggest that they were derived from basalts and gabbros. The metamorphic mineral assemblage of the Farmington Canyon Complex is of amphibolite grade. In the southern part of the area, silli-

manite and muscovite form a stable mineral assemblage in pelitic layers. In a broad belt to the north, coexisting muscovite, sillimanite, and microcline are characteristic of the schists. In the northern part of the complex, sillimanite and microcline occur in the pelitic rocks, and this zone is approximately geographically coincident with the area of migmatite development. As Winkler (1979) pointed out, the conditions of the sillimanite-microcline zone are favorable for incipient melting in rocks rich in quartz and feldspar. A few orthopyroxenes found in mafic inclusions in the quartz monzonite gneiss and in adjacent migmatitic gneiss may indicate an earlier metamorphism under conditions of the granulite facies, but these pyroxenes are not in equilibrium with the main mineral assemblage.

The above observations suggest the following history for the Farmington Canyon Complex. A thick sequence of dominantly continentally derived sediments containing small bodies of basalt and gabbro was highly deformed, in part migmatized, and highly metamorphosed. Before the height of the metamorphism, quartz monzonite was intruded. A long and complex structural and lower temperature thermal history occurred after this main metamorphism.

A pervasive shearing and partial retrogressive metamorphism affected large parts of the Farmington Canyon Complex possibly in Proterozoic time or during the Sevier orogeny or both. The complex was buried to a depth of about 8 km by Late Cretaceous time, just before the Sevier orogeny. That orogeny produced thrust faults and uplift. The rocks of the complex were again uplifted several kilometers in late Tertiary and Quaternary times.

The geology of the Farmington Canyon Complex on Antelope Island resembles that in the Wasatch Mountains. Migmatitic gneiss and schist were metamorphosed to sillimanite grade and intruded, perhaps synkinematically, by granite. Later shearing and retrogressive metamorphism occurred before deposition of diamictite of late Proterozoic age (Larsen, 1957; Bryant and Graff, 1980). The granite is now a pyroxene-hornblende gneiss with foliation parallel to that of the metamorphic rocks. Granite gneiss of contrasting mineralogy and texture was encountered in a drill hole at 10,391 ft by Amoco Production Company about 8 km west of Antelope Island. That rock is a blastomylonitic garnet-bearing biotite-muscovite granite gneiss that apparently was subjected to cataclasis and recrystallization under medium-grade metamorphic conditions.

Only minerals from the Farmington Canyon Complex have previously been dated. K-Ar ages of biotite range from 224 m.y. on Antelope Island through 487 to 559 m.y. above the Ogden thrust in the Wasatch Mountains to 1079 m.y. below the thrust (Hashad, 1964). K-Ar ages of hornblende are similar on both sides of the Ogden thrust and range from 1364 to 1700 m.y. An Rb-Sr age of muscovite from a pegmatite above the thrust is 1580 m.y. (Giletti and

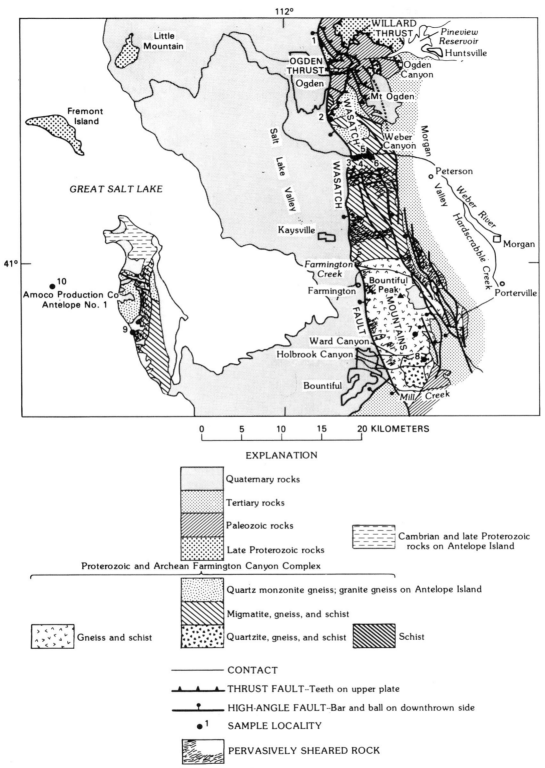

Figure 2. Geologic map of the Farmington Canyon Complex showing sample localities. From Bryant (1979), Larsen (1957), and Bryant and Graff (1980).

TABLE 1. SAMPLE LOCALITIES, DESCRIPTIONS, AND Rb-Sr DATA

Locality No.	Sample No.	Rock type	Rb (ppm)	Sr (ppm)	$^{87}Rb/^{86}Sr$	$^{87}Sr/^{86}Sr$

Wasatch Mountains, quartz monzonite gneiss

Spoil from north end of water tunnel, sec. 3, T.6N., R.1W. Coarse-grained granite gneiss with locally irregular distribution of mafics.

1	NO-1-A	Biotite granite gneiss	239	39.6	18.45	1.2501
1	NO-1-B	Biotite-hornblende granite gneiss	194	46.9	12.43	1.0946
1	NO-1-C	Biotite-hornblende granite gneiss	190	55.2	10.27	1.0335

Outcrops and blocks about 200 m north of the mouth of Beus Creek. Quartz monzonite gneiss with some mafic segregations and pegmatite veinlets and pods. Two 5-m-thick mylonitic zones cut the rock, and the samples came from these zones because they weather less easily.

2	0-41-A	Mylonitic biotite-hornblende quartz monzonite gneiss	115	79.3	4.271	0.8814
2	0-41-D	Cataclastic biotite-hornblende quartz monzonite gneiss	125	80.8	4.550	0.8864

Migmatitic gneiss

Typical of migmatitic gneiss from cuts along pipeline on the south side of Weber Canyon south and southwest of powerhouse. Migmatitic gneiss contains lenses of amphibolite and lenses, pods, and dikes of pegmatite.

3	0-65	Biotite-garnet-microcline-plagioclase gneiss	129	124	3.039	0.8141
4	0-64	Garnet-hornblende quartz monzonite gneiss	86.6	97.1	2.605	0.7994

Biotite-quartz-feldspar gneiss with lenses of amphibolite and pegmatite from outcrops adjacent to I-80 in Weber Canyon opposite Broad Canyon.

5	0-57-A	Partly chloritized amphibolite	90.4	136	1.933	0.7904
5	0-57-B	Migmatitic hornblende quartz monzonite gneiss	96.0	96.4	2.923	0.8521
5	0-57-C	Migmatitic biotite-hornblende quartz monzonite gneiss	108	92.5	3.419	0.8382

Migmatitic layered gneiss with pods of amphibolite and pegmatite from road cut on I-80 opposite Devil's Gate in Weber Canyon.

6	S-28-A	Hornblende-biotite granite gneiss	110	92.2	3.511	0.8447
6	S-28-B	Hornblende-biotite-plagioclase-microcline-quartz-gneiss	140	84.8	4.870	0.8825

Gneiss and schist

Roadcut west of BM 7946 on road from Bountiful to Bountiful Peak.

7	B-118-A	Quartz-bearing amphibolite	10.1	163	0.1797	0.7086
7	B-118-Z	Garnet-quartz-biotite-plagioclase gneiss	115	200	1.682	0.7567
7	B-118-H	Garnet-biotite-plagioclase-quartz gneiss	59.4	158	1.090	0.7451

Quartzite, gneiss, and schist

Amphibolite cut by pegmatite pods from spoil along pipeline at 8,350-ft altitude on ridge south of Ward Canyon.

8	B-122-A	Amphibolite	4.8	111	0.1250	0.7100

Antelope Island and vicinity, granite gneiss

Outcrop at 4,430-ft altitude near center sec. 7, T.2N., R.2W.

9	G-1-4	Partly epidotized granite gneiss	..	..	..	..

Core from depth of 10,391 ft, Amoco Production Co., Antelope Island State of Utah well.

10	Am-1	Garnet-bearing biotite-muscovite granite gneiss	..	..	..	..

TABLE 2. Sm-Nd DATA FOR WHOLE-ROCK SAMPLES OF FARMINGTON CANYON COMPLEX

Sample No.	Sm (ppm)	Nd (ppm)	$^{147}Sm/^{144}Nd$	$^{143}Nd/^{144}Nd$	Model age*
0-57-A	2.73	11.3	0.1472	0.51176	2740 m.y.
0-57-B	20.0	97.7	0.1248	0.51136	3150 m.y.
0-64	18.1	86.2	0.1280	0.51112	3430 m.y.

*Model age calculated from:

$$\text{Age} = 1.529 \times 10^{-11} \ln \left[1 + \left(\frac{.51264 - ^{143}Nd/^{144}Nd}{.19584 - ^{147}Sm/^{144}Nd} \right) \right]$$

Gast, 1961). This range indicates that the ages are best interpreted as having been reset to varying degrees. All that can be concluded from these ages is that the Farmington Canyon Complex is at least 1700 m.y. old.

RESULTS AND DISCUSSION

The rocks of the Farmington Canyon Complex yield a confusing array of radiometric age data, and the results do not lend themselves to simple interpretations. In an attempt to unravel the long and complex history, we have analyzed the following: 16 whole-rock Rb-Sr samples (Table 1), 3 Sm-Nd whole-rock samples (Table 2), and 18 U-Pb zircon fractions from 6 different samples (Table 3). Because the quartz monzonite gneiss apparently intrudes the layered metamorphic rocks, separate interpretations of the data from these respective units will be made.

The Rb-Sr data for the layered metamorphic rocks do not define an age. On an isochron diagram (Fig. 3), the data points define a crude cone, with an upper boundary of 3600 m.y. and a lower boundary of 2600 m.y. One sample (B-

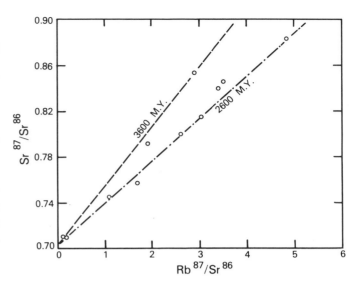

Figure 3. Rb-Sr isochron diagram for samples of the layered metamorphic rocks of the Farmington Canyon Complex.

118Z) plots slightly below this cone. The scatter is such that no unique interpretation can be made from these data. It is apparent, however, that very old material was incorporated into what is now the Farmington Canyon Complex.

Similarly, the Sm-Nd data do not define an age (Table 2), but the data do support an interpretation that crustal material as old as 2800 to 3600 m.y. was included in the complex.

Several interpretations of the Rb-Sr and Sm-Nd data of the layered metamorphic rocks are possible: (1) the rocks are very old (3500 to 3600 m.y.) and later metamorphisms have more or less lowered the apparent ages of many sam-

TABLE 3. U-Pb DATA FOR ZIRCON SEPARATES

Sample No.	Size fraction	Concentrations (ppm)			Atomic ratios				Age estimates (m.y.)				Blank corrected
		Pb	U	Th	$\frac{^{206}Pb}{^{238}U}$	$\frac{^{207}Pb}{^{235}U}$	$\frac{^{207}Pb}{^{206}Pb}$	$\frac{^{208}Pb}{^{232}Th}$	$\frac{^{238}U}{^{206}Pb}$	$\frac{^{235}U}{^{207}Pb}$	$\frac{^{207}Pb}{^{206}Pb}$	$\frac{^{232}Th}{^{208}Pb}$	$\frac{^{206}Pb}{^{204}Pb}$
0-41-D	+150	306	1274	163	0.21220	3.1794	0.10881	0.0907	1241	1452	1779	1755	459
	-150+250 M4°	409	1360	175	0.20431	3.0453	0.10824	0.1103	1198	1419	1770	2115	142
	-150+200 M2°	162	1367	499	0.21755	3.2636	0.10927	0.1422	1296	1475	1787	2687	104
	-200+250 NM8°	286	1226	137	0.22418	3.3883	0.10976	0.0855	1304	1502	1795	1714	1245
	-150+250 Clear	298	1644	..	0.20734	2.9967	0.10482	..	1215	1407	1711	..	950
0-57-B	+250 NM	208	724	107	0.27162	4.3321	0.11567	0.10628	1549	1699	1890	2041	1276
	+250 M	255	937	347	0.24504	3.7474	0.11092	0.0541	1413	1582	1814	1065	735
	-250+325 NM5°	218	784	85	0.26779	4.1293	0.11183	0.1062	1539	1660	1829	2040	1388
	-325 NM5°	251	966	63	0.26160	3.8982	0.10807	0.10010	1498	1613	1767	1928	3126
0-64	+150 NM	253	875	..	0.27708	4.4722	0.11706	..	1577	1726	1912	..	2590
	-150+250	231	825	..	0.27753	4.2782	0.11180	..	1579	1689	1829	..	4936
	-325 NM	246	892	..	0.27758	4.1593	0.10860	..	1579	1666	1777	..	5394
B-118	+200	305	828	..	0.28064	5.3119	0.13727	..	1595	1871	2193	..	233
	-200+250 NM	193	688	..	0.30664	6.0710	0.14359	..	1724	1986	2271	..	288
GA-1-4	+100 NM	222	1672	212	0.12744	1.9799	0.11268	..	773	1109	1843	1074	1560
	+100 M	232	1986	339	0.10986	1.6071	0.11026	..	672	997	1804	753	1091
AM-1	+100	328	1414	..	0.21901	3.7111	0.12290	..	1277	1574	1999	..	1710
	-400	193	740	..	0.30308	5.1835	0.12404	..	1707	1850	2015	..	5270

ples, thus producing the observed scatter; (2) the rocks are no older than 2600 m.y., but contain varying amounts of inherited sedimentary components that were derived from a very old terrane; and (3) the true age is some intermediate value (3000 to 3200 m.y.), and the scatter of the data is the result of a combination of inheritance and metamorphism.

The quartz monzonite gneiss was interpreted in the field as having been originally an intrusive rock. The gradational contacts with migmatitic gneiss and the abundant large inclusions of wall rocks within the quartz monzonite suggest that the quartz monzonite was derived relatively locally and that the wall rocks were hot. The whole rock Rb-Sr data (Fig. 4) from samples of the quartz monzonite gneiss both from one locality above the Ogden thrust and one locality below the thrust yield an age of 1808 ± 34 m.y., and a very high initial Sr^{87}/Sr^{86} ratio of 0.769. The isochron passes through the upper part of the field of data for the layered metamorphic rocks. This data array is consistent with an interpretation that the quartz monzonite magma was derived by melting of the more leucocratic parts of the layered complex.

The U-Pb data for zircons from the Farmington Canyon Complex are plotted on Figure 5. The points for the quartz monzonite gneiss lie on a chord with an upper intercept of 1780 ± 20 m.y. This age is within analytical uncertainty of the Rb-Sr age of 1808 ± 34 m.y., but the data scatter more than can be attributed to analytical error. There are several possible explanations for this scatter, but since we have no criteria for evaluating them, we have simply combined the U-Pb and Rb-Sr results to say that the best age of the quartz monzonite gneiss is about 1790 ± 20 m.y.

U-Pb data for zircons from the layered metamorphic complex yield a spectrum of Pb^{207}/Pb^{206} ages from 1770 to 2271 m.y. These are interpreted as indicating that there is, in fact, an old component of radiogenic lead in the zircons

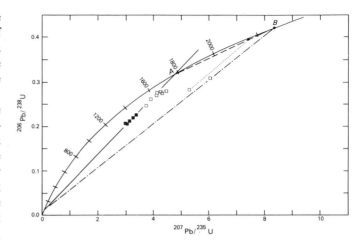

Figure 5. Concordia diagram of U-Pb data for zircons from the Farmington Canyon Complex. Quartz monzonite samples are solid squares and layered rocks are open squares. The chord has an upper intercept of 1973 ± 19 (2σ) m.y. and a lower intercept of 70 m.y.

from the metamorphic rocks (something that may be much older than 2271 m.y.). However, the zircons were partially to totally reset during the intense metamorphism that accompanied the emplacement of the quartz monzonite.

In Figure 5, the chord A-B represents one possible distribution of the zircon data points in the metamorphic complex immediately after the intrusion of the quartz monzonite gneiss. The present location of the data points on the Concordia diagram is the result of further lead loss since late Mesozoic time. Dilatancy as an effect of uplift of the region and the consequent leaching of Pb from the zircon crystals constitute a likely cause of this second episode of lead loss (Goldich and Mudrey, 1972).

In Figure 5, chord A-B is drawn so that the lower end B defines the age of intrusion of the quartz monzonite gneiss and the upper end A represents the minimum possible age for the metamorphic complex. This latter age (2270 m.y.) is the Pb^{207}/Pb^{206} age for the oldest zircon fraction, B118 (−200 +250 NM). Both size fractions from sample B118, the only sample of gneiss from outside the zone of migmatization, have older apparent ages than any of the fractions from the migmatitic gneiss. The actual age of the metamorphics could be considerably older than this—in which case the point B would lie somewhere to the right of its location shown in Figure 5. Stacey and others (1968), in their study of galenas from the Park City mining district in the Wasatch Mountains, concluded from a well-defined $^{207}Pb/^{204}Pb-^{206}Pb^{204}Pb$ "isochron" that the basement age in that region must be 2345 ± 50 m.y. This is somewhat less than the 2600-m.y. minimum age inferred from the strontium data, but one point in Figure 3 does lie below the 2600-m.y. line and may indicate a slightly younger event.

While the 1790-m.y. event is not precisely dated, the

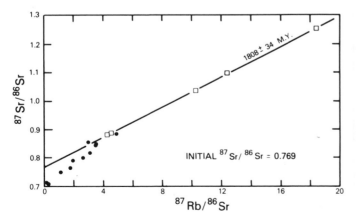

Figure 4. Rb-Sr isochron diagram for samples of the quartz monzonite gneiss (squares) and the layered metamorphic rocks (dots). The line corresponds to an age of 1808 ± 34 (2σ) m.y., and the initial $^{87}Sr/^{86}Sr$ is 0.769.

isotopic data do confirm the field evidence in testifying to its intensity. The very high initial Sr^{87}/Sr^{86} ratio of the quartz monzonite indicates a shallow melting, and the high degree of resetting of the zircon ages from the metamorphic rocks argues for extreme conditions. It is, therefore, remarkable that the Rb-Sr data for the layered metamorphic rocks do not reflect the 1790-m.y. event. This apparent paradox can be explained if the rocks had gone through an earlier metamorphism (probably at about 2400 to 2600 m.y.) that created a high temperature mineral assemblage and, thus, made them relatively resistant to later resettings. This interpretation is also consistent with the Rb-Sr data for the amphibolites. These rocks have mineralogies and chemical compositions indicating tholeiitic basalt precursors. The Rb contents, however, are high for tholeiitic basalt composition, and since they have correspondingly high Sr^{87}/Sr^{86} ratios, this Rb must have been added much before 1790 m.y. ago.

The granitic gneiss from Antelope Island is a relatively uniform and nonlayered rock, suggesting a metaigneous origin (Bryant and Graff, 1980), as is the core from the Amoco hole in Great Salt Lake (Fig. 1). However, the rock from the core apparently had a somewhat different metamorphic history. We have attempted to date these rocks in order to determine whether they are the same age as the quartz monzonite gneiss from the Wasatch Mountains.

The samples from Antelope Island were not fresh enough for Rb-Sr dating, but two zircon fractions indicate an age of 1993 ± 22 m.y., about 200 m.y. older than that of the quartz monzonite gneiss (Fig. 6). An age of 2023 ± 4 m.y. was obtained from the core sample. Because samples from Antelope Island and the core have only marginally different upper intercepts ages and because the Antelope Island zircons are highly discordant and from relatively

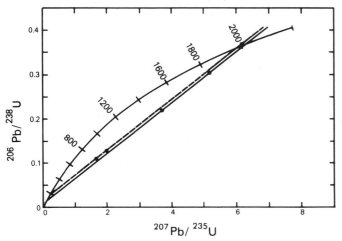

Figure 6. Concordia diagram of U-Pb data for zircons from Antelope Island (circles) and the Amoco core (dots). The Antelope Island chord (dashed line) has an upper intercept of 1993 ± 22 m.y., and the chord for the core sample (solid line) has an upper intercept of 2023 ± 4 m.y.

weathered rocks, we suggest that the rocks are essentially the same age. More important, this age (2020 m.y.) is clearly much older than that of the quartz monzonite gneiss from the Wasatch Mountains. The 2020-m.y. age is particularly interesting because it was not found in the Wasatch Mountains, and, in fact, is virtually unknown in the Western United States. The only previous indication of 2000-m.y.-old rocks in this part of the world was found in the lead-isotope study of the ores from the Tintic mining district (Stacey and others, 1968).

CONCLUSIONS

The rocks of the Farmington Canyon Complex originated as a pile of dominantly sedimentary rocks intruded by basalt and gabbro sometime prior to 2600 m.y. ago. Some or all of the material in this pile had a crustal history in excess of 3000 m.y. A high-grade metamorphic event about 2600 m.y. ago affected the Rb-Sr systems in the rocks. An intense metamorphism accompanied by anatectic melting occurred about 1790 m.y. ago, and this event is responsible for the present mineral assemblages in the rocks that have not been subject to shearing and retrogressive metamorphism. U-Pb systems were re-equilibrated in the anatectic melt, and they were nearly completely reset in the metamorphic rocks. These isotopic systems were somewhat less severely reset in the gneiss outside the area subject to migmatization and melting. Whether the sparse relict monoclinic pyroxene remains from the 2600-m.y.-old metamorphism or formed during an early stage of the 1790-m.y.-old metamorphism cannot be determined from out data. Mineral ages were reset to varying degrees during cooling and other events after these metamorphisms. The rocks on Antelope Island were intruded by granitic plutons 2020 m.y. ago, and this event is not recognized in the rocks of the Wasatch Mountains.

The history of the Farmington Canyon Complex resembles that of rocks at the southern edge of the Archean craton in Wyoming (Hills and others, 1968; Hills and Armstrong, 1974), in western Wyoming (Reed and Zartman 1973; Peterman, 1979), and in southwestern Montana (Vitaliano and others, 1979; James and Hedge 1980). However, the early Proterozoic event reflected by the present mineral assemblages in the metamorphic rocks and the migmatization and melting was more drastic in the Farmington Canyon Camplex than in these other areas of the Archean craton exposed in the United States.

ACKNOWLEDGMENTS

We thank Amoco Production Company and Vincent Matthews III for core samples of granite gneiss from west of Antelope Island. We achnowledge the technical assistance of Gerry Cebula, Jack Groen, and Becky Hopper in mineral separations, Kiyoto Futa in the Rb-Sr and Sm-Nd analyses, and Lynn Fischer in the U-Pb analyses.

REFERENCES CITED

Bryant, Bruce, 1979, Reconnaissance geologic map of the Precambrian Farmington Canyon Complex and surrounding rocks in the Wasatch Mountains between Ogden and Bountiful, Utah: U.S. Geological Survey Open-File Report 79-709.

Bryant, Bruce, 1980, Metamorphic and structural history of the Farmington Canyon Complex, Wasatch Mountains, Utah [abs.]: Geological Society of America Abstracts with Programs, v. 12, no. 5, p. 269.

Bryant, Bruce, and Graff, Paul, 1980, Metaigneous rocks on Antelope Island [abs.]: Geological Society of America Abstracts with Programs, v. 12, no. 5, p. 269.

Crittenden, M. D., Jr., 1972, Willard thrust and the Cache allochthon, Utah: Geological Society of America Bulletin, v. 83, p. 2871-2880.

Eardley, A. J., and Hatch, R. A., 1940, Pre-cambrian crystalline rocks of north-central Utah: Journal of Geology, v. 48, p. 58-72.

Giletti, B. J., and Gast, P. W., 1961, Absolute age of Precambrian rocks in Wyoming and Montana: New York Academy of Science, Ann. 91, p. 454-458.

Goldich, S. S., and Mudrey, M. G., 1972, Dilatancy model for discordant U-Pb zircon ages, in Tugarinov, A. I., ed., Recent contributions to geochemistry and analytical chemistry: New York, John Wiley and Sons, p. 466-470.

Hashad, A. H., 1964, Geochronologic studies in the central Wasatch Mountains, Utah [Ph. D. thesis]: University of Utah.

Hill, F. A., and Armstrong, R. L., 1974, Geochronology of Precambrian rocks in the Laramie Range and implications for the tectonic framework of Precambrian southern Wyoming: Precambrian Research, v. 1, p. 213-225.

Hills, F. A., Gast, P. W., Houston, R. S., and Swainbank, I. G., 1968, Precambrian geochronology of the Medicine Bow Mountains, southeastern Wyoming: Geological Society of America Bulletin, v. 79, p. 1757-1784.

James, H. L., and Hedge, C. E., 1980, Age of basement rocks of southeast Montana: Geological Society of America Bulletin, pt. 1, v. 91, p. 11-15.

Larsen, W. N., 1957, Petrology and structure of Antelope Island, Davis County, Utah [Ph. D. thesis]: University of Utah, 142 p.

Peterman, Z. E., 1979, Geochronology and the Archean of the United States: Economic Geology, v. 74, p. 1544-1562.

Reed, J. C., Jr., and Zartman, R. E., 1973, Geochronology of Precambrian rocks of the Teton Range, Wyoming: Geological Society of America Bulletin, v. 84, p. 561-582.

Royce, F., Jr., Warner, M. A., and Reese, D. L., 1975, Thrust belt structural geometry and related stratigraphic province, Wyoming-Idaho-northern Utah, in Bolyard, D. W., ed., Deep drilling frontiers of the Central Rocky Mountains: Rocky Mountain Association of Geologists, Geologic Symposium Volume, p. 41-54.

Stacey, J. S., Zartman, R. E., and Nkomo, I. T., 1968, A lead isotope study of galenas and selected feldspars from mining districts in Utah: Economic Geology, v. 63, p. 796-814.

Stokes, W. L., 1963, Geologic map of northwestern Utah: Utah Geological and Mineralogical Survey, scale 1:250,000.

Vitaliano, C. J., Cordoa, W. S., Burger, H. R., Hanley, T. B., Hess, D. F., and Root, F. K., 1979, Geology and structure of the southern part of the Tobacco Root Mountains, southwestern Montana map summary: Geological Society of America Bulletin, pt. I, v. 90, p. 712-715.

Winkler, H.G.F., 1979, Petrogenesis of metamorphic rocks (fifth edition): Springer-Verlag, 348 p.

MANUSCRIPT ACCEPTED BY THE SOCIETY AUGUST 20, 1982

Geological Society of America
Memoir 157
1983

Overthrusts and salt diapirs, central Utah

Irving J. Witkind
U.S. Geological Survey
Box 25046, Denver Federal Center
Denver, Colorado, 80225

ABSTRACT

At three separate localities in central Utah, elongate, diapiric folds, formed by and above salt diapirs, appear to have deformed parts of the upper plate of the Charleston-Nebo thrust fault. The causative salt diapirs, contained within the Twelvemile Canyon Member of the Arapien Shale (Middle Jurassic), have surged upward at least three and possibly four times since Middle Jurassic time. Each time they forced the overlying mudstones to intrude and bow up the overlying sedimentary rocks. Wherever the thrust plate overlay these folds it has been broken and locally tilted. In the Gardner Canyon–Red Canyon area, near Nephi, the thrust plate is an overturned, almost recumbent anticline, and the upward push of the core of the Levan diapiric fold has punched up a wedge of the beds that form part of the thrust plate. In the Pigeon Creek area, near Levan, mudstones of the same diapiric fold have intruded an overturned anticline and locally bowed up and broken off its east limb. In the Thistle area, the upwelling of an offshoot of the Fairview diapiric(?) fold has domed the thrust plate and arched younger sedimentary rocks that mantle and conceal much of it. Tenuous evidence suggests that the thrust plate was emplaced after Middle Jurassic (Bathonian) time and eroded before Late Cretaceous (Campanian) time; parts of it were deformed during the third diapiric episode—tentatively dated as late(?) Oligocene or Miocene. If oil pools did form in the overturned anticlinal fold that forms the upper plate of the Charleston-Nebo thrust, or in other comparable overturned folds, the pools may have been displaced laterally as a result of the diapiric tilting, or they may have been dissipated because the reservoir rocks were too fractured to retain them.

INTRODUCTION

A narrow zone marked by intense deformation trends southward through central Utah. This zone, commonly referred to as the "transition zone," delineates the join between the eastern edge of the Basin and Range province and the western margin of the Colorado Plateau. The zone has attracted the attention of geologists since the days of Gilbert and Dutton, and its fascination stems as much from structural complexity as it does from the fact that the field evidence lends itself to conflicting interpretations. A recently published cartoon (Spearing, 1981) showing two geologists arguing fiercely and captioned "Don't worry—it's a typical field trip—one outcrop, two geologists and three interpretations!" is only too true for this sector of central Utah.

The late Prof. E. M. Spieker and his many Ohio State University graduate students, attracted by the structural complexity, mapped and studied much of the area in detail. Spieker (1946, 1949) explained the many diverse structures as the result of multiple episodes of orogeny—in all, he postulated 14 sequences of crustal disturbance (1949, p. 78). Subsequently, his views were challenged by Prof. W. L. Stokes of the University of Utah who proposed in two abstracts (1952, 1956) that much of the deformation could as readily be explained by salt tectonics. Regrettably, Stokes never fully developed his ideas in the published literature. During the past decade, as interest in the oil potential of central Utah has increased, seismic surveys have criss-crossed the transition zone. A test well in the transition zone (Phillips Petroleum Company's Price-N well, sec. 29, T. 15 S., R. 3 E.) penetrated as much as 610 m (2,000 ft) of rock salt. Workers in the area, impressed by the seismic and test-well data, reemphasized the possibility that salt diapirism may have played a significant role in the structural evolution of central Utah (Moulton, 1975; Baer, 1976).

In this article I present evidence suggesting that diapiric folds, formed by and above salt diapirs, have intruded, broken, and locally arched parts of the upper plate of the Charleston-Nebo thrust fault (hereafter referred to as the Charleston-Nebo thrust plate)—a major structural element in this part of central Utah (Fig. 1). The plate, now much dissected, forms the southern Wasatch Mountains which lie athwart the north end of the transition zone. The plate and the underlying thrust fault have been mapped and described previously (Black, 1965; Johnson, 1959; and Eardley, 1933a, 1933b).

STRATIGRAPHY

The sedimentary rocks that crop out in this part of central Utah consist of two sequences; one sequence comprises the autochthonous plate that underlies much of the Sanpete–Sevier Valley area (Table 1); the second forms much of the Charleston-Nebo thrust plate in the Mount Nebo area (Table 2). No attempt is made here to describe either of these sequences in detail. The rocks of the Sanpete–Sevier Valley area have been described repeatedly and thoroughly by Spieker and his many graduate students (Spieker, 1946, 1949; Gilliland, 1948, 1951; Hardy, 1962; Schoff, 1951; Hardy and Zeller, 1953; and McGookey, 1960, among others). The rocks of the thrust plate have been described by Eardley (1933a), Black (1965), Johnson (1959), and Hintze (1962).

The Twelvemile Canyon Member of the Arapien Shale

The rock salt (halite) responsible for the recurrent diapiric episodes is contained within one of the most unusual units in central Utah—the Twelvemile Canyon Member of the Arapien Shale. Originally named and defined by Spieker (1946, p. 123–125), the Arapien Shale of Middle Jurassic age consists of an upper member, the Twist Gulch, underlain by a lower member, the Twelvemile Canyon (Table 3). The Twist Gulch Member consists of beds of reddish-brown shaly siltstone and fine-grained sandstone with sparse, thin interbeds of light-gray sandstone. It may be equivalent to the Entrada, Curtis, and Summerville Formations of the San Rafael Group (Hardy, 1949, p. 36).

The salt- and evaporite-rich Twelvemile Canyon Member, chiefly a calcareous mudstone, appears in a variety of colors; commonly it is light gray, mottled here and there by blotches of pale red. In a few places, it is uniformly reddish brown, and elsewhere drab gray. In addition to the

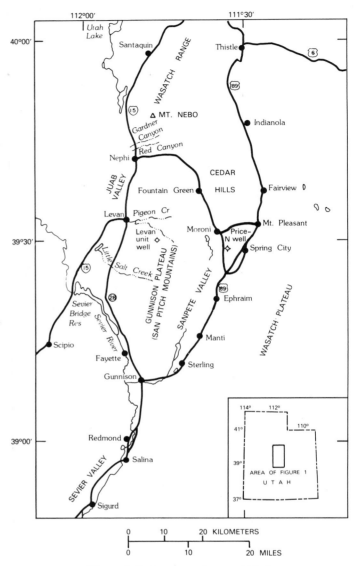

Figure 1. Index map of central Utah.

calcareous mudstone, the Twelvemile Canyon Member contains gypsiferous shaly siltstone, sparse shale, and some fine-grained sandstone. Some gray, thin-bedded limestone beds may be in the lower part of the member. Evaporites are common and include rock salt (halite)—hereafter referred to as "salt"—gypsum, anhydrite, and calcite. Many outcrops of the Twelvemile Canyon Member are littered with selenite crystals. Salt was and is being mined near Redmond; large amounts of it probably underlie central Utah (Moulton, 1975, p. 93). It is likely that much salt has been dissolved and removed from the area by surface and ground water in the past (Pratt and others, 1966, p. 55; Mitchell, 1979, p. 505, 508).

Previous workers (Picard, 1980, p. 131; Spieker, 1949, p. 17; Moulton, 1975, p. 9) have suggested that the Twelvemile Canyon Member rests on the Navajo Sand-

TABLE 1. SOME STRATIGRAPHIC UNITS EXPOSED IN THE TRANSITION ZONE OF CENTRAL UTAH

SYSTEM	SERIES	UNIT		APPROXIMATE THICKNESS		LITHOLOGY
				Meters	Feet	
Tertiary	Miocene?	"Lava beds" (Basalt)		152+	500+	Dark-gray to black, thick-bedded to massive, fine-grained basalt
		—?—				
	Oligocene	Gray Gulch Formation of Spieker (1949)		0 to 213	0 to 700	Light-gray, thin-bedded to massive, friable volcaniclastic sediments
		Bald Knoll Formation of Gilliland (1951)		305?	1000?	Light-gray to tan, thin-bedded mudstone, siltstone, and sandstone
		Crazy Hollow Formation of Spieker (1949)		0 to 304	0 to 1000	Light-gray to reddish-brown, thin- to medium- bedded sandstone, shaly siltstone, and some conglomerate
		—?—				
	Eocene	Green River Formation		365	1200	Consists of a lower light-green, thin-bedded, fissile shale unit overlain by a light-brown, thin- and even-bedded limestone unit
		Colton Formation		137 to 487	450 to 1600	Variegated red and gray mudstone, reddish-brown siltstone, sandstone, and conglomerate
		Flagstaff Limestone		15 to 548	50 to 1800	Light-gray, thin- to thick-bedded, even-bedded, fine-grained limestone
	Paleocene	North Horn Formation		45 to 914	150 to 3000	Light-yellow, to reddish-brown, thin- to thick-bedded mudstone, sandstone, conglomeratic sandstone and sparse limestone
Cretaceous	Upper Cretaceous	Price River Formation		6 to 609	20 to 2000	Light-gray to gray, thin- to thick-bedded conglomerate, conglomeratic sandstone, and sandstone
		West of Sanpete Valley / East of Sanpete Valley — Indianola Group undivided		915 to 2130	3000 to 7000	Red to reddish-brown, thick-bedded to massive, well-cemented conglomerate
			Sixmile Canyon Formation	830	2725	Brown sandstone, conglomeratic sandstone, carbonaceous shale, and some coal
			Funk Valley Formation	685	2250	Light-brown sandstone and interbedded shale
			Allen Valley Shale	182 to 245	600 to 800	Dark-gray to black, thin- and even-bedded shale
			Sanpete Formation	410	1350	Brown, thin- to medium-bedded, sandstone and conglomeratic sandstone
Jurassic	Upper Jurassic	Morrison(?) Formation		90 to 610	300 to 2000	Reddish-brown shaly siltstone and mudstone
	Middle Jurassic	Arapien Shale	Twist Gulch Member	915	3000	Reddish-brown, thin- and even-bedded shaly siltstone and sandstone
			Twelvemile Canyon Member	1220 to 3960	4000 to 13,000	Variegated red and gray calcareous mudstone, shaly siltstone, and shale with much salt and other evaporites
		Units of the Twin Creek Limestone		100 to 137	320 to 450	Dominantly light-gray, thin-bedded limestones; includes some reddish siltstone and fine-grained sandstone
Triassic	Lower Jurassic and Upper Triassic	Navajo Sandstone or Nugget Sandstone		150 to 305	500 to 1000	Light-brown, medium- to thick-bedded, fine- to medium-grained, quartzose sandstone

TABLE 2. SOME STRATIGRAPHIC UNITS EXPOSED IN THE CHARLESTON-NEBO THRUST PLATE, MOUNT NEBO AREA, CENTRAL UTAH

SYSTEM	SERIES	UNIT	APPROXIMATE THICKNESS		LITHOLOGY
			Meters	Feet	
Jurassic	Middle Jurassic	Twin Creek Limestone	?	?	Chiefly light-gray, thin-bedded limestone; includes interbedded reddish-brown shaly siltstone members
	Jurassic and Triassic(?)	Navajo Sandstone: Nugget Sandstone	183?	600?	Orange-brown to reddish-brown to light-brown, thick-bedded, fine- to medium-grained quartzose sandstone
Triassic	Upper to Lower Triassic	Ankareh Formation	122	400	Reddish-brown shaly siltstone and cross-bedded sandstone with some intercalated, thin, conglomerate beds
	Lower Triassic	Thaynes Limestone	91-305	300-1,000	Chiefly light-gray limestone with some reddish-brown to light-gray shaly siltstone and sandstone beds
		Woodside Sandstone	122	400	Reddish-brown shaly siltstone and cross-bedded, fine- to medium-grained sandstone
Permian	Lower Permian	Park City Formation	198	650	Chiefly light-gray to pale-red, thin- to thick-bedded limestone; includes beds of brownish-black cherty limestone
		Diamond Creek Sandstone	122	400	Reddish-brown to light-brown cross-bedded sandstone; some intercalated limestone
		Kirkman Limestone	113	370	Light- to medium-gray, thin- to thick-bedded limestone; contains chert
Pennsylvanian	Upper to Lower Pennsylvanian	Oquirrh Formation	3,505	11,500	Gray to brownish-gray, thin- to thick-bedded limestone and interbedded light-brown, fine- to medium-grained sandstone

⌐ Thicknesses from Eardley, 1933a; Johnson, 1959; and Black, 1965

TABLE 3. CORRELATION CHART FOR SOME MIDDLE JURASSIC FORMATIONS IN CENTRAL UTAH

Series	Stage	Charleston-Nebo thrust plate		Sanpete-Sevier Valley area, Gunnison Plateau		Wasatch Plateau, San Rafael Swell	
Middle Jurassic	Callovian	Not exposed		Arapien Shale	Twist Gulch Member	Entrada Sandstone (May also include Summerville and Curtis Formations)	
		? — ? — ? — ? — ?					
	Bathonian	Twin Creek Limestone	Leeds Creek(?) Member		Twelvemile Canyon Member	Carmel Formation	Upper part
			Watton Canyon Member		Carbonate beds of the Twin Creek Formation (in the subsurface)		Lower part
			Boundary Ridge Member				
	Bajocian		Rich Member				
			Sliderock Member				
			Gypsum Spring Member				
Lower Jurassic	Toarcian and older	Navajo Sandstone: Nugget Sandstone of Thistle area					

Modified slightly from correlation charts prepared by Imlay (1980, p. 74), and D. A. Sprinkel (Placid Oil Company, 1981, written commun.)

stone of Jurassic and Triassic age. In Red Canyon, near Nephi (Fig. 1), however, the Twelvemile Canyon Member is separated from the underlying Navajo Sandstone by a sequence of beds, chiefly carbonate, that are correlative with the lower and medial parts of the Twin Creek Limestone. Exploratory wells drilled in and near the Sevier and Juab Valleys by Placid Oil Company confirm that, at least in those areas, the Twelvemile Canyon Member is separated from the Navajo by this intercalated Twin Creek sequence (Sprinkel, 1982). The Twelvemile Canyon Member, thus, may be correlative with the Leeds Creek Member of the Twin Creek Limestone (Table 3). Likely the Twelvemile Canyon Member rests on units of the Twin Creek Limestone throughout the Sanpete–Sevier Valley area, although locally it may rest directly on the Navajo Sandstone. I suspect that this intercalated sequence of Twin Creek beds is correlative with the lower part of the Carmel Formation of the San Rafael Group. Table 3, modified from correlation charts prepared by Imlay (1980, p. 74) and by D. A. Sprinkel (Placid Oil Company, 1981, written commun.), summarizes the stratigraphic relations between the Twin Creek Limestone and the Twelvemile Canyon Member of the Arapien Shale.

Wherever exposed, the Twelvemile Canyon Member is greatly deformed; crumpled and contorted beds are widespread, small folds accompany larger ones, overturned folds are common. Bewildering lateral and vertical changes in lithology and appearance are common.

The Twelvemile Canyon Member has been so deformed that its original thickness is uncertain; previous estimates range from 1,220 to 3,960 m (4,000 to 13,000 ft) (Spieker, 1949, p. 17; Gilliland, 1948, p. 30; Hardy, 1949, p. 16, 17; Eardley, 1933a, p. 331). I doubt whether any of these thicknesses is a reliable indicator of original thickness. The member is probably thickest near the axes of the diapiric folds and thinnest between them.

CHARLESTON–NEBO THRUST PLATE

In central Utah the Charleston–Nebo thrust plate is probably best exposed in the Mount Nebo area (Fig. 1) where it consists of a large, overturned, almost recumbent anticline. The west edge of the exposed thrust plate, cut by the Wasatch fault, forms the steep, imposing west face of the southern Wasatch Mountains overlooking Juab Valley. The east edge of the plate, much eroded and more or less mantled by younger rocks, trends northeastward toward Thistle and beyond. The configuration of the east edge of the thrust plate in the subsurface can only be surmised; it may be an overturned anticline.

An overturned anticline (Ritzma, 1972, p. 78) that consists of Jurassic and Triassic strata occurs east of Levan and south of Nephi—beyond what is commonly accepted as the southern margin of the exposed Charleston–Nebo thrust plate. This overturned anticline is here considered to be an eroded outlier of the Charleston–Nebo thrust plate.

The thrust plate is composed of rocks that range in age from Precambrian to Jurassic, but in the three specific areas discussed in this article, the age range is only from Permian and Pennsylvanian (Oquirrh Formation) to Jurassic (Twin Creek Limestone).

The Charleston–Nebo thrust plate was eroded after it was emplaced, and its dissected remnants were then concealed beneath younger Mesozoic and Tertiary strata. I propose that at some time after this cover of younger rocks was emplaced, one or more episodes of salt diapirism deformed parts of both the plate and the mantle of younger rocks. Since then, some of these younger rocks have been eroded, and the thrust plate now is either very well exposed, as in the Mount Nebo area (Fig. 4), or mostly concealed beneath the younger rocks, as in the Thistle area (Fig. 7). The exposed, eroded east margin of the thrust plate is probably near, but not necessarily at, the leading edge of the plate.

The time of emplacement of the thrust plate is unclear. The youngest beds involved in the overturned anticline that forms the thrust plate in the Mount Nebo area appear to be part of the Watton Canyon(?) Member of the Twin Creek Limestone of Middle Jurassic (Bathonian) age. The thrust plate, then, must have been emplaced at some time after Middle Jurassic time. Parts of the eroded east margin of the thrust plate are concealed beneath conglomerate beds assigned to the Price River Formation of Late Cretaceous (Campanian) age (Schoff, 1951, p. 627), implying that the plate was emplaced before Price River time. But, as noted by Black (1965, p. 78), it is uncertain whether these conglomerates are truly equivalent to the type Price River Formation. On the assumption that they are, the thrust plate must have been emplaced, folded, and partly eroded before Price River sediments were deposited. It seems reasonable to conclude, therefore, that the thrust plate was emplaced at some time after Middle Jurassic (Bathonian) time but before Late Cretaceous (Campanian) time.

DIAPIRIC DEFORMATION

Diapiric Concept

Many of the complex structures in the Sanpete–Sevier Valley sector of central Utah can be explained by salt diapirism. In brief, the structural pattern impressed on the area appears to stem from the repeated growth and collapse of diapiric folds formed by and above salt diapirs.

A more exhaustive discussion of the salt diapir concept, including some of the geologic evidence that has led me to the views expressed here, is contained in another article (Witkind, 1982). For purposes of this discussion, only the major concepts advanced in that paper are included here.

Diapiric Folds

I believe the explanation for the intense deformation of the Twelvemile Canyon Member lies in the fact that it is an intrusive sedimentary body. Wherever exposed, it appears to be in intrusive contact with all overlying units older than Holocene. I propose that the salt component of the Twelvemile Canyon Member is the intrusive agent and that it has repeatedly pushed the mudstone component both vertically and laterally. The salt has been moving continuously but at widely varying speeds almost since it was deposited; probably it is moving today. Some of this movement has been a slow, almost imperceptible upwelling, but at times the salt seems to have surged upward rapidly and sporadically. During these surges the salt deformed the overlying mudstones and shaly siltstones and literally forced them to intrude, bow up, and fold back the overlying consolidated sedimentary rocks to form large, elongate, gently curving, diapiric folds, most of which, I suspect, were fan-shaped in cross section. Partial removal of the salt, either by extrusion, dissolution, or lateral flowage, removed support from the folds and caused subsidence of the mudstones. Expectably, the folds then failed, either along high-angle faults that formed along the flanks of the folds, or by broad, general downwarp.

The mudstones of the Twelvemile Canyon Member seemingly have been pushed up time and again by the salt, and time and again have subsided as salt was removed. This has resulted in the repeated growth and collapse of diapiric folds whose trends mirrored the underlying causative salt diapirs. The structural complexity that marks the transition zone between Richfield on the south and Nephi on the north is a direct reflection of this recurrent growth and collapse of ever younger diapiric folds which occupied the same structural zones and had the same trends as those folds that were formed during the previous diapiric episodes. It is much as if the salt and mudstone, confined to northerly trending, near-vertical linear conduits, repeatedly used these conduits each time the salt was reactivated.

Diapiric Episodes

Geologic evidence in central Utah suggests that the mudstones of the Twelvemile Canyon Member, forced upward by the salt, deformed the overlying rocks into diapiric folds at least three and possibly four times in the past (Witkind, 1982). Each of these diapiric episodes is arbitrarily divided into three recognizable, but related and continuous stages. The first stage is an *intrusive stage* during which an upward surge of the salt forces the calcareous mudstones to bow up and fold back the overlying sedimentary rocks to form the diapiric folds. An *erosional stage* begins with the destruction of the fold as the salt is removed. This failure of the fold is expressed either by collapse of the crest of the fold along high-angle faults, or by a general downwarp of the uplifted country rocks. The remnants of the fold are then rapidly eroded to form a surface of low relief. Sediments from adjacent highlands are spread across the newly formed erosional surface and mark the beginning of the *depositional stage*. The fact that the beds deposited on this surface are much thinner near the axes of the folds than elsewhere suggests that the slow but persistent upwelling of the salt resumed at this time. The depositional stage ends as the salt surges upward once again marking the onset of the intrusive stage of the next diapiric episode.

The first recognizable diapiric episode began during Late Cretaceous time and ended during early Paleocene time. The second episode began during early Paleocene time and probably ended during late(?) Oligocene time. The third episode extended from late(?) Oligocene time into Pliocene or Pleistocene time; when it ended is uncertain. A fourth episode, marked by localized deformation, appears to have begun and ended at some time during late Pliocene or Pleistocene time.

The fact that the Twelvemile Canyon Member, which was responsible for these huge upwarps, moved repeatedly—once during Late Cretaceous time, again during early Paleocene time, and yet a third time during the late(?) Oligocene or Miocene—suggests that it has different "ages." Its *depositional age* is Middle Jurassic; it contains Middle Jurassic fossils. Its *emplacement age*, however, has changed repeatedly throughout geologic time. To show this repeated change in emplacement age, I have, on a published map (Witkind, 1981) and on figures in this and other articles (Witkind, 1982), used the symbols "T(Jat)" or "K(Jat)" to indicate the times (T=Tertiary; K=Cretaceous) during which the country rocks were deformed by the upward movement of the Twelvemile Canyon Member (J=Jurassic; a=Arapien Shale; t=Twelvemile Canyon Member).

PATTERN OF FOLDS

At least three, and probably as many as seven, major diapiric folds are exposed or lie within the transition zone of central Utah (Witkind, 1982). Their general map pattern is shown on Figure 2. The Sanpete-Sevier Valley fold (No. 1) appears to be the master fold. Apparently branching off its southern end is the Redmond fold (No. 2) which extends at least from Redmond (possibly Salina or Sigurd) to near Gunnison. It is unknown whether the Redmond fold continues northward to join the Levan fold (No. 3). Near Fayette, the Sevier Bridge Reservoir fold (No. 4) either branches off the Redmond fold or is a separate fold wholly independent of the Redmond fold. West Hills (No. 5), a north-trending "anticline," has been drilled and much salt

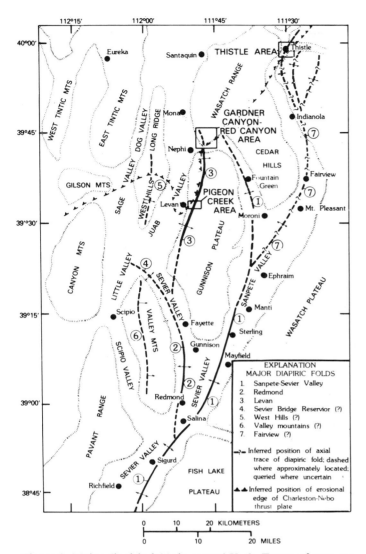

Figure 2. Major diapiric folds in central Utah. Traces of axes are approximately located.

the fold's flanks (Fig. 3). In most places, however, the core of the fold is concealed beneath surficial deposits, commonly an alluviated valley floor, and the long, linear to gently curving bands of upturned strata are the only exposed parts of the fold.

DIAPIRIC DEFORMATION OF THE CHARLESTON-NEBO THRUST PLATE

Of the seven folds shown in Figure 2, only two (the Levan fold, No. 3, and the Fairview fold, No. 7) are discussed here, chiefly because both of these folds appear to pass below and to deform the Charleston-Nebo thrust plate.

Levan Diapiric Fold

The Levan fold extends for at least 38 km (24 mi) along the west flank of the Gunnison Plateau where its core, composed of Twelvemile Canyon Member mudstones, appears as an intricately dissected range of low hills that trends about N. 15° E. Southeast of Nephi, the fold is concealed beneath enormous landslides. Directly south of these landslides both flanks of the fold are expressed by tilted strata which dip outward; the west flank of the diapiric core is overlain for much of its length by bowed-up beds of the Green River Formation which dip moderately westward. The east flank is mantled by bowed-up beds, chiefly of the Twist Gulch Member of the Arapien Shale, which dip eastward, also at moderate angles.

The core of the Levan fold, expressed as exposures of the Twelvemile Canyon Member, extends from near Little Salt Creek on the south to Red Canyon on the north, where it disappears beneath a more northerly segment of the Charleston-Nebo thrust plate. The core reappears a short distance to the north in Gardner Canyon where it seemingly has disrupted part of the thrust plate.

Gardner Canyon-Red Canyon Area. In Red Canyon (Fig. 4), about 4 km (2.5 mi) northeast of Nephi, the Charleston-Nebo thrust plate appears as an overturned anticline composed of beds that range in age from Permian and Pennsylvanian (Oquirrh Formation) to Jurassic (Twin Creek Limestone). In and near Red Canyon, these overturned beds rest on contorted mudstones of the Twelvemile Canyon Member. As the overturned beds of the thrust plate are traced northeastward, the relations between them and the underlying beds are masked by younger sedimentary and volcaniclastic rocks.

About 0.8 km (0.5 mi) north of Red Canyon, in Gardner Canyon, the overturned beds of the thrust plate are disrupted above a mass of Twelvemile Canyon strata that are much deformed. In one place beds of the Twelvemile Canyon are warped into a small S-fold, implying vertical uplift. These strata, exposed along the lower flanks

found in its core. Possibly the Valley Mountains (No. 6), an eastward-tilted block, may have a salt core, although recent drilling has failed to confirm this. Collapsed diapiric structures east of Indianola (Runyon, 1977) and the downwarp of those beds that form the west flank of the Wasatch Plateau (Witkind, 1982) suggest the presence of the Fairview fold (No. 7). The fold may extend from near Ephraim northeastward to near Indianola. Likely it splits near Indianola to form a series of north-trending minor diapiric folds.

In a few places these diapiric folds appear as outcrops of the mudstones of the Twelvemile Canyon Member flanked on one or both sides by upturned strata. Locally, these tilted strata are vertical or overturned; elsewhere they are gently to moderately inclined. I interpret the Twelvemile Canyon outcrops to be part of the core of a diapiric fold and the upturned strata to be the eroded remnants of

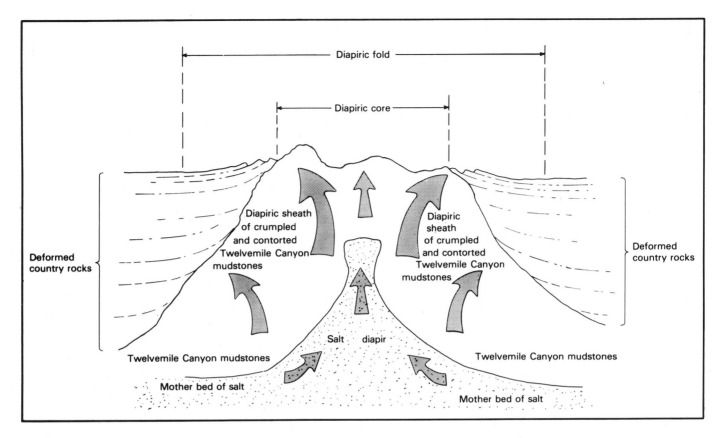

Figure 3. Possible geologic relations between a *salt diapir,* the *diapiric sheath* of Twelvemile Canyon Member mudstones, and the upturned country rocks. All form the diapiric fold. The salt diapir and its diapiric sheath compose the *diapiric core* of the fold. Arrows denote general direction of movement of the plastic and mobile salt and mudstones. In places, vertical forces, stemming from the intrusive salt diapir, are translated laterally into horizontal compressional forces. Not to scale.

of the canyon walls, abut strata of the Oquirrh Formation, and the contact between them is steep to near-vertical. The Twelvemile Canyon beds are overlain by light-gray, thin limestone beds of the Leeds Creek(?) Member of the Twin Creek Limestone which crop out as a narrow band, about 0.8 km (0.5 mi) wide. Both the east and west flanks of this band of Leeds Creek(?) beds is in near-vertical fault contact with overturned, northwest-dipping beds of the Oquirrh Formation. Farther to the north, in Birch Creek, Navajo Sandstone beds overlain by Twin Creek strata abut overturned Oquirrh beds, and here, too, the fault contact between these units is near vertical. This band of Navajo and Twin Creek strata extends for about 1 km (0.75 mi) north of Birch Creek before it disappears, near Long and Wide Canyons, beneath Paleozoic strata that form part of the Charleston-Nebo thrust plate. In this general area, then, a narrow, north-trending band of Jurassic strata is in fault contact, along both sides, with overturned Permian and Pennsylvanian beds.

Although these geologic relations may represent a window in a folded thrust sheet, as interpreted by Black (1965, cross section C-C'), I believe that the steep to near-vertical contacts between the Twelvemile Canyon Member and the Oquirrh Formation and between the Twin Creek Limestone and the Oquirrh, plus the evidence of vertical compression (S-fold), are more reasonably explained by vertical uplift. Consequently, as suggested by cross section A-A', Figure 4, I interpret the Twelvemile Canyon strata to be a plug of the Levan diapiric core. This plug of Twelvemile Canyon strata uplifted, tilted, and juxtaposed this band of Leeds Creek(?) limestone beds against overturned Oquirrh strata, which in this specific area form the main mass of the Charleston-Nebo thrust plate.

The mudstones of the Twelvemile Canyon Member exposed in front of the Charleston-Nebo thrust plate give every sign of intense compression. It is possible that this compression was induced during emplacement of the Charleston-Nebo thrust plate (Black, 1965, p. 80). There is, however, a lack of uniformity in the axial trends of the folds formed in the Twelvemile Canyon Member. This lack

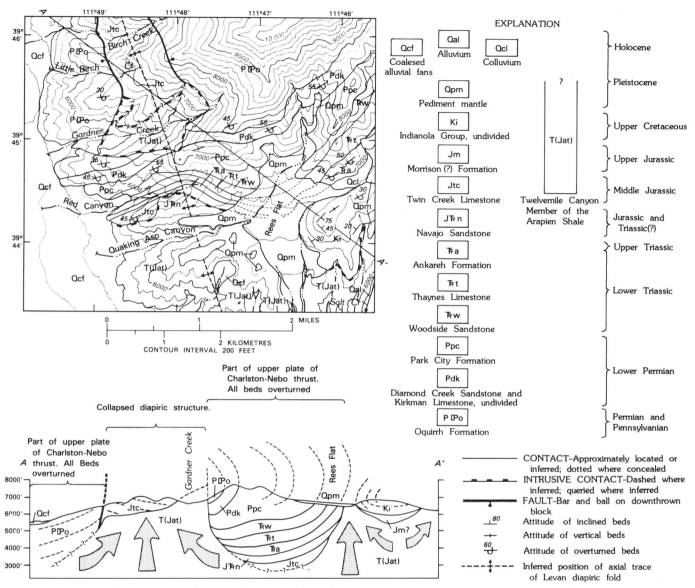

Figure 4. Geologic map and cross section of the Gardner Canyon–Red Canyon area showing relations between the upper plate (here an overturned fold) of the Charleston-Nebo thrust fault and the north end of the Levan diapiric fold. As interpreted, the Twelvemile Canyon Member of the Arapien Shale, T(Jat), both intruded and bowed up the thrust plate. In the Gardner Canyon area, mudstones and contained evaporites of the Twelvemile Canyon Member intruded, broke, and raised a small wedge of the thrust plate, thus juxtaposing beds of the Jurassic Twin Creek Limestone (Jtc) against overturned beds of the Permian and Pennsylvanian Oquirrh Formation (PⅠPo).

suggests that the compression more likely stems from the later, vertical movement of the Twelvemile Canyon Member as it was forced upward by its salt component.

Some evidence in support of this interpretation is exposed farther east near Rees Flat (Fig. 4). As the contact between the sole of the thrust plate and the Twelvemile Canyon mudstones is traced eastward up Red Canyon, it eventually disappears beneath a pediment mantle of sand and gravel that floors Rees Flat. The eastern edge of Rees Flat is bounded on the north by overturned rocks (Ankareh Formation) of the thrust plate, and on the south by a small, irregular-shaped basin formed by upturned Indianola conglomerate beds. The Indianola beds nearest the thrust plate are vertical or overturned; those along the east flank of the basin dip westerly, and those along the south flank dip northerly. I suspect that the basin is encircled and underlain by the Twelvemile Canyon Member, which in its upward movement (prior to deposition of the surficial deposits that now conceal much of it) bowed up the Indianola conglomerates irregularly. The vertical to overturned beds that mark the north edge of the basin imply that the Twelvemile Canyon Member, confined between the edge of the thrust plate and the once near-horizontal Indianola beds, pushed up and bowed back the Indianola beds. Indeed, if this deformation is due to vertical uplift of the Twelvemile Canyon Member rather than to eastward movement of the thrust plate (Black, 1965, p. 80), the mudstones of the Twelvemile Canyon Member likely underlie the toe of the plate and may have bowed up part of the plate even as they deformed the adjacent Indianola conglomerates.

Pigeon Creek Area. A second area in which the Charleston-Nebo thrust plate has been deformed by the core of the Levan diapiric fold is east of Levan, about 23 km (14 mi) south of the Gardner Canyon–Red Canyon area. Field relations are best exposed in the valley of Pigeon Creek (Fig. 5).

Pigeon Creek cuts through the Charleston-Nebo thrust plate which in this area consists of a north-trending anticline. This anticline appears to have been intruded and deformed by the mudstone and gypsum beds of the Twelvemile Canyon Member that make up the core of the Levan diapiric fold. The anticline appears as a broad upwarp of gray, thin limestone beds of the Leeds Creek(?)

Member of the Twin Creek Limestone. A test well, Standard Oil of California's Levan Unit in the NE1/4NW1/4 sec. 17, T. 15 S., R. 1 E., some 6 km (4 mi) south of Pigeon Creek, demonstrated that the fold is overturned (Ritzma, 1972, p. 78). Near the mouth of Pigeon Creek Canyon, the limestone beds that form the west flank of the north-trending anticline dip moderately westward (Fig. 5). As one proceeds eastward into the canyon and toward the axis of the fold, the beds become nearly horizontal, and then beyond the fold axis they gradually roll over and dip moderately eastward. At the very eastern edge of the fold, the eastward-dipping beds are flexed up sharply to westward dips and end abruptly against the Twelvemile Canyon mudstone beds that form the core of the Levan diapiric fold. Locally, these westward-dipping limestone beds are bowed up, folded back, and overturned; in these localities, thus, they dip eastward. It seems clear that the intrusive action of the Twelvemile Canyon Member bowed up beds that originally dipped eastward, forced them into a westward dip, locally overturned them, and then broke them off.

Further evidence of the intrusive and deforming action of the Twelvemile Canyon Member is exposed near the center of this north-trending, overturned anticline. There, the limestone beds appear to be intruded by several vertical

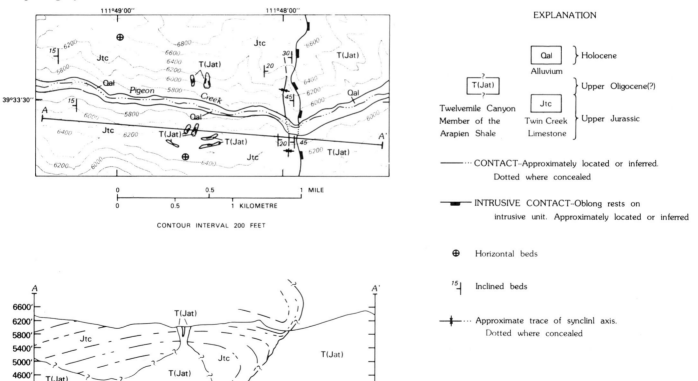

Figure 5. Geologic map and cross section of the Pigeon Creek area showing relations between an overturned anticline (composed of limestone beds of the Twin Creek Limestone) and the intrusive mudstone and gypsum beds of the Twelvemile Canyon Member of the Arapien Shale. The anticline is interpreted to be an erosional outlier of the Charleston-Nebo thrust plate.

veins of gypsum, which range in thickness from about 15 to 76 m (50 to 250 ft), and by red mudstone beds of the Twelvemile Canyon Member. Locally, seams of gypsum, about about 15 m (50 ft) thick, are intercalated between limestone beds to form elongate, near-horizontal pods conformable with the bedding.

In the past, these limestone beds of the Leeds Creek(?) Member have been included with the Twelvemile Canyon Member (Spieker, 1949, p. 17) and mapped with the latter as a single unit (Hunt, 1950). Thus, published descriptions of the Twelvemile Canyon Member have included limestone as one of the lithologic units that make up the member (Hardy, 1952, p. 18). The evidence of deformation and intrusion of the limestones by the mudstones, however, suggests that, at least in this area, units of two totally different "ages" are involved even though both contain Jurassic fossils. In the Pigeon Creek area, the Twelvemile Canyon Member is best viewed as an intrusive unit that has a depositional age of Middle Jurassic and an emplacement age of late(?) Oligocene or Miocene. As such, it has intruded and deformed an overturned anticline composed of the Middle Jurassic Twin Creek Limestone, a structure that formed at some time after Middle Jurassic time but before Late Cretaceous time. Grouping the Leeds Creek(?) limestone beds with the Twelvemile Canyon Member mudstones not only has masked critical structural relations but also has led to the erroneous assumption that bedding attitudes of the Twelvemile Canyon Member mudstones reflect the structure of the concealed, underlying Navajo Sandstone, one of the principal oil reservoir rocks in central Utah.

Fairview Diapiric(?) Fold

In the Thistle area (Fig. 1), a minor fold—a northern offshoot of the Fairview diapiric(?) fold (No. 7 of Fig. 2)—has deformed part of the Charleston-Nebo thrust plate.

Whereas the Levan fold is represented both by its core of Twelvemile Canyon Member mudstones and by its flanks of tilted sedimentary rocks, much of the evidence supporting the existence of the Fairview fold, and thus of its inferred northern offshoot, is speculative. The presence of the Fairview diapiric(?) fold is based on the westward downwarp of those beds that border the east side of Sanpete Valley—the Wasatch monocline (Witkind, 1982). It is postulated that partial removal of a salt diapir that underlies Sanpete Valley resulted in the westward subsidence of once near-horizontal beds to form the Wasatch monocline. The causative salt diapir is thought to be compound, consisting of the Sanpete–Sevier Valley salt diapir in the southern part of Sanpete Valley and the Fairview salt diapir(?) in the northern part of the valley.

As interpreted, then, the Fairview diapiric fold branches off the main Sanpete–Sevier Valley fold near Ephraim and continues northeastward past Mount Pleasant and Fairview, concealed beneath the valley fill, to some indefinite point near Indianola (Fig. 2). There, the fold seems to split into three small folds, the westernmost one of which extends through Thistle to underlie the eroded eastern margin of the Charleston-Nebo thrust plate.

Thistle Area. In the Thistle area, the erosional edge of the Charleston-Nebo thrust plate appears as an imposing ridge that consists of steeply tilted beds of orange to tan Navajo Sandstone (Jurassic and Triassic?) overlain chiefly by gray, thin limestone beds of the Twin Creek Limestone (Jurassic). All beds are upright and dip about 60° to the east; they are particularly well exposed where U.S. Highway 6 and 50 cuts through the ridge (Fig. 6). This outcrop of the thrust plate is mantled by younger units; along both its east and west flanks the thrust plate is overlain by beds of the North Horn Formation (Paleocene and Cretaceous) and Flagstaff Limestone (Eocene and Paleocene). East of the ridge, the North Horn–Flagstaff sequence dips gently eastward; west of the ridge, it dips gently westward (Fig. 7). Seemingly, the Navajo and Twin Creek beds have been exhumed from beneath an arched mantle of North Horn and Flagstaff beds.

Directly east of the ridge and north of U.S. Highway 6 and 50, the North Horn–Flagstaff sequence has been intruded and deformed by a wedge of contorted Twelvemile Canyon Member strata, most of which is concealed beneath a mantle of pediment gravels. This intrusive mass appears to have bowed up the Flagstaff Limestone to a vertical position, then to have broken through it and spread laterally westward across its top. Mudstones of the Twelvemile Canyon Member now abut the eastward-tilted Twin Creek Limestone beds that form the erosional, eastern edge of the thrust plate. The exposure is significant in that it indicates (1) that the Twelvemile Canyon Member underlies the Thistle area, (2) that the Twelvemile Canyon Member has deformed all exposed units to some extent, and (3) that at least part of this deformation occurred at some time after deposition of the Flagstaff Limestone and long after the emplacement of the Charleston-Nebo thrust plate. This deformation by Twelvemile Canyon strata probably stems from the intrusive stage that marked the beginning of the third diapiric episode, which began during or at some time after the late(?) Oligocene

STRUCTURAL IMPLICATIONS

The geologic evidence in the three areas discussed above indicates that the thrust plate was deformed by the upward movement of the cores of several diapiric folds after it was emplaced. Salt in the Twelvemile Canyon Member surged upward and through narrow, elongate, near-vertical conduits, forcing the overlying diapiric sheath, composed of mudstones of the Twelvemile Canyon

Figure 6. View looking north in the Thistle area at the exposed, erosional, east edge of the Charleston-Nebo thrust plate. At left (west) are steeply tilted, eastward-dipping thick beds of the Navajo (Nugget) Sandstone (JRn). These are conformably overlain by beds of reddish-brown, shaly siltstone and sandstone and light-gray, thin limestone beds of the Twin Creek Limestone (Jtc).

Member, to intrude, break, raise, and locally arch parts of the overlying Charleston-Nebo thrust plate.

The resultant folds, of which the diapiric cores are but a part (Fig. 3), display an unusual regularity in distribution and orientation, which suggests some form of structural control. This concept of structural control is supported by the fact that in place after place in central Utah the stacks of sedimentary rocks that form the flanks of the diapiric folds display similar patterns of crustal disturbance. Younger diapiric folds with the same trends as the older folds appear to have formed repeatedly in the same sites. An additional implication of this pattern is that these repeated episodes of deformation must have occurred more or less in unison; that is, all the salt diapirs seemingly surged upward at about the same time, and new diapiric folds formed simultaneously in response to this concerted movement of the salt diapirs.

I believe that the uniformity in distribution and orientation of the diapiric folds reflects deep-seated, fundamental faults that were formed before the salt was deposited. It is possible that autonomous, isostatic movement of the salt ("halo-kinesis" of Trusheim, 1957) led to the development of the parental salt diapirs above these faults. Subsequent tectonic impulses may have stimulated the development of daughter diapirs which formed in the same sites and with the same trends as the parental diapirs. I doubt, however, whether all the daughter diapirs would have formed simultaneously in all the diapiric sites as a result of self-contained salt movement. As Sannemann (1968) and Trusheim (1960) have pointed out, the pattern of development of the diapirs in the Zechstein salt of northern Germany appears to be that of ever-younger generations of diapirs forming centrifugally away from the parental diapirs. The end result, when the mother bed of salt is depleted, is very young diapirs at the outermost margins of the diapiric belt and ever older generations of diapirs inward toward the parental diapir, which is in the center. This is not the diapiric pattern in central Utah. As far as I can determine, the same rocks are deformed in the outermost diapiric folds as in the central ones—the pattern appears to be one of repeated, simultaneous diapiric development.

This had led me to speculate (Witkind, 1982) that this pattern may result from the repeated reactivation of the deep-seated breaks ("halo-tectonism" of Trusheim, 1957). When these faults reactivated, the salt, under great static load, surged up each newly formed fault plane pushing the overlying mudstones ahead of it. As a result the fault planes were obliterated by the rising salt and mudstones. The end result was a series of diapiric sheaths, each of which was cored by an intrusive salt diapir and flanked by upturned

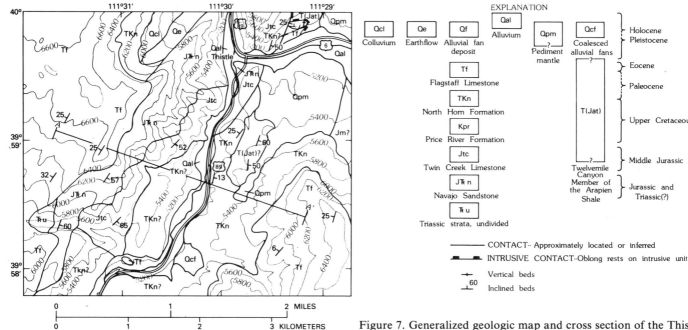

EXPLANATION

Qcl	Qe	Qf	Qal	Qpm	Qcf	Holocene / Pleistocene
Colluvium	Earthflow	Alluvial fan deposit	Alluvium	Pediment mantle	Coalesced alluvial fans	Eocene / Paleocene

Tf — Flagstaff Limestone

TKn — North Horn Formation

Kpr — Price River Formation

Jtc — Twin Creek Limestone

Jℝn — Navajo Sandstone

ℝu — Triassic strata, undivided

T(Jat) — Twelvemile Canyon Member of the Arapien Shale — Upper Cretaceous / Middle Jurassic / Jurassic and Triassic(?)

———— CONTACT- Approximately located or inferred

◄▬▬► INTRUSIVE CONTACT-Oblong rests on intrusive unit

Vertical beds

Inclined beds

Figure 7. Generalized geologic map and cross section of the Thistle area, compiled and modified from maps prepared by H. D. Harris (1954) for the area west of U.S. Highway 89, and by M. L. Pinnell (1972) for the area east of U.S. Highway 89.

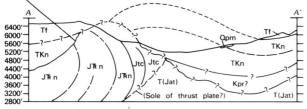

Unconformity. Price River (Kpr) and younger rocks (Tf, TKn) were deposited on the flanks of a pre-existing ridge composed of older rocks (Jtc, Jℝn).

beds—the diapiric folds. Each fold presumably reflects the trend of the causative fault. Repeated reactivation of the faults resulted in repeated renewal of the salt diapirs and their overlying diapiric folds, which formed in the same sites and with the same trends as the older diapirs and their overlying folds. This pattern also implies that this repeated deformation occurred more or less in unison—all the faults reactivated at about the same time resulting in the development of new diapiric folds.

Inherent uncertainties flaw this concept. It is unknown whether such deep-seated faults do exist, and if they do, whether they have moved during the times postulated. At present, little is known about older structures that underlie this part of central Utah. Two major fault zones in central Utah—the Sevier and Wasatch fault zones—appear either to align with or to parallel major diapiric folds. Even as the Sevier fault zone ends northward near Richfield, the Sanpete–Sevier Valley diapiric fold (No. 1 of Fig. 2) begins and extends northeastward, essentially collinear with the projected trace of the fault zone. In similar fashion, the southern extension of the Wasatch fault zone aligns with the south-trending Redmond fold (No. 2 of Fig. 2) (Wit-

kind, 1981). And the projected trend of the Wasatch fault zone, along the east side of Juab Valley, parallels the Levan fold (No. 3 of Fig. 2).

Basin and Range extensional tectonics may have begun in southwestern Utah some 22 to 21 m.y. ago (Miocene) (Rowley and others, 1978), and this may date the beginning of the third episode of diapirism. But documentation supporting the earlier fault movements is missing.

It may be significant, however, that the first workers in the Paradox Basin, also impressed by the uniformity of distribution and orientation of the "salt anticlines" there, attributed this uniformity to a sequence of deep-seated breaks in the basement rocks (Stokes, 1948, p. 14; Cater, 1955, p. 125), even though little evidence supported their views. Since then, drilling and seismic surveys in Paradox Basin have demonstrated that such faults have, indeed, influenced the distribution of the salt anticlines (Baars and Stevinson, 1981, p. 28). The concept that deep-seated faults may have determined the diapiric pattern in central Utah has still to be tested.

How many times the Charleston-Nebo thrust plate was deformed by the upward movement of the salt diapirs is unknown. I have been unable to find any evidence to indicate that the thrust plate was deformed during either the first or second diapiric episode (Late Cretaceous to early Paleocene; early Paleocene to late(?) Oligocene). Almost all available evidence suggests that the major deformation of the thrust plate occurred during the intrusive stage of the third diapiric episode—tentatively dated as late(?) Oligocene or Miocene. The youngest rocks deformed along the flanks of the Levan diapiric fold belong to the Crazy Hollow Formation of Oligocene age. I assume, therefore, that the disruption of the thrust plate in the Gardner Canyon area and the warpage and intrusion of the overturned fold in the Pigeon Creek area occurred at some time after deposition of the Crazy Hollow Formation, presumably during this late(?) Oligocene or Miocene intrusive stage.

In the Thistle area the youngest beds seemingly bowed up as a result of intrusion by a salt diapir are part of the Green River Formation of Eocene age. The intrusion and warpage of the Flagstaff Limestone (Eocene and Paleocene) in that area therefore would appear to have occurred at some time after the Eocene, presumably also during the intrusive stage of the third diapiric episode, in essence, the late(?) Oligocene or Miocene.

In summary, then, the Charleston-Nebo thrust plate was likely emplaced after Middle Jurassic (Bathonian) time, and eroded before Campanian time of the Late Cretaceous, for its eroded remnants are overlain by Price River strata of Campanian age. Deformation of the thrust plate seems to have occurred during the intrusive stage of the third diapiric episode—Late Oligocene(?) or Miocene.

ECONOMIC IMPLICATIONS

Overturned folds have been a prime target in the ever increasing search for new oil fields. As one of these anticlinal folds—the upper plate of the Charleston-Nebo thrust—has been intruded, tilted, and deformed by the Twelvemile Canyon Member mudstones, it would seem likely that other anticlinal folds in the area are similarly deformed. Whatever oil pools are contained within the reservoir rocks of these tilted anticlines may have been shifted and displaced. And if the oil has had time to migrate, drilling on the original structural crests of the anticlines may be fruitless. But perhaps of even greater concern is the possibility that these tilted folds have been so pervasively broken during one or more episodes of diapirism that their reservoir rocks may have long since lost the ability to retain whatever oil had pooled in them.

It would seem prudent for most petroleum geologists considering exploration in this part of central Utah to determine what effects salt diapirism has had on overturned anticlines, their reservoir rocks, and whatever oil pools might be contained within these reservoir rocks.

ACKNOWLEDGMENTS

All articles reflect the efforts of one's colleagues and editors. This article is no exception; I thank D. A. Sprinkel of Placid Oil Company for some stimulating field discussions, E. R. Cressman, E. R. Verbeek, and Victoria R. Todd of the Geological Survey, and W. R. Muehlberger of the University of Texas for their thoughtful and friendly technical reviews. And finally my thanks go to Editor Dorothy Merrifield for her helpful and constructive, but devastating, attack on my commas.

REFERENCES CITED

Baars, D. L., and Stevinson, G. M., 1981, Tectonic evolution of the Paradox Basin, Utah and Colorado: Rocky Mountain Association of Geologists Guidebook, 1981 Field Conference, Geology of the Paradox Basin, p. 23–31.

Baer, J. L., 1976, Structural evolution of central Utah—Late Permian to Recent: Rocky Mountain Association of Geologists Guidebook, 27th Annual Field Conference, Geology of Cordilleran Hingeline, p. 37–45.

Black, B. A., 1965, Nebo overthrust, southern Wasatch Mountains, Utah: Brigham Young University Geology Studies, v. 12, p. 55–89.

Cater, F. W., Jr., 1955, The salt anticlines of southwestern Colorado and southeastern Utah: Four Corners Geological Society Guidebook, Geology of Parts of Paradox, Black Mesa, and San Juan Basins, p. 125–131.

Eardley, A. J., 1933a, Stratigraphy of the southern Wasatch Mountains, Utah: Michigan Academy of Science, Arts, and Letters, v. 18, p. 307–344.

——1933b, Structure and physiography of the southern Wasatch Mountains, Utah: Michigan Academy of Science, Arts, and Letters, v. 19, p. 377–400.

Gilliland, W. N., 1948, Geology of the Gunnison quadrangle, Utah [Ph.D. thesis]: Ohio State University, 178 p.

——1951, Geology of the Gunnison quadrangle, Utah: University of Nebraska Studies, no. 8.

Hardy, C. T., 1949, Stratigraphy and structure of the Arapien Shale and the Twist Gulch Formation in Sevier Valley, Utah [Ph.D. thesis]: Ohio State University, 85 p.

——1952, Eastern Sevier Valley, Sevier and Sanpete Counties, Utah: Utah Geological and Mineralogical Survey Bulletin 43, 98 p.

——1962, Mesozoic and Cenozoic stratigraphy of north-central Utah: Brigham Young University Geology Studies, v. 9, Pt. 1, p. 50–64.

Hardy, C. T., and Zeller, H. D., 1953, Geology of the west-central part of the Gunnison Plateau, Utah: Geological Society of America Bulletin, v. 64, p. 1261–1278.

Harris, H. D., 1954, Geology of the Birdseye area, Thistle Creek Canyon, Utah: The Compass, v. 31, no. 3, p. 189–208.

Hintze, L. F., 1962, Precambrian and Lower Paleozoic rocks of north-central Utah: Brigham Young University Geology Studies, v. 9, Pt. 1, p. 8–16.

Hunt, R. E., 1950, Geology of the northern part of the Gunnison Plateau [Ph.D. thesis]: Ohio State University.

Imlay, R. W., 1980, Jurassic paleobiogeography of the conterminous United States in its continental setting: U.S. Geological Survey Professional Paper 1062, 134 p.

Johnson, K. D., 1959, Structure and stratigraphy of the Mount Nebo–Salt Creek area, southern Wasatch Mountains, Utah [M.S. thesis]: Brigham Young University, 49 p.

McGookey, D. P., 1960, Early Tertiary stratigraphy of part of central Utah: American Association of Petroleum Geologists Bulletin, v. 44, no. 5, p. 589–615.

Mitchell, G. C., 1979, Stratigraphy and regional implications of the Argonaut Energy No. 1 Federal, Millard County, Utah: Rocky Mountain Association of Geologists–Utah Geological Association Guidebook, Basin and Range Symposium and Great Basin Field Conference, p. 503–514.

Moulton, F. C., 1975, Lower Mesozoic and Upper Paleozoic petroleum potential of the hingeline area, central Utah: Rocky Mountain Association of Geologists Guidebook, Deep Drilling Frontiers in the Central Rocky Mountains, p. 87–97.

Picard, M. D., 1980, Stratigraphy, petrography, and origin of evaporites, Jurassic Arapien Shale, central Utah: Utah Geological Association Publication 8, Henry Mountains Symposium, p. 129–150.

Pinnell, M. L., 1972, Geology of the Thistle quadrangle, Utah: Brigham Young University Geology Studies, v. 19, Pt. 1, p. 89–130.

Pratt, A. R., Heylmun, E. B., and Cohenour, R. E., 1966, Salt deposits of Sevier Valley, Utah: *in* Second Symposium on Salt: Northern Ohio Geological Society, Pt. I, p. 48–58.

Ritzma, H. R., 1972, Six Utah "hingeline" wells: Utah Geological Association Publication 2, Basin and Range Transition Zone, Central Utah, p. 75–80.

Rowley, P. D., Anderson, J. J., Williams, P. L., and Fleck, R. J., 1978, Age of structural differentiation between the Colorado Plateaus and Basin and Range provinces in southwestern Utah: Geology, v. 6, p. 51–55.

Runyon, D. M., 1977, Collapsed diapiric structures and their potential economic significance, Indianola, Utah: Wyoming Geological Association Guidebook, 29th Annual Field Conference, Rocky Mountain Thrust Belt Geology and Resources, p. 479–486.

Sannemann, D., 1968, Salt-stock families in northwestern Germany: American Association of Petroleum Geologists Memoir 8, Diapirism and diapirs, p. 261–270.

Schoff, S. L., 1951, Geology of the Cedar Hills, Utah: Geological Society of America Bulletin, v. 62, p. 619–645.

Spearing, C., 1981, Cartoon: Rocky Mountain Association of Geologists Guidebook, Geology of the Paradox Basin, p. 186.

Spieker, E. M., 1946, Late Mesozoic and early Cenozoic history of central Utah: U.S. Geological Survey Professional Paper 205-D, p. 117–161.

——1949, The transition between the Colorado Plateaus and the Great Basin in central Utah: Utah Geological Society, Guidebook to the Geology of Utah, no. 4, 106 p.

Sprinkel, D. A., 1982, Twin Creek Limestone–Arapien Shale relations in central Utah: Utah Geological Association Publication 9, The Overthrust Belt of Utah.

Stokes, W. L., 1948, Geology of the Utah-Colorado salt dome region with emphasis on Gypsum Valley: Guidebook to the Geology of Utah No. 2, Utah Geological and Mineralogical Survey.

——1952, Salt-generated structures of the Colorado Plateau and possible analogies [abs.]: American Association of Petroleum Geologists Bulletin, v. 36, no. 5, p. 961.

——1956, Tectonics of Wasatch Plateau and near-by areas [abs.]: American Association of Petroleum Geologists Bulletin, v. 40, no. 4, p. 790.

Trusheim, F., 1957, Uber halokinese and ihre bedeutung fur die strukturelle entwicklung Norddeutschlands: Zeitschrift der Deutschen Geologischen Gesellschaft, v. 109, p. 111–151.

——1960, Mechanism of salt migration in northern Germany: American Association of Petroleum Geologists Bulletin, v. 44, no. 9, p. 1519–1540.

Witkind, I. J., 1981, Reconnaissance geologic map of the Redmond quadrangle, Sanpete and Sevier Counties, Utah: U.S. Geological Survey Miscellaneous Investigations Map I-1304-A.

——1982, Salt diapirism in central Utah: Utah Geological Association Publication 9, The overthrust belt of Utah.

MANUSCRIPT ACCEPTED BY THE SOCIETY AUGUST 20, 1982

Geological Society of America
Memoir 157
1983

Variations in structural style and correlation of thrust plates in the Sevier foreland thrust belt, Great Salt Lake area, Utah

Edwin W. Tooker

U.S. Geological Survey
345 Middlefield Road
Menlo Park, California 94025

ABSTRACT

Analysis of geologic data from the Salt Lake City region provides new understanding of the Sevier thrust belt foreland. Faulted segments of six major thrust plates, the Charleston-Nebo, North Oquirrh, Midas, Stockton, Tintic Valley, and Skull Valley, have been identified and correlated on the basis of their structural position and their stratigraphic and structural characteristics. Substantial movement is indicated on the sole Charleston-Nebo thrust, and smaller but unknown amounts of movement are inferred along the later, younger, and discontinuous thrusts farther to the west, which successively telescoped Paleozoic and Mesozoic miogeoclinal strata. Thrust plates were transported east-northeast to eastward, and their emplacement was affected by a buttress consisting of an uplifted Uinta-Cortez axial zone and adjacent northern Utah highland. Segmentation of plates by intraplate thrusts and by transcurrent (tear) faults was intensified particularly along the westward trace of the Uinta-Cortez axis. Plates thrust against the buttress are generally narrow and elongate north-south, and the easternmost plates are wrapped more tightly around the buttress. Fold style seems to vary systematically with thickness of plate strata and distance of a plate from the buttress. The North Oquirrh thrust plate, which apparently moved southeastward, presumably was derived from an unknown hinterland of the western Wyoming salient and stands in contrast with the main group thrust plates. The western edge of the thrust belt, or the root zone, lies an unknown distance west of the study area, probably in eastern Nevada.

INTRODUCTION

Additional geologic information has been developed in the 17 years since Roberts and others (1965) described the occurrence of multiple allochthonous thrust plates in the vicinity of Salt Lake City, Utah (Fig. 1). The present paper examines these new data in the area of the Salt Lake recess (Fig. 1) in order to establish regional correlations between the numerous segmented thrust plates.

During Mesozoic time, a regional system of thrust faults called the Sevier orogenic belt (Armstrong, 1968) originated in the western Utah–eastern Nevada region and was moved eastward. The genesis for these thrust plates is still controversial, ranging from gravity-propelled overthrusting resulting from uplift in the hinterland (Roberts and Crittenden, 1973) to underthrusting resulting from compressional stresses originating at depth (Armstrong, 1972; Lowell, 1977; Allmendinger and Jordan, 1981). The characteristics of the thrust belt foreland, particularly the development of curvature along its trace, were considered by Beutner (1977). The objective of this paper is to focus on a segment of the thrust belt in the vicinity of Salt Lake City, Utah, which Beutner (1977) called the Uinta reentrant. It is renamed the Salt Lake recess in recognition of the broader

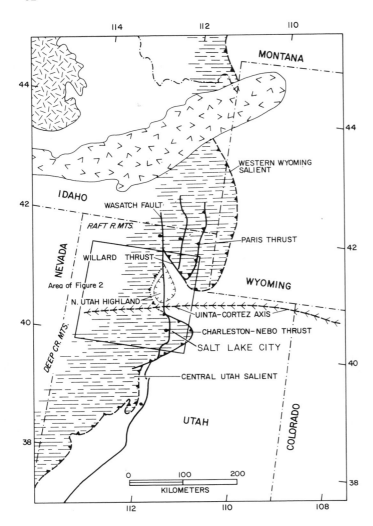

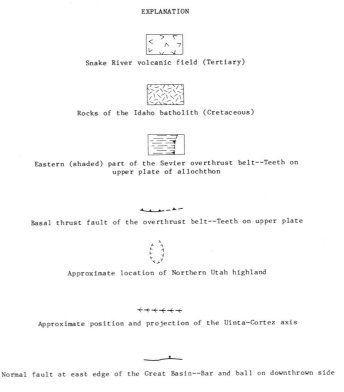

EXPLANATION

Snake River volcanic field (Tertiary)

Rocks of the Idaho batholith (Cretaceous)

Eastern (shaded) part of the Sevier overthrust belt--Teeth on upper plate of allochthon

Basal thrust fault of the overthrust belt--Teeth on upper plate

Approximate location of Northern Utah highland

Approximate position and projection of the Uinta-Cortez axis

Normal fault at east edge of the Great Basin--Bar and ball on downthrown side

Figure 1. Index map showing the locations of the foreland of the Sevier thrust belt and of Figure 2, the area of the Salt Lake recess.

involvement of northern Utah highland as well as the Uinta-Cortez axis in the development of structures in and between the thrust plates.

The Salt Lake recess is located in an area extending from Willard, Utah, south to the vicinity of Twelvemile Pass, south of Tooele, Utah, and from Park City, Utah, westward to the Newfoundland Mountains (Fig. 2a). In this general region, the overthrust belt jogs westward around the Uinta-Cortez axis (Roberts and others, 1965) and the adjacent remnant of the northern Utah highland (Eardley 1968), which together separate the western Wyoming salient to the north from the central Utah salient to the south. The Sevier belt is here composed of a number of imbricate thrust plates that commonly are offset into segments by transcurrent, or tear, faults. Recognition of thrust faults is difficult because only short segments of some crop out in the mountain ranges of the Great Basin. Most of the major thrust faults must be inferred to lie between ranges, generally beneath Tertiary and Quaternary deposits. Their existence is not inferred; the structural and stratigraphic evidence exposed in the mountain ranges requires their

presence. The western margin of the Sevier thrust belt lies an unknown distance west of the Salt Lake area and is not identified in this report. The southern continuation of the Sevier thrust belt into the central Utah salient is considered by Morris (1983).

The still-evolving geologic models for Sevier thrust faulting have benefited from numerous studies by Max Crittenden. His career was closely identified with the development of geologic data and concepts that are fundamental to our understanding of the thrust belt in western Utah and its northward continuation into the western Wyoming salient. Geologic maps and stratigraphic reports for areas along the western margin of the Wasatch Mountains by Crittenden have provided a basis for correlating thrust plates with the autochthon and for making geologic estimates of allochthon displacement.

REGIONAL STRATIGRAPHIC CHARACTERISTICS

At the beginning of Paleozoic time, a cratonal landmass in eastern Utah contributed material that formed shallow-marine detrital and carbonate sediments nearshore and that thickened westward as deep-water deposits

(Stewart and Poole, 1974). East-west facies changes caused by depth of deposition in the miogeocline provided sedimentational contrasts that Jordan (1979) described in upper Paleozoic rocks. Local north-south facies changes and depositional unconformities along the shelf in the Utah area resulted from the uplift along the Uinta-Cortez axis at intervals during the Paleozoic (Roberts and others, 1965).

Lower Paleozoic (Cambrian through Devonian) rocks consist of miogeoclinal carbonate and transitional assemblage rocks and deep-water eugeosynclinal siliceous assemblage rocks whose characteristic lithologies and thickness are summarized by Stewart and Poole (1974). These are lithologically distinctive and areally extensive units consisting mainly of quartzite and siltstone, local conglomerate, carbonate, and shale. Cross-bedded strata indicate westerly transport. The section thickens westward from a few hundred to more than 6,000 m. The stratigraphic section is interrupted locally by a depositional unconformity that cuts out Ordovician, Silurian, and some Devonian strata. Uplift is recorded in the limited appearance of an Upper Devonian conglomerate/quartzite facies.

The Mississippian Antler orogeny, which included the accretion of the Roberts Mountain tectonostratigraphic terrane to the craton on the west (Albers, 1982), changed the pattern of sedimentation. Materials were derived from both the craton and the accreted terrane to the west (Rose, 1976). The latter provided a source of siliceous eugeosynclinal sediments while the craton continued to supply detrital sands and carbonate.

Upper Paleozoic (Mississippian through Permian) rocks therefore are of mixed composition reflecting their dual source and shallow- and deep-water deposition. Roberts and others (1965) observed changes in lithology and a general thinning of strata to the west. Mississippian rocks are predominantly alternating carbonate, sandstone, and shale facies in the east; siliceous shales and quartzite increase westward. The lower part of Pennsylvanian deposits consists of clastic limestone, sandy limestone, and minor shale. These are succeeded by cyclically bedded limestone and sandstone. The Permian rocks are primarily sandstone, chert, shale, and minor carbonate rocks. For the most part these sediments seem to have been deposited in deeper water than the earlier rocks (Roberts and others, 1965). An exception to this trend is thick limestones of Permian age in the central Utah salient (Morris and others, 1977).

Mesozoic rocks are only observed as fragments of sections in allochthonous sequences of the eastern Great Basin. These rocks are well represented on the Wasatch Mountain autochthon where Crittenden and others (1952) described the characteristic shale, sandstone, limestone, quartzite succession of Triassic and Jurassic age. A small thrust sliver of Triassic rocks was reported by Jordan and Allmendinger (1979) in the Stansbury Mountains, and by Stifel (1964) in the Terrace Mountains.

AUTOCHTHON OF THE WASATCH MOUNTAINS

A thin sequence of autochthonous cratonic rocks of Precambrian and Paleozoic ages lies in the Wasatch Mountains immediately east of Salt Lake City, Utah, and on Antelope Island in the Great Salt Lake (Crittenden and others, 1952; Crittenden, 1959, 1969). The Paleozoic part of the stratigraphic section (Fig. 3), which unconformably overlaps Archean and Proterozoic rocks, is attenuated in comparison with allochthonous rocks of comparable age and contains a number of stratigraphic unconformities representing greatly thinned and intermittent nearshore sediment deposition. Broad, open to asymmetrical folds in the Paleozoic and Mesozoic rocks trend northeast on the north side of the Uinta uplift. Compression of these strata from the northwest resulted in the development of thrust faults (Crittenden and others, 1952) such as the Mount Raymond thrust (Fig. 2a). The stratigraphic and structural relations south of the uplift in the area shown in Figure 2a are concealed by the Charleston-Nebo thrust plate. The Charleston-Nebo thrust fault and its northward equivalent, the Willard thrust (Crittenden, 1972), form the south and north sides of the autochthon, respectively. An inferred thrust west of exposures of autochthonous rocks on Antelope Island (identified as part of the Charleston-Nebo system in Fig. 2a) forms the western boundary of the autochthon.

MAJOR ALLOCHTHONOUS THRUST PLATES

At least six thrust plates are believed to have been imbricately stacked on the Wasatch autochthon sequence in the Salt Lake recess (Figs. 2b, 3). These are, from east to west, the Charleston-Nebo, North Oquirrh, Midas, Stockton, Tintic Valley, and Skull Valley thrusts. A major thrust plate is defined here as a structural block containing a laterally continuous sedimentary sequence that is stratigraphically distinct from sequences in other blocks and whose structural configuration and continuity can be demonstrated. The name of the thrust plate is derived from the underlying thrust fault. The recess plates commonly are segmented and offset into several parts along steep-dipping transcurrent faults. Some plates contain subordinate thrust faults, but the stratigraphic composition of the component parts of plates overall is similar. The correlation of plate segments is based on their relation to bounding thrust faults, comparability of the rock units forming the stratigraphic sections, and continuity in the geometry of fold structures in relation to the Uinta-Cortez axis and northern Utah highland buttress.

Charleston-Nebo Plate

The Charleston-Nebo plate is defined as the unique stratigraphic sequence lying on the Charleston-Nebo thrust

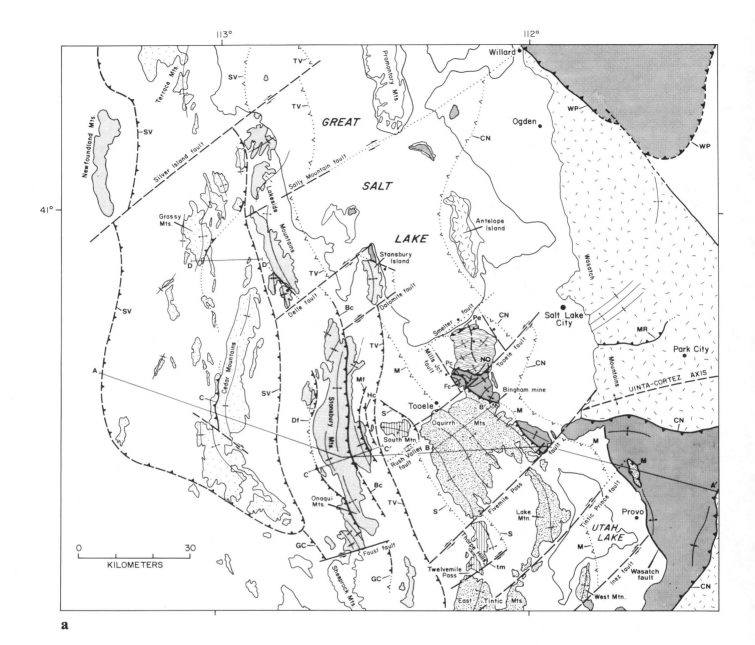

a

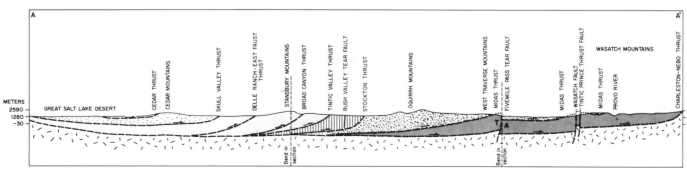

b

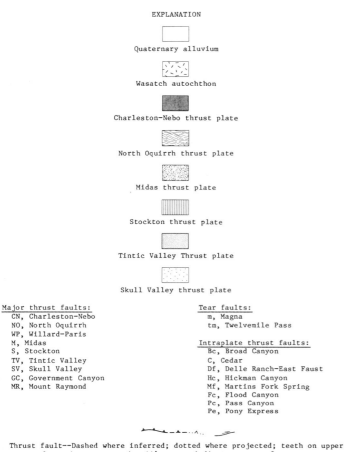

EXPLANATION

Quaternary alluvium

Wasatch autochthon

Charleston-Nebo thrust plate

North Oquirrh thrust plate

Midas thrust plate

Stockton thrust plate

Tintic Valley Thrust plate

Skull Valley thrust plate

Major thrust faults:
  CN, Charleston-Nebo
  NO, North Oquirrh
  WP, Willard-Paris
  M, Midas
  S, Stockton
  TV, Tintic Valley
  SV, Skull Valley
  GC, Government Canyon
  MR, Mount Raymond

Tear faults:
  m, Magna
  tm, Twelvemile Pass

Intraplate thrust faults:
  Bc, Broad Canyon
  C, Cedar
  Df, Delle Ranch-East Faust
  Hc, Hickman Canyon
  Mf, Martins Fork Spring
  Fc, Flood Canyon
  Pc, Pass Canyon
  Pe, Pony Express

Thrust fault--Dashed where inferred; dotted where projected; teeth on upper plate; in cross section AA', arrow indicates sense of movement

Transcurrent or tear and normal fault--Dashed where inferred with certainty; dotted where projected; arrows indicate direction of displacement; bar and ball indicate direction of vertical displacement; in section AA', T indicates movement toward observer; A, away from observer

Anticline   Syncline

Figure 2a (this and facing page). Simplified map of the Salt Lake recess area in the vicinity of Salt Lake City, Utah, showing the location and correlation of exposures of major thrust plates and plate segments, the principal fold axes, and the main known and inferred transcurrent or tear and thrust faults. The approximate locations of cross sections AA', shown in Figure 2b, and BB', CC', and DD', shown in Figure 4, are indicated. Basin-and-range faults and Tertiary intrusive and extrusive rocks are not shown, and unstudied ranges are unpatterned.

Figure 2b (this and facing page). General cross section showing the proposed patterns of imbrication of thrust plates along AA' of Figure 2a.

fault. This thrust plate crops out mainly in the central Wasatch Mountains, east of Provo, Utah (Baker and others, 1949). Correlative thrust plate segments occur in the central Utah salient (Morris, 1983). Rocks of the Charleston-Nebo plate were also recognized on the east side of the central Oquirrh Mountains by Welch and James (1961). A series of underlying, very tight to isoclinally folded, overturned,

sheared and altered, undated rocks, which are the lower plates of an intraplate (Flood Canyon) thrust on the west side of the range (E. W. Tooker, 1978, unpub. data), are considered to be a sliver of the Charleston-Nebo plate. This correlation is based on gross stratigraphic and structural similarity with less deformed Charleston-Nebo plate rocks that unconformably overlie them in thrust-fault contact. Moore (1973) showed an unnamed thrust fault in the West Traverse Mountains which I now believe to be the Midas thrust fault. It separates the Charleston-Nebo and Midas plates. The Charleston-Nebo plate also includes all or parts of several islands in the Great Salt Lake (Crittenden, 1961).

Sedimentary rocks in the Charleston-Nebo thrust plate range in age from Precambrian to Permian and are well exposed in the Wasatch Mountains (Baker and others 1949; Crittenden 1959). The stratigraphic section comprises a very thick sequence of representative rocks of Proterozoic and Paleozoic age, except for a marked nondepositional break in Ordovician, Silurian, and part of Devonian time. The section is not equally represented in all segments of the plate owing to folding, proximity with the overlapping thrust fault, and variations in the amount of subsequent local uplift and erosion near the Uinta-Cortez axis and northern Utah highland buttress. Proterozoic rocks crop out on several small islands in the Great Salt Lake and in a fault block at the north end of Stansbury Island (Hintze, 1980). In the central Oquirrh Mountains the rocks of the plate are of Permian age, and in the West Traverse Mountains a partial section of Pennsylvanian strata is exposed (Table 1).

Large-amplitude asymmetrical folds in the Wasatch Mountains form a west-facing concave arc south of the Cortez-Uinta axis and locally are overturned eastward. Folds in the Oquirrh Mountains form a westward-convex arc around the westward-plunging Uinta-Cortez axis; tight to isoclinal folds are overturned to the east and segmented by steep faults. Rocks in the upper plate north of Willard, which have been considered to be correlative with those in the Charleston-Nebo plate, are folded into north-trending, broad, open folds (Crittenden, 1972).

The northern and southern extent of the Charleston-Nebo thrust plate and breaks in continuity within the plate are determined by several transcurrent faults. Within the recess the plate ends against or is offset to the east along the Sally Mountain fault. The stratigraphic section in the Promontory Mountains seems to represent a sequence of rocks unlike those of the Charleston-Nebo or Midas plates. The south end of the Charleston-Nebo plate is bounded by a major transcurrent (Leamington) fault in the central Utah salient (Morris, 1983).

North Oquirrh Plate

An enigmatic stratigraphic and structural sequence of

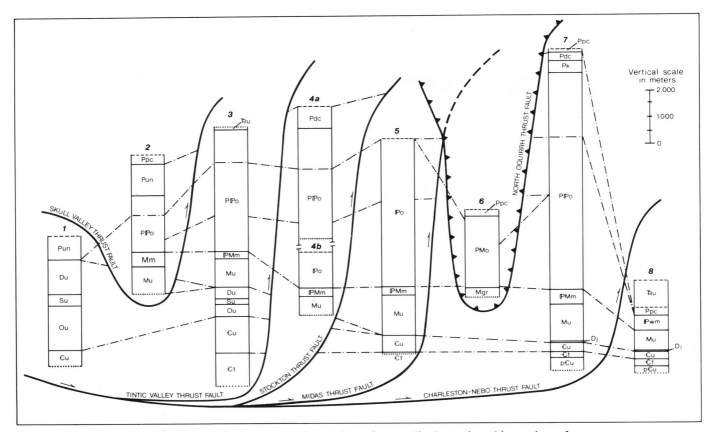

Figure 3 (this and facing page). Correlation of generalized stratigraphic section of pre-Cretaceous rocks of the autochthon and major allochthonous thrust plates composing the foreland of the Sevier orogenic belt in the Salt Lake recess. Comparative thicknesses of major lithologic units are based on the following references. Allochthonous stratigraphic sections: 1. Newfoundland Mountains (Tintic Valley plate), Paddock (1956). 2. Cedar Mountains (Skull Valley plate), Maurer (1970). 3. Stansbury Mountains (Tintic Valley plate), Rigby (1958). 4a. South Mountain (Stockton plate), Welch and James (1961). 4b. Thorpe Hills (Stockton plate), Disbrow (1957). 5. Oquirrh Mountains (Midas plate), Bingham sequence of Tooker and Roberts (1970). 6. Oquirrh Mountains (North Oquirrh plate), Rogers Canyon sequence of Tooker and Roberts (1970). 7. Wasatch Mountains (Charleston-Nebo plate), Provo section of Baker and others (1949). Autochthonous stratigraphic section: 8. Wasatch Mountains, Midway section of Baker and others (1949).

rocks defined as the upper plate of the North Oquirrh thrust fault is located in the northern Oquirrh Mountains, west of Salt Lake City, Utah. Its southern end is along the North Oquirrh thrust and Tooele tear faults. Along the shore of the Great Salt Lake, the north end of the plate is broken by several intraplate thrusts having small displacements, but is finally truncated by a concealed strong east-northeast normal fault inferred from geophysical evidence (Mabey and others, 1963). The east and west sides of the plate are normal faults of the basin and range system (Tooker and Roberts, 1970). Folds within the North Oquirrh thrust plate trend northeast, suggesting movement from the northwest. The folds are broad, open, locally asymmetric, and overturned to the southeast at the north end of the range.

While most of the rocks are generally similar to those in the Midas plate, studies by Tooker and Roberts (1970) showed that the rocks are sufficiently different in detail to propose that they formed in a part of the geosyncline different from the rocks of the Charleston-Nebo and Midas plates. Studies by M. D. Crittenden, Jr. (1981, oral commun.) in the Promontory Mountains and northern ranges in the state may provide evidence for more logical stratigraphic correlation and northern counterparts of the plate. Only the upper Paleozoic part of the stratigraphic section is exposed in the North Oquirrh thrust plate. The lowermost Upper Mississippian part, however, contains a limestone sequence in place of the Manning Canyon Shale (Fig. 3), which otherwise is a persistent stratigraphic feature of recess plates. In this respect the section is more comparable with that on the autochthon in which a limestone facies is present in this interval (Crittenden, 1959). The

EXPLANATION

Erosion surface at top of section

....

Covered base of section

- - - - -

Unconformity

— · — · — ·

Formation or Member correlation

_____

Formation contact

Thrust fault—Dotted where concealed;
arrows indicate sense of motion

Thrust fault—Movement normal to page

pЄu, undivided Precambrian rocks
Єt, Tintic Quartzite of Cambrian age
Єu, undivided post-Tintic Quartzite Cambrian rocks
Ou, undivided Ordovician rocks
Su, undivided Silurian rocks
Du, undivided Devonian rocks
Dj, Jefferson(?) Dolomite of Devonian age
Mgr, Green Ravine Formation of Mississippian age
Mu, undivided Mississippian rocks
Mm and ℙMm, Manning Canyon Shale of Mississippian and
    Pennsylvanian and Mississippian ages, respectively
ℙwm, Weber and Morgan Formations of Pennsylvanian age
ℙMo, ℙo, ℙℙo, and Po, Oquirrh Formation or Group of
    Permian to Mississippian, Pennsylvanian, Permian and
    Pennsylvanian, and Permian ages, respectively
Pu, post-Oquirrh Permian rocks
Pun, unnamed Permian formation
Pk, Pdc, and Ppc, Kirkman Limestone, Diamond Creek
    Sandstone, and Park City Formation, respectively,
    all of Permian age
Ṟu, undivided Triassic rocks

thickness of Pennsylvanian rocks on the North Oquirrh plate is less than that in the Midas plate, but the characteristic types of interbedded carbonate, quartzite, and shale are similar. The North Oquirrh thrust plate contains a thin section of probable Permian rocks and is capped in turn by a thinned section of Permian Park City Formation. Comparable rocks are not present in the Midas plate, probably because of erosion.

Midas Plate

The Midas thrust plate is defined by a sequence of

rocks that overlies the Midas thrust fault, located mostly in ranges west of the Wasatch Mountains. The Charleston-Nebo and Midas plates seem to have been overlapped by the North Oquirrh thrust fault. The Midas thrust plate is exposed only on the south and west sides of the Uinta-Cortez axis and is overlapped by the Tintic Valley and Stockton thrust faults. The northernmost exposures of the Midas thrust plate are truncated by the Tooele fault, which seems to have left-lateral displacement. The southern end of the plate terminates against the Inez fault, having a right-lateral displacement. Two major intervening faults, the Five-mile Pass and Tintic Prince faults, also displace segments of the Midas plate in the Lake Mountains and West and East Tintic Mountains (Morris, 1983). One small easternmost klippe of the plate is shown in the Wasatch Mountains (Fig. 2a). The intraplate thrust faults seem to represent local slippage along tightly overturned parts of fold axes.

The folds in Midas plate rocks (Fig. 4a) are broad, of large amplitude, and locally asymmetrical to overturned eastward. They lie along a west-facing concave arc that abuts the Uinta-Cortez uplift on the north and follows the curve of the salient southward. Where overturning in the fold was tight, local intraplate thrusts formed. These structures occur mainly along the projection of the Uinta-Cortez axis. Here also the folds are offset by closely spaced, northeast-trending, steep normal faults of small displacement (Tooker, 1971).

The stratigraphic section of the Midas thrust plate (Fig. 3) closely resembles that of the Charleston-Nebo thrust plate, both in composition and thickness (Baker and others, 1949; Gilluly, 1932; Tooker and Roberts, 1970; Moore, 1973). The Ordovician, Silurian, and Lower Devonian unconformity is recognized in both plates. The Mississippian and Pennsylvanian parts are closely similar, but Permian rocks are missing in the Midas plate in the Oquirrh Mountains owing to the Midas thrust and erosion. Permian rocks are present in the Midas plate in the central Utah salient (Morris and others, 1977).

Stockton Plate

The inferred Stockton thrust fault, which underlies the Stockton plate, has been proposed to explain the structural and stratigrpahic discontinuities evident between the rocks exposed on South Mountain and in the Thorpe Hills and the rocks in the adjoining Midas and Tintic Valley thrust plates (Figs. 2a, 3). The Stockton plate overlaps the Midas thrust plate on the east and is overlapped by the Tintic Valley thrust on the north and west. The plate is truncated by the Twelvemile Pass transcurrent fault on the south, which juxtaposes rocks of the Midas and overriding Stockton plates. Exposures on South Mountain, southwest of Tooele, Utah, which contain parts of the Oquirrh Group of

TABLE 1. PROPOSED CORRELATION OF EXPOSED SEQUENCES OF THE OQUIRRH GROUP AND ADJOINING FORMATIONS IN THE CHARLESTON-NEBO PLATE, IN THE WASATCH, OQUIRRH, AND WEST TRAVERSE MOUNTAINS AND ISLANDS IN THE GREAT SALT LAKE

	Wasatch Mountains (Baker and others, 1949; Baker, 1972, 1973)	Islands in the Great Salt Lake (Hintze, 1980)	Oquirrh Mountains (Welch and James, 1961; Swensen, 1975)	West Traverse Mountains (Moore, 1973)
	MIDAS THRUST FAULT			
Permian	Diamond Creek Sandstone and Kirkman Limestone		Diamond Creek Sandstone and Kirkman Limestone	
Permian	Oquirrh Formation:		Freeman Peak and Curry Peak Formations	
Pennsylvanian	Upper and middle parts, Shingle Mill Limestone Member, Unnamed Member, and Bridal Veil Limestone Member			Bingham Mine Formation of Oquirrh Group
Mississippian	Manning Canyon Shale Great Blue Limestone, Humbug Formation, Deseret Limestone, Gardison Limestone, and Fitchville Formation			
Devonian	Jefferson(?) Dolomite			
Cambrian	Maxfield Limestone, Ophir Formation, and Tintic Quartzite			
Precambrian	Mineral Fork Tillite and Big Cottonwood Formation	Mutual Formation, Mineral Fork Tillite, and Big Cottonwood Formation		
	CHARLESTON-NEBO THRUST FAULT			

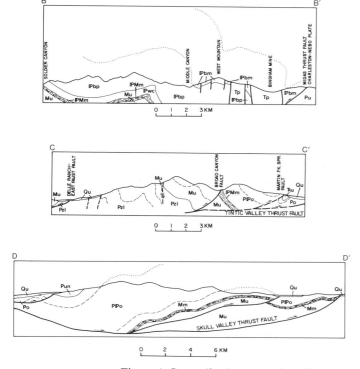

Figure 4. Generalized cross sections illustrating the characteristic fold and fault patterns in the Midas, Tintic Valley, and Skull Valley thrust plates in the Oquirrh, Stansbury, and Grassy Mountains located on Figure 2a. Section BB' is from Tooker (1970, unpub. data); CC' is modified from Tooker and Roberts (1971); DD' is modified from data in Doelling (1964).

Pennsylvanian and Permian age, were mapped by Gilluly (1932) and studied by Jordan (1979) and E. W. Tooker and R. J. Roberts (1960–1970, unpub. data). Welch and James (1961) have correlated the uppermost part of the South Mountain section with the Diamond Creek Sandstone of Permian age of the Wasatch Mountains. The Thorpe Hills segment, south of Fivemile Pass, was mapped by Disbrow (1957), and the stratigraphic section, which includes the lower and middle parts of the Oquirrh Group and the underlying Mississippian rocks, was measured by Mackenzie Gordon Jr., H. M. Duncan, and E. W. Tooker (1962, unpub. data).

The nearly normal structural discontinuity across the Tintic Valley thrust between South Mountain and the Stansbury Mountains is more evident than the structural break between South Mountain and the Oquirrh Mountains. The marked differences in stratigraphy originally described as a sharp facies change (Welch and James, 1961) is not borne out by stratigraphic or structural relations in subsequent detailed mapping by E. W. Tooker and R. J. Roberts (1970, unpub. data).

Tintic Valley Plate

The Tintic Valley thrust plate was originally recog-

nized and named from exposures in the central Utah salient (Roberts and others, 1965, p. 1946). The plate extends into the recess area in the Stansbury Mountains, where the rocks were described by Rigby (1958) and Moore and Sorenson (1979). I propose that stratigraphic sequences on Stansbury Island (Chapusa, 1969; Palmer, 1970), in the Onaqui (Armin and Moore, 1981), Lakeside (Young, 1955; Doelling, 1964), and Newfoundland Mountains (Paddock, 1956), are correlative faulted segments of the Tintic Valley plate. The Tintic Valley thrust fault underlies the thrust plate. It is overlapped at the southern end of the Onaqui Mountains by the Government Canyon thrust fault, which is defined by Morris (1983), and by the Skull Valley thrust fault on the north and west.

The stratigraphic section in the Tintic Valley plate ranges in age from Cambrian through Triassic. The section differs in detail from units in the Stockton, Midas, and Skull Valley plates. First, the lower Paleozoic strata include those Ordovician, Silurian, and Devonian units missing in the Midas plate and therefore resemble stratigraphic sections found farther west (Stewart and Poole, 1974). The Devonian section also contains a quartzite conglomerate formation that occurs throughout the plate. Conglomerate seems to be thickest along the trace of the Uinta-Cortez axis, becoming a quartzite farther from the axis. It is the presence of this unit, particularly (Roberts and Tooker, 1969), that prompts considering the Newfoundland Mountains to be a more faulted western window exposure of the Tintic Valley thrust plate. Second, Pennsylvanian rocks exposed in the Tintic Valley plate contrast with comparable thicker units in the Midas and Stockton plates and thinner units in the Skull Valley thrust plate. Permian rocks are not known in most exposures of the Tintic Valley thrust plate (Hintze, 1980); however, a very thin Permian section on the east side of the Stansbury Mountains is conformable with the Pennsylvanian section. It is an intraplate thrust contact overlying Triassic sandy, silty, cherty limestones (Jordan and Allmendinger, 1979).

The major folds on the Tintic Valley plates are open, generally asymmetric, and locally overturned along the trace of the Uinta-Cortez axis (Fig. 4b). The folds (Fig. 2a) trend sinuously north, describing a gentle arc about the Uinta-Cortez axis and northern Utah highland.

The Tintic Valley plate is segmented and offset by steep transcurrent right- and left-lateral faults such as the Faust, Dolomite, Delle, Sally Mountain, and Silver Island faults. The stratigraphic section of the plate is also segmented by the Hickman Canyon, Martin Fork Spring, Broad, and Delle Ranch–East Faust intraplate thrust faults, and possibly several lesser thrusts. The Broad Canyon thrust emplaces a lower Paleozoic section, which contains a fine-grained Devonian clastic unit, against lower and upper Paleozoic rocks that contain the coarse Devonian clastic unit. These relations suggest that substantial thrust displacement is possible within the Tintic Valley thrust plate. The Hickman Canyon thrust places Permian rocks in contact with a small sliver of the Devonian conglomerate.

The Promontory Mountains, recently under study by M. D. Crittenden, Jr., contain a structurally disturbed stratigraphic section that seems to be more closely related to Tintic Valley thrust plate units than to those of the Midas or Charleston-Nebo plates. Determination of the northern limit or an extension of Tintic Valley rocks on the north side of the Salt Lake recess awaits further study. The marked break in continuity of stratigraphy and structure north and west of the buttress prompts me to extend the Sally Mountain tear fault toward Willard. There has also been a later large vertical component impressed on many of what initially were transcurrent faults.

Skull Valley Plate

The Skull Valley plate is bounded at its base by the Skull Valley thrust fault, which overlaps both the Government Canyon thrust plate on the south and the Tintic Valley thrust plate on the east. The Tintic Valley plate crops out in the Newfoundland Mountains on the west edge of the recess area, an occurrence that supposes the trace of the Skull Valley thrust fault overlies this range. The extent of the Skull Valley plate north of the Terrace Mountains is not determined, but is most likely cut off by a later Raft River Mountains uplift (Todd, 1980). Outcrops of the Skull Valley plate have not yet been traced west of the Newfoundland Mountains in Utah.

The stratigraphic section in the Skull Valley plate was mapped and described in the Cedar Mountains by Maurer (1970), in the Grassy Mountains by Doelling (1964), and in the Terrace and Hogup Mountains by Stifel (1964). The stratigraphic section includes rocks ranging from Mississippian to Permian ages in the Cedar and Grassy Mountains and also includes Triassic rocks in the Terrace Mountains. The section is somewhat thinner than the comparable units in Tintic Valley plate and more closely resembles rock sequences in the Confusion Range and eastern Nevada areas (Hose and Repenning, 1959).

The structural style of the Skull Valley plate contrasts strongly with those styles of plates to the east. Numerous small-amplitude (Fig. 4c) open folds trend in sinuous fashion around the Uinta-Cortez axial zone (Fig. 2a). The plate is as much as 50 km wide and more than 140 km long in north-south extent. It is cut by the Silver Island transcurrent fault at the north end. Extension of the Skull Valley plate north of the Terrace Mountains has not been determined. The thrust appears to wrap around the south end of the Cedar Mountains, where the plate may overlie the Government Canyon thrust plate of the central Utah salient, discussed by Morris (1983). Remnants of the plate have not yet been recognized in the region west of the Deep Creek–Raft River uplifted zone. This domal uplift, attributed to intrusion of a metamorphic-plutonic core complex (Coney, 1980), may have arched the sole Charleston-Nebo thrust plate across the state line region. Comparability of sedimentary facies, fossils, and thicknesses of strata between the eastern Nevada (Ely) and Skull Valley plate (Oquirrh) rocks is evidence for a possible original connection between them, which may have been erased by erosion and concealed by Quaternary sediments.

CORRELATION OF THRUST FAULTS

The foregoing thrust faults have been correlated by consideration of individual and collective stratigraphic and structural characteristics: (1) gross relations observed along the frontal edge of the thrust belt, (2) identity of the major thrust faults separating the belt into individual plates, and (3) distinctive styles within plates. Beutner (1977) concluded that the causes for the distinctive curvature forming recesses and salients along the frontal edge of the Sevier thrust belt (Fig. 1) lay in the interaction of the allochthons with stratigraphic and structural irregularities of the craton margin. Thus, thrust plates that moved into a structurally low or unbuttressed region in western Wyoming and southeastern Idaho formed a series of broad east-facing, convex, folded, and locally segmented imbrications. The structurally high buttressed zone, formed by the persistent remnant of northern Utah highland (Eardley, 1968) and the adjoining uplifted Uinta-Cortez axial zone (Roberts and other, 1965), deflected thrust plates to form the complexly folded and faulted irregularly stacked imbricate sequence of plates in the Salt Lake recess seen in Figure 2a.

The Sevier foreland thrust faults in Utah, formed during Cretaceous time (Armstrong, 1968), apparently moved in an easterly to northeasterly direction on the basis of fold and tear-fault evidence. Thicker parts of the miogeoclinal basin deposits were moved 130 to 170 km eastward over the thinner parts of the craton shelf (Baker and others, 1949; Crittenden, 1961). Whether a thrust plate moved as a single plate, breaking into discrete plates and plate segments at the end, or whether different parts moved successively has not been determined. A systematic thinning of Pennsylvanian basin rocks and the repetition in successively younger thrust plates containing younger parts of the stratigraphic section suggest that the plates were derived from a single plate. The apparent mixing of deep- and shallow-water Pennsylvanian and Permian sedimentary deposits (Jordan, 1979, 1981) suggests that following breakup, plates moved in somewhat separate paths resulting in overlap of shallow- and deep-water facies.

Differences in the structural style and stratigraphic composition along the Sevier belt north and south of the recess may pose correlation problems. Plate transition from the western Wyoming salient into the Salt Lake recess is angular, and the plates seem to interfinger abruptly. This feature raises the possibility that the family of thrusts overlying the Charleston-Nebo thrust fault represents a separate central Utah lobe of the thrust belt that may not have been coupled with a western Wyoming–southeastern Idaho lobe of the belt. In view of this hypothesis, the presently uncertain correlation of thrust plates immediately north of the recess becomes important. In contrast, the transition into the central Utah salient is smooth. Plates and structures are continuous from the recess into an asymmetrically narrowing structural low of the salient. The style of recess folding and faulting and the stratigraphy observed in most plates permit reconstruction of the center and western part of a depositional basin that has been compressed northeastward and eastward. The exception to this pattern observed in the North Oquirrh thrust plates seems to indicate derivation of that plate from a separate lobe derived from another part of the miogeocline in a separate undatable movement.

Differences in structural style between individual thrust plates of the recess imply that there were different components of motion within the central Utah lobe. The Charleston-Nebo, Midas, and Stockton thrust plates, which are stratigraphically comparable and derived from the thickest part of the miogeocline, are characterized by strongly directed motion from the southwest. The plates were lodged in an arc against the buttress on the north and the salient lobe on the south; there is no evidence to suppose that the plates extend north of the Sally Mountain tear fault or south of the Leamington fault. The overlying Tintic Valley and Skull Valley thrust plates extend north of the Sally Mountain tear fault, and the correlative part of the Tintic Valley thrust plates (Morris, 1983) extends south into the central Utah salient, overlapping the Leamington fault. The strongest component of movement on these plates, based on fold axes, seems to be eastward; however, the strong northeast-trending tear faults are characteristic of the central Utah lobe segment the Tintic Valley and Skull Valley plates and may have been initiated by recurrence of movement on faults in the underlying, earlier plates.

The suggestion has been made (Crittenden, 1974) that the Mount Raymond thrust fault on the autochthon (parautochthon of Beutner, 1977) is the southwestward extension of one of the easternmost thrusts of the western Wyoming salient. The thrust and its fold structures were formed by impact against the Uinta-Cortez axis buttress. Projections of the Mount Raymond and Charleston-Nebo thrust faults and their upper plate structures intersect at a near-normal angle in the Salt Lake recess and represent motion from different directions. The characteristic stratigraphic sections of the respective plates also support the interpretation that they represent separate events.

Correlation of the Proterozoic rocks on the islands in the Great Salt Lake west of Antelope Island with those of the Charleston-Nebo plate is based mostly on the analysis of Crittenden (1961). The type of structural and stratigraphic evidence used elsewhere in this report is sparse here. One must assume that erosion of the Paleozoic rocks in the Charleston-Nebo plate was most intense in the buttress area of the northern Utah highland and that all that remains are the lowermost Precambrian strata. Subsequent basin-and-range faulting has further obscured evidence for more positive correlation.

TIMING OF THRUSTING

The evidence of a Cretaceous age for the period of Sevier thrust faulting (Armstrong, 1968; Morris, 1983) is clear, but the ages and sequences of thrust faults in the Salt Lake recess have not been well established previously. Oriel and Armstrong (1966) established that early thrust movement in the Idaho-Wyoming area was from west to east and that easternmost thrusts are younger than those to the west. Recurrent movement undoubtedly occurred on some of these thrusts. The age of movement on those parts of the Sevier belt west of the Paris thrust was not resolved by Oriel and Armstrong. A different timing sequence (younger to the west) has been established in the central

Utah salient (Morris, 1983). He shows that the Charleston-Nebo thrust fault, which is the oldest in the allochthonous sequence, ends along the Leamington transcurrent fault. South of that fault the oldest exposed foreland thrust is correlated with the Tintic Valley thrust. This correlation suggests that within the central Utah salient, the basal plate in the northern part was emplaced earlier than the comparable basal plate of the thrust belt in the southern part and probably did not extend south of the Leamington transcurrent fault. This faurther suggests that the frontal edge of the thrust belt was propagated sequentially southward in separate episodes, a conclusion that is consistent with the broader regional age relations between the western Wyoming and central Utah salients.

If the Charleston-Nebo and Willard-Paris thrust faults are age correlatives (Crittenden, 1972) and if the Mount Raymond thrust on the autochthon is directly related to one of the easternmost thrusts in the western Wyoming salient, earliest thrust movement in the western Wyoming salient must predate thrusting in the Salt Lake recess. Present data thus strongly suggest that the Idaho-Wyoming thrust activity was decoupled from a later thrusting phase in the Salt Lake recess and central Utah salient.

The timing of movement of the North Oquirrh thrust plate is still uncertain. The plate moved into place some time after the Charleston-Nebo plate, which it overlies, and probably postdates Midas movement. There is no way to compare the age of North Oquirrh plate motion with later recess thrust movements.

CONCLUSIONS

1. Six stratigraphically and structurally distinctive thrust-plate and plate segments in the foreland recess of the Sevier belt near Salt Lake City, Utah, are identified and correlated.

2. Fold structures, tear faults, intraplate thrust faults, and plate segmentation may be directly correlated with the characteristic stratigraphic properties of individual plates, the surface over which the plates moved, and the angle of plate impact on the craton margin.

3. The style of thrust emplacement, as reconstructed, suggests that a buttress in northern Utah, formed by a combination of a remnant northern Utah highland and a mobile uplifted Uinta-Cortez axis, was at the south edge of a spreading lobe of early thrust activity in Idaho and Wyoming. In this region, direction of movement was mainly from the north and west. The same buttress had different impact on a later central-Utah salient lobe of the Sevier belt that moved from the southwest and west.

4. Still unanswered is the mechanism for detachment, movement, and stacking of these thin-skinned structures; however, present data impose constraints upon future genetic models. The westward continuity of stratigraphic and

structural styles in the foreland thrust plates, and their resemblance to correlative western rocks, is support for proposing that the Ely and Oquirrh "basin" sediments of Pennsylvanian and Permian ages were deposited in adjacent parts of a common geosyncline and that they were subsequently separated by an intervening uplift of the core-complex terrane lying along the Utah-Nevada border. If sustained by further study, this hypothesis may provide an important clue to location of the hinterland western margin of the Sevier thrust belt.

REFERENCES CITED

Albers, J. P. 1982, Distribution of mineral deposits in accreted terranes and cratonal rocks of western United States: Canadian Journal of Earth Sciences (in press).

Allmendinger, R. W., and Jordan, T. E., 1981, Mesozoic evolution, hinterland of the Sevier orogenic belt: Geology, v. 9, p. 308–313.

Armin, R. A., and Moore, W. J., 1981, Geology of the southeastern Stansbury Mountains and southern Onaqui Mountains, Tooele County, Utah: U.S. Geological Survey Open-File Report 81-247, 28 p.

Armstrong, R. L., 1968, Sevier orogenic belt in Nevada and Utah: Geological Society of America Bulletin, v. 79, p. 429–458.

——1972, Low-angle (denudation) faults, hinterland of the Sevier orogenic belt, eastern Nevada and western Utah: Geological Society of America Bulletin, v. 83, p. 1729–1754.

Baker, A. A., 1972, Geologic map of the Bridal Veil Falls quadrangle Utah: U.S. Geological Survey Geologic Quadrangle Map GQ-998, scale 1:24,000.

——1973, Geologic map of the Springville quadrangle, Utah County, Utah: U.S. Geological Survey Geologic Quadrangle Map GQ-1103, scale 1:24,000.

Baker, A. A., Huddle, J. W., and Kinney, D. M., 1949, Paleozoic geology of north and west sides of Uinta Basin, Utah: American Association of Petroleum Geologists Bulletin, v. 33, p. 1161–1197.

Beutner, E. C., 1977, Causes and consequences of curvature in the Sevier orogenic belt, Utah to Montana: Wyoming Geological Association Guidebook, 29th Annual Field Conference, p. 353–365.

Chapusa, F.W.P., 1969, Geology and structure of Stansbury Island [M.S. thesis]: Salt Lake City, University of Utah, 81 p.

Coney, P. J., 1980, Cordilleran metamorphic core complexes: An overview: Geological Society of America Memoir 153, p. 7–34.

Crittenden, M. D., Jr., 1959, Mississippian stratigraphy of the central Wasatch and western Uinta Mountains, Utah: Intermountain Association of Petroleum Geologists Guidebook, 10th Annual Field Conference, p. 63–74.

——1961, Magnitude of thrust faulting limits in northern Utah: U.S. Geological Survey Professional Paper 424-D, p. D128–D131.

——1969, Interaction between Sevier orogenic belt and Uinta structures near Salt Lake City, Utah [abs.]: Geological Society of America Abstracts with Programs, pt. 5, p. 18.

——1972, Willard thrust and Cache allochthon: Geological Society of America Bulletin, v. 83, p. 2871–2880.

——1974, Regional extent and age of thrusts near Rockport reservoir and relation to possible exploration targets in northern Utah: American Association of Petroleum Geologists Bulletin, v. 58, p. 2428–2435.

Crittenden, M. D., Jr., Sharp, B. V., and Calkins, F. C., 1952, Geology of the Wasatch Mountains east of Salt Lake City—parley's Canyon to Traverse Range: Utah Geological Society Guidebook to the Geology of Utah, no. 8, p. 1–37.

Disbrow, A. E., 1957, Preliminary geologic map of the Fivemile Pass quadrangle, Tooele and Utah Counties, Utah: U.S. Geological Survey Mineral Investigations Field Studies Map, MF-131, scale 1:24,000.

Doelling, H. H., 1964, Geology of the northern Lakeside Mountains and Grassy Mountains and vicinity [Ph.D. thesis]: Salt Lake City, University of Utah, 354 p.

Eardley, A. J., 1968, Major structures of the Rocky Mountains of Colorado and Utah: University of Missouri Journal, no. 1, p. 71–99.

Gilluly, James, 1932, Geology and ore deposits of the Stockton and Fairfield quadrangles, Utah: U.S. Geological Survey Professional Paper 173, 171 p.

Hintze, L. F., compiler, 1980, Geologic map of Utah: Utah Geological and Mineral Survey, scale 1:500,000.

Hose, R. K., and Repenning, C.A., 1959, Stratigraphy of Pennsylvanian Permian and Lower Triassic rocks of Confusion Range, west-central Utah: American Association of Petroleum Geologists Bulletin, v. 43, p. 2167–2196.

Jordan, T. E., 1979, Lithofacies of the upper Pennsylvanian and lower Permian western Oquirrh Group, northwest Utah: Utah Geology, v. 6, p. 41–56.

—— 1981, Enigmatic deep water depositional mechanisms, upper part of the Oquirrh Group, Utah: Journal of Sedimentary Petrology, v. 51, p. 879–894.

Jordan, T. E., and Allmendinger, R. W., 1979, Upper Permian and lower Triassic stratigraphy of the Stansbury Mountains, Utah: Utah Geology, v. 6, p. 69–74.

Lowell, J. D., 1977, Underthrusting origin for thrust-fold belts with application to the Idaho-Wyoming belt: Wyoming Geological Association Guidebook, 29th Annual Field Conference, p. 449–455.

Mabey, D. R., Tooker, E. W., and Roberts, R. J., 1963, Gravity and magnetic anomalies in the northern Oquirrh Mountains, Utah: U.S. Geological Survey Professional Paper 450 E, p. E28–E31.

Maurer, R. E., 1970, Geology of the Cedar Mountains, Tooele County, Utah [Ph.D. thesis]: Salt Lake City, Unversity of Utah, 184 p.

Moore, W. J., 1973, Preliminary geologic map of western Traverse Mountains and northern Lake Mountains, Salt Lake and Utah Counties, Utah: U.S. Geological Survey Miscellaneous Field Studies Map, MF-490, scale 1:24,000.

Moore, W. J., and Sorensen M. L., 1979, Geologic map of the Tooele 1° by 2° quadrangle, Utah: U.S. Geological Survey Miscellaneous Investigations Map I-1132, scale 1:250,000.

Morris, H. T., 1983, Interrelations of thrust and transcurrent faults in the central Sevier orogenic belt near Leamington, Utah, *in* Miller, D. M., Todd, V. R., and Howard, K. A., eds., Tectonic and stratigraphic studies in the eastern Great Basin: Geological Society of America memoir 157 (this volume).

Morris H. T., Douglass, R. C., and Kopf, R. W., 1977, Stratigraphy and microfaunas of the Oquirrh Group in the southern East Tintic Mountains, Utah: U.S. Geological Survey Professional Paper 1025, 22 p.

Oriel, S. S., and Armstrong, F. C., 1966, Times of thrustng in the Idaho-Wyoming thrust belt—Reply: American Association of Petroleum Geologists Bulletin, v. 50, p. 2614–2621.

Paddock, R. E., 1956, Geology of the Newfoundland Mountains, Box Elder County, Utah [M.S. thesis]: Salt Lake City, University of Utah, 101 p.

Palmer, D. E., 1970, Geology of Stansbury Island, Tooele County Utah: Brigham Young University Geological Studies, v. 17, pt. 2, p. 3–30.

Rigby, J. K., 1958, Geology of the Stansbury Mountains, eastern Tooele County, Utah, *in* Rigby, J. K., ed., Geology of the Stansbury Mountains, Tooele County, Utah: Utah Geological Society Guidebook to the Geology of Utah, no. 13, p. 1–133.

Roberts, R. J., and Crittenden, M. D., Jr., 1973, Orogenic mechanisms, Sevier orogenic belt, Nevada and Utah, *in* Dejong, K. A., and Sholten, Robert, eds., Gravity and tectonics: New York, John Wiley and Sons, p. 409–428.

Roberts, R. J., and Tooker, E. W., 1969, Age and regional significance of conglomerate in the Newfoundland and Silver Island Mountains, Utah [abs.]: Geological Society of America Abstracts with Programs, 1969, pt. 5, p. 69.

Roberts, R. J., Crittenden, M. D., Jr., Tooker, E. W., Morris, H. T., Hose, R. K., and Cheney, T. M., 1965, Pennsylvanian and Permian basins in northwestern Utah, northeastern Nevada, and south-central Idaho: American Association of Petroleum Geologists Bulletin, v. 49, p. 1926–1956.

Rose, P. R., 1976, Mississippian carbonate shelf margins, Western United States: U.S. Geological Survey Journal of Research, v. 4, p. 449–466.

Stewart, J. H., and Poole, F. G., 1974, Lower Paleozoic and uppermost Precambrian Cordillerian Miogeocline, Great Basin, Western United States, *in* Dickinson, W. R., ed., Tectonics and sedimentation: Society of Economic Paleontologists and Mineralogists Special Publication no. 22, p. 28–57.

Stifel, P. B., 1964, Geology of the Terrace and Hogup Mountains, Box Elder County, Utah [Ph.D. thesis]: Salt Lake City, University of Utah, 173 p.

Swensen, A. J., compiler, 1975, Geologic map of the Bingham district, *in* Bray, R. E., and Wilson, J. C., eds., Guidebook to the Bingham mining district, Bingham Canyon Utah: Society of Economic Geologists, scale 1:24,000.

Todd, V. R., 1980, Structure and petrology of a Tertiary gneiss complex in northwestern Utah: Geological Society of America Memoir 153, p. 349–384.

Tooker, E. W., 1971, Regional structural controls of the deposits, Bingham mining district, Utah, USA.: Geological Society of Japan Special Issue 3, p. 76–81.

Tooker, E. W., and Roberts, R. J., 1970, Upper Paleozoic rocks in the Oquirrh Mountains and Bingham mining district, Utah: U.S. Geological Survey Professional paper 629-A, 76 p.

—— 1971, Structures related to thrust faults in the Stansbury Mountains, Utah, *in* U.S. Geological Survey Research 1971: U.S. Geological Survey Professional paper 750-B, p. B1–B12.

Welch, J. E., and James, A. H., 1961, Pennsylvanian and Permian stratigraphy of the central Oquirrh Mountains, Utah, *in* Cook, D. R., ed., Geology of the Bingham mining district and northern Oquirrh Mountains, Utah: Utah Geological Society Guidebook 16, p. 1–16.

Young, J. C., 1955, Geology of the southern Lakeside Mountains, Utah: Geological and Mineral Survey Bulletin, v. 56, 116 p.

MANUSCRIPT ACCEPTED BY THE SOCIETY AUGUST 20, 1982

Geological Society of America
Memoir 157
1983

Interrelations of thrust and transcurrent faults in the central Sevier orogenic belt near Leamington, Utah

Hal T. Morris
U.S. Geological Survey
345 Middlefield Road
Menlo Park, California 94025

ABSTRACT

The structural and stratigraphic interrelationships between thrust plates and transcurrent faults provide a means to establish the relative ages and regional correlations of imbricate thrust plates in the central Sevier orogenic belt. The key structure in this area is the Leamington transcurrent fault, which defines the southern boundary of the Charleston-Nebo thrust plate and which appears to have undergone at least two periods of renewed movement after the initial emplacement of this thrust plate. Detailed evaluation of these recurrent movements, coupled with stratigraphic comparisons and structural analysis, indicates the apparent correlation of the Tintic Valley and Pavant thrusts and the Sheeprock and Canyon Range thrusts, and it also highlights the great regional extent of the Wah Wah–Frisco thrust sheet. The westward concealment of the Leamington transcurrent fault and possibly the Canyon Range thrust fault beneath the Wah Wah–Frisco plate further indicates that the Sevier orogeny apparently took place in two relatively independent episodes, a conclusion originally suggested by the general relationships of the Indianola Group and the Price River Formation.

INTRODUCTION

The relative ages of emplacement of the imbricate thrust plates of many orogenic belts are only imperfectly known. It is commonly assumed that the overlapping or superimposed plates are successively younger upward or toward their hinterland, obeying a general law of superposition. It is equally possible that the lower plates in an imbricate sequence moved late in an orogenic sequence, carrying overlying plates with them in a pickaback fashion. Most of these conclusions, however, are merely speculation. In the central part of the Sevier orogenic belt (Armstrong, 1968; Morris, 1978) the interrelationships between thrust and transcurrent faults appear to provide reasonably reliable information concerning thrust-plate relationships that elsewhere is not available. Of particular importance are the structural relations between the Leamington transcurrent fault and the four or more thrust plates with which it is directly associated. Other transcurrent faults of possible similar usefulness and interest are the Inez and Tintic Prince faults of the East Tintic Mountains (Morris and

Shepard, 1964), the Indian Springs fault of the Sheeprock Mountains (Morris and Kopf, 1970a), and possibly the inferred Fivemile Pass fault of the southern Oquirrh Mountains (Tooker, 1983).

GEOLOGIC SETTING

The dominant feature of the central part of the northerly trending Sevier orogenic belt in central Utah is the Charleston-Nebo lobe, the southern half of which is shown on Figure 1. This eastward-projecting salient underlies an area of about 2,900 km^2 in the general vicinity of Utah Lake. The Paleozoic rocks of which this thrust plate is dominantly composed are prominently exposed in the Wasatch Mountains and in the Oquirrh, East Tintic, and adjacent ranges in the east-central part of the Great Basin. Along the eastern margin of the thrust plate, however, the plate is overlapped and partly concealed by orogenic and postorogenic deposits of late Mesozoic and early Tertiary

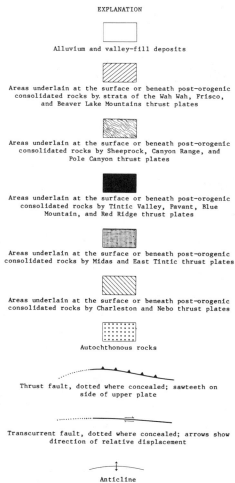

EXPLANATION

Alluvium and valley-fill deposits

Areas underlain at the surface or beneath post-orogenic consolidated rocks by. strata of the Wah Wah, Frisco, and Beaver Lake Mountains thrust plates

Areas underlain at the surface or beneath post-orogenic consolidated rocks by Sheeprock, Canyon Range, and Pole Canyon thrust plates

Areas underlain at the surface or beneath post-orogenic consolidated rocks by Tintic Valley, Pavant, Blue Mountain, and Red Ridge thrust plates

Areas underlain at the surface or beneath post-orogenic consolidated rocks by Midas and East Tintic thrust plates

Areas underlain at the surface or beneath post-orogenic consolidated rocks by Charleston and Nebo thrust plates

Autochthonous rocks

Thrust fault, dotted where concealed; sawteeth on side of upper plate

Transcurrent fault, dotted where concealed; arrows show direction of relative displacement

Anticline

Syncline

Figure 1. Generalized map of thrust and tear faults and associated major folds in west-central Utah. Patterns indicate bedrock areas that are underlain at surface or beneath postorogenic deposits by specific thrust plates.

THRUST FAULTS

BLM	Beaver Lake Mountains
BM	Blue Mountain
CN	Charleston-Nebo
CR	Canyon Range
ET	East Tintic
F	Frisco
M	Midas
P	Pavant
PC	Pole Canyon
RR	Red Ridge
S	Sheeprock
TV	Tintic VAlley
WW	Wah Wah
WWF	Wah Wah-Frisco

TRANSCURRENT FAULTS

H	Homansville
I	Inez
IS	Indian Springs
LTF	Leamington
TP	Tintic Prince

age. The northern and northeastern boundary of the thrust lobe, about 30 km northeast of Provo, is defined by the trace of the Charleston thrust fault (Baker, 1947; Baker and Crittenden, 1961). The southern boundary is defined by the northeast-trending Leamington transcurrent fault (LTF), which is largely concealed except in the southern Gilson Mountains but which apparently extends northeastward to a point near Nephi at the base of the Wasatch Mountains. This transcurrent feature has been described by Christiansen (1952, p. 717), Costain (1960, p. 111–112), Morris and Shepard (1964, p. C21), and Wang (1970, p. 92–97). Although originally termed a tear fault (Morris, 1977), the Leamington is here designated by the more general term of transcurrent fault because of its suspected multiple origin and more complicated history.

In the area north of the LTF, the Charleston-Nebo thrust plate is overlain successively westward by six addi-

tional thrust plates. Five of these are shown in Figures 1 and 2, including: (1) Midas, (2) East Tintic, (3) Tintic Valley, (4) Sheeprock, and (5) Pole Canyon. Of these structures, the East Tintic thrust may be considered to be a subordinate feature in the folded plate of the Midas thrust

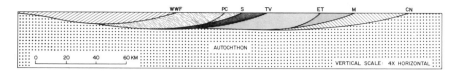

Figure 2. Diagramatic cross section approximately at the northern boundary of Figure 1 showing general relations of thrust plates and underlying decollement. Letter symbols are the same as on Figure 1.

(Morris and Lovering, 1979), and the Pole Canyon thrust a subordinate structure in the plate of the Sheeprock thrust (Blick, 1979). The sixth thrust plate, the Government Canyon plate, crops out in the northern part of the Sheeprock Mountains north of Figure 1 (Cohenour, 1959), but like the Pole Canyon thrust it also may be only a subsidiary feature of the Sheeprock thrust. Of the three main thrust plates recognized north of the LTF, only two of them, the Charleston-Nebo and the Tintic Valley, can be projected southward to a contact with the LTF with confidence. It is possible, however, that the Sheeprock plate also extends to the LTF, but this cannot be confirmed on the basis of bedrock exposures.

In the area immediately south of the LTF, only two major thrust plates are recognized: (1) Pavant, which like the Charleston-Nebo is the apparent sole thrust of the area; and (2) Canyon Range thrust, which is sharply compressed into a north-trending syncline. Farther southward along the Sevier belt the Pavant thrust appears to have its counterpart in the Blue Mountain thrust, and the Canyon Range thrust appears to have been overridden by the Wah Wah–Frisco thrust plate or to have been removed by erosion. It is also possible that the Canyon Range thrust is correlative with the Beaver Lake Mountains thrust, which is exposed a short distance north of Milford, but on Figure 1 this thrust is interpreted to be a subsidiary structure in the lower plate of the Wah Wah–Frisco thrust similar to the subsidiary Red Ridge thrust in the lower plate of the Pavant thrust (Steven and Morris, 1981).

The Wah Wah–Frisco thrust, which apparently overrode the Blue Mountain thrust plate in the general area of Milford, has in the past been correlated with the Canyon Range thrust (Morris and Lovering, 1979, p. 73), but this now appears not to be the case, and its regional relations are discussed in detail later in this paper.

LEAMINGTON TRANSCURRENT FAULT (LTF)

The LTF crops out on the north side of the Sevier River about 3 km northwest of the community of Leamington, which is 95 km southwest of Provo. The fault strikes about N. 60° E. and at its outcrop appears to dip 50° to 70° southeast; the fault zone is relatively narrow, consisting of about 1 to 3 m of poorly cemented gouge and breccia. Although actual exposures of the fault zone are limited, the nearly straight-line termination of the strata within the plates of the Tintic Valley and Charleston-Nebo thrusts between Leamington and Nephi allow the fault to be projected beneath younger sedimentary deposits for more than 35 km with some confidence. The obvious straightness of the exposed and concealed fault zone also suggests that despite the apparently moderate dips at the surface, the fault plane is essentially vertical and the shallower observed dips are merely parts of a corrugated fault plane similar to the planes of other large strike-slip faults that have been exposed in deep mine workings (Wallace and Morris, 1979, p. 79–100).

Neither the eastward nor westward limits of the LTF are known with certainty. Near Nephi, at the western base of the Wasatch Mountains, no comparable high-angle structure is recognized in either the upper plate or autochthon of the Nebo thrust, and it is assumed that the LTF flattens and merges with the thrust near this locality. Similar terminations of transcurrent faults at thrusts have been observed in the East Tintic mining district (Morris and Lovering, 1979, p. 81). Southwest of Leamington the LTF is concealed by Lake Bonneville deposits, and it does not reappear.

If the LTF were a simple tear fault, thus being a lateral shear fault within a clearly recognized thrust plate, correlations of strata and minor thrust faults across it would seem to be a fairly straightforward matter. However, no direct correlations are obvious. This relationship suggests that the LTF might best be interpreted as a transcurrent fault, or perhaps merely the southern edge of the Charleston-Nebo thrust plate, that during the earlier part of the thrusting episode may have separated two relatively distinct areas containing separate stress fields that were at least unequal or even more or less independent of each other. Similar relationships have been described for some of the wrench faults of the Jura Mountains of Europe (Aubert, 1959). Alternatively, it may be assumed that the LTF is a simple tear fault and that a southern extension of the Charleston-Nebo plate lies beneath the Pavant, Canyon Range, and Wah Wah–Frisco thrust plates. If this is so, however, the remnant plate is totally concealed and its existence based only on speculation.

Theoretically, wrench faults resulting from horizontal compression commonly form pairs of conjugate faults, or at least conjugate directions of faulting. For this reason it may be important to note that the transcurrent faults of the Sevier belt all trend northeast, and the LTF in particular does not appear to have a northwest-trending left-lateral companion fault.

CHARLESTON-NEBO AND PAVANT THRUSTS

Although the Nebo and Pavant thrusts both are sole thrusts of the Sevier belt on opposite sides of the LTF, and both override Mesozoic strata, marked differences in the stratigraphy of their upper plates suggest that they are not directly correlative. The major differences in the plates are found in a comparison of their Ordovician, Silurian, and particularly their Devonian strata. In all of the exposures of the Charleston-Nebo plate north of the LTF, no strata assignable with confidence to the Devonian Sevy, Simonson, or Guilmette Formations are recognized, and in many areas only a thin sequence of Upper Devonian strata overlies an unconformity cut into lower Paleozoic beds. In addition, in areas where post-Cambrian rocks are exposed on the Charleston-Nebo plate, no units equivalent to the Eureka Quartzite or the Kanosh Shale are present. In contrast, the upper plate of the Pavant thrust, which is prominently exposed in the southeastern Canyon Range and the adjacent Pavant Range, contains relatively thick and well-developed sections of both the Sevy and Simonson Dolomites, as well as an extensive sequence of Ordovician strata that includes both the Eureka and Kanosh. Differences also have been noted between the Cambrian sequences of the Pavant and Charleston-Nebo thrust plates, but these have not been studied in detail and are not discussed here.

TINTIC VALLEY THRUST

A much more direct comparison—or general correlation—may be made between the sedimentary strata of the Pavant thrust and the Tintic Valley thrust. Both plates contain similar if not identical stratigraphic sections, including most of the same formations of Ordovician, Silurian, and Devonian age. If a structural correlation between these two thrust plates is valid, it must be assumed that the Charleston-Nebo thrust plate was emplaced during the early part of the thrusting episode and that the LTF or a similar protostructure may have been its southern boundary. Shortly following this emplacement, a continuous plate containing a regionally extensive, characteristic sequence of Ordovician to Devonian strata was then thrust over the Charleston-Nebo plate and, as well, over the LTF or its protostructure and the autochthonous Mesozoic rocks south of it. Later movements on the LTF may then have broken the continuous allochthonous mass, separating it into the Tintic Valley plate and the Pavant plate. Such displacement on the LTF most logically would have resulted from minor renewed eastward movement on the Charleston-Nebo thrust followed by moderate to strong vertical displacement as described later in this report.

CANYON RANGE THRUST

The most obvious discontinuity across the LTF is the pronounced synclinal folding of the plate of the Canyon Range thrust, which also requires similar, although not as extensive, folding of at least part of the underlying Pavant thrust plate. The axis of the Canyon Range syncline trends nearly due north; the east limb is partly concealed by Mesozoic orogenic conglomerates but appears to dip moderately westward. In comparison, the southern part of the west limb dips steeply eastward, but within 15 km south of the LTF it is overturned as much as 20°, dipping about 70° westward or northwestward into the fault zone.

No similar folding is recognized immediately north of the LTF, although folding of the upper plate of the Midas thrust fault has yielded the great anticlines and synclines of the Bingham and Tintic mining districts. The general similarities of the great folds of the Midas and Canyon Range plates offer some temptations to generally correlate these plates, but if a correlation of the Tintic Valley and Pavant thrusts is valid, it becomes readily apparent from regional mapping that the Midas plate *underlies* the Tintic Valley thrust, whereas the Canyon Range thrust *overlies* the Pavant, and therefore the folded plates cannot be correlative. In addition, the Midas thrust probably did not ever extend as far south as the LTF, but terminated against the Inez transcurrent fault of the central East Tintic Mountains (Morris and Shepard, 1964).

The termination of the Canyon Range syncline against the LTF and the apparent strong eastward drag of the upper-plate formations against the fault suggest eastward-directed right-lateral displacement on the LTF during or after emplacement of the Canyon Range thrust plate. The magnitude and direction of the drag folding, as well as the overturning of the syncline, indicate that this movement would logically have been a late surge of eastward movement on the Charleston-Nebo thrust plate that bent the axis of the Canyon Range eastward and also dragged the formations of both its western and eastern limbs until they were nearly parallel to the Leamington fault. The apparent *left-lateral* offset of the Tintic Valley thrust from the projected position of the Pavant thrust as shown in Figure 1, however, indicates that the earlier right-lateral displacement that doubtless resulted from the near-horizontal movement was apparently followed by vertical displacement of the southern block of the LTF. This vertical displacement is also confirmed by the preservation of the Mesozoic conglomerates south of the LTF and the wide separation of the exposed remnants of the apparently correlative Sheeprock and Canyon Range thrust plates.

SHEEPROCK THRUST

Correlation of the Sheeprock thrust with the Canyon Range thrust is tempting for several reasons. Both plates are chiefly composed of stratified Precambrian rocks, some of which, like the Mutual Formation and Inkom Shale, are distinctively colored. In detail, however, some differences between the strata of the two plates are noted, particularly the occurrence of Precambrian limestone in the Canyon Range sequence (Christiansen, 1952, p. 721), whereas none has been recognized in the Sheeprock plate (Morris and Knopf, 1970a, 1970b; Blick, 1979, p. 30–40). Also, no Precambrian diamictites have been recognized in the Canyon Range plate, although this may only be the result of shallower exposures in the Canyon Range area.

Despite the minor stratigraphic differences between the Canyon Range and Sheeprock thrust plates, their general similarity appears to warrant a general correlation. If they are correlative, the late right-lateral displacement that occurred on the LTF also broke the originally continuous thrust plate and moved the Sheeprock segment east of the relative position of the Canyon Range segment. Evidence for this right-lateral displacement, however, was completely obliterated by the relative uplift and possible northward tilting of the western part of the northern block of the LTF and the concomitant downward displacement of the southern block. It is important to note that the late vertical displacement on the LTF preceded the eruption of the Oligocene volcanic rocks of the area (Morris, 1977) and is not related to basin-and-range normal faulting.

POLE CANYON THRUST

The Pole Canyon thrust of the Sheeprock Mountains (Cohenour, 1959; Morris and Kopf, 1970a; Blick, 1979) is unique among the thrusts of the central Sevier orogenic belt in the mode of its origin. It is limited to the area north of the Indian Springs tear fault and is underlain by an exposed stratigraphic sequence more than 5,000 m thick that is completely overturned. It apparently originated as a large anticline within the plate of the Sheeprock thrust, which became asymmetric and then overturned as the thrust plate moved eastward. As this local fold became recumbent, the upper limb, which was nearly horizontal and contained a normal, right-side-up section, apparently began to crumple and then sheared off, creating the Pole Canyon thrust plate. This mode of origin indicates that the Pole Canyon thrust may have been contemporaneous with the origin and development of the Canyon Range syncline.

WAH WAH AND FRISCO THRUSTS

The most interesting and instructive aspect of the LTF in relation to the overthrust sheets is its westward disap-

pearance. West of the village of Leamington, the fault and adjacent thrust sheets are concealed by the alluvial fill of the Sevier Desert. West of the desert, where the consolidated rocks are moderately well exposed, the geologic features consist of essentially through-going stratigraphic sequences and structural belts that extend from the traces of the Wah Wah and Frisco thrusts northward to the mountain ranges of the southern Great Salt Lake Desert. An example is the relatively undisturbed structural continuity of the northern Wah Wah, House, and Fish Springs Ranges, which are unbroken by major thrust or transcurrent faults (see Hintze, 1980). The southwestward projection of the LTF carries it into the center of this zone of continuous geology, most probably into the area of the middle part of the House Range. The absence of any transcurrent fault in this area perhaps can best be explained by the assumption that the LTF is overridden by the regionally extensive plate of the Wah Wah–Frisco thrust. Examination and evaluation of the stratigraphy and internal structures of the plate of the Wah Wah–Frisco thrust give no clues to the position of the LTF beneath it, and no similar fault is recognized in the adjacent part of Nevada.

The Wah Wah–Frisco thrust crops out in the central Frisco Range and southern Wah Wah Mountains from which the separate exposures are individually named. In both areas the faulted plates contain Precambrian strata including the Mutual and Inkom Formations and older stratified units, as well as a sequence of younger Cambrian quartzites and limestones. Regionally, as in the Confusion and Needle Ranges, the Cambrian strata are succeeded by an essentially complete sequence of Paleozoic formations. One of the most distinctive features of the Wah Wah–Frisco plate is its Mississippian sequence including the Pilot Shale, Joana Limestone, and Chainman Shale. These units are not present in the lower plate of the Wah Wah–Frisco thrust, but rocks of the same age consist of lithologically distinctive units that are more readily correlated with formational units whose type areas are in the general vicinity of Las Vegas, Nevada.

The full extent of the Wah Wah–Frisco thrust plate is not completely known because of limited exposures. On the basis of nearly continuous outcrops the thrust plate can be identified with some confidence to the northern limits of the Cricket, Fish Springs, and Confusion Ranges. In view of the general similarity of the Cambrian sections of the House Range with those of the Drum and Dugway Mountains, the buried trace is interpreted to pass east of these latter mountain ranges and beneath the general area of the Keg Mountains. A short distance north of the Kegs it appears to merge downward into the décollement of western Utah and eastern Nevada (Misch, 1960).

The southern and southwestern extent of the Wah Wah–Frisco plate is also imperfectly known. Relatively isolated exposures of Mississippian strata in the southern

Needle or Indian Peaks Range indicate that the concealed thrust extends westward through the area of the southern part of this range and then probably extends into Nevada, possibly through the area of the Clover Mountains.

TECTONIC HISTORY

On the basis of the stratigraphic and structural relations of the thrust and transcurrent faults of the area, the tectonic history of the central Sevier orogenic belt can be summarized as follows:

1. Emplacement of the Charleston-Nebo thrust plate, which was bounded on the south by the LTF or a proto-structure, and which may be part of a regionally extensive thrust sheet in northern Utah and southern Idaho (Crittenden, 1961);

2. Emplacement and compressional folding of the Midas thrust plate north of the Inez transcurrent fault;

3. Emplacement of a continuous thrust plate, now represented by the separate Tintic Valley and Pavant thrust plates, across the southern part of the Charleston-Nebo thrust plate, the LTF, and the Mesozoic autochthonous rocks south of it;

4. Emplacement of another continuous thrust plate now represented by the Sheeprock and Canyon Range plates across the LTF and onto the Tintic Valley and Pavant thrust plates;

5. During the latter part of the period of emplacement of the Sheeprock-Canyon Range thrust plate, renewed right-lateral movement on the LTF and the Charleston-Nebo thrust, causing rupture and separation of the Sheeprock and Canyon Range thrusts and similar separation of the Tintic Valley and Pavant thrusts. Localized compressional folding, possibly limited to the Canyon Range plate south of the LTF, resulted in the synclinal deformation of this plate; local stresses in the Sheeprock plate also caused the development of a recumbent anticline and the development of the Pole Canyon thrust. Continued compressional forces then caused drag-folding of the Canyon Range syncline and its constituent strata. This general structural episode was followed by erosion of the general area of the thrust plates, producing the clastic sedimentary rocks of the Indianola Group east of the Canyon Range (Christiansen, 1952) and elsewhere in east-central Utah (Schoff, 1951);

6. Following the compressional stage of deformation of the Canyon Range and Sheeprock plates, normal displacement took place on the LTF, dropping the southern block and preserving large areas of conglomerate of the Indianola Group south of fault. This normal displacement obviously postdates the Cretaceous Indianola Group and predates the Oligocene volcanic rocks of the area (Morris, 1977) and thus was not part of the basin-and-range orogenic episode;

7. After the rupture and separation of the Canyon Range—Sheeprock plate, emplacement and local internal deformation of a great thrust plate that is underlain in western Utah by the Frisco and Wah Wah thrust faults. Since the emplacement of this great sheet of rock there has been no apparent movement on either the Charleston-Nebo thrust or the LTF, and this younger regional thrust plate effectively conceals the Leamington transcurrent fault. This event was followed by erosion of the Wah Wah–Frisco plate and associated structures and uplifts, apparently producing the clastic debris of the Price River Formation, which unconformably overlies the Indianola Group (Schoff, 1951).

CONCLUSIONS

The relative ages and regional correlations of the imbricate thrust sheets of the central Sevier orogenic belt can be determined with reasonable success by their relationships to the intermittently active Leamington transcurrent fault and by detailed stratigraphic comparisons. Other parts of the orogenic belt should be examined for similar transcurrent faults, which also may be useful in determining the sequence and timing of major structural events during the Sevier orogeny.

REFERENCES CITED

Armstrong, R. L., 1968, Sevier orogenic belt in Nevada and Utah: Geological Society of America Bulletin, v. 79, p. 429–458.

Aubert, D., 1959, Le décrochement de Pontarlier et L'orogenese du Jura: Memoires de la Societe Vaudoise des Sciences Naturelles, v. 12, p. 93–152.

Baker, A. A., 1947, Stratigraphy of the Wasatch Mountains in the vicinity of Provo, Utah: U.S. Geological Survey Oil and Gas Chart OC-30.

Baker, A. A., and Crittenden, M. D., Jr., 1961, Geologic map and cross sections of the Timpanogos Cave quadrangle, Utah: U.S. Geological Survey Geologic Quadrangle Map GQ-132, scale 1:24,000.

Blick N. H., 1979, Stratigraphic, structural, and paleogeographic interpretation of upper Proterozoic glaciogenic rocks in the Sevier orogenic belt, southwestern Utah [Ph.D. thesis]: Santa Barbara, University of California, 636 p.

Christiansen, F. W., 1952, Structure and stratigraphy of the Canyon Range, central Utah: Geological Society of America Bulletin, v. 63, no. 7, p. 717–740.

Cohenour, R. E., 1959, Sheeprock Mountains, Tooele and Juab Counties [Utah]: Utah Geological and Mineral Survey Bulletin 81, 201 p.

Costain, J. K., 1960, Geology of the Gilson Mountains and vicinity, Juab County, Utah [Ph.D. thesis]: Salt Lake City, University of Utah, 139 p.

Crittenden, M. D., Jr., 1961, Magnitude of thrust faulting in northern Utah: U.S. Geological Survey Professional Paper 424-D, art. 35, p. D128–D131.

Hintze, L. F., 1980, Geologic map of Utah: Utah Geological and Mineral Survey, scale 1:500,000.

Misch, Peter, 1960, Regional structural reconnaissance in central-northeast Nevada and some adjacent areas: observations and interpretations: Intermountain Association of Petroleum Geologists and Eastern Nevada Geological Society, 11th Annual Field Conference Guidebook to the Geology of Central Nevada, p. 17–42.

Morris, H. T., 1977, Geologic map and sections of the Furner Ridge quadrangle, Juab County, Utah: U.S. Geological Survey Miscellaneous Geologic Investigations Map I-1045, scale 1:24,000.

——1978, Preliminary geologic map of the Delta 2° quadrangle, west-central Utah: U.S. Geological Survey Open-File Report 78-705, scale 1:125,000.

Morris, H. T., and Kopf, R. W., 1970a, Preliminary geologic map and cross sections of the Cherry Creek quadrangle and adjacent part of the Dutch Peak quadrangle, Juab County, Utah: U.S. Geological Survey Open-File Map, February 3, 1970, scale 1:24,000.

——1970b, Preliminary geologic map and cross section of the Maple Peak quadrangle and adjacent part of the Sabie Mountain quadrangle, Juab County, Utah: U.S. Geological Survey Open-File Map, February 3, 1970, scale 1:24,000.

Morris, H. T., and Lovering, T. S., 1979, General geology and mines of the East Tintic mining district, Utah and Juab Counties, Utah: U.S. Geological Survey Professional Paper 1024, 203 p.

Morris, H. T., and Shepard, W. M., 1964, Evidence for a concealed tear fault with large displacement in the central East Tintic Mountains, Utah: U.S. Geological Survey Professional Paper 501C, p. C19–C21.

Schoff, S. L., 1951, Geology of the Cedar Hills, Utah: Geological Society of America Bulletin, v. 62, no. 6, p. 619–646.

Steven, T. A., and Morris, H. T., 1981, Geologic map [and cross sections] of the Cove Fort quadrangle, west-central Utah: U.S. Geological Survey Open-File Map 81-1093, scale 1:50,000.

Tooker, Edwin W., 1983, Thrust plates of the eastern Sevier overthrust belt, Salt Lake recess, Utah, *in* Miller, D. M., Todd, V. R., and Howard, K. A., eds., Tectonic and stratigraphic studies in the eastern Great Basin: Geological Society of America Memoir 157, (this volume).

Wallace, R. E., and Morris, H. T., 1979, Characteristics of fault and shear zones as seen in mines at depths as much as 2.5 km below the surface, *in* Analysis of actual fault zones in bedrock: U.S. Geological Survey Open-File Report 79-1239.

Wang, Yun Fei, 1970, Geological and geophysical studies of the Gilson Mountains and vicinity, Juab County, Utah [Ph.D. thesis]: Salt Lake City, University of Utah, 126 p.

MANUSCRIPT ACCEPTED BY THE SOCIETY AUGUST 20, 1982

Geological Society of America
Memoir 157
1983

Microfacies of the Middle Pennsylvanian Part of the Oquirrh Group, central Utah

Calvin H. Stevens
Department of Geology
San Jose State University
San Jose, California 95192

Richard A. Armin
U.S. Geological Survey
Menlo Park, California 94025
and
Department of Geosciences
University of Arizona
Tucson, Arizona 85721

ABSTRACT

Middle Pennslvanian rocks of the immensely thick Oquirrh Group in the Stansbury and Onaqui Mountains, Tooele County, Utah, consist primarily of thick, fine-grained siliciclastic rock units with relatively thin interbedded limestone units. A total of 12 microfacies are recognized; 3 are primarily siliciclastic, and 9 are primarily carbonate. The relatively pure siliciclastic rock composed of coarse silt and very fine sand-size quartz and feldspar is here called the "siltstone microfacies." Two other microfacies, composed primarily of quartz and feldspar grains, are (1) peloidal siltstone (siltstone and very fine grained sandstone containing micritic peloids), and (2) fossiliferous siltstone (siltstone and very fine grained sandstone containing rounded fossil fragments). The carbonate microfacies, assigned to modified standard microfacies, are shallow spiculite (smf-1-s); boundstone (smf-7); whole-fossil wackestone (smf-8); bioclastic wackestone (smf-9); fusulinid-bearing bioclastic wackestone (smf-9-f); coated bioclasts in sparite, grainstone (smf-11); coquina, bioclastic grainstone or rudstone, shell hash (smf-12); laminated to bioturbated pelleted lime mudstone-wackestone (smf-19); and coarse lithoclastic-bioclastic rudstone or floatstone (smf-24).

The vertical arrangement of microfacies, fauna, flora, and sedimentary structures indicates that each carbonate sequence, from base to top, represents a transgression of the sea over shoreline deposits, followed some time later by reestablishment of these environments. The thick sequences of siltstone sandwiched between marine limestone sequences represent deposition in various shoreline and, probably, terrestrial environments.

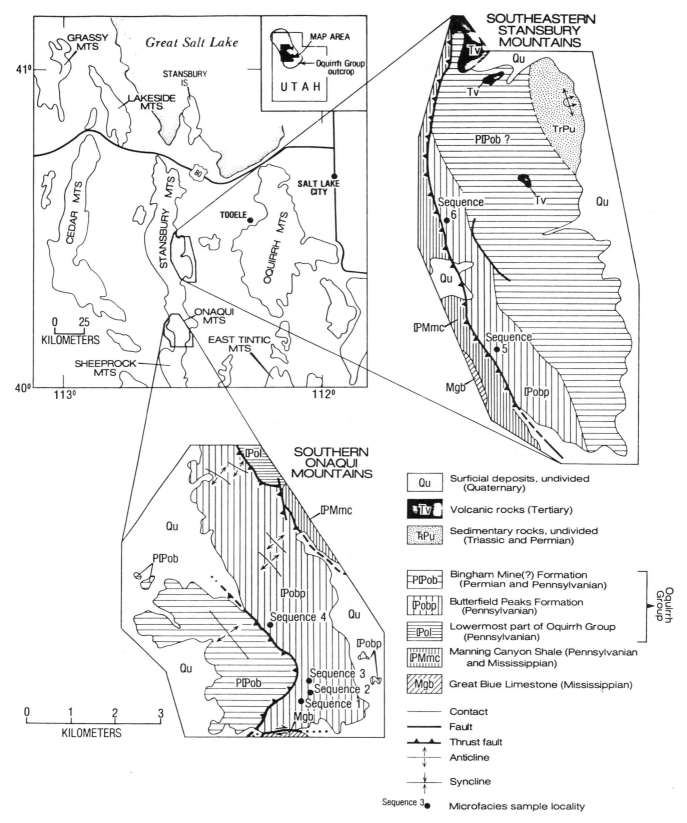

Figure 1. Central Utah showing locations of measured sequences in the Butterfield Peaks Formation. Mapping by Armin (1979) and Jordan and Allmendinger (1979).

INTRODUCTION

The middle part of the Oquirrh Group (the Butterfield Peaks Formation) in the Onaqui and Stansbury Mountains, central Utah (Fig. 1), was selected for microfacies analysis because field reconnaissance suggested the presence of diverse microfacies and several fossil communities resembling those described by Stevens (1965, 1966) in rocks of similar age in Colorado. Thus, microfacies and fossil communities in the Oquirrh Group might be compared to rocks of the same age, the depositional environments of which had been determined previously. Also, the rocks in the middle part of the Oquirrh Group seemed ideal for such a study because the limestone units are relatively thin (from several meters to several tens of meters thick), and thus are relatively easily studied; in addition, deposition appears to have been relatively rapid and more or less continuous so that complications arising from periods of erosion or nondeposition could be avoided.

Previously, few data have been presented to allow a detailed interpretation of the environments of deposition of Middle Pennsylvanian rocks in this part of the Oquirrh basin; in fact, even the outline of the basin is uncertain. Welsh (1979) included the Onaqui Mountains within an area that he designated a subaqueous delta, although he did not present data for this designation and went no further in interpreting subenvironments.

Here we describe the microfacies and fossil content of the Butterfield Peaks Formaton in the Onaqui and Stansbury Mountains, central Utah (Fig. 1), interpret the environments of deposition of the unit, and present a depositional model based on the nature of the microfacies and the contained fossil communities. This information and these interpretations should be widely applicable to similar rocks of late Paleozoic age, and perhaps other ages, far beyond the confines of the Oquirrh basin.

STRATIGRAPHY

The Oquirrh Group, comprising as much as 8,000 m of limestone and fine-grained siliciclastic rocks (Roberts and others, 1965), is exposed over a large area of northwestern Utah. The name "Oquirrh Formation" was first applied by Nolan (1930) to Pennsylvanian and Permian rocks in the Gold Hill mining district of west-central Utah, but later the Oquirrh Mountains (Fig. 1) were considered to embrace the type area. Welsh and James (1961) proposed raising the Oquirrh Formation to group status and restricted the name "Oquirrh Group" to the Pennsylvanian strata. In the southern Oquirrh Mountains, Tooker and Roberts (1970) recognized three formations in the Oquirrh Group: the lower West Canyon Limestone, the middle Butterfield Peaks Formation, and the upper Bingham Mine Formation. The sequences of interbedded limestone and siliciclas-

tic rocks reported on here in the Onaqui and Stansbury Mountains correlate lithologically and faunally with the Butterfield Peaks Formation of the southern Oquirrh Mountains (Armin, 1978; Armin and Moore, 1981). In the Stansbury Mountains, the Butterfield Peaks Formation is about 1,850 m thick, of which 60% to 75% is composed of siltstone and sandstone, and 25% to 40% of limestone. Although the thickness of the Butterfield Peaks Formation in the Onaqui Mountains is difficult to determine owing to structural complexity, the unit is lithologically similar to the Butterfield Peaks Formation in the Stansbury Mountains.

Six predominantly limestone sequences (Fig. 1) in the Butterfield Peaks Formation, representing a total thickness of 167 m, were measured and sampled. These six limestone sequences, the exact stratigraphic position of which is uncertain, were chosen for study because they are well exposed and appear to be typical of those within the Butterfield Peak Formation. The thickest limestone sequence is about 50 m thick, and all the sequences grade vertically from fine-grained siliciclastic rocks, into limestone, and then back into siliciclastic rocks. A total of 84 hand specimens were collected at an average stratigraphic interval of about 2 m throughout the predominantly limestone sequences (Fig. 2), as well as random samples from the thicker siliciclastic intervals. Interspecimen distance varies according to the changes in lithology; closer sampling was done in intervals that megascopically showed relatively rapid lithologic change. Thin sections were prepared from surfaces oriented perpendicular to the bedding of samples from sequences 2, 4, 5, and 6 and of the siliciclastic rocks; acetate peels and polished surfaces were made of samples from sequences 1 and 3.

Fusulinids show that sequences 2 and 6 are Desmoinesian and that sequence 5 is Atokan (Armin and Moore, 1981). Although sequences 1, 3, and 4 do not contain fusulinids, on the basis of other faunal elements and stratigraphic position they also are believed to be Atokan or Desmoinesian (Middle Pennsylvanian).

MICROFACIES ANALYSIS

In all, 12 microfacies have been recognized in the samples of which thin sections and peels were prepared. One, composed almost entirely of siliciclastic grains, is here called the "siltstone microfacies (ss), although some samples are composed of very fine sand. Two other primarily siliciclastic microfacies containing substantial amounts of carbonate debris are here referred to as the "peloidal siltstone microfacies" (ssp) and the "fossiliferous siltstone microfacies" (ssf), although some samples are composed of very fine sand; these two microfacies occur primarily near contacts with the siltstone microfacies. The nine primarily carbonate microfacies are assigned to the standard micro-

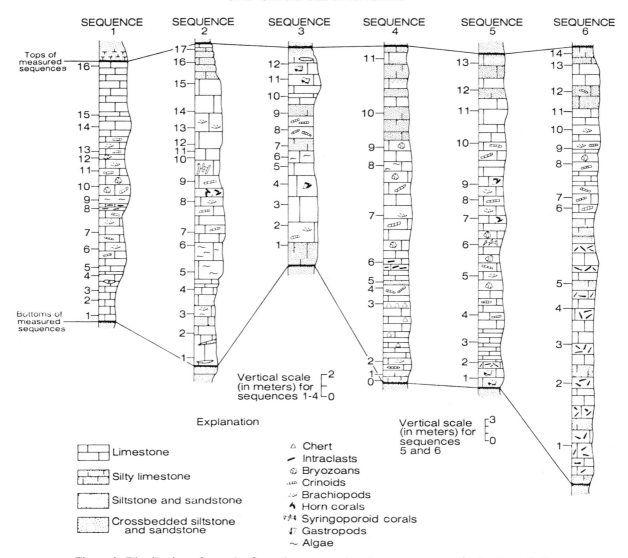

Figure 2. Distribution of samples from the measured carbonate sequences in the Butterfield Peaks Formation. Correlation of sequences is not implied.

facies of Wilson (1975), although a few minor modifications of names have been made. These microfacies are here designated: shallow spiculite (smf-1-s); boundstone (smf-7); whole-fossil wackestone (smf-8); bioclastic wackestone (smf-9); fusulinid-bearing bioclastic wackestone (smf-9-f); coated bioclasts in sparite, grainstone (smf-11); coquina, bioclastic grainstone or rudstone, shell hash (smf-12); laminated to bioturbated pelleted lime mudstone-wackestone (smf-19); and coarse lithoclastic-bioclastic rudstone or floatstone (smf-24).

DESCRIPTION OF MICROFACIES

Siltstone Microfacies (ss)

The siltstone microfacies (ss, Fig. 3A) typically is com-

posed of yellowish-gray- to grayish-brown-weathering, thick-bedded to less commonly laminated, coarse-grained siliciclastic siltstone to very fine grained sandstone. Most rocks of this type are subarkosic (terminology of Pettijohn and others, 1973) and consist of subangular, well-sorted, loosely to moderately packed quartz and feldspar grains. Commonly, the cement is calcareous, although few clastic carbonate grains and no fossils are present. No evidence of bioturbation was noted. This microfacies, which normally is rather poorly exposed, forms most of the Butterfield Peaks Formation.

Peloidal Siltstone Microfacies (ssp)

The peloidal siltstone microfacies (ssp, Fig. 3D) typically is composed of yellowish-gray- to grayish-brown-

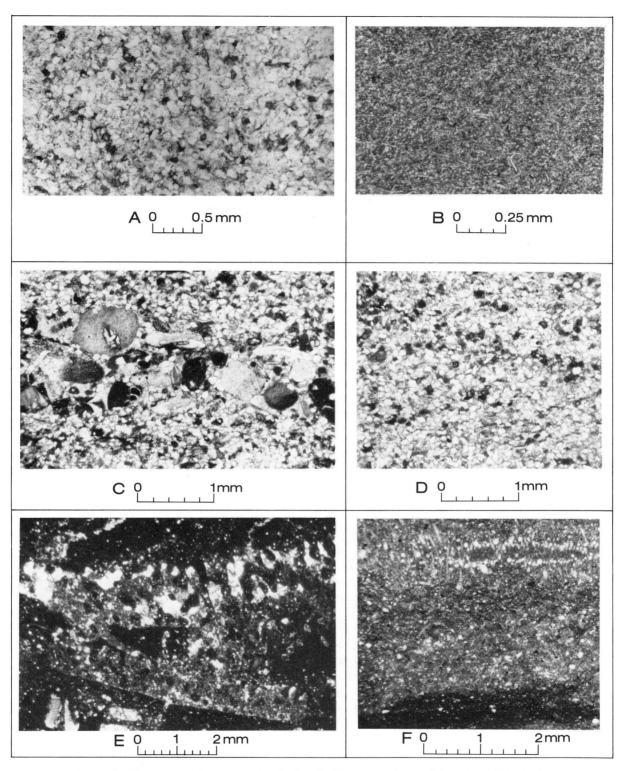

Figure 3. Siltstone and limestone microfacies: A, Siltstone (ss) containing about 1% micrite particles (sample 1-17). B, Shallow spiculite (smf-1-s) composed primarily of spicules (sample 4-3). C, Fossiliferous siltstone (ssf) (sample 6-4). D, Peloidal siltstone (ssp) containing micritic peloids (dark grains) (sample 2-2). E, Boundstone (smf-7) containing calcareous sponges (sample 2-6). F, Shallow spiculite (smf-1-s) containing sponge (at top) and dark intraclast with indistinct boundaries at the bottom (sample 2-4).

weathering, laminated to medium-bedded, calcareous, coarse-grained siliciclastic siltstone to very fine grained sandstone. Parallel laminae are typical, and low-angle tabular-, wedge-, and lens-shaped crosslaminated sets, 5 to 15 cm thick, are common; high-angle avalanche-type cross-bedding occurs more rarely. Contrasts in the amounts of carbonate peloids and minor micritic matrix commonly define the laminae in thin section. Subangular, well-sorted quartz and felspar grains form most of the rock, complemented by abundant silt-size to very fine sand-size, micritic peloids. Fecal pellets, generally about 0.5 mm in diameter and composed of micrite, and sand- to silt-size siliciclastic grains also occur. Two samples show considerable evidence of bioturbation. Fossils are rare and average only about one group per sample; the only fossils noted are siliceous sponge spicules, productid brachiopod spines, and echinoderm parts. The rock is cemented by microspar. Secondary dolomite, dedolomitized calcite rhombs, and iron oxide pseudomorphs after pyrite are scattered throughout the rock.

Occurrence. 1-4?, 2-2, 2-17, 4-0, 4-1, 6-2, 6-14. (In this and all other occurences listed, the first number is the sequence number; the second is the sample number in that sequence; the query indicates that assignment of a given sample to this microfacies is questionable. Some apparent discrepancies with columns shown in Figure 2 are due to the generalized field identification of rocks compared to thin-section study of small samples from those units.)

Fossiliferous Siltstone Microfacies (ssf)

The fossiliferous siltstone microfacies (ssf, Fig. 3C), which commonly occurs at the top and bottom of the limestone sequences, typically consists of thin-bedded to laminated, yellowish-brown, subarkosic, very coarse grained siliciclastic siltstone to very fine grained sandstone and interlaminated, dusky-blue-weathering, bioclastic layers composed of medium-sand-size, rounded, well-sorted bioclasts containing few peloids and intraclasts. The layers are commonly 1 to 4 clasts thick. A few medium-sand-size quartz grains also are present. Low-angle ($< 10°$) crosslaminated sets, as thick as 20 cm, are common and are accentuated by differential weathering of the siliciclastic siltstone and bioclastic laminae. Layers within the crossbeds commonly are 1.5 to 5 mm thick. Generally, about five different groups of fossils are distinguishable, all worn and obviously transported. Echinoids and fewer pelmatozoans dominate; foraminifers, brachiopods, and bryozoans are less common; and immature fusulinids are rare. Rounded intraclasts with enclosed foraminifers and fossil fragments are moderately rare to common constituents. Some of the echinoderms are enveloped by syntaxial cement.

Occurrence. 1-2, 1-16, 3-12, 4-11, 5-12, 6-4.

Shallow Spiculite microfacies (smf-1-s)

The shallow spiculite microfacies (smf-1-s, Fig. 3B, 3F) includes wackestone and mudstone, which form moderately to weakly resistant outcrops that are laminated to medium bedded and weather light and medium grayish yellow and pale reddish purple with local mottling. Fresh surfaces are brownish gray to dark gray and pinkish gray. Bedding contacts commonly are wavy. The beds are sparsely fossiliferous except for sponge spicules. Commonly these spicules are scattered, but locally they make up as much as 60% of this microfacies. Several *in situ* flattened sponges approximately 1.5 by 5 mm, occur in two samples. Some ostracods were noted, and pelmatozoan and brachiopod remains are rare. In addition, a few planispiral gastropods were seen in outcrop. The petrographic fossil diversity (Smosna and Warshauer, 1978) is very low, generally about 2.

Silt and very fine sand, consisting of well-sorted, subangular quartz and feldspar, make up as much as 25% of this microfacies. A few waferlike intraclasts (0.4 by 4 mm) and fecal pellets (0.1 mm) also are present.

The texture of the matrix ranges from homogeneous micrite, to patchy microspar and micrite, to pervasively recrystallized microspar after micrite. Some slides exhibit undulatory pinch-and-swell laminae, commonly 2 to 7 mm thick, with silty swirls probably caused by bioturbation. Disseminated small iron oxide pseudomorphs after pyrite, tiny euhedral laths of muscovite, and patches of tiny dolomite rhombohedra are present in trace amounts in about one-third of the samples. Dolomite is abundant in one sample.

The presence of gastropods, which appear to be planispiral bellerophontids, suggests that this fossil community is related to the euphemitid fossil community of Stevens (1965, 1966).

The shallow spiculite microfacies is here designated "smf-1-s" because in many ways it resembles miocrofacies smf-1 of Wilson (1975). However, as is shown later, this microfacies of the Butterfield Peaks Formation is believed to represent shallow-water deposition (whence the "s"), in contrast to the deep-water microfacies described by Wilson.

Occurrence. 1-1, 1-3, 1-5, 1-7, 1-14, 1-15, 2-1, 2-4, 3-3, 3-5, 3-7, 3-10, 3-11, 4-3, 4-5, 4-8, 4-10, 5-1?, 5-3, 6-11.

Boundstone Microfacies (smf-7)

Rock of the boundstone microfacies (smf-7, Figs. 3E, 4A) generally is medium to thick bedded in outcrop and laterally lenticular on a scale of several meters to several tens of meters. Most samples consist of a dark-gray-weathering organic framework of encrusting calcareous sponges 2 by 10 mm, surrounded by light-gray-weathering lime mud. Virtually no siliciclastic grains occur in the rock. Abundant encrusting calcareous sponges are characteristic

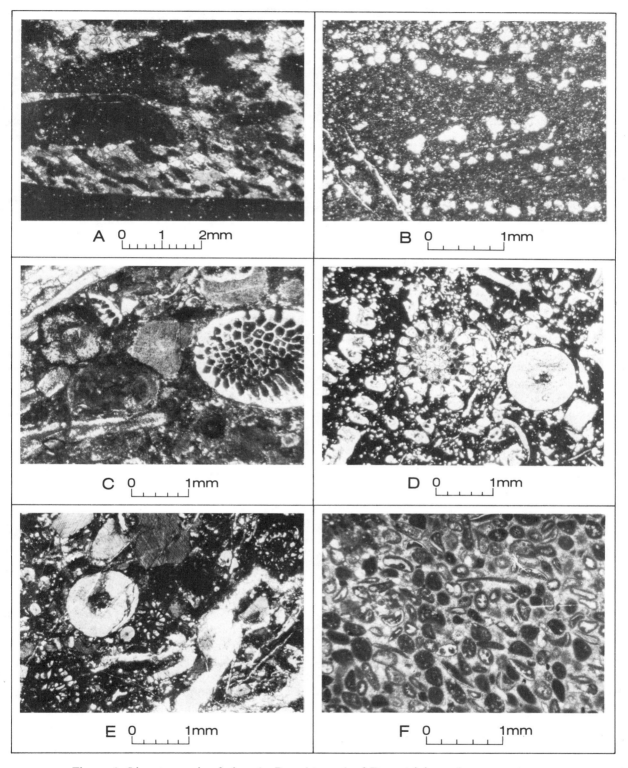

Figure 4. Limestone microfacies. A, Boundstone (smf-7) containing calcareous sponges (sample 2-5). B, Whole-fossil bioclastic wackestone (smf-8) showing transverse sections of fenestrate bryozoans (sample 5-6). C, D, Bioclastic wackestone (smf-9) (samples 5-5, 2-13). E, Fusulinid-bearing bioclastic wackestone (smf-9-f) (sample 6-10). F, Coated bioclasts in sparite, grainstone (smf-11) containing coated shells and fragments of dasycladacean algae (sample 2-10).

of this microfacies. Other abundant encrusting organisms include *Tubiphytes,* cyclostome bryozoans, *Tuberitina, Tetrataxis,* other foraminifers, and a *Komia*-like form. Delicate-shelled brachiopods, ostracods, pelmatozoans, sponge spicules, and small coiled foraminifers are scarce. The micrite matrix is mostly homogeneous and locally clotted or pelleted. Spar commonly occurs under umbrellas of bivalves and other fossils. In one sample, several apparently unbroken fenestrate bryozoans formed a baffle that trapped lime mud. The petrographic fossil diversity is comparatively high, about 9—the highest observed.

This fossil community probably is allied to the productid-*Composita* fossil community of Stevens (1965, 1966).

Occurrence. 1-9, 2-3, 2-5, 2-6, 3-6, 4-7.

Whole-Fossil Wackestone Microfacies (smf-8)

Outcrops of the whole-fossil wackestone (smf-8) microfacies generally form resistant ledges of medium-gray- and grayish-yellow-weathering, medium- to thick-bedded, fossiliferous limestone. Fresh surfaces are dark gray and dark brownish gray. Subangular silt-size and very fine sand-size siliciclastic detritus averages less than 1% and is localized within swirls and streaks in the rock. Most of the matrix is micrite, locally replaced by microspar. Some samples show a clotted texture suggestive of bioturbation. Minor matrix-replacing and rare void-filling spar also is present.

This microfacies (Fig. 4B) is a wackestone containing unsorted, mostly whole, solitary rugose corals, brachiopods, and fenestrate bryozoans. Most of the corals lie approximately parallel to the bedding planes, but some stand in living position. A few pelmatozoan ossicles commonly are suspended in the micritic matrix. Palaeotextulariid foraminifers, *Tetrataxis,* other unidentified foraminifers, and *Komia* are scarce; monaxonic sponge spicules are rare to abundant. The petrographic fossil diversity is about 6.

Many skeletal parts have been replaced by silica and sparry calcite. Iron oxide pseudomorphs after pyrite are scarce and are restricted to the matrix. Some small rhombs of dolomite are present.

The fossil community represented in this microfacies resembles that of the productid-*Composita* fossil community of Stevens (1965, 1966).

Occurrence. 3-4, 5-6, 5-7, 5-9.

Bioclastic Wackestone Microfacies (smf-9)

Rock of the bioclastic wackestone microfacies (smf-9, Fig. 4C, D) forms resistant ledges to platy, talus-covered slopes that weather bluish gray, medium gray, and pale reddish purple; fresh surfaces generally are medium to dark gray.

This microfacies is dominantly a wackestone and packstone with some mudstone that contains abundant fossils in a matrix ranging from homogeneous or recrystallized micrite to biogenically churned micrite containing finely comminuted fossil fragments. The larger particles are mostly unsorted fossils exhibiting varying degrees of disarticulation and fragmentation. In one sample, a trail of fecal pellets extending from the matrix through the pedicle opening of a slightly unhinged brachiopod provides evidence that burrowers have caused some of the disarticulation. Fecal pellets and peloids, about 0.3 mm in diameter, locally are present, particularly in zones containing a high percentage of fossil debris. From 5% to 15% moderately sorted, subangular silt- to very fine sand-size quartz and feldspar detritus is present in almost all thin sections of the rock.

The bioclastic wackestone microfacies is distinguished by its texture and by the abundance of brachiopods, echinoids, bryozoans, and pelmatozoans. Fenestrate, ramose, and encrusting bryozoans all are common, although the fenestrate types dominate. Most fossils lie parallel to bedding, and the bivalves are so oriented that either the convex or concave side is up. Foraminifers (including *Tuberitina, Bradyina,* and other, unidentified coiled forms) and a few trilobites and corals occur in this microfacies. Monaxonic sponge spicules and ostracods are rare to common components; several chonetid brachiopods and a few fragments of possible phylloid algae also occur. The petrographic fossil diversity is about 7.

Fossils commonly are silicified or replaced by sparry calcite. Rarely, some brachiopod valves are replaced by sodic plagioclase. Tiny iron oxide crystals after pyrite and, rarely, dolomite selectively replace parts of echinoderms and brachiopods. Syntaxial rim cement around pelmatozoan plates and microstylolites also are moderately common.

Characteristics that distinguish the bioclastic wackestone (smf-9) from the somewhat similar whole-fossil wackestone microfacies (smf-8) are that the latter contains more whole fossils, generally forms thicker, less fossiliferous beds, and contains more micritic matrix and less siliciclastic material.

The fossils, which include most components of the productid-*Composita* fossil community of Stevens (1965, 1966), probably represent *in situ* accumulations, as indicated by their suspension in micrite, although they have been broken and highly disturbed, presumably during bioturbation.

Occurrence. 1-6, 1-11, 1-12, 2-9, 2-13, 3-2, 3-9, 4-2, 4-4, 4-9, 5-4, 5-5, 5-11, 6-6, 6-8, 6-12.

Fusulinid-Bearing Bioclastic Wackestone Microfaces (smf-9-f)

The fusulinid-bearing bioclastic wackestone microfacies (smf-9-f) consists of moderately resistant wackestone and minor packstone that weather medium and light gray to pale reddish purple and are laminated to medium bedded; fresh surfaces are medium to dark gray and brownish gray. Beds contain abundant poorly sorted fossils and some peloids. Outcrops with mottled markings and siliciclastic silt stringers along bedding planes are common. Limestone of this microfacies ranges from mud-rich wackestone to packstone. Moderately well sorted, subangular siliciclastic silt is unimportant overall and averages less that 5% of the rock. The matrix is chiefly micrite, locally replaced by microspar.

Echinoderms, especially pelmatozoan remains, are the most abundant fossils, but the common occurrence of mature fusulinids serves to differentiate this microfacies from the others. Fenestrate, ramose, and encrusting bryozoans, brachiopods, *Tubiphytes,* possible phylloid algae, and ostracods are present in minor amounts. Besides fusulinids, other foraminifers present in some samples include palaeotextulariids, *Tuberitina, Tetrataxis,* and other coiled and encrusting forms. The fossils generally lie parallel to the bedding and appear to be randomly oriented in that plane. The petrographic fossil diversity averages slightly higher than 8.

The outer fusulinid walls commonly are broken or crumpled at contacts with other organic debris, suggesting that considerable compaction of the sediment has occurred. Syntaxial rim cement around pelmatozoan ossicles was noted rarely, and many skeletal parts are silicified. Small burrows and associated fecal pellets record biogenic activity.

The fusulinid-bearing bioclastic wackestone (smf-9-f, Fig. 4E) differs from the bioclastic wackestone microfacies (smf-9) in its content of mature fusulinids, generally more pelmatozoan stems, and fewer brachiopods and bryozoans. The siliciclastic material also is slightly finer grained and less abundant.

This microfacies contains fossils characteristic of the fusulinid fossil community of Stevens (1965, 1966). Fusulinids were shown by Stevens (1965) to constitute a very important, easily identified biofacies; therefore, they are here employed to designate a microfacies separate from the bioclastic wackestone.

Occurrence. 2-7, 2-8, 5-8, 6-10.

Coated Bioclasts in Sparite, Grainstone Microfacies (smf-11) with Interbedded Laminae of the Coated and Worn Bioclasts in Micrite, Packstone-Wackestone Microfacies (smf-10)

The coated-bioclasts-in-sparite, grainstone microfacies, composed dominantly of very fine grained grainstone (Figs. 4F, 5A), forms resistant outcrops that are medium gray weathering, locally light gray mottled, and laminated

to medium bedded with wispy streaks of quartz and feldspar silt and sand; a fresh surface is medium to dark gray. Megafossils are absent. This microfacies consists chiefly of pelletal grainstone (pellets 0.1 to 0.4 mm across), although different textures commonly are displayed in different laminae within a single thin section. Most of the grains (0.25-mm diameter) are superficial to true oolites. They consist of nuclei of pellets, peloids, and fragments of fossils coated with one to several layers of fine-grained fibroradiating carbonate material. Coated grains are rarely nucleated by siliciclastic detritus. Common molded grain contacts suggest that plastic deformation occurred during compaction or that the coatings were pliable at the time of deposition. Textures present include grainstone composed of pellets, tiny fossil fragments, and coated grains cemented by spar (smf-11); silty packstone containing patches of micrite and interstitial sparry cement (smf-10); and wavy, spiculiferous micrite layers. Laminae generally are 1 to 5 mm thick; micro-crosslaminae were noted rarely. Fairly well sorted, subangular siliciclastic silt ranges from a trace to 15% of the samples. Calcite cement, including rare sparry syntaxial overgrowths, makes up about 50% of the rock.

A diverse assemblage of highly fragmented, moderately well sorted to well-sorted, commonly coated fossils occurs in this microfacies, although ostracods and foraminifers, including *Tuberitina,* dominate. Some productid spines, bryozoans, rare pelmatozoan parts, a few echinoid spines, and many thin, small shells also are present. Stringers of pellets cutting vertically through disrupted bedding document burrowing activity by undetermined organisms. Some shells appear to have algal coatings, and fragments of dasycladacean algae were noted in three samples. The petrographic fossil diversity is about 7.

Occurrence. 2-10, 2-11, 2-12, 2-14, 2-15, 3-1.

Coquina, Bioclastic Grainstone or Rudstone, Shell Hash Microfacies (smf-12)

The coquina, bioclastic grainstone or rudstone, shell hash microfacies, composed of grainstone and packstone (smf-12, Fig. 5B, 5C), generally is thin to medium bedded, weathers gray to bluish gray, and consists mostly of a diverse assemblage of poorly sorted normal marine fossils in spar cement with minor micritic matrix. Micritic intraclasts, pellets, and small amounts of siliciclastic silt are present locally. Patches of matrix composed of grumous micrite and comminuted allochems may represent material that filtered in after deposition. Pressure solution and stylolitization between allochems are particularly noticeable where large skeletal fragments are concentrated. Many fossils show partial micritization, and many others are silicified. Rare secondary dolomite and iron oxide pseudomorphs after pyrite have formed in echinoid plates.

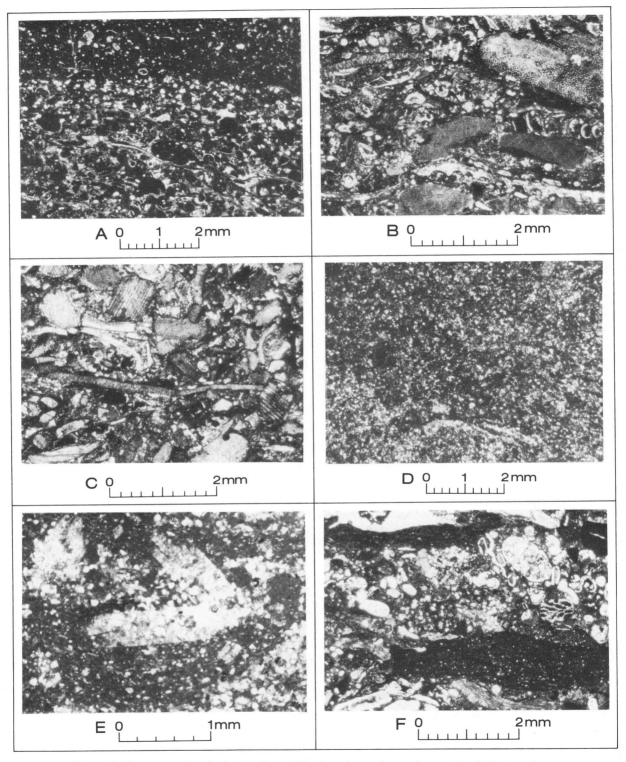

Figure 5. Limestone microfacies. A, Coated bioclasts in sparite, grainstone (smf-11), showing interlayering of grainstone and wackestone layers (sample 2-15). B, C, Coquina, bioclastic grainstone or rudstone, shell hash (smf-12) (samples 6-9, 6-3). D, E, Laminated to bioturbated pelleted lime mudstone-wackestone (smf-19): D, containing numerous peloids (sample 5-13); E, containing a swallowtail calcite pseudomorph after gypsum (sample 2-16). F, Coarse lithoclastic-bioclastic rudstone or floatstone (smf-24) containing two bent and intruded flakes of the shallow spiculite microfacies (sample 5-2).

Echinoderms (especially pelmatozoans), bryozoans (especially fenestrate forms), and brachiopods are the dominant fossils, but several types of foraminifers included *Tetrataxis* also are common. The fossils occur mostly as fragments, commonly as long as 10 mm and few larger than 30 mm across. The petrographic fossil diversity is about 6.

Occurrence. 1-10, 1-13, 3-8, 5-10?, 6-3, 6-7, 6-9.

Laminated to Bioturbated Pelleted Lime Mudstone-Wackestone Microfacies (smf-19)

The laminated to bioturbated pelleted lime mudstone-wackestone microfacies (smf-19, Fig. 5D, 5E) consists of thin- to thick-bedded, medium-gray-weathering mudstone and wackestone that lack easily recognizable megafossils. Bioturbation features are common. Micrite and pellets (many as large as 0.25-mm diameter) are the main rock components, but as much as 10% siliciclastic silt also occurs, and layers of fecal pellets are rarely interbedded with micrite and siliciclastic silt. In one sample (2-16), calcite pseudomorphs after swallowtail gypsum twins, as long as 2.5 mm, were observed.

This microfacies typically contains abundant scattered ostracods and siliceous sponge spicules. Two or three types of foraminifers, brachiopods, echinoids, pelmatozoans, and trilobites rarely are present. The petrographic fossil diversity is about 4.

Occurrence. 2-16, 5-13, 6-13?.

Coarse Lithoclastic-Bioclastic Rudstone or Floatstone Microfacies (smf-24)

The coarse lithoclastic-bioclastic rudstone or floatstone microfacies (smf-24) shown on Figure 5F is medium gray to bluish gray and yellowish gray weathering, is laminated to medium bedded, and contains abundant poorly sorted, tabular intraclasts and fossil fragments. This lithology occurs in beds in which lag-filled scours are common. The presence of dark-colored intraclasts makes this microfacies easy to recognize in the field.

This microfacies consists mostly of packstone to grainstone that locally has a matrix of comminuted allochems and, rarely, patches of partially recrystallized micrite. Sparry calcite has filled the original void space in fossils and in some samples preserved geopetal structures. The framework commonly is densely packed and consists of various intraclasts, fossils, and siliciclastic silt and very fine sand (Fig. 5F). Intraclasts oriented subparallel to bedding make up as much as half the thin-section areas, although in the field some beds were estimated to contain even more. Elongate, lensoidal, and tabular intraclasts range up to 30 mm long, although generally they are 3 to 10 mm long and 0.4 to 1.4 mm thick. Most of these clasts are composed of dark micrite containing abundant sponge

spicules (pieces of shallow spiculite, smf-1-s). Clasts composed of micrite with enclosed skeletal fragments, small coated grains cemented by microspar, and sparse siliciclastic silt also were observed. Many intraclasts are bent and pinched and are intruded by other materials, indicating that they were only partially lithified when deposited. Moderately well sorted, subangular siliciclastic silt and very fine sand commonly make up as much as 10% of the rock. The matrix has been replaced by microspar except for numerous clots of micrite. Blocky spar and microspar form the cement, which averages 5% to 10% of the rock. Pressure solution has affected some grain boundaries, especially those between larger crinoid plates. Silica or blocky calcite spar has partially to totally replaced many fossils, particularly the bivalves. Tiny iron oxide crystals locally replace echinoid plates, and secondary dolomite rarely is present in some intraclasts.

Disarticulated and fragmented echinoid, pelmatozoan, brachiopod, and rhomboporoid bryozoan parts are common in all thin sections of this microfacies. Encrusting foraminifers, coated fossil debris, ostracods, and peloids are present in some slides, and one immature fusulinid was observed. The petrographic diversity is about 3.

Occurrence. 1-8, 4-6?, 5-2, 6-1, 6-5.

INTERPRETATION OF ENVIRONMENTS OF DEPOSITION

Introduction

Each microfacies in the limestone sequences occurs either (1) near the center or (2) at the tops and/or bottoms of sequences (Fig. 6); none occurs in all positions. In general, the boundstone, whole-fossil wackestone, bioclastic

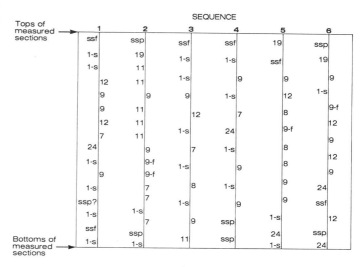

Figure 6. Distribution of microfacies in limestone sequences. Microfacies interpreted as nearshore are on left side of line, and offshore microfacies are on right side. Not to scale.

wackestone, and fusulinid-bearing bioclastic wackestone microfacies (smf-7, smf-8, smf-9, and smf-9-f, respectively) occur near the middles of sequences, whereas the peloidal siltstone, fossiliferous siltstone, shallow spiculite, laminated to bioturbated pelleted mudstone-wackestone, and rudstone-floatstone microfacies (ssp, ssf, smf-1-s, smf-19, smf-24, respectively) generally occur near the tops and/or bottoms of sequences (Fig. 6). Of all of the two lowest and two highest samples in each of the six sequences (24 samples), the latter five microfacies comprise all except those at the base of sequence 3. Other occurrences of some of these microfacies are scattered within the sequences, although these occurrences are relatively few.

Major differences between the two groups of microfacies suggest that the middles of the limestone sequences were deposited offshore, whereas the tops and bottoms were formed nearshore. The first group of microfacies (those occurring in the middles of sequences) is characterized by abundant normal marine organisms, relatively high petrographic fossil diversity (Table 1), and relatively sparse siliciclastic silt or sand (Table 1). These characteristics are representative of shallow, open-shelf conditions, an environment of deposition confirmed by the occurrence within this group of microfacies of the fusulinid fossil community of Stevens (1965, 1966), shown to have been deposited in offshore, open, normal marine environments.

The second group of microfacies (those generally occurring near the tops and/or bottoms of sequences), in contrast, contains few normal marine organisms, has a low petrographic fossil diversity (Table 1), may contain a fossil community related to the very nearshore euphemitid fossil community (Stevens, 1965, 1966), and has relatively abundant siliciclastic silt or sand (Table 1). These are characteristics of very nearshore, somewhat restricted environments. In addition, the laminated to bioturbated pelleted mudstone-wackestone microfacies (smf-19) and lithoclastic-bioclastic rudstone-floatstone microfacies (smf-24) are considered to be characteristic of restricted lagoons (Wilson, 1975), and the presence of calcite pseudomorphs after gypsum in one sample of smf-19 demonstrates that the pore waters had achieved very high salinity. Occurrence of the microfacies of this second group (ssp, ssf, smf-1-s, smf-19, and smf-24) is consistent with an interpretation of very shallow water environments except, perhaps, for the shallow spiculite. Rocks that could possibly be assigned to this microfacies (smf-1-s) generally have been considered to represent a relatively deep water microfacies (Heath and others, 1967; Rich, 1969; Wilson, 1975), although this interpretation was disputed (Folk and McBride, 1976) for the pelleted spiculitic chert part of the Caballos Novaculite, which lacks radiolarians. The shallow spiculite (smf-1-s) in the Butterfield Peaks Formation is considered to be a very shallow water microfacies because it is most commonly associated with other very shallow nearshore microfacies and also because many of the intraclasts in the coarse lithoclastic-bioclastic rudstone or floatstone microfacies (smf-24) are bent and intruded wafers of the shallow spiculite. The nature of these clasts shows that this sediment was partially lithified before erosion, a condition suggestive of desiccation in the upper part of an intertidal zone.

The nature, stratigraphic distribution, and fossil content of the microfacies in the sequences studied clearly show that each limestone sequence represents a transgressive-regressive event during which deeper water, farther offshore deposits were intercalated between very shallow nearshore ones. The fusulinid-bearing bioclastic wackestone (smf-9-f) is uncommon but, when present, appears near the middle of the limestone sequence. The position of this microfacies suggests that it represents one of the farthest offshore microfacies, and the fusulinid fossil community representing this microfacies previously was interpreted in this manner (Stevens, 1969). Because no deeper water microfacies has been recognized, it can reasonably be assumed that the water depth during deposition of the Butterfield Peaks Formation rarely exceeded that necessary for the development of this fossil community (>13 m).

General Interpretations

Combining information on the petrographic fossil diversity, the abundance of siliciclastic silt, and the actual position of microfacies in the stratigraphic columns, we can

Table 1. AVERAGE PETROGRAPHIC FOSSIL DIVERSITY AND SILICICLASTIC CONTENT OF MICROFACIES OF THE BUTTERFIELD PEAKS FORMATION

Microfacies*	Petrographic fossil diversity	Siliciclastic content
ss	0	high
ssp	1	high
smf-1-s	2	moderate
smf-24	3	moderate
smf-19	4	moderate
ssf	5	high
smf-8	6	low
smf-12	6	low
smf-11	7	moderate
smf-9	7	moderate
smf-9-f	8	low
smf-7	9	low

*ss	siltstone
ssp	peloidal siltstone
ssf	fossiliferous siltstone
smf-1-s	shallow spiculite
smf-7	boundstone
smf-8	whole-fossil wackestone
smf-9	bioclastic wackestone
smf-9-f	fusulinid-bearing bioclastic wackestone
smf-11	coated bioclasts in sparite, grainstone
smf-12	coquina, bioclastic grainstone or rudstone, shell hash
smf-19	laminated to bioturbated pelleted lime mudstone-wackestone
smf-24	coarse lithoclastic-bioclastic rudstone or floatstone

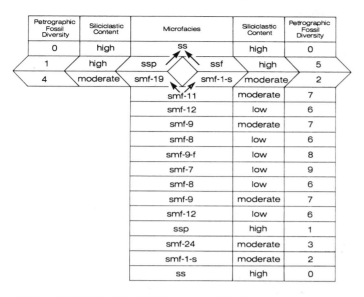

Petrographic Fossil Diversity	Siliciclastic Content	Microfacies		Siliciclastic Content	Petrographic Fossil Diversity
0	high	ss		high	0
1	high	ssp	ssf	high	5
4	moderate	smf-19	smf-1-s	moderate	2
		smf-11		moderate	7
		smf-12		low	6
		smf-9		moderate	7
		smf-8		low	6
		smf-9-f		low	8
		smf-7		low	9
		smf-8		low	6
		smf-9		moderate	7
		smf-12		low	6
		ssp		high	1
		smf-24		moderate	3
		smf-1-s		moderate	2
		ss		high	0

Figure 7. Idealized vertical sequence of microfacies in the limestone sequences in the Middle Pennsylvanian part of the Oquirrh Formation.

derive a rough, idealized sequence of microfacies (Fig. 7). This ideal, although never actually realized, is useful for a detailed interpretation of the environments of microfacies deposition. In general, this sequence is asymmetrical about the fusulinid-bearing bioclastic wackestone microfacies (smf-9-f) which is believed to represent the maximum marine transgression. The siltstone microfacies (ss), which is farthest from smf-9-f, is therefore interpreted as representing the nearest shore environment.

INTERPRETATION OF MICROFACIES

Siltstone Microfacies (ss)

Interpretation of the mode of deposition of the thick siliciclastic units that overlie and underlie the relatively thin limestone intervals in the Butterfield Peaks Formation is difficult because of the paucity of sedimentary structures and the absence of fossils. However, the vertical distribution of microfacies in the limestone sequences puts limits on the possibilities; because the limestone sequences record shallow- to deep- to shallow-water episodes of deposition, the siltstone that occurs above and below the limestone sequences apparently was deposited in more shallow and presumably more landward environments. Thus, the siltstone microfacies (ss) probably represents very nearshore sublittoral sandsheets, beaches, subaerial to submarine parts of deltaic systems, and eolian deposits that accumulated contemporaneously with limestone in farther offshore marine waters. A similar phenomenon is occurring today along the North African Mediterranean coastline (Caulet, 1973).

Peloidal Siltstone Microfacies (ssp)

The well-sorted, laminated, peloidal siltstone microfacies (ssp), containing a few fragments of marine fossils, probably was well worked in a shallow marine environment. Sedimentary structures such as common parallel laminae, low-angle tabular-, wedge-, or lens-shaped cross-laminated sets are suggestive of beach-accretion deposits (Conybeare and Crooke, 1968). The less common high-angle, avalanche-type crosslaminae typically form in migrating sand waves or on tidal bars (Klein, 1970). Because the peloidal siltstone microfacies commonly overlies the possibly subaerial siltstone microfacies and is overlain by various of the marine limestone microfacies, this microfacies probably represents relatively low energy beach to slightly subtidal environments.

Fossiliferous Siltstone Microfacies (ssf)

Interlaminated coarse-silt-size siliciclastic grains and coarser rounded fossil fragments of moderate diversity within crossbedded units in the fossiliferous siltstone microfacies (ssf) suggest a moderately high energy marine environment. Bedding features similar to those of the peloidal siltstone microfacies (ssp) suggest a similar, though probably slightly higher energy, environment.

The fossiliferous siltstone microfacies occurs near the tops and bottoms of the limestone sequences and indicates extensive physical reworking; these observations suggest that this microfacies represents a moderately high energy nearshore bar or possibly beach deposit.

Shallow Spiculite Microfacies (smf-1-s)

The shallow spiculite microfacies (smf-1-s), essentially identical to that which has been interpreted as a relatively deep, offshore facies in some former studies (Heath and others, 1967; Rich, 1969), is here interpreted to represent one of the most nearshore restricted environments, for the following reasons: (1) it commonly is associated with other microfacies considered to represent very nearshore environments; (2) it contains a very limited fauna of mostly euryhaline forms, including elements of the nearshore euphemitid fossil community; (3) it resembles other nearshore microfacies in that it commonly contains abundant siliciclastic grains; and (4) it evidently was subjected to erosion because flakes of this microfacies have been redeposited into the coarse lithoclastic-bioclastic rudstone or floatstone microfacies (smf-24), a probable tidal channel deposit.

This microfacies is interpreted to represent deposition in extremely shallow coastal embayments generally formed during early stages of marine transgression or in the last stages of shoreline progradation. In part, deposition may have been at the margins of tidal flats where flakes of this microfacies were removed and redeposited in the coarse lithoclastic-bioclastic rudstone or floatstone microfacies (smf-24).

Boundstone Microfacies (smf-7)

Rocks of the boundstone microfacies (smf-7) contain a wide diversity of marine fossils adapted to relatively shallow marine waters. Evidently the absence of siliciclastic material, perhaps due to weak current or wave action, favored the development of carbonate buildups formed partly by encrusting organisms, especially calcareous sponges. These organisms also probably trapped lime mud carried by weak currents. Distribution of this microfacies in the limestone sequences suggests deposition in water slightly more shallow than for the fusulinid-bearing bioclastic wackestone microfacies (smf-9-f), which is interpreted to have been deposited in water deeper than 13 m, the minimum depth at which fusulinids flourished (Stevens, 1969).

Whole-Fossil Wackestone Microfacies (smf-8)

The whole-fossil wackestone microfacies (smf-8) contains a relatively large population of pelmatozoans, brachiopods, and bryozoans; solitary rugose corals, some in growth position, also are abundant. The high percentage of micritic matrix, the numerous articulated fossils, the spar-filled voids under bivalve roofs, and the very low percentage of siliciclastic particles suggest that the fauna is in place and that the water was not highly agitated.

The presence of elements of the productid-*Composita* fossil community and the occurrence of this microfacies both above and below (and, thus, presumably shoreward of) the fusulinid-bearing bioclastic wackestone microfacies (smf-9-f) suggest a quiet, relatively shallow water environment of deposition. Presumably, the absence of fusulinids is due to a water depth less than 13 m.

Bioclastic Wackestone Microfacies (smf-9)

The bioclastic wackestone microfacies (smf-9) contains a relatively large population of pelmatozoans, brachiopods, and bryozoans. The presence of these suspension feeders indicates a quiescent, yet oxygenated, normal marine environment. The moderately low concentration of siliciclastic grains also suggests low-energy conditions. The presence of most elements of the productid-*Composita* fossil community and the occurrence of this microfacies adjacent to the fusulinid-bearing bioclastic wackestone microfacies (smf-9-f) within two sequences suggest a moderately shallow, nearshore environment of deposition. The absence of fusulinids is attributed to water depths shallower than those ordinarily inhabited by those organisms.

Fusulinid-Bearing Bioclastic Wackestone Microfacies (smf-9-f)

Recognition of the fusulinid-bearing bioclastic wackestone microfacies (smf-9-f) depends on the presence of moderately abundant mature fusulinids, which are utilized because of their demonstrated importance in interpreting paleobathymetry (Stevens, 1969). Detailed study of their occurrence in Middle Pennsylvanian rocks of Colorado indicates that these fusulinids lived in normal marine waters greater than 13 m deep (Stevens, 1969), and Permian fusulinids in Nevada have been shown to have ranged in water depth to at least 35 m (Yancey and Stevens, 1981). Intervals containing fusulinids occur, without exception, in the middle parts of the limestone sequences studied. Because no indications have been found in the Butterfield Peaks Formation in either the Stansbury or Onaqui Mountains to suggest that water depths exceeded those in which the fusulinids lived, it seems reasonable to surmise that water depths were never greater than about 35 m during deposition of these Middle Pennsylvanian rocks. Furthermore, because the fusulinid-bearing bioclastic wackestone microfacies represents maximum marine inundation and is poorly represented within the few sequences where it has been found, it seems probable that most deposition of this microfacies was close to the minimum depth for development of the fusulinid fossil community.

Coated-Bioclasts-in-Sparite, Grainstone Microfacies (smf-11) with Interbedded Laminae of the Coated and Worn Bioclasts in Micrite, Packstone-Wackestone Microfacies (smf-10)

The coated-bioclasts-in-sparite, grainstone microfacies was deposited in a shallow marine environment in which energy levels fluctuated, as indicated by the presence of mudstone and grainstone layers within the same samples. Fossil diversity is relatively high, but the high degree of fragmentation and abrasion indicates that all these fossils have been transported. The dominant lithology of well-sorted, fine-sand-size, coated allochems cemented by spar shows that the water was sufficiently agitated to winnow out smaller particles. In addition, the grains must have been disturbed regularly so that complete envelopes developed around most grains.

The average grain size of the rock in this microfacies is considerably smaller than the common maximum grain diameter (1 mm) of oolites. Bathurst (1967) suggested that "young embryonic ooids" similar to those reported here may form in water depths of 10 m or more, but it is just as likely that the small grain size is due to diminished wave energy in very shallow water. This latter interpretation is favored here because in sequence 2, where this microfacies is well represented, the rock is overlain by the restricted shallow laminated to bioturbated pelleted lime mudstone-wackestone (smf-19) and peloidal siltstone microfacies (ssp), which, in turn, are overlain by the nearshore to subaerial siltstone microfacies (ss).

The rapidly changing energy conditions thus indicated could be obtained on the shoreward side of a bar or shoal and be a function of tide levels; during low tide the shallow water shoreward of the shoal might have been calm so that micritic material settled from suspension, but during high tide, when the bottom was no longer protected from the waves, fine material was winnowed out, leaving a well-washed grainstone lamina behind. This general environmental setting is consistent with the presence of dasycladacean algae, which typify extremely shallow water sediment deposited in tidal bars or lagoonal channels (Wilson, 1975). A modern analog, not unlike this microfacies, is the interior sand on the lee of barriers in Bahamian-type carbonate deposits (Ball, 1967). This microfacies is well represented in sequence 2, where it makes up about 6 m of section, the top of which is about 2 m below the possibly subaerially deposited siltstone microfacies.

Coquina, Bioclastic Grainstone or Rudstone, Shell Hash Microfacies (smf-12)

The coquina, bioclastic grainstone or rudstone, shell hash microfacies (smf-12), consisting of grainstone and packstone containing a diverse assemblage of reworked normal marine fossils with micritized rims, and the position of this microfacies within the limestone sequences (Figs. 6, 7) suggest deposition under relatively high energy conditions above wave base on a shallow marine shelf. This microfacies is interpreted to represent a shallower, higher energy marine environment than the boundstone, whole-fossil wackestone, bioclastic wackestone, and fusulinid-bearing bioclastic wackestone microfacies (smf-7, smf-8, smf-9, smf-9-f, respectively).

Laminated to Bioturbated Pelleted Lime Mudstone-Wackestone Microfacies (smf-19)

The laminated to bioturbated pelleted lime mudstone-wackestone microfacies (smf-19) consists of largely unfossiliferous, pelleted mudstone and wackestone. Pseudomorphs after gypsum demonstrate that the pore waters once had a very high salinity, a condition common in a tidal flat environment.

This microfacies occurs only at or near the tops of the limestone sequences. Thus, the interpretation that this microfacies was deposited in restricted bays and ponds (Wilson, 1975), formed during the last stages of marine deposition, is favored here. This microfacies probably represents the final filling in of depressions as the shoreline prograded across the area.

Coarse Lithoclastic-Bioclastic Rudstone or Floatstone Microfacies (smf-24)

The coarse lithoclastic-bioclastic rudstone or floatstone microfacies (smf-24), composed chiefly of intraclasts and skeletal fragments, indicates fairly rapid deposition in a high-energy environment. Organic remains with mud-fill, but spar cement, suggest that the fossils were deposited first in a muddy matrix and were reworked later by currents that left a concentration of relatively large, well-washed clasts. The waferlike intraclasts, some clearly pieces of the shallow spiculite (smf-1-s), are similar to mud chips derived from desiccated surfaces. Laporte (1967) described reworked flat-pebble limestone clasts and shelly material which he thought accumulated where tidal flow was concentrated, that bear a close resemblance to this microfacies. Point bars in tidal channels on Andros Island, composed of coquinas and soft-mud clasts (Shinn and others, 1969), support this conjecture.

This microfacies occurs primarily near the bottoms of the limestone sequences and most commonly is associated with the shallow spiculite microfacies. This relation suggests that deposition as point bars in tidal channels is a realistic interpretation. This is especially attractive because the shallow spiculite microfacies probably represents deposition on the margins of, and in, bays and tidal channels. Probably many of the meandering tidal channels undercut banks and thus created the intraclasts that were incorporated into this microfacies.

SEDIMENTARY MODELS

Depositional models can be constructed from the inferred environments of deposition of each microfacies and their vertical distribution. Using Walthers's law to explain their superposition, the microfacies in the columnar sections are related to their distribution within schematic block diagrams (Fig. 8). Relatively small geomorphic features on the Oquirrh shelf, such as lagoons, tidal channels, and oolite shoals, probably migrated parallel as well as perpendicular to the strandline through time, or were discontinuous in time and space; therefore, it should not be assumed that all the microfacies representing these features are as laterally persistent as shown in Figure 8. Common pinching out of individual beds in the Butterfield Peaks Formation within a few tens to hundreds of meters, in fact, demonstrates the lack of persistence of some of the units parallel to depositional strike. Sequences 2 and 5 are illustrated here because they show relatively simple patterns of deposition. Only those microfacies deposited during the transgression are shown in the block diagrams; later deposits would follow a similar pattern, in reverse.

Figure 9 models the vertical arrangement of the major microfacies as they would appear in the transgressive part

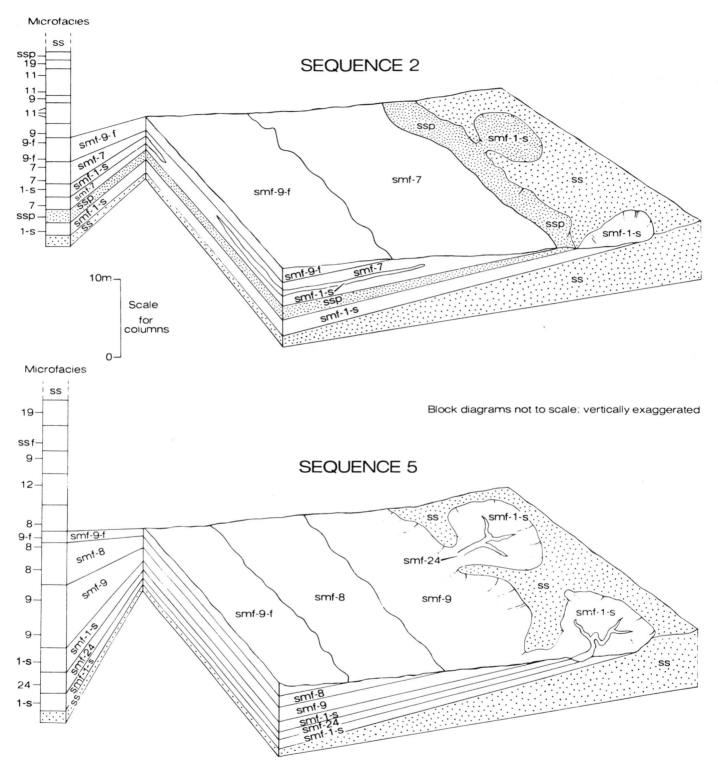

Figure 8. Depositional models of two limestone sequences in the Butterfield Peaks Formation. Columns on the left represent the carbonate sequences; tic marks with microfacies numbers show location of thin sections. Block diagrams show how microfacies are superimposed during a marine transgression. In each sequence, the tops of blocks represent the time of maximum marine transgression in that sequence. Upper part of each column formed during a regression is a crude mirror image of lower part.

Microfacies

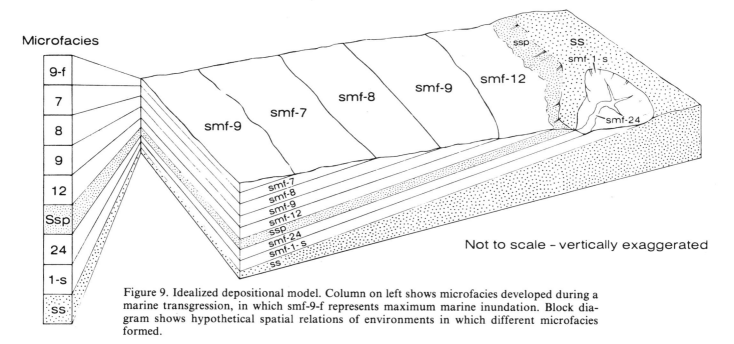

Figure 9. Idealized depositional model. Column on left shows microfacies developed during a marine transgression, in which smf-9-f represents maximum marine inundation. Block diagram shows hypothetical spatial relations of environments in which different microfacies formed.

of the idealized sequence. Several microfacies not shown on the block diagram are those present only in the upper parts of the limestone sequences, because of evolution of the geomorphology.

CONCLUSIONS

The results of the present study of microfacies of the Butterfield Peaks Formation indicate that at least 12 microfacies are present: siltstone (ss); peloidal siltstone (ssp); fossiliferous siltstone (ssf); shallow spiculite (smf-1-s); boundstone (smf-7); whole-fossil wackestone (smf-8); bioclastic wackestone (smf-9); fusulinid-bearing bioclastic wackestone (smf-9-f); coated bioclasts in sparite, grainstone (smf-11); coquina, bioclastic grainstone or rudstone, shell hash (smf-12); laminated to bioturbated pelleted lime mudstone-wackestone (smf-19); and coarse lithoclastic-bioclastic rudstone or floatstone (smf-24).

Within individual limestone sequences in the Butterfield Peaks Formation the vertical sequence of microfacies, fossil communities, sedimentary structures, and abundance of siliciclastic detritus are arranged in a somewhat imperfect mirror image about the middle of the sequence. The succession of microfacies and all other data from the lower parts of the limestone sequences suggest a relatively simple transgressive episode; the upper part, in contrast, is most easily explained by progradation of nearshore and, later, subaerial environments. The limestone sequences commonly are capped by the laminated to bioturbated pelleted lime mudstone-wackestone microfacies (smf-19) which probably filled depressions left behind by the prograding

shoreline before these deposits were covered by the siltstone microfacies, deposited at least partly under subaerial conditions.

ACKNOWLEDGMENTS

We are grateful to David Andersen, San Jose State University, California, and to Gus Armstrong and Ken Bird, U.S. Geological Survey, Menlo Park, California, for discussions on environments of deposition and for critical readings of the manuscript.

REFERENCES CITED

Armin, R. A., 1978, Correlation and paleoenvironment of the Oquirrh Group in the Stansbury and Onaqui Mountains, central Utah: Geological Society of America Abstracts with Programs, v. 10, p. 210.

—— 1979, Geology of the southeastern Stansbury Mountains and southern Onaqui Mountains, Tooele County, Utah, with a paleoenvironmental study of part of the Oquirrh Group [M.S. thesis]: San Jose, California, San Jose State University, 105 p.

Armin, R. A., and Moore, W. J., 1981, Geology of the southeastern Stansbury Mountains and southern Onaqui Mountains, Tooele County, Utah: U.S. Geological Survey Open-File Report 81-247, p. 1–28.

Ball, M. M., 1967, Carbonate sand bodies of Florida and the Bahamas: Journal of Sedimentary Petrology, v. 37, p. 556–591.

Bathurst, R.G.C., 1967, Oolitic films on low energy carbonate sand grains, Bimini Lagoon, Bahamas: Marine Geology, v. 5, p. 89–109.

Caulet, J. P., 1973, Recent biogenic calcareous sedimentation on the Algerian continental shelf, *in* Stanley, D. J., ed., The Mediterranean Sea: A natural sedimentary laboratory: Stroudsburg, Pennsylvania, Dowden, Hutchinson Ross, p. 261–277.

Conybeare, C.E.B., and Crook, K.A.W., 1968, Manual of sedimentary structures: Australia Department of National Development, Bureau of Mineral Resources, Geology and Geophysics Bulletin 102, 327 p.

Folk, R. L., and McBride, E. F., 1976, The Caballos Novaculite revisited. Part 1: Origin of Novaculite Member: Journal of Sedimentary Petrology, v. 46, p. 659–669.

Heath, C. P., Lumsden, D. N., and Carozzi, A. V., 1967, Petrography of a carbonate transgressive-regressive sequence—the Bird Spring Group (Pennsylvanian), Arrow Canyon Range, Clark County, Nevada: Journal of Sedimentary Petrology, v. 37, p. 377–400.

Jordon, T. E., and Allmendinger, R. W., 1979, Upper Permian and Lower Triassic stratigraphy of the Stansbury Mountains: Utah Geology, v. 6, no. 2, p. 69–74.

Klein, G. D., 1970, Depositional and dispersal dynamics of intertidal sand bars: Journal of Sedimentary Petrology, v. 10, p. 1095–1127.

Laporte, L. F., 1967, Carbonate deposition near mean sea level and resultant facies mosaic—Manlius Formation (Lower Devonian) of New York State: American Association of Petroleum Geologists Bulletin, v. 51, p. 73–101.

Nolan, T. B., 1930, Paleozoic formations of the Gold Hill quadrangle, Utah: Washington Academy of Sciences Journal, v. 20, p. 421–432.

Pettijohn, F. J., Potter, P. E., and Siever, R., 1973, Sand and Sandstone: New York, Springer-Verlag, 619 p.

Rich, Mark, 1969, Petrographic analysis of Atokan carbonate rocks in central and southern Great Basin: American Association of Petroleum Geologists Bulletin, v. 53, p. 340–366.

Roberts, R. J., Jr., Crittenden, M. D., Jr., Tooker, E. W., Morris, H. T., Hose, R. K., and Cheney, T. M., 1965, Pennsylvanian and Permian basins in northwestern Utah, northeastern Nevada, and south-central Idaho: American Association of Petroleum Geologists Bulletin, v. 49, p. 1926–1956.

Shinn, E. A., Lloyd, R. M., and Ginsberg, R. N., 1969, Anatomy of a modern carbonate tidal-flat, Andros Island, Bahamas: Journal of Sedimentary Petrology, v. 39, p. 1202–1228.

Smosna, Richard, and Warshauer, S. M., 1978, Fossil diversity in thin section: Journal of Sedimentary Petrology, v. 48, p. 331–336.

Stevens, C. H., 1965, Faunal trends in near-shore Pennsylvanian deposits near McCoy, Colorado: Mountain Geologist, v. 2, p. 71–77.

——1966, Paleoecologic implications of Early Permian fossil communities in eastern Nevada and western Utah: Geological Society of America Bulletin, v. 77, p. 1121–1130.

——1969, Water depth control of fusulinid distribution: Lethaia, v. 2, p. 121–132.

Tooker, E. W., and Roberts, R. J., 1970, Upper Paleozoic rocks in the Oquirrh Mountains and Bingham mining district, Utah, with a section on biostratigraphy and correlation, by Mackenzie Gordon, Jr., and H. M. Duncan: U.S. Geological Survey Professional Paper 629-A, p. A1–A76.

Welsh, J. E., 1979, Paleogeography and tectonic implications of the Mississippian and Pennsylvanian in Utah, in Newman, G. W., and Goode, H. D., eds., Basin and Range Symposium and Great Basin Field Conference: Denver, Colorado, Rocky Mountain Association of Geologists, p. 93–106.

Welsh, J. E., and James, A. H., 1961, Pennsylvanian and Permian stratigraphy of the central Oquirrh Mountains, in Cook, D. R., ed., Geology of the Bingham mining district and northern Oquirrh Mountains, Utah: Utah Geological Society Guidebook to the Geology of Utah, no. 16, p. 1–16.

Wilson, J. L., 1975, Carbonate facies in geologic history: New York, Springer-Verlag, 471 p.

Yancey, T. E., and Stevens, C. H., 1981, Early Permian fossil communities in northeastern Nevada and northwestern Utah, in Gray, J., Boucot, A. J., and Berry, W.B.N., eds., Communities of the past: Stroudsburg, Pennsylvania, Hutchinson Ross Publishing Company, p. 243–269.

MANUSCRIPT ACCEPTED BY THE SOCIETY AUGUST 20, 1982

Geological Society of America
Memoir 157
1983

Structural geology of the southern Sheeprock Mountains, Utah: Regional significance

Nicholas Christie-Blick
Exxon Production Research Co.
P.O. Box 2189
Houston, Texas 77001

ABSTRACT

The Sheeprock Mountains are part of a horst of Proterozoic, Paleozoic and Cenozoic sedimentary and igneous rocks located in the transitional region between the Cordilleran fold-thrust belt and the hinterland in the Basin-Range province of west-central Utah.

Prominent structural elements in the Sheeprock Mountains are the Sheeprock Thrust, juxtaposing Proterozoic rocks above Paleozoic ones with a stratigraphic separation exceeding 10 km; the Pole Canyon Thrust, thought to be an upper plate imbrication of the Sheeprock Thrust; the Pole Canyon Anticline, a recumbent fold vergent to the northeast and cut by the Pole Canyon Thrust; the east-northeast-striking Indian Springs (tear) Fault; and two low-angle normal faults (the Harker and Lion Hill Faults) which together account for stratigraphic omission of several kilometres.

The Pole Canyon Anticline is thought to have developed in the late Mesozoic during propagation of the thrusts parallel to the Indian Springs Fault, and this transport direction is corroborated by minor structures. Fault geometry suggests that the Harker and Lion Hill Faults are younger than the thrusts and probably of late Cenozoic age, although some mid-Cenozoic or even earlier displacement cannot be entirely ruled out.

My preferred interpretation of the structural history of the Sheeprock Mountains is consistent with minimal regional extension before the mid-Cenozoic and with the view that crustal shortening in the fold-thrust belt is for the most part unrelated to hinterland extension.

INTRODUCTION

The Sheeprock Mountains are part of a northwest- to north-trending horst of little-metamorphosed Upper Proterozoic[1] to Paleozoic and Cenozoic sedimentary and igneous rocks in the Basin-Range province of west-central Utah (Fig. 1). The internal structure of the range is dominated by low-angle faults. Some place older rocks on younger ones, and with one possible exception are thrusts.

Other low-angle faults place younger rocks on older ones, with stratigraphic omission, and are inferred to have been produced by extension. The Sheeprock Mountains are therefore transitional, both geographically and structurally, between the Cordilleran fold-thrust belt to the east (see King, 1969) and a hinterland of open folds, metamorphic core complexes and predominantly extensional structures to the west (Armstrong, 1972; Coney, 1980).

This paper has three objectives: (1) to provide an up-to-date description of the structure of the southern Sheeprock Mountains; (2) to present new interpretations of the structural history; and (3) to consider how structural rela-

[1]Upper Proterozoic is used in the sense of Harrison and Peterman (1980), referring to the interval between 900 and 570 Ma ago. Symbols used on figures in this paper are based on the approximately equivalent term of the U.S. Geological Survey, Proterozoic-Z (James, 1972, 1978), which embraces the interval from 800 to 570 Ma ago.

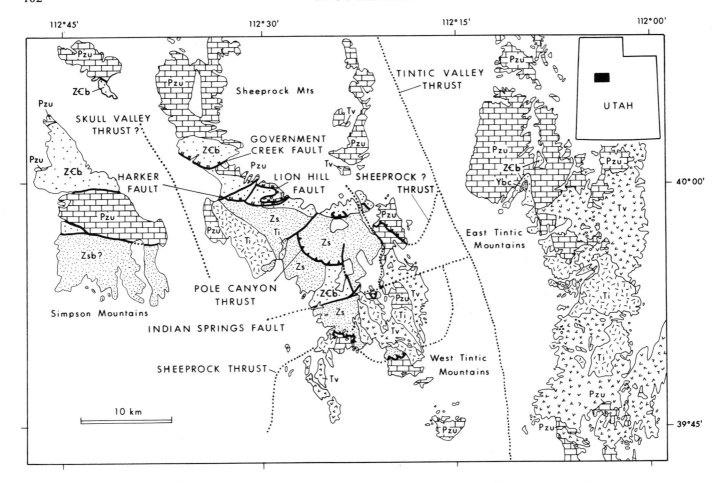

Figure 1. Location map of the Sheeprock Mountains in west-central Utah showing major faults and gross rock units (modified from Cohenour, 1959; Stokes, 1963; Morris, 1977; Moore and Sorensen, 1979). Faults (bold contacts, dotted where concealed): thrust (sawteeth on upper plate); low-angle extension fault (double ticks on hanging-wall block); high-angle fault (no ornament). Rock units: Ybc, Big Cottonwood(?) Formation (Proterozoic-Y); Zs, Sheeprock Group (Proterozoic-Z); ZCb, Brigham Group (Proterozoic-Z to Cambrian); Zsb, undifferentiated Sheeprock and Brigham Groups; Pzu, undifferentiated Paleozoic rocks; Ti, Tertiary intrusive rocks including intrusion breccia (large body on southwestern flank of Sheeprock Mountains is Sheeprock Granite); Tv, Tertiary extrusive rocks; unpatterned area, undifferentiated Tertiary and Quaternary sediments.

tions in the Sheeprock Mountains bear on the continuing controversy about the relation between thrusts and low-angle normal faults in the Basin-Range province (e.g., Armstrong, 1972; Crittenden, 1979; Compton and Todd, 1979; Coney, 1980; Allmendinger and Jordan, 1981; Wernicke, 1982).

Conclusions presented here are based largely on geologic mapping at 1:24,000 scale of the southern Sheeprock Mountains between latitudes 40°00′ and 39°54′N. by Blick (1979), and of the southernmost part of the range and adjacent West Tintic Mountains by Morris and Kopf (1970a, 1970b).

TERMINOLOGY OF LOW-ANGLE FAULTS

It is now widely recognized that low-angle faults can originate in both extensional and contractional regimes, but there is no consensus about terminology appropriate for such faults. Purely descriptive terms are useful in the initial stages of geologic mapping, but genetic terms are desirable as we formulate kinematic interpretations of structural history. This is especially the case where rocks have been folded and faulted at different times and where faults initiated under one stress regime have been reactivated under a different regime (Dahlstrom, 1970; Armstrong, 1972). In these cases, the relative ages of hanging-wall and footwall blocks at a particular locality may be a fallible guide to fault genesis. For example, Figure 2, based on faults in the Sheeprock Mountains, illustrates how reverse fault geometry[2] can locally arise from normal slip (and hypothetically, vice versa) if reference surfaces are appropriately folded or tilted prior to fault movement. However, crustal extension and contraction can generally be distinguished from observations of structural relations over a large area, providing the timing of deformation is known.

GEOMETRY   GENESIS

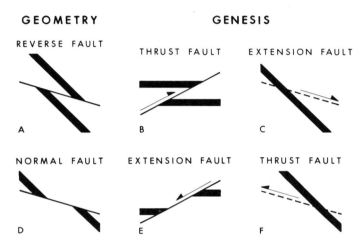

Figure 2. Vertical views of a low-angle reverse fault (A) and a low-angle normal fault (D) transverse to fault strike, and their genesis as either thrusts (B and F) or extension faults (C and E). It is assumed that in each case all separation is achieved on the fault illustrated by slip in the plane of the page. In cases B and E, faulted beds are subsequently folded or tilted, whereas in C and F, folded or tilted beds are subsequently offset by a fault.

Among descriptive terms for gently dipping faults, those indicating the relative age of juxtaposed rocks are probably most useful, although somewhat unwieldy (e.g., low-angle reverse fault, low-angle normal fault, older-over-younger fault, younger-over-older fault). Other terms used recently in the Basin-Range province are imprecisely defined or not particularly descriptive, and some have assumed genetic implications that may be inappropriate in the Sheeprock Mountains (R. G. Bohannon, 1982, personal commun.). For example, "dislocation surface" (Rehrig and Reynolds, 1980) is practically synonymous with fault; "detachment fault" and its synonym "décollement" imply that rocks above and below the fault are characterized by independent styles of deformation (de Sitter, 1964; Bates and Jackson, 1980; Davis and others, 1980); the term "denuda-

tion fault" (Moores and others, 1970; Armstrong, 1972) seems to imply erosion or exposure of lower plate rocks as a result of fault movement.

In this paper, conspicuous low-angle faults in the Sheeprock Mountains are interpreted as "thrusts" and "extension faults," defined kinematically as suggested by McClay (1981).[3] A thrust is a map scale contraction fault which shortens an arbitrary datum, commonly but not necessarily bedding. An extension fault extends an arbitrary datum such as bedding. As noted above, not all thrust segments are reverse faults (e.g., Fig. 2D, 2F) and not all extension faults are normal faults (Fig. 2A, 2C), because bedding may not be an appropriate datum in rocks that are already deformed. Observations, other than the relative age of juxtaposed rocks, such as fold vergence and the timing of deformation, may bear on fault interpretation. However, in spite of possible ambiguities, the terms "thrust" and "extension fault" are valuable, because they allow two genetically distinct classes of fault to be distinguished in interpretations without invoking any particular regional model.

The kinematic terms "normal-slip fault" and "reverse-slip fault" (Hill, 1959), though valid, are avoided in this paper, because where applied to tilted or folded faults, it is unclear whether normal and reverse refer to the orientation of the fault during displacement or to its present orientation.

PREVIOUS WORK

Early descriptions of rocks in the Sheeprock and adjacent West Tintic Mountains are by Loughlin (1920), Eardley and Hatch (1940), Stringham (1942) and Gardner (1954). The stratigraphy was established in a definitive work by Cohenour (1959), and his scheme was largely followed by Groff (1959) and by Morris and Kopf (1967, 1970a, 1970b). Harris (1958) independently established an alternative terminology for Proterozoic rocks in the vicinity of Dutch Peak (see Fig. 3 for location), but there are difficulties with his nomenclature as a result of structural complications, which he did not fully understand (Christie-Blick, 1982). The stratigraphy of Upper Proterozoic and Lower Cambrian rocks, which underlie much of the southern Sheeprock Mountains (Fig. 1), has been revised extensively by Christie-Blick (1982), in part, based on new understanding of the geologic structure. This paper therefore constitutes a companion to the stratigraphic summary presented in Christie-Blick (1982).

Essential features of the structure of the Sheeprock Mountains have been known for many years. Loughlin (1920) recognized that in the northern part of the Cherry Creek Quadrangle (Fig. 3), Proterozoic quartzite is juxtaposed by a thrust above Paleozoic limestone. The fault,

---

[2]Following Hill (1959) and Crowell (1959), in this paper "reverse fault" and "normal fault" are geometric terms describing dip separation. A fault that dips less steeply than bedding is here termed a normal fault if younger rocks are juxtaposed on older ones, and a reverse fault if the opposite is true.

---

[3]The terms "thrust" and "extension fault" may be applied to faults of any dip.

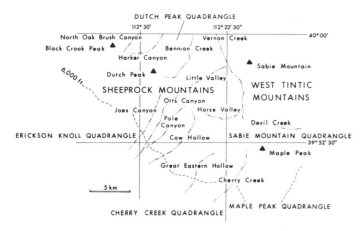

Figure 3. Topographic features of the southern Sheeprock Mountains and adjacent part of the West Tintic Mountains mentioned in this paper. The ranges are separated by Vernon Creek and Cherry Creek.

named the Sheeprock Thrust by Eardley (1939), was subsequently delineated in mapping by Stringham (1942), Gardner (1954), Cohenour (1959), Groff (1959), and Morris and Kopf (1970a, 1970b) (Figs. 1, 4). Near the drainage divide of the southern Sheeprock Mountains, Cohenour (1959) mapped and named the Pole Canyon Thrust (Figs. 1, 4), which he regarded as structurally above the Sheeprock Thrust. However, Harris (1958) and later Armstrong (1968) interpreted the two faults as one. Cohenour (1959) was the first to recognize that beds beneath the Pole Canyon Thrust are overturned and dip toward the west, whereas upper plate rocks are mostly upright and dip to the northeast. He also established the presence of a major steeply dipping fault between the traces of the Pole Canyon and Sheeprock Thrusts, which Morris and Kopf (1967) termed the Indian Springs Fault. Groff (1959) and Morris and Kopf (1970a) interpreted this fault as a tear, although they disagreed about its slip sense, whereas Armstrong (1968) considered it another segment of a folded Sheeprock Thrust. In the northern part of the Sheeprock Mountains, Cohenour (1959) recognized and named the Government Creek Fault (Fig. 1), a third fault that juxtaposes older rocks above younger ones, but a fault that he considered to have moved by strike slip. In addition to these faults that repeat the stratigraphic section, Cohenour (1959) also mapped several (including the Lion Hill "Thrust"; Figs. 1, 4) that attenuate it. However, as was standard procedure at the time, he regarded all gently dipping faults (except the Government Creek Fault) as thrusts whatever the relative age of hanging-wall and footwall blocks.

STRATIGRAPHY

The Sheeprock Mountains are underlain by a miogeoclinal section of little-metamorphosed Upper Protero-

zoic to Lower Cambrian clastic sedimentary rocks and Paleozoic carbonate rocks with an aggregate thickness of over 12,500 m (Fig. 5; Hintze, 1973; Christie-Blick, 1982). The southern flank of the range and much of the adjacent West Tintic Mountains are overlain and intruded by Oligocene(?) to Miocene volcanic and plutonic rocks (Fig. 1; Morris and Kopf, 1967, 1970a, 1970b).

The Upper Proterozoic to Lower Cambrian beds are as thick as 7,200 m in the southern Sheeprock Mountains, and Christie-Blick (1982) has suggested detailed correlations with formations established in northern Utah and southeastern Idaho by Crittenden and others (1971). The Sheeprock sequence begins at the base with 2,700 to 4,300 m of phyllite, quartzite, glaciomarine diamictite and shale assigned to the Otts Canyon, Dutch Peak, and Kelley Canyon Formations of the Sheeprock Group (Fig. 5). These units are overlain by 1,950 to 4,000 m of quartzite and minor shale assigned to the Caddy Canyon Quartzite, Inkom and Mutual(?) Formations and Prospect Mountain Quartzite, which together constitute the Brigham Group. Details about regional and local lateral variations of thickness and facies within these rocks may be found in Blick (1979), Christie-Blick (1982), and Crittenden and others (1983).

The Proterozoic stratigraphy in the Sheeprock Mountains, although relatively coherent for more than 350 km to the north, is markedly different from that in the East Tintic Mountains, only 20 km to the east (Fig. 1). There, Lower Cambrian Tintic Quartzite overlies 510 m of quartzite and shale, correlated by Morris and Lovering (1961) with the Middle Proterozoic Big Cottonwood Formation. Upper Proterozoic rocks, thicker than 6,000 m in the Sheeprock Mountains, are apparently missing at a contact interpreted to be a regional unconformity. This stratigraphic contrast, along with differences in the thickness and facies of Upper Ordovician to Devonian carbonate rocks (Morris and Kopf, 1969), suggests several tens of kilometres displacement on the intervening but mostly concealed Tintic Valley Thrust (Morris, 1977).

Proterozoic rocks of the Sheeprock Mountains also differ from those of the southern Simpson Mountains, about 10 km to the west (Fig. 1), where Thomas (1958) described a sequence estimated by H. T. Morris (1978, personal commun.) to be about 5,000 m thick. The age of this sequence is uncertain, but at least two interpretations are possible (Blick, 1979). One is that the rocks of the Simpson Mountains are temporally equivalent to the Sheeprock and Brigham Groups. Lateral facies changes implicit in this correlation are somewhat abrupt, suggesting the presence of an intervening thrust, perhaps the Skull Valley Thrust of Roberts and others (1965) (Fig. 1). For example, graywacke of unit I of Thomas (1958) may be equivalent to the upper part of the Dutch Peak Formation; unit II (siltstone and shale), to the Kelley Canyon Formation; and unit V

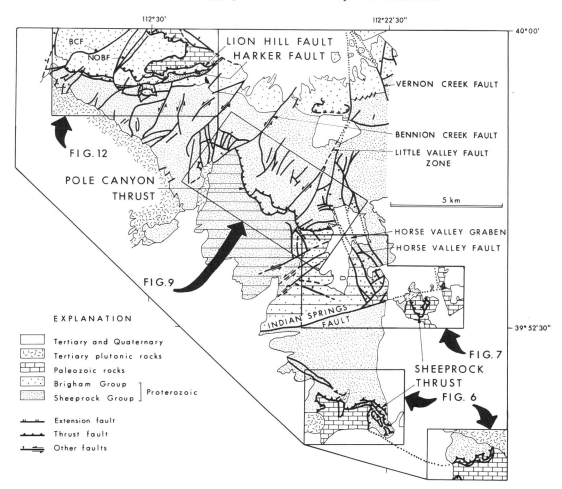

Figure 4. Simplified fault map of the southern Sheeprock Mountains, indicating the location of more detailed geologic maps illustrated in Figures 6, 7, 9, 12 (after Blick, 1979, and unpublished mapping; Morris and Kopf, 1970a, 1970b). Horizontal ruling indicates overturned rocks between the Sheeprock and Pole Canyon Thrusts. Abbreviations for faults: BCF, Black Crook Fault; NOBF, North Oak Brush Fault.

(quartzite), to the Caddy Canyon Quartzite. Siltstone and sandstone of unit IV resemble the Papoose Creek Formation of Crittenden and others (1971), but that formation is poorly developed in the Sheeprock Mountains (Christie-Blick, 1982). Units III (quartzite) and VI (mainly argillite) are unknown in the Sheeprock Mountains. A second interpretation is that the Simpson Mountains sequence is correlative with the Middle Proterozoic Big Cottonwood Formation as in the East Tintic Mountains. However, such a correlation implies exhumation to a structurally very deep level in comparison with adjacent ranges. If Middle Proterozoic, the Simpson Mountains sequence should either stratigraphically underlie a thick section of Upper Proterozoic and Paleozoic rocks, comparable to that in the Sheeprock Mountains, or occur structurally beneath the Tintic Valley and/or Sheeprock Thrusts. Note that this lat-

ter possibility might involve unrealistically large displacements on these thrusts (see Fig. 1 for scale).

Paleozoic beds overlying the Prospect Mountain Quartzite in the Sheeprock Mountains are at least 5,300 m thick even if the incomplete section of Pennsylvanian-Permian Oquirrh Group is excluded (Hintze, 1973). Paleozoic rocks are best exposed in the northern Sheeprock Mountains (Fig. 1; Cohenour, 1959), but they also occur in the southernmost part of the range, in the lower plate of the Sheeprock Thrust (Morris and Kopf, 1970a, 1970b). The Cambrian beds consist of thin shale and limestone with minor quartzite (340 m), overlain by thick limestone and dolomite (1,400 m), and have been subdivided into 12 formations (Fig. 5; Cohenour, 1959; Hintze, 1973). The Ordovician rocks in the northern Sheeprock Mountains include, from oldest to youngest, argillaceous and sandy limestone

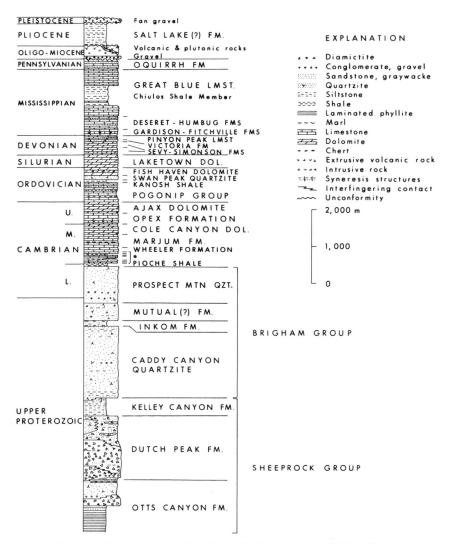

Figure 5. Composite stratigraphy of the Sheeprock and West Tintic Mountains (simplified from Hintze, 1973; Christie-Blick, 1982). *Unlabeled Cambrian formations between the Pioche Shale and the Wheeler Formation, from oldest to youngest: Tatow Formation, Millard-Howell Limestone, Chisholm Shale, Dome Limestone, Whirlwind Formation, and Swasey Limestone. Note that the Cambrian stratigraphy may require revision in the light of the work of Hintze and Robison (1975) in the House Range, 100 km southwest of the Sheeprock Mountains.

of the Pogonip Group (520 m), the Kanosh Shale (75 m), the Swan Peak Quartzite (100 to 140 m), and the Fish Haven Dolomite (215 m). A comparable but incomplete and faulted Ordovician section, totaling 1,165 m, occurs in the lower plate of the Sheeprock Thrust (Groff, 1959; Morris and Kopf, 1970a, 1970b). The Silurian is thin (275 to 430 m) and is represented by a single formation, the Laketown Dolomite. From oldest to youngest, the Devonian consists of the Sevy Dolomite and Simonson Dolomite (together about 250 to 450 m), quartzite of the Victoria Formation and the Pinyon Peak Limestone, together

about 175 m thick in the northern Sheeprock Mountains, but only 70 m thick beneath the Sheeprock Thrust (Hintze, 1973; Morris and Kopf, 1970a, 1970b). The Mississippian is thick (1,800 m) and mostly limestone, except for the Chiulos Shale Member (550 m) of the Great Blue Formation (Fig. 5). The Pennsylvanian section is not complete in the Sheeprock Mountains.

Deformed Proterozoic and Paleozoic strata are unconformably overlain and intruded by various volcanic and plutonic rocks which from regional arguments are of probably Oligocene to Miocene age (McKee, 1971; Lipman and others, 1972; Lindsey and others, 1975). These igneous rocks include latite; quartz latite; monzonite and quartz monzonite porphyry; quartz diorite porphyry; aplite in dikes, sills and plugs; olivine basalt dikes and one flow; rhyolitic tuff; and intrusion breccias of both rhyolitic and andesitic composition (Morris and Kopf, 1967, 1970a, 1970b). The total thickness of extrusive volcanic rocks is several hundred metres. The Sheeprock Granite, which intrudes the southwest side of the Sheeprock Mountains (Fig. 1; Cohenour, 1959) is of Early Miocene age (17 to 19 Ma by the K-Ar method on biotite and 15 to 20 Ma from lead-alpha determinations on zircon; Armstrong, 1970, and quoted in Cohenour, 1970). These igneous rocks are unconformably overlain by lacustrine silt, marl, and bentonitic tuff of the Salt Lake(?) Formation (600 m?) and by Pleistocene to Holocene alluvial fan gravel, mudflow deposits, alluvium, lake deposits, and aeolian sand (Morris and Kopf, 1970a, 1970b).

STRUCTURE

The internal structure of the southern Sheeprock Mountains is dominated by low-angle faults. Those with significant stratigraphic separation are the Sheeprock and Pole Canyon Thrusts[4] and the Harker and Lion Hill (normal) Faults. Several steeply dipping faults are inferred to be tears. Of these, the Indian Springs Fault is most conspicuous. With the exception of overturned beds beneath the Pole Canyon Thrust, the Pole Canyon Anticline of Cohenour (1959), folds are subordinate and for the most part of mesoscopic scale.

Sheeprock Thrust

The Sheeprock Thrust is best exposed in the Cherry Creek and Maple Peak Quadrangles (Figs. 3, 4, 6; Morris and Kopf, 1970a, 1970b), where it was first recognized by Loughlin (1920). It consists of several anastomosing strands with fault dips ranging from near zero to as much as 50° northward. Between Great Eastern Hollow and Cherry Creek (located in Fig. 3), the thrust juxtaposes gently dipping rocks of the Proterozoic Otts Canyon and Dutch Peak Formations above steeply dipping and overturned strata of the Lower Ordovician Pogonip Group through Upper Mississippian Deseret Limestone (Fig. 5). About 2 km east of Cherry Creek in the southern West Tintic Mountains, the Otts Canyon and Dutch Peak Formations are faulted against Upper Mississippian Great Blue Limestone, a stratigraphic separation exceeding 10 km.

Two small klippen, apparently separated by the concealed eastward extension of the Indian Springs Fault,

---

[4]I continue to use the term "thrust" in these well-established names in spite of genetic connotations, because there is little doubt that these faults were responsible for significant crustal shortening.

occur in the southwestern corner of the Sabie Mountain Quadrangle (Figs. 3, 4, 7). The southern klippe consists of gently dipping Paleozoic carbonate rocks in thrust contact above overturned Deseret Limestone that dips gently to the north and northwest (Morris and Kopf, 1970b). Slices of Dutch Peak Formation are included in the fault zone, along with quartzite here interpreted as Otts Canyon Formation, by analogy with exposures of the Sheeprock Thrust in the Cherry Creek and Maple Peak Quadrangles. Note, however, that Morris and Kopf (1970b) assigned the quartzite to the upper Sheeprock Series (Caddy Canyon Quartzite in this paper). The northern klippe illustrated in Figure 7 consists of quartzite, similarly interpreted as Otts Canyon Formation, faulted above Great Blue Formation. The fault beneath both klippen is thought to be the Sheeprock Thrust for reasons discussed below (see the cross section in Fig. 8).

In the vicinity of Sabie Mountain in the northern West Tintic Mountains (Figs. 1, 3), another thrust strand entirely within the Paleozoic section is possibly also part of the Sheeprock Thrust (Groff, 1959; Morris, 1977). I have not studied the geology of that area.

Pole Canyon Thrust

The trace of the Pole Canyon Thrust extends for about 6 km from Bennion Creek to the northern part of Horse Valley across the drainage divide of the Sheeprock Mountains (Figs. 3, 4, 9). For most of its exposed length, the thrust is relatively planar, and its dip ranges from 12° north-northeast in Pole Canyon to 20° east-northeast on the ridge south of Bennion Creek. Between the northern parts of Otts and Pole Canyons, a subsidiary upper strand dips to the northeast at 40° to 50° (Fig. 10). However, contrary to mapping by Cohenour (1959, cross section AA'), this inclination does not reflect the true dip of the main strand of the thrust.

In the vicinity of Bennion Creek, Cohenour (1959) thought that the Pole Canyon Thrust was folded and locally near-vertical, whereas I interpret the thrust to be offset by a steeply dipping fault, here named the Bennion Creek Fault (Fig. 9). The new interpretation is based on careful mapping of fault traces and the observation that lower plate rocks are not correspondingly deformed along with the supposedly folded thrust. Only upper plate rocks are exposed on the west side of the Bennion Creek Fault, except in a small window intruded by the Sheeprock Granite at the head of Joes Canyon (Fig. 9).

Where exposed, the Pole Canyon Thrust cuts only Proterozoic rocks and juxtaposes Otts Canyon Formation above rocks as young as Caddy Canyon Quartzite, a stratigraphic separation of only 2 to 3 km. As noted by Cohenour (1959), upper plate rocks for the most part dip steeply to the north and northeast, whereas strata of the lower

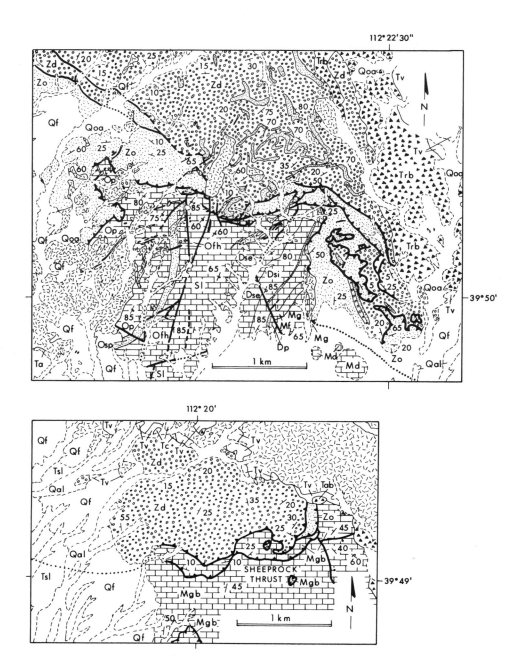

Figure 6. Geologic maps of parts of the Cherry Creek and Maple Peak Quadrangles, showing the trace of the Sheeprock Thrust (simplified from Morris and Kopf, 1970a, 1970b). See Figure 4 for location. Faults (bold contacts, dashed where approximately located, dotted where concealed): thrust (sawteeth on upper plate); high-angle fault (no ornament). Stratigraphic units: Zo, Otts Canyon Formation; Zd, Dutch Peak Formation, with quartzite beds and lenses indicated by stipple pattern; Op, Pogonip Group; Osp, Swan Peak Formation; Ofh, Fish Haven Dolomite; Sl, Laketown Dolomite; Dse, Sevy Dolomite; Dsi, Simonson Dolomite; Dp, Pinyon Peak Limestone, with quartzite member indicated by stipple pattern; Mf, Fitchville Formation; Mg, Gardison Limestone; Md, Deseret Limestone; Mgb, Great Blue Formation; Tc, Tertiary conglomerate; Tv, latite and quartz latite volcanic rocks; Ta, andesite or latite; Tab, andesite intrusion breccia; Trb, rhyolitic intrusion breccia; random dash pattern, undifferentiated Tertiary plutonic rocks; Tsl, Salt Lake Formation; Qf, Qoa, and Qal, Quaternary fan gravel, older alluvium, and alluvium.

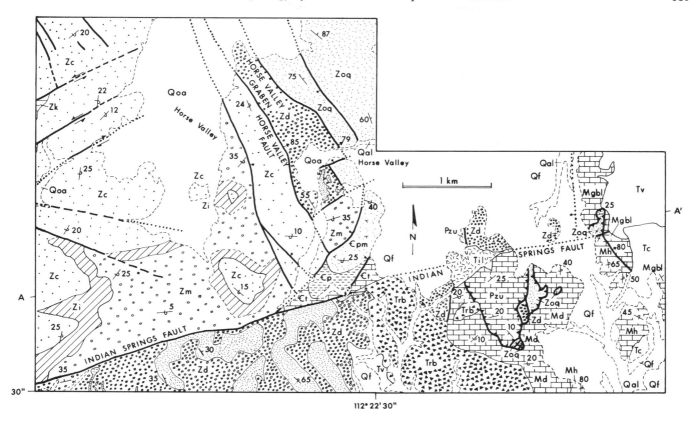

Figure 7. Geologic map of parts of the Dutch Peak and Sabie Mountain Quadrangles (simplified from Morris and Kopf, 1970a, 1970b; Blick, 1979). See Figure 4 for location. Prominent structural features are two klippen of the Sheeprock(?) Thrust, the Indian Springs (tear) Fault and the Horse Valley Graben. Rocks west of the Horse Valley Fault are overturned and in the lower plate of the Pole Canyon Thrust; to the east, rocks are for the most part upright and belong to the upper plate of this thrust. Faults (bold contacts, dashed where approximately located, dotted where concealed): thrust (sawteeth on upper plate); high-angle fault (no ornament; ball on downthrown side; arrows signify inferred lateral slip). Stratigraphic units: Zd, Md, Tc, Tv, Trb, Qf, Qoa, and Qal as in Figure 6; Zoq, upper (quartzite) member of Otts Canyon Formation; Zk, Kelley Canyon Formation; Zc, Caddy Canyon Quartzite; Zi, Inkom Formation; Zm, Mutual(?) Formation; €pm, Prospect Mountain Quartzite; €p, Pioche Shale; €t, Tatow Formation; Mh, Humbug Formation; Mgbl, lower member of Great Blue Formation; Pzu, undifferentiated Paleozoic rocks; Til, intrusive latite porphyry. Cross section AA' is shown in Figure 8.

plate are overturned and dip gently to the west (Figs. 9, 10). Equal-area plots of poles to bedding in the Dutch Peak Formation and Caddy Canyon Quartzite from both plates are illustrated in Figure 11.

Rocks in the lower plate of the thrust are more deformed than those in the upper plate. Pebbles in lower plate conglomerate and diamictite are more strongly flattened, and the Sheeprock Group is correspondingly thinner than in the upper plate (Christie-Blick, 1982). Cleavage and bedding are subparallel beneath the Pole Canyon Thrust, whereas they typically intersect with a large dihedral angle above it (Fig. 11). Cleavage is deformed close to the thrust and therefore predates latest movement on it.

Indian Springs Fault

The Indian Springs Fault extends east-northeast for more than 8 km from Cow Hollow to the divide between Horse Valley and Devil Creek (Figs. 3, 4, 7). The fault is close to vertical and its trace is relatively straight. It appears to offset the Sheeprock Thrust (Fig. 7), and thus, contrary to the interpretation of Armstrong (1968), it is not a folded segment of the thrust.

Several additional observations suggest that the Indian Springs Fault is a tear, as proposed by Groff (1959) and Morris and Kopf (1967, 1970a, 1970b). It juxtaposes different facies of the Dutch Peak Formation (Blick, 1979); it

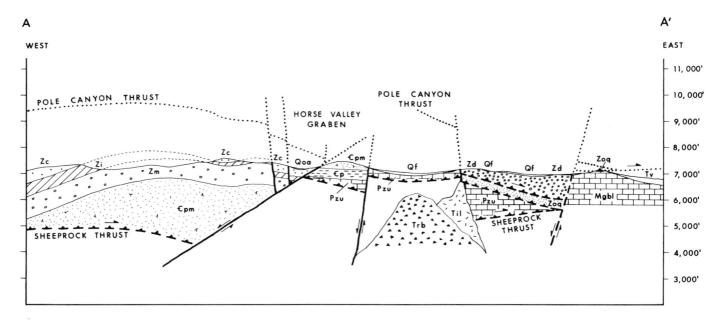

Figure 8. Geologic cross section AA′ through parts of the Dutch Peak and Sabie Mountain Quadrangles (located in Fig. 7). Geologic symbols and abbreviations for stratigraphic units as in Figure 7. The Sheeprock Thrust is only approximately located for much of the cross section.

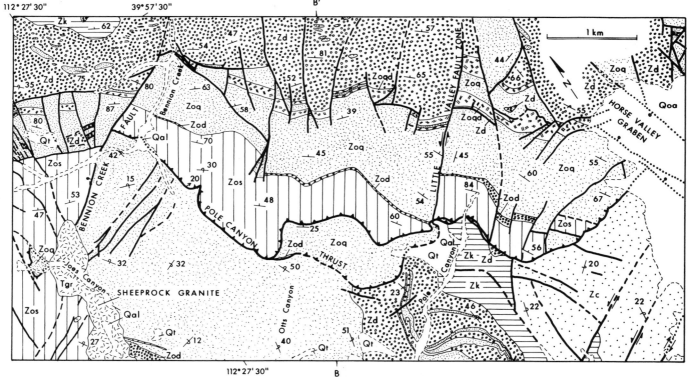

Figure 9. Geologic map of part of the Dutch Peak Quadrangle (simplified from Blick, 1979). See Figure 4 for location. Prominent structural features are the Pole Canyon Thrust, Bennion Creek Fault, Little Valley Fault Zone, and Horse Valley Graben. Fault symbols as in Figure 7. Stratigraphic units: Zos, Zod, Zoq, and Zoqd, lower (slate), middle (diamictite), upper (quartzite) members, and diabase sills of Otts Canyon Formation; Zd, Dutch Peak Formation, with quartzite beds and lenses indicated by stipple pattern; Zk, Kelley Canyon Formation; Zc, Caddy Canyon Quartzite; Tgr, Tertiary granite; Qoa, Qal, and Qt, Quaternary older alluvium, alluvium, and talus. Cross section BB′ is shown in Figure 10.

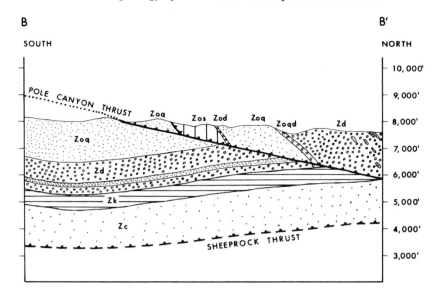

Figure 10. Geologic cross section BB′ through part of the Dutch Peak Quadrangle (located in Fig. 9). Geologic symbols and abbreviations for stratigraphic units as in Figure 9.

separates upright beds on the south side from overturned strata on the north side; and slip is thought to have occurred before Basin-Range deformation, because the Indian Springs Fault is cut by igneous rocks of probable Oligocene age (Fig. 7; and Morris and Kopf, 1970a).

Lion Hill Fault and Harker Fault

Structurally above the Pole Canyon Thrust is a complexly faulted zone cropping out between Harker Canyon and the western flank of the Sheeprock Mountains (Figs. 4, 12). The major feature of this zone is a set of faults that, for the most part, dip at between 10° and 30° to the north and northeast and attenuate the stratigraphic section.

The uppermost low-angle fault may be traced continuously from the entrance of Harker Canyon to the entrance of North Oak Brush Canyon, where it is probably truncated by the North Oak Brush Fault (new name; Fig. 12). The normal fault, named the Lion Hill "Thrust" by Cohenour (1959), juxtaposes Lower Cambrian Pioche Shale through Middle Cambrian Marjum Formation above Upper Proterozoic Caddy Canyon Quartzite (Fig. 5), a stratigraphic separation of about 1,500 to 3,500 m. Cohenour (1959) interpreted much of the footwall block as Lower Cambrian Tintic Quartzite (Prospect Mountain Quartzite in this paper), but I have assigned these rocks to the Caddy Canyon Quartzite on the basis of the lithology of interbeds within the quartzite (Christie-Blick, 1982). Rocks of both hanging-wall and footwall blocks dip moderately steeply to the north or northeast (Figs. 12, 13), and beds above the Lion Hill Fault are locally overturned toward the north close to the fault surface.

Beneath the Lion Hill Fault there are at least two other important low-angle fault strands, here together termed the Harker Fault. The Harker fault extends from the entrance of Harker Canyon, where its trace is subparallel to that of the Lion Hill Fault, to the western side of the range (Fig. 12). It cuts only Proterozoic rocks at the present level of exposure and unlike the Lion Hill Fault is offset in several places by cross-faults. The hanging wall of the upper strand consists of lower Caddy Canyon Quartzite east of North Oak Brush Canyon and upper Caddy Canyon Quartzite to the west. The footwall of the lower strand ranges from lower Caddy Canyon Quartzite at the entrance of Harker Canyon, through uppermost Dutch Peak Formation west of North Oak Brush Canyon. The combined stratigraphic separation thus increases from a few hundred metres in the east to more than 1,000 m in the west. The relative displacement of blocks may be as much as several kilometres because the fault juxtaposes different facies of the Caddy Canyon Quartzite (Blick, 1979). The attitude of bedding in hanging-wall and footwall blocks ranges from subparallel to the Harker Fault or slightly steeper to markedly discordant (Figs. 12, 13).

Both strands of the Harker Fault are offset by the North Oak Brush Fault, which has a stratigraphic separation of at least 2,000 m in North Oak Brush Canyon, but farther south does not appear to displace contacts within the Sheeprock Group (Fig. 12). This fault is therefore in part older than the Harker Fault or genetically related to it.

Other Faults

Several other faults are of interest in a consideration of

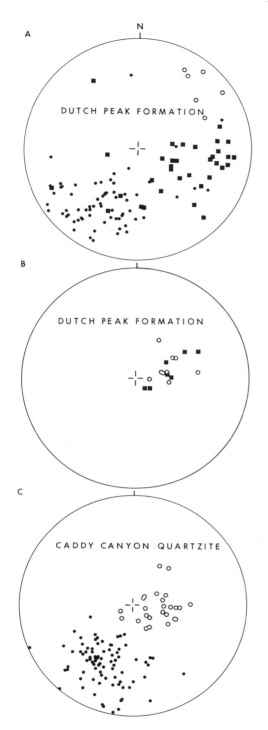

Figure 11. Equal-area, lower hemisphere plots of poles to bedding and cleavage in the Dutch Peak Formation and Caddy Canyon Quartzite. Symbols: dots, poles to upright beds; open circles, poles to overturned beds; squares, poles to cleavage. A. Poles to bedding (73) and cleavage (34), Dutch Peak Formation, upper plate of Pole Canyon Thrust, between Vernon Creek and Harker Canyon. B. Poles to bedding (9) and cleavage (7), Dutch Peak Formation, lower plate of Pole Canyon Thrust in Pole Canyon. C. Poles to bedding (109), Caddy Canyon Quartzite, upper plate

the structural history of the Sheeprock Mountains. The upper plate of the Pole Canyon Thrust is segmented by a set of steeply dipping faults that strike northeast to north-northeast (Figs. 4, 9). Many of these are characterized by left separation. One fault, a strand of the Little Valley Fault Zone (new name), with left separation of about 750 m, clearly terminates against the Pole Canyon Thrust. Several faults with less separation also appear to terminate at the thrust. I conclude that they are tears, active before and/or during movement on the Pole Canyon Thrust. Faults sub-parallel to the Little Valley Fault Zone, west of and including the Bennion Creek Fault, may also have been initiated as upper plate tears.

The lower plate of the Pole Canyon Thrust is similarly segmented by steeply dipping faults. These strike northeast to east-northeast and are characterized by both right and left separation. However, their significance is uncertain because none of them intersects the thrust.

In the vicinity of Horse Valley four near-vertical faults define what is here named the Horse Valley Graben (Figs. 4, 7, 9). The faults strike north-northwest and terminate against the Little Valley Fault Zone in the north and the Indian Springs Fault in the south. One of the graben faults, the Horse Valley Fault (Morris and Kopf, 1970a, is thought to cut the Pole Canyon Thrust, because it juxtaposes upright beds of the Dutch Peak Formation against overturned Caddy Canyon Quartzite (Fig. 7).

In addition to the Lion Hill and Harker Faults, there is in the upper part of the Harker Canyon a bewildering array of high-angle normal faults that break the rocks into a large number of small blocks (Fig. 12). The faults occur in at least three sets. Some dip to the north, some to the south, and some are approximately vertical and north-striking. Several are cut by the Harker Fault, whereas others offset this fault.

KINEMATIC INTERPRETATION OF MAJOR STRUCTURES

The Sheeprock Thrust is thought to occur at depth beneath much of the southern Sheeprock Mountains. The Pole Canyon Thrust is interpreted as a structurally higher fault that occurs only north of the Indian Springs Fault. The thrusts probably merge to both east and west. The now-dismembered Pole Canyon Anticline is thought to have developed during propagation of one or both of the thrusts. The recumbent attitude of this fold and the near-horizontal to eastward dips of the thrusts are explained chiefly by late Cenozoic tilting and propagation of the thrusts from ramps to a flat within the Upper Mississippian

(upright beds) and lower plate (overturned beds) of Pole Canyon Thrust, from approximately the same areas as in A and B. See Figures 3 and 4 for locations.

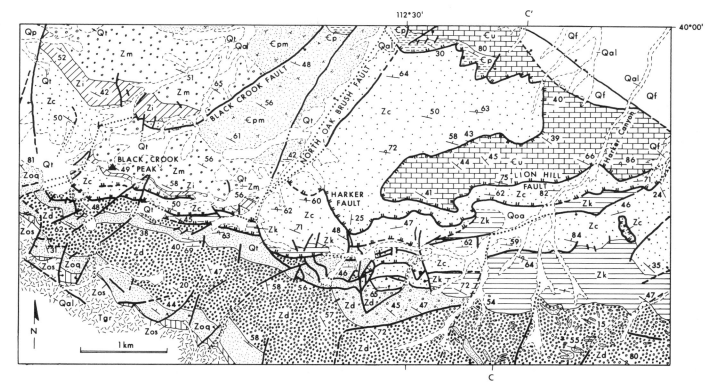

Figure 12. Geologic map of parts of the Dutch Peak and Erickson Knoll Quadrangles (modified from Blick, 1979). See Figure 4 for location. Prominent structural features are the Harker, Lion Hill, Black Crook, and North Oak Brush Faults. Faults (bold contacts, dashed where approximately located; dotted where concealed): low-angle extension fault (double ticks on hanging-wall block); high-angle fault (no ornament; ball on downthrown side). Stratigraphic units: Zos, Zoq, Zd, Zk, and Zc, as in Figure 9, Zi, Inkom Formation; Zm, Mutual(?) Formation; €pm, Prospect Mountain Quartzite; €p, Pioche Shale; €u, undifferentiated Cambrian rocks; Tgr, Tertiary granite; Qp, Qf, Qoa, Qal, and Qt, Quaternary pediment gravel, fan gravel, older alluvium, alluvium, and talus. Cross section CC′ is shown in Figure 13.

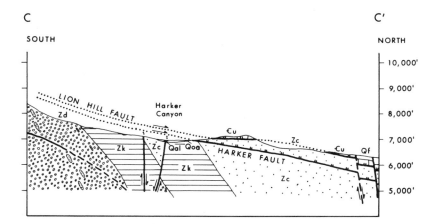

Figure 13. Geologic cross section CC′ through part of the Dutch Peak Quadrangle (located in Fig. 12). Geologic symbols and abbreviations for stratigraphic units as in Figure 12.

section. East-directed tectonic transport, parallel to the Indian Springs Fault (075° azimuth), is corroborated by the vergence of the Pole Canyon Anticline and mesoscopic folds, by the orientation of slickensides on the Pole Canyon Thrust, and by the direction in which the thrusts climb stratigraphically.

The Harker and Lion Hill Faults are interpreted as late Cenozoic extension faults younger than the thrusts, although the possibility of earlier displacement cannot be ruled out. In spite of its reverse separation, the Government Creek Fault of the northern Sheeprock Mountains may also be an extension fault.

Relation Between Sheeprock and Pole Canyon Thrusts

The geometric relation between the Sheeprock and Pole Canyon Thrusts is uncertain, because they are exposed in different parts of the range and are separated in map view by the Indian Springs Fault (Fig. 4). Two interpretations are possible. In one, favored by Armstrong (1968), the Sheeprock and Pole Canyon Thrusts are considered to be the same fault. In the other interpretation, favored by Morris (1977) and in this paper, the Pole Canyon Thrust is viewed as structurally higher than and subsidiary to the Sheeprock Thrust. In this view the Pole Canyon Thrust is located only north of the Indian Springs Fault, whereas the Sheeprock Thrust occurs on both sides of it. Possible variants of the second interpretation hold that the Pole Canyon and Sheeprock Thrusts merge to the west, or to both east and west. Morris (1977) thought that both thrusts also merge northward with the Tintic Valley Thrust, the trace of which is concealed beneath alluvium of the Tintic Valley, east of the Sheeprock and West Tintic Mountains (Fig. 1).

Critical to the distinction between the two-thrust and single-thrust hypotheses is a comparison of the maximum stratigraphic separation (and probable slip magnitude) of the Pole Canyon Thrust and the separation associated with the small thrust klippe shown in Figure 7 north of the Indian Springs Fault. Stratigraphic separation on the Pole Canyon Thrust probably increases toward the southeast, but is no more than 6 km, or perhaps 4 km when allowances are made for tectonic flattening beneath the thrust. Assuming that the upper plate was displaced toward the east-northeast (discussed below) and that the offset of formational boundaries in this direction is a measure of the amount of slip, I estimate the slip to be less than 10 km. In comparison, if the klippe is correctly mapped as Otts Canyon Formation faulted above Great Blue Limestone, the stratigraphic separation on that thrust exceeds 10 km, and the slip magnitude is probably considerably greater (at least 16 km, according to Morris, 1977).

In addition, the klippe is only about 3 km east of the Horse Valley Graben, where overturned rocks in the lower plate of the Pole Canyon Thrust are at least as old as the Caddy Canyon Quartzite, stratigraphically more than 5 km beneath the Great Blue Limestone (Fig. 5). Even if I have underestimated slip on the Pole Canyon Thrust, it is unlikely that a low-angle thrust could cut up through 5 km of gently dipping beds in a horizontal distance of 3 km. It is for these reasons that I favor the interpretation of the Pole Canyon and Sheeprock Thrusts as different faults, and interpret the fault beneath the klippe as the Sheeprock Thrust.

In spite of these arguments, there are three circumstances in which the single-thrust hypothesis might still be correct. One arises if the Indian Springs Fault projects or is offset north of the klippe and nearby outcrops of Dutch Peak Formation, which probably occur within the same thrust plate. However, elsewhere the Indian Springs Fault is relatively straight and is not offset by younger faults. A second possibility is that the klippe consists of Caddy Canyon Quartzite or even younger Prospect Mountain Quartzite, rather than Otts Canyon Formation, thus reducing the stratigraphic separation on the thrust. Countering this idea is the close association of the quartzite with Dutch Peak Formation and the fact that Caddy Canyon quartzite does not crop out nearby in the upper plate of the Pole Canyon Thrust. A third possibility is that between the Horse Valley Graben and the klippe the thickness of rocks between the Caddy Canyon Quartzite and Great Blue Formation is substantially less than 5 km owing to pronounced tectonic flattening.

Where exposed, the Sheeprock Thrust dips gently and probably occurs in the shallow subsurface beneath much of the Dutch Peak Quadrangle. Assuming that the Pole Canyon Thrust is a different fault, there seems to be insufficient vertical space between the thrusts to accommodate a complete upright fold limb beneath the overturned beds (Figs. 7, 8). In Figure 8, I have shown the Pole Canyon and Sheeprock Thrusts merging toward the east to account for the occurrence of the Otts Canyon Formation above the Sheeprock Thrust near location A'.

Relation Between the Thrusts and Pole Canyon Anticline

Cohenour (1959) thought that the recumbent Pole Canyon Anticline formed during eastward displacement above the Sheeprock Thrust and that it was subsequently truncated and overridden by the Pole Canyon Thrust plate moving toward the south or southwest. The slip direction of the Pole Canyon Thrust was inferred from "drag and compressional folds observed along the thrust plane" and presumably based, in part, on the prevailing assumption that thrust plates move up-slope.

Working in the Sheeprock Mountains after Cohenour, Harris (1958) recognized the steeply dipping Bennion Creek Fault southwest of its intersection with the Pole Canyon

Thrust (Fig. 9). He interpreted this segment of the Bennion Creek Fault as a tear and inferred northeastward displacement on the thrust parallel to it. Harris (1958) also suggested a genetic relation between folding and thrusting, although he did not recognize that lower plate rocks are overturned, and he incorrectly concluded that the Pole Canyon Thrust dips to the southwest rather than to the northeast. As a result, Harris's (1958) autochthon (i.e., lower plate) is actually the upper plate of the thrust.

Morris (1977) suggested that the Pole Canyon Thrust developed by dislocation of the limbs of the Pole Canyon Anticline during east-directed displacement above the Sheeprock Thrust. Recently, in a sketch cross section,

H. T. Morris (1982, personal commun.) interpreted the Pole Canyon Anticline as a recumbent fold older than and only indirectly related to the Pole Canyon Thrust, for which he proposed several tens of kilometres of slip. A possible criticism of this idea is that it implies a substantial change in overall crustal shortening across the Indian Springs Fault.

Here I suggest two kinematic interpretations, different from and simpler than that of Morris, but consistent with observations in the Sheeprock Mountains (Fig. 14A) and also with mechanisms of folding and thrust propagation established in well-known fold-thrust belts such as the Canadian Rocky Mountains. In the absence of direct evidence for large slip on the Pole Canyon Thrust, I assume

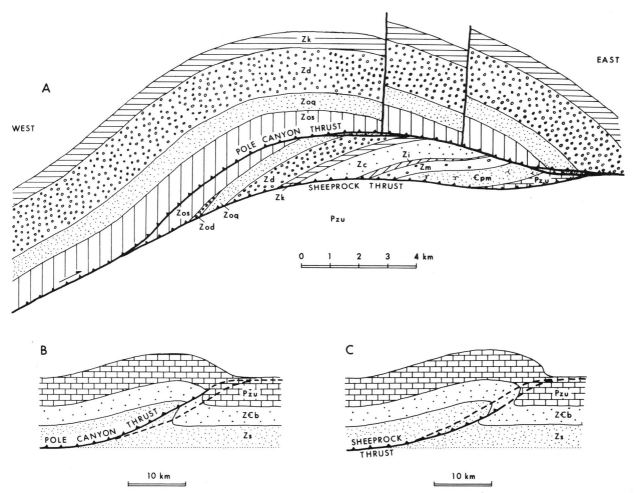

Figure 14. A. Conceptual cross section showing the relation between prominent structures in the southern Sheeprock Mountains (Sheeprock and Pole Canyon Thrusts, the recumbent Pole Canyon Anticline and tear faults in the upper plate of Pole Canyon Thrust). Late Cenozoic faults and the hypothetical Skull Valley Thrust (Fig. 1) have been omitted for simplicity. B and C. Proposed origin of the Pole Canyon Anticline as a fault propagation fold related to the Pole Canyon Thrust (B) or Sheeprock Thrust (C). Future thrust traces are shown as dashed lines in each cross section. The present recumbent attitude of the Pole Canyon Anticline is explained by a combination of late Cenozoic tilting (see Fig. 16) and propagation of the thrusts from ramps to a flat within the Great Blue Formation (Mississippian). Stratigraphic units: Zos, Zod, and Zoq, lower (slate), middle (diamictite), and upper (quartzite) members of Otts Canyon Formation; Zd, Dutch Peak Formation; Zk, Kelley Canyon Formation; Zc, Caddy Canyon Quartzite; Zi, Inkom Formation; Zm, Mutual (?) Formation; €pm, Prospect Mountain Quartzite; Zs, Sheeprock Group; Z€b, Brigham Group; Pzu, undifferentiated Paleozoic rocks.

the minimum required to explain the observed stratigraphic separation. In addition, I assume that the thrusts and the Pole Canyon Anticline are genetically related.

One possible interpretation, illustrated in Figure 14B, is that the Pole Canyon Anticline was generated by propagation of the Pole Canyon Thrust and that the Sheeprock Thrust is a slightly younger fault which subsequently displaced part of the overturned fold limb. My colleague, T. R. Bultman, has suggested the term "fault propagation fold" for folds generated in this manner. A second possibility (Fig. 14C), which I prefer, is that the Pole Canyon Anticline is a fault propagation fold primarily related to the Sheeprock Thrust and the Pole Canyon Thrust is a minor upper plate imbrication.

The interpretation of the Pole Canyon Anticline as a fault propagation fold is consistent with observed geometry, although not demonstrated conclusively. The limbs of the fold, as they are now juxtaposed, are modeled in Figure 15. In the model, one limb (X) is upright, dipping homoclinally at 50° toward the northeast. The other limb (Y) is overturned and dips at 20° toward the west. Figure 11 indicates the degree of scatter of bedding attitudes in the actual fold. The axial surface of the model fold (S in Fig. 15) is near-horizontal and almost parallel to the Pole Canyon Thrust (T). Note, however, that the geometry of the Pole Canyon Anticline probably changed progressively during deformation as a result of flattening of the overturned limb and displacement on a non-planar fault.

South of the Indian Springs Fault, rocks beneath the Sheeprock Thrust are locally overturned and may constitute the lower limb of a fault propagation fold similar to the Pole Canyon Anticline. In contrast to the region north of the tear, bedding in upper plate rocks is only slightly oblique to the thrust. These rocks may therefore represent the back limb of the proposed fold (compare the left side of Fig. 14A). Such an interpretation is consistent with greater displacement on the Sheeprock Thrust south of the Indian Springs Fault than north of it, where some shortening was accommodated by the Pole Canyon Thrust. Another possibility is that south of the Indian Springs Fault, rocks beneath the Sheeprock Thrust were locally overturned by drag.

Explanation for Present Attitude of Thrusts

The recumbent attitude of the Pole Canyon Anticline and near-horizontal to eastward dips of the thrusts in the Sheeprock Mountains are anomalous and are thought to be due chiefly to two factors: late Cenozoic tilting and folding above the Sheeprock Thrust where it passed from a ramp to a flat within the Upper Mississippian section. Suppe (1979) has suggested the term "fault-bend folding" for the latter type of deformation.

The magnitude and direction of late Cenozoic tilting

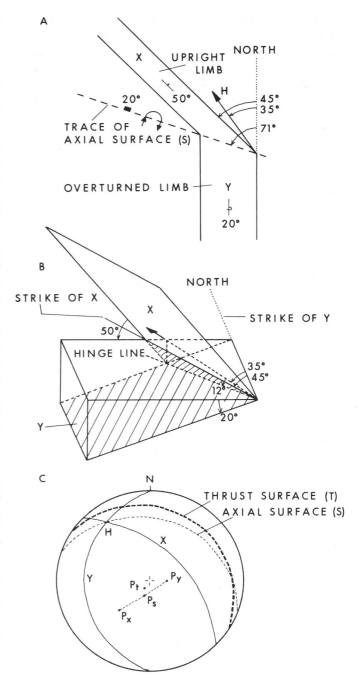

Figure 15. Model of the Pole Canyon Anticline, a prominent recumbent fold, prior to detachment of upright (X) and overturned (Y) limbs by the Pole Canyon Thrust. Bedding attitudes are generalized from the areas shown in Figures 7 and 9. A. Map view. H indicates trend of hinge line. B. Block diagram. C. Equal-angle, lower hemisphere plot of surfaces (and poles to surfaces): X (Px) and Y (Py), limbs of Pole Canyon Anticline; S (Ps), axial surface (dip = 20°N, strike = 109°); T (Pt), Pole Canyon Thrust (dip = 12°N, strike = 123°). H is hinge line (plunge = 12°, trend =325°).

are difficult to estimate, because in the Sheeprock Mountains there are no layered rocks that postdate the thrusts and predate Basin-Range extension. The attitude of widespread Oligocene(?) volcanic flows in the adjacent West Tintic Mountains is variable (Morris and Kopf, 1970b) and presumably influenced locally by initial dip, by independent tilting of small blocks, and by associated intrusions. However, away from faults and intrusions, extrusive volcanic rocks, exposed over several square kilometres of the Maple Peak Quadrangle (Fig. 3), dip relatively uniformly toward the northeast at about 25°. This direction of tilting is consistent, perhaps fortuitously, with the northward to eastward dips of the Proterozoic and Paleozoic rocks in much of the Sheeprock Mountains. However, the magnitude of tilting is reasonable when compared with that observed elsewhere in the Basin-Range province of Utah and Nevada. Regionally, range tilts average 15° to 20°, and few initially horizontal Cenozoic rocks in Utah and Nevada dip at more than 32° (Stewart, 1980). For these reasons, I tentatively estimate late Cenozoic tilting of the Sheeprock Mountains to have been about 25° toward the northeast. Although the evidence is admittedly weak and the direction of tilt may be revised by future observations, the magnitude of tilt is unlikely to be substantially more than 25° and may be considerably less.

For want of a better value, the estimated tilt has been used in Figure 16 to correct the attitudes of the Pole Canyon Thrust and bedding depicted in Figure 15. The corrected attitude of the Pole Canyon Thrust (13° dip to the southwest) is still near horizontal. If the Pole Canyon Anticline is a fault propagation fold, it is likely that segments of the associated thrusts originally dipped more steeply to the southwest or west. In the Maple Peak Quadrangle, the Sheeprock Thrust is preferentially located within the upper and Chiulos (shale) Members of the Mississippian Great Blue Formation (Morris and Kopf, 1970b; and Fig. 6). I therefore propose that the upper plate of the Sheeprock Thrust (including the Pole Canyon Anticline and Thrust north of the Indian Springs Fault) was folded during displacement over a bend in the fault where it passed from a ramp in much of the Proterozoic and Paleozoic section to a flat within the Upper Mississippian carbonate and shale section (Fig. 14). Fault-bend folding may also be associated with ramps in structurally lower faults such as the Tintic Valley Thrust, although this has not yet been documented.

Direction of Slip on Pole Canyon and Sheeprock Thrusts

Evidence for the direction of tectonic transport associated with the Pole Canyon Thrust and Anticline is summarized in Table 1. If the Indian Springs Fault is a major tear as I and others have supposed, it is reasonable to infer tectonic transport parallel to it (075° azimuth). This is cor-

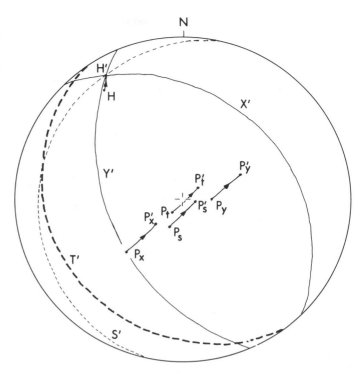

Figure 16. The attitudes of fold limbs (Px and Py), axial surface (Ps), Pole Canyon Thrust (Pt), and hinge line (H) from Figure 15 corrected for late Cenozoic tilting, estimated as 25° toward the northeast: X' (P'x), dip = 25°NE, strike = 135°; Y' (P'y), dip = 42°W overturned, strike = 156°; S' (P's), dip = 9°W, strike = 013°; T' (Pt') dip = 13°W, strike = 143°; H', plunge = 7°, trend = 329°.

roborated by the direction of vergence of the Pole Canyon Anticline (Figs. 15, 16), by a separation arc plot (Hansen, 1971) of 20 asymmetrical folds in the upper plate, close to the thrust (Fig. 17), and by the trend of slickensides on several square metres of thrust surface on the ridge south of Bennion Creek (located in Fig. 9). Azimuths are not significantly changed by a correction for late Cenozoic tilting of the Sheeprock Mountains.

TABLE 1. EVIDENCE FOR THE DIRECTION OF TECTONIC TRANSPORT ASSOCIATED WITH THE POLE CANYON THRUST AND ANTICLINE

	Azimuth
Indian Springs tear fault (strike; Fig. 7)	075°
Modeled recumbent Pole Canyon anticline	
(normal to fold axis trend; Figs. 15, 16)	055° (059°)
Separation arc plot of asymmetrical folds	
(bisector of separation angle; Fig. 17)	086° ± 43°
Slickensides on Pole Canyon thrust (trend)	069° (070°)
Upper plate tear faults (strike; Fig. 9)	~ 037°
Lower plate tear (?) faults (strike; Fig. 7)	~ 060°

Note: correction for late Cenozoic tilting of the Sheeprock Mountains changes the azimuths given by less than 5° (values in parentheses).

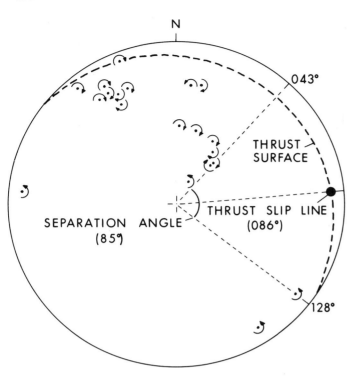

Figure 17. Equal-area, lower hemisphere separation arc plot of 20 minor asymmetrical folds in the lower member of the Otts Canyon Formation, in the upper plate of the Pole Canyon Thrust, close to the thrust surface. Fold axes are indicated by dots. Clockwise arrows signify Z-folds and counterclockwise arrows, S-folds. Separation angle = 85°, measured in a horizontal plane. Inferred slip line of Pole Canyon Thrust is approximately 086°. Note that a great circle approximating the fold-axis distribution is inclined to the Pole Canyon Thrust.

It is not clear why minor upper plate tear faults (average strike, 037°) are oblique to the Indian Springs Fault. Three possible explanations are that (1) they represent longitudinal shortening parallel to the axis of the Pole Canyon Anticline (but see Dahlstrom, 1970), (2) the Pole Canyon Thrust experienced an earlier phase of northeastward displacement, and (3) the tear faults originated parallel to the Indian Springs Fault and assumed their present orientation through counterclockwise rotation.

Eastward displacement on the Sheeprock Thrust, parallel to the Indian Springs Fault, is consistent with the direction in which the thrust climbs stratigraphically and the sense of overturning of beds beneath the thrust.

The inferred direction of tectonic transport in the Sheeprock Mountains is also consistent with regional geologic relationships (Armstrong, 1968). In particular, the Indian Springs Fault is approximately parallel to inferred tear faults in the adjacent East Tintic Mountains (located in Fig. 1) and in the Gilson Mountains at the southern end of the Tintic Valley (055° to 065° azimuth; Morris and Shepard, 1964; Roberts and others, 1965; Morris, 1977).

Sense of Slip on Indian Springs Fault

Groff (1959) inferred left slip on the Indian Springs Fault based on left separation of juxtaposed Proterozoic rocks. However, separation is a poor indicator because beds are overturned to the north but not to the south and because the Dutch Peak Formation changes facies across the fault (Blick, 1979). Morris and Kopf (1970a) and Morris (1977) indicated right slip, presumably on the assumption that the Pole Canyon Thrust and Anticline were responsible for additional shortening in the Sheeprock Thrust plate north of the tear.

In my opinion, the sense of slip on the Indian Springs Fault is probably variable, depending on structural position and on any changes in crustal shortening across it. For example, assuming equivalent overall shortening on each side of the fault, right slip would be favored between rocks above the Sheeprock Thrust on the north side and lower plate rocks to the south (and vice versa). Segments of the tear which juxtapose upper plate rocks that are overturned on the north side but upright to the south would be characterized by left slip.

Relation Between Thrusts and Low-Angle Normal Faults

Fault geometry in Harker Canyon (Figs. 3, 12, 18) sug-

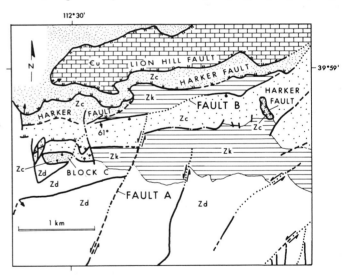

Figure 18. Simplified fault map of the central part of Harker Canyon abstracted from Figure 12. (See Fig. 12 for explanation of symbols.) Fault A probably originated as a tear fault and terminates against normal fault B. Fault B is truncated by the Harker Fault (see also cross section in Fig. 13). Latest movement on the Harker Fault is therefore thought to postdate the Pole Canyon Thrust. The Harker Fault is also cut by younger normal faults. For example, based on a comparison of facies of the Caddy Canyon Quartzite (Zc), block C belongs in the hanging-wall block of the Harker Fault (dense stipple pattern). The Caddy Canyon Quartzite of the footwall block is indicated by a sparse stipple pattern.

gests that latest movement on the Harker and Lion Hill Faults postdates the Sheeprock and Pole Canyon Thrusts. Fault A in Figure 18 probably originated as a tear, because it parallels tear faults in the upper plate of the Pole Canyon Thrust and, like them, displays left separation. However, the following argument holds even if it is younger than the thrusts. It terminates against normal fault B, which dips steeply to the south and juxtaposes lower Caddy Canyon Quartzite against Kelley Canyon Formation. Fault B is clearly truncated by the north-dipping Harker Fault in the eastern part of Harker Canyon (Figs. 12, 13, 18). The Harker Fault is thus thought to be younger than the Pole Canyon Thrust.

The slip direction of the Harker and Lion Hill Faults is not well constrained, but the vergence of a few asymmetrical folds close to the faults suggests movement to the north or northeast. This tentative slip direction, transverse to that inferred for the thrusts, is consistent with a different age of deformation, but it poses a geometrical problem.

In Harker Canyon, the Harker and Lion Hill Faults dip to between north and northeast less steeply than bedding in juxtaposed rocks (Fig. 13), and northward slip should therefore have led to stratigraphic repetition, not to the observed attenuation (e.g., Fig. 2A, 2C). At least two kinds of explanation are possible. One is that stratigraphic omission was achieved by a component of slip to the southwest during an earlier phase of deformation. In this view, the Harker and Lion Hill Faults are (1) back-thrusts (analogous to Fig. 2F); or (2) expressions of extension in the hinterland of the advancing thrust wedge; or (3) extension faults either older than or younger than the thrusts, but active before Basin-Range block-faulting and tilting. The back-thrust explanation is geometrically possible but mechanically unlikely, because thrust faults do not ordinarily propagate down-section (Dahlstrom, 1970). Explanations invoking earlier extension are discussed below in the context of regional geology. However, local evidence against any postulated slip to the southwest is that there is no support for such a slip direction in minor structures close to the faults.

My preferred explanation for stratigraphic omission on the Harker and Lion Hill Faults involves prior displacement on a hypothetical high-angle (normal?) fault or faults located south or southwest of the Sheeprock Mountains, and requires movement on the low-angle faults only toward the north or northeast, down the present dip. The hypothesis is illustrated with reference to the Lion Hill Fault by means of a sketch cross section in Figure 19, and a similar scenario could be envisaged for the Harker Fault. Locally, the Harker Fault is parallel to bedding in the Kelley Canyon Formation (shale), the oldest unit exposed in the hanging-wall block. The oldest beds exposed above the Lion Hill Fault belong to the Pioche Shale. Thus, the Harker and Lion Hill Faults may have originated as

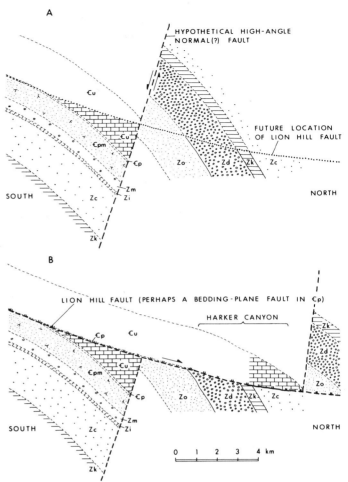

Figure 19. Explanation for stratigraphic omission on the Lion Hill Fault. Large normal separation is first achieved on a hypothetical high-angle fault (or faults) south or southwest of the Sheeprock Mountains (A). Subsequently, hanging-wall and footwall blocks are juxtaposed by northward displacement on the Lion Hill Fault (B). The cross sections are approximately to scale, but only conceptual (solid lines indicate parts constrained by observation). The high-angle fault may be listric at depth. The interpretation of the Lion Hill Fault as a bedding-plane fault in the Pioche Shale ($\mathcal{C}$p) is not a necessary aspect of the model. Other stratigraphic units: Zo, Otts Canyon Formation; Zd, Dutch Peak Formation; Zk, Kelley Canyon Formation; Zc, Caddy Canyon Quartzite; Zi, Inkom Formation; Zm, Mutual (?) Formation; $\mathcal{C}$pm, Prospect Mountain Quartzite; $\mathcal{C}$u, undifferentiated Cambrian rocks.

bedding-plane faults within these incompetent formations, although such an origin is not a necessary aspect of the hypothesis.

Three implications of the hypothesis are that (1) the Harker and Lion Hill Faults were active primarily after pronounced block-faulting, consistent with a late Cenozoic age; (2) the stratigraphic units cut out by the low-angle faults occur in the hanging-wall blocks of these faults beneath the valley fill north of the Sheeprock Mountains; and (3) high-angle faults with large separation terminate against the low-angle faults in both hanging-wall and footwall blocks. The North Oak Brush Fault and Black Crook Fault (Fig. 12) and fault B in Figure 18 may be examples of such faults, although the first two were reactivated after displacement by the Harker Fault. The second and third implications cited above suggest possible tests of this model through deep drilling and the acquisition of seismic reflection profiles for the region north of the Sheeprock Mountains.

A significant problem evident in Figure 19 is the magnitude of slip indicated on the postulated high-angle fault(s) (about 10 km), although the slip could be reduced using different geometric assumptions. It is also conceivable, if unlikely, that rocks in the hanging-wall block of the Lion Hill Fault were derived from south of the Indian Springs Fault or from the lower plate of the Sheeprock Thrust.

Government Creek Fault

The Government Creek Fault (Cohenour, 1959) occurs north of the area on which this paper is primarily focused (Fig. 1), but it is of interest because it is approximately parallel to the Harker and Lion Hill Faults and may be a related structure.

Figure 20 is a cross section through the Government Creek Fault, modified from section DD′ of Cohenour (1959). The fault dips to the north at 20° to 25° and juxtaposes rocks as old as the Caddy Canyon Quartzite above the Middle Cambrian Marjum Formation. Hanging-wall and footwall rocks both dip to the north at about 45°.

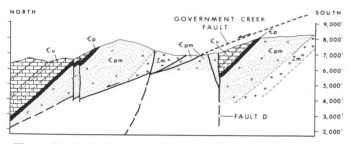

Figure 20. Geologic cross section through the Government Creek Fault (modified from cross section DD′ of Cohenour, 1959). According to Cohenour, fault D terminates against the Government Creek Fault, which is therefore interpreted by me to be younger. Stratigraphic units: Zm, €pm, €p and €u, as in Figure 19.

Cohenour (1959) thought the Government Creek Fault was a strike-slip fault, associated with the Pole Canyon and Lion Hill "Thrusts." However, I propose an alternative interpretation, that it is a Cenozoic extension fault geometrically analogous to the one shown in Figure 2A and 2C. Cohenour's (1959) geologic map indicates that fault D in Figure 20 terminates against the Government Creek Fault in the same way as fault B (Fig. 18) terminates against the Harker Fault. The segment of fault D inferred to be displaced by the Government Creek Fault is not exposed in

the hanging-wall block because another low-angle fault intervenes.

AGE OF DEFORMATION

The timing of deformation is poorly constrained in the Sheeprock Mountains but may be summarized as follows if my preferred interpretations are assumed.

The Sheeprock and Pole Canyon Thrusts and the Indian Springs (tear) Fault deformed Pennsylvanian rocks and are cut by igneous rocks of probable Oligocene to Miocene age. Basin-Range extension, between Miocene and present time, produced both high-angle faults (e.g., the North Oak Brush Fault, the Horse Valley Graben, and range-bounding faults) and low-angle ones (e.g., the Harker, Lion Hill, and Government Creek Faults). Some Mesozoic (tear?) faults, such as the Bennion Creek Fault, were reactivated, possibly during mid- to late Cenozoic time. However, all activity on the Bennion Creek Fault predates intrusion of the Early Miocene Sheeprock Granite.

Regional arguments suggest additional constraints on the age of the Sheeprock and Pole Canyon Thrusts. Stratigraphic evidence in central Utah indicates major orogenic activity during Late Cretaceous time, deformation beginning in latest Jurassic or Early Cretaceous time (Armstrong, 1968; Burchfiel and Hickcox, 1972; Crittenden, 1976). In several well-studied parts of the Cordilleran fold-thrust belt, there is a tendency for the principal thrusts to initiate sequentially in the direction of tectonic transport. Examples are in the Canadian Rocky Mountains (Dahlstrom, 1970; Price, 1981), Wyoming and Idaho (Armstrong and Oriel, 1965; Oriel and Armstrong, 1966; Royse and others, 1975), and in southern South America (Winslow, 1981). With the possible exception of the hypothetical Skull Valley Thrust (Fig. 1), the Sheeprock and Pole Canyon Thrusts are the westernmost thrusts in the foreland belt at the latitude of the Sheeprock Mountains (Fig. 21). The Sheeprock and Pole Canyon Thrusts are therefore likely to be of latest Jurassic to Early Cretaceous age, as proposed by Armstrong (1968).

REGIONAL SIGNIFICANCE OF LOW-ANGLE FAULTS IN SHEEPROCK MOUNTAINS

The relative ages and significance of low-angle faults in the Basin-Range province continue to be controversial. Available evidence suggests that the thrusts are for the most part older than and unrelated to low-angle extension faults, and my preferred interpretation of the structure of the Sheeprock Mountains is consistent with this view. However, because there is still no consensus, it is worth reviewing the spectrum of possible interpretations.

Thrusts in the Cordilleran fold-thrust belt of Utah,

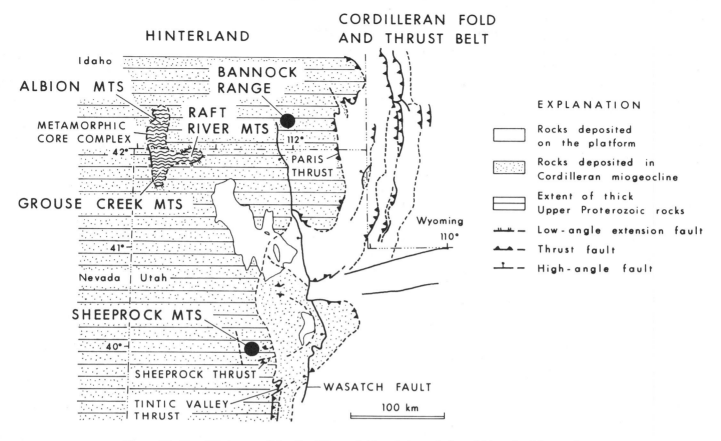

Figure 21. Simplified map of the Cordilleran fold and thrust belt and hinterland in northern Utah and adjacent Idaho and Wyoming (modified from Crittenden, 1976; Allmendinger and Jordan, 1981). Filled circles indicate location of extension faults, discussed in the text, in the southern Bannock Range and Sheeprock Mountains.

Idaho, and Wyoming range in age from Late Jurassic to earliest Eocene (Armstrong and Oriel, 1965; Oriel and Armstrong, 1966; Armstrong, 1968). In contrast, low angle extension faults of the hinterland are thought to be mainly middle to late Cenozoic structures for two reasons. First, many of the extension faults juxtapose rocks of very different metamorphic facies in regions such as the Raft River and Grouse Creek Mountains (Fig. 21), where metamorphism is known to have persisted until latest Oligocene time (Compton and others, 1977; Compton and Todd, 1979). In that area, the horizontal separation of high-grade rocks suggests as much as 30 km of eastward transport after metamorphism. In addition, geobarometry suggests that metamorphism occurred at depths of perhaps 10 to 20 km (Wernicke, 1982), thus limiting the amount of earlier tectonic denudation. A second reason for inferring that the extension faults are predominantly Cenozoic is that there is no evidence for significant surface faulting prior to the deposition of Eocene-Oligocene volcanic and sedimentary rocks, because Cenozoic rocks consistently overlie relatively undeformed and unmetamorphosed Upper Paleozoic beds (Armstrong, 1972; Wernicke, 1982). Many workers

thus now regard low-angle faults of the so-called metamorphic core complexes south of the Snake River Plain for the most part as only slightly older than the Miocene to Holocene Basin-Range faults (Crittenden, 1980). The flat-lying detachment into which the latter are presumed to merge is viewed as structurally lower than detachments associated with the metamorphic core complexes.

However, some pre-Cenozoic extension in the metamorphic terranes is not precluded, and some (older-over-younger) faults may be thrusts rather than extension faults. Available dates suggest that metamorphism, igneous activity, and associated deformation were widespread although possibly sporadic in the hinterland during Jurassic and Cretaceous times (summarized by Allmendinger and Jordan, 1981). For example, premetamorphic to synmetamorphic ductile low-angle faults in the Albion Mountains (Fig. 21) may be as old as Jurassic and thus predate the foreland thrusts (Miller, 1980). In addition, although pure gravity gliding models for the foreland thrusts (e.g., Hose and Daneš, 1973) are untenable (Armstrong, 1972), some workers continue to interpret the principal hinterland extension faults as reactivated Mesozoic structures (Crittenden, 1979; Allmendinger and Jordan, 1981). Allmendinger and Jordan (1981) attributed some of the observed stratigraphic omission to translation of the foreland thrust wedge over independent thermal domes.

In view of the regional evidence for significant Cenozoic extension before the development of the present Basin-Range topography and the possibility that some hinterland extension may have occurred still earlier before or during thrusting, I here reconsider the possibility of pre–late Cenozoic movement on the Harker and Lion Hill Faults of the Sheeprock Mountains. Assuming that the Sheeprock Mountains tilted about 25° to the northeast during late Cenozoic time, these faults, if they existed, would have been near-horizontal in mid-Cenozoic time and southwest- to west-dipping in Jurassic time. Such attitudes would have permitted a component of movement toward the southwest during earlier extensional events (see Fig. 2D, 2E). On the other hand, the suggestion of substantial Mesozoic extension in the Sheeprock area is arguably at odds with the apparent lack of corresponding surface faulting elsewhere in the hinterland (e.g., Wernicke, 1982). Significant Cretaceous extension is particularly unlikely if the fold-thrust belt is primarily a response to regional horizontal compression (e.g., Burchfiel and Davis, 1975; Chapple, 1978; Price, 1981). The possibility of appreciable mid-Cenozoic movement on the Harker and Lion Hill Faults is perhaps more defensible if not demonstrable.

These deliberations about extension faults in the Sheeprock Mountains bear on the interpretation of similar structures in the southern Bannock Range of southeastern Idaho (Fig. 21). That range occurs west of the trace of the Paris Thrust, and like the Sheeprock Mountains it occupies a transitional position between the fold-thrust belt and the hinterland. In the Bannock Range, Oriel and Platt (1979) and Link (1980, 1981, 1982) have recognized three major lithologically distinct "thrust plates" separated by low-angle normal faults. The lowermost plate consists of the Scout Mountain (diamictite) and Bannock Volcanic Members of the Pocatello Formation (equivalent to the Dutch Peak Formation of the Sheeprock Mountains; Christie-Blick, 1982), the middle plate consists of quartzite rocks of the Brigham Group, and the upper plate is composed of Middle Cambrian through Ordovician carbonate rocks. Oriel and Platt (1979) implied that the carbonate plate is structurally continuous with a plate of correlative strata similarly bounded by low-angle normal faults in the Raft River and Albion Mountains 100 km to the west (Fig. 21). A provocative observation is that low-angle faults in the Bannock Range preferentially eliminate predominantly shaly units that are stratigraphically equivalent to strata eliminated or thinned by extension faults in the Sheeprock Mountains (the Kelley Canyon Formation and the Lower Cambrian Pioche Shale; Fig. 5). One interpretation is that prior to late Cenozoic disruption, the "thrust plates" identi-

fied by Oriel and Platt (1979) extended continuously from southern Idaho to central Utah. However, my preferred interpretation is that low-angle faults of the southern Bannock Range and Sheeprock Mountains are only locally developed extension faults of probable late Cenozoic (and/or perhaps mid-Cenozoic) age. In different places, the faults may independently have initiated within mechanically weak stratigraphic units that happen to be regionally persistent (see Fig. 19).

CONCLUSIONS

The internal structure of the southern Sheeprock Mountains is dominated by gently dipping thrusts and extension faults. The Sheeprock Thrust juxtaposes Upper Proterozoic strata above beds of Paleozoic age with a stratigraphic separation exceeding 10 km, and underlies much of the range. The Pole Canyon Thrust is interpreted as an imbrication in the upper plate of the Sheeprock Thrust, and occurs only north of the east-northeast-striking Indian Springs (tear) Fault. The recumbent Pole Canyon Anticline originated through propagation of the thrusts. Tectonic transport inferred parallel to the Indian Springs Fault is corroborated by the vergence of the Pole Canyon Anticline, the direction in which the Sheeprock Thrust ramps upward, a separation arc plot of asymmetrical folds, and the orientation of slickensides on the Pole Canyon Thrust.

The principal extension faults are the Harker and Lion Hill Faults, which together account for stratigraphic omission of as much as several kilometres. Their slip direction is poorly constrained as approximately toward the north or northeast.

The thrust faults deform Pennsylvanian rocks and are cut by igneous rocks of probable Oligocene-Miocene age. Regional arguments suggest deformation in latest Jurassic to Early Cretaceous time. Fault geometry suggests that the extension faults are younger than the thrusts and probably of late Cenozoic age. However, the possibility of mid-Cenozoic or even earlier displacement on the extension faults cannot be entirely ruled out. My preferred interpretation of these faults could be tested through deep drilling or acquisition of seismic reflection profiles north of the range.

The suggested timing of deformation in the southern Sheeprock Mountains is consistent with the view that there was minimal regional extension before middle Cenozoic time and that crustal shortening in the fold-thrust belt is for the most part unrelated to hinterland extension.

ACKNOWLEDGMENTS

This paper is a by-product of a regional study of Upper Proterozoic glacial deposits in northwestern Utah, supported by National Science Foundation grants ATM 74-24201, EAR 77-06008, and EAR 78-15194 to J. C. Crowell.

Although the project was aimed primarily at reconstructing the ancient glacial facies, mapping in the Sheeprock Mountains led to new interpretations of both the Proterozoic stratigraphy and the geologic structure of that range.

It is particularly appropriate that a paper on the latter topic should appear in a volume dedicated to Max Crittenden, to whom I am indebted for many stimulating discussions in the field. He is to be credited with the idea that puzzling stratigraphic and structural relations that I had observed along the northeast flank of the Sheeprock Mountains are due to substantial stratigraphic omission on low-angle faults, an interpretation borne out by subsequent mapping. I also thank T. R. Bultman and C. C. Wielchowsky for discussions about deformational mechanisms in fold-thrust belts, and R. G. Bohannon, S. H. Lingrey, D. M. Miller, and H. T. Morris for helpful criticisms that measurably improved the manuscript. All drafts were typed by Donna Kelly. Publication has been approved by Exxon Production Research Company.

REFERENCES CITED

Allmendinger, R. W., and Jordan, T. E., 1981, Mesozoic evolution, hinterland of the Sevier orogenic belt: Geology, v. 9, p. 308–313.

Armstrong, F. C., and Oriel, S. S., 1965, Tectonic development of Idaho-Wyoming thrust belt: American Association of Petroleum Geologists Bulletin, v. 49, p. 1847–1866.

Armstrong, R. L., 1968, Sevier orogenic belt in Nevada and Utah: Geological Society of America Bulletin, v. 79, p. 429–458.

—— 1970, Geochronology of Tertiary igneous rocks, eastern Basin and Range province, western Utah, eastern Nevada, and vicinity, U.S.A.: Geochimica et Cosmochimica Acta, v. 34, p. 203–232.

—— 1972, Low-angle (denudation) faults, hinterland of the Sevier orogenic belt, eastern Nevada and western Utah: Geological Society of America Bulletin, v. 83, p. 1729–1754.

Bates, R. L., and Jackson, J. A., eds., 1980, Glossary of geology (second edition): Virginia, American Geological Institute, 749 p.

Blick, N. H., 1979, Stratigraphic, structural and paleogeographic interpretation of upper Proterozoic glaciogenic rocks in the Sevier orogenic belt, northwestern Utah [Ph.D. dissertation]: Santa Barbara, University of California, 633 p.

Burchfiel, B. C., and Davis, G. A., 1975, Nature and controls of Cordilleran orogenesis, western United States: Extensions of an earlier synthesis: American Journal of Science, v. 275-A, p. 363–396.

Burchfiel, B. C., and Hickcox, C. W., 1972, Structural development of central Utah, *in* Baer, J. L., and Callaghan, E., eds., Plateau—Basin and Range transition zone, central Utah: Utah Geological Association Publication 2, p. 55–66.

Chapple, W. M., 1978, Mechanics of thin-skinned fold-and-thrust belts: Geological Society of America Bulletin, v. 89, p. 1189–1198.

Christie-Blick, N., 1982, Upper Proterozoic and Lower Cambrian rocks of the Sheeprock Mountains, Utah: Regional correlation and significance: Geological Society of America Bulletin, v. 93, p. 735–750.

Cohenour, R. E., 1959, Sheeprock Mountains, Tooele and Juab Counties: Precambrian and Paleozoic stratigraphy, igneous rocks, structure, geomorphology, and economic geology: Utah Geological and Mineralogical Survey Bulletin 63, 201 p.

—— 1970, Sheeprock Granite, *in* Whelan, J. A., compiler, Radioactive and isotopic age determinations of Utah rocks: Utah Geological and Mineralogical Survey Bulletin 81, p. 31.

Compton, R. R., and Todd, V. R., 1979, Oligocene and Miocene metamorphism, folding, and low-angle faulting in northwestern Utah: Reply: Geological Society of America Bulletin, v. 90, p. 307–309.

Compton, R. R., Todd, V. R., Zartman, R. E., and Naeser, C. W., 1977, Oligocene and Miocene metamorphism, folding, and low-angle faulting in northwestern Utah: Geological Society of America Bulletin, v. 88, p. 1237–1250.

Coney, P. J., 1980, Cordilleran metamorphic core complexes: An overview, *in* Crittenden, M. D., Jr., Coney, P. J., and Davis, G. H., eds., Cordilleran metamorphic core complexes: Geological Society of America Memoir 153, p. 7–31.

Crittenden, M. D., Jr., 1976, Stratigraphic and structural setting of the Cottonwood area, Utah, *in* Hill, J. G., ed., Geology of the Cordilleran hingeline: Rocky Mountain Association of Geologists, Symposium, p. 363–379.

—— 1979, Oligocene and Miocene metamorphism, folding, and low-angle faulting in northwestern Utah: Discussion: Geological Society of America Bulletin, v. 90, p. 305–306.

—— 1980, Metamorphic core complexes of the North American Cordillera: Summary, *in* Crittenden, M. D., Jr., Coney, P. J., and Davis, G. H., eds., Cordilleran metamorphic core complexes: Geological Society of America Memoir 153, p. 485–490.

Crittenden, M. D., Jr., Schaeffer, F. E., Trimble, D. E., and Woodward, L. A., 1971, Nomenclature and correlation of some upper Precambrian and basal Cambrian sequences in western Utah and southeastern Idaho: Geological Society of America Bulletin, v. 82, p. 581–602.

Crittenden, M. D., Jr., Christie-Blick, N., and Link, P. K., 1983, Evidence for two pulses of glaciation during the late Proterozoic in northern Utah: Geological Society of America Bulletin, in press.

Crowell, J. C., 1959, Problems of fault nomenclature: American Association of Petroleum Geologists Bulletin, v. 43, p. 2653–2674.

Dahlstrom, C.D.A., 1970, Structural geology in the eastern margin of the Canadian Rocky Mountains: Bulletin of Canadian Petroleum Geology, v. 18, p. 332–406.

Davis, G. A., Anderson, J. L., Frost, E. G., and Shackelford, T. J., 1980, Mylonitization and detachment faulting in the Whipple-Buckskin-Rawhide Mountains terrane, southeastern California and western Arizona, *in* Crittenden, M. D., Jr., Coney, P. J., and Davis, G. H., eds., Cordilleran metamorphic core complexes: Geological Society of America Memoir 153, p. 79–129.

de Sitter, L. U., 1964, Structural geology: New York, McGraw-Hill, 551 p.

Eardley, A. J., 1939, Structure of the Wasatch–Great Basin region: Geological Society of America Bulletin, v. 50, p. 1277–1310.

Eardley, A. J., and Hatch, R. A., 1940, Proterozoic(?) rocks in Utah: Geological Society of America Bulletin, v. 51, p. 795–843.

Gardner, W. C., 1954, Geology of the West Tintic mining district and vicinity, Juab County, Utah [M.S. thesis]: Salt Lake City, University of Utah, 43 p.

Groff, S. L., 1959, Geology of the West Tintic Range and vicinity, Tooele and Juab Counties, Utah [Ph.D. thesis]: Salt Lake City, University of Utah, 183 p.

Hansen, E., 1971, Strain facies: New York, Springer-Verlag, 208 p.

Harris, D., 1958, The geology of Dutch Peak area, Sheeprock Range, Tooele County, Utah: Brigham Young University, Research Studies, Geology Series, v. 5, no. 1, 82 p.

Harrison, J. E., and Peterman, Z. E., 1980, North American Commission on Stratigraphic Nomenclature Note 52—A preliminary proposal for a chronometric time scale for the Precambrian of the United States and Mexico: Geological Society of America Bulletin, Part I, v. 91, p. 377–380.

Hill, M. L., 1959, Dual classification of faults: American Association of Petroleum Geologists Bulletin, v. 43, p. 217–237.

Hintze, L. F., 1973, Geologic history of Utah: Brigham Young University

Geology Studies, v. 20, pt. 3, 181 p.

Hintze, L. F., and Robison, R. A., 1975, Middle Cambrian stratigraphy of the House, Wah Wah, and adjacent ranges in western Utah: Geological Society of America Bulletin, v. 86, p. 881–891.

Hose, R. K., and Daneš, Z. F., 1973, Development of the late Mesozoic to early Cenozoic structures of the eastern Great Basin, in de Jong, K. A., and Scholten, R., eds., Gravity and tectonics: New York, John Wiley and Sons, p. 429–441.

James, H. L., 1972, Stratigraphic Commission Note 40—Subdivision of Precambrian: An interim scheme to be used by U.S. Geological Survey: American Association of Petroleum Geologists Bulletin, v. 56, p. 1128–1133.

—— 1978, Subdivision of the Precambrian—a brief review and a report on recent decisions by the Subcommission on Precambrian Stratigraphy: Precambrian Research, v. 7, p. 193–204.

King, P. B., 1969, The tectonics of North America—a discussion to accompany the tectonic map of North America, scale 1:5,000,000: U.S. Geological Survey Professional Paper 628, 95 p.

Lindsey, D. A., Naeser, C. W., and Shawe, D. R., 1975, Age of volcanism, intrusion, and mineralization in the Thomas Range, Keg Mountain, and Desert Mountain, western Utah: U.S. Geological Survey Journal of Research, v. 3, p. 597–604.

Link, P. K., 1980, Younger-over-older thrust fault reactivated as landslide surface, Oxford Mountain, Bannock Range, southeastern Idaho [abs.]: Geological Society of America Abstracts with Programs, v. 12, no. 3, p. 116.

—— 1981, Geometry of younger-over-older thrusts and high-angle normal faults, Oxford Mountain, Bannock Range, southeast Idaho [abs.]: Geological Society of America Abstracts with Programs, v. 13, no. 2, p. 67.

—— 1982, Geology of the Upper Proterozoic Pocatello Formation, Bannock Range, southeastern Idaho [Ph.D. dissertation]: Santa Barbara, University of California, 131 p.

Lipman, P. W., Prostka, H. J., and Christiansen, R. L., 1972, Cenozoic volcanism and plate-tectonic evolution of the western United States. I. Early and Middle Cenozoic: Philosophical Transactions of the Royal Society of London A, v. 271, p. 217–248.

Loughlin, G. F., 1920, Sheeprock Mountains, in Butler, B. S., Loughlin, G. F., Heikes, V. C., and others, The ore deposits of Utah: U.S. Geological Survey Professional Paper 111, p. 423–444.

McClay, K. R., 1981, What is a thrust? What is a nappe?, in McClay, K. R., and Price, N. J., eds., Thrust and nappe tectonics: Geological Society of London Special Publication 9, p. 7–9.

McKee, E. H., 1971, Tertiary igneous chronology of the Great Basin of western United States—Implications for tectonic models: Geological Society of America Bulletin, v. 82, p. 3497–3502.

Miller, D. M., 1980, Structural geology of the northern Albion Mountains, south-central Idaho, in Crittenden, M. D., Jr., Coney, P. J., and Davis, G. H., eds., Cordilleran metamorphic core complexes: Geological Society of America Memoir 153, p. 399–423.

Moore, W. J., and Sorensen, M. L., 1979. Geologic map of the Tooele 1° by 2° quadrangle, Utah: U.S. Geological Survey Miscellaneous Investigations Series, Map I-1132, scale 1:250,000.

Moores, E. M., Scott, R. B., and Lumsden, W. W., 1970, Tertiary tectonics of the White Pine–Grant Range region, east-central Nevada, and some regional implications: Reply: Geological Society of America Bulletin, v. 81, p. 323–330.

Morris, H. T., 1977, Geologic map and sections of the Furner Ridge Quadrangle, Juab County, Utah: U.S. Geological Survey, Map I-1045, scale 1:24,000.

Morris, H. T., and Kopf, R. W., 1967, Breccia pipes in the West Tintic and Sheeprock Mountains, Utah: U.S. Geological Survey Professional Paper 575-C, p. 66–71.

—— 1969, Tintic Valley thrust and associated low-angle faults, central Utah [abs.]: Geological Society of America Abstracts with Programs,

v. 1, no. 5, p. 55–56.

—— 1970a, Preliminary geologic map and cross section of the Cherry Creek Quadrangle and adjacent part of the Dutch Peak Quadrangle, Juab County, Utah: U.S. Geological Survey Open-File Map, scale 1:24,000.

—— 1970b, Preliminary geologic map and cross section of the Maple Peak Quadrangle and adjacent part of the Sabie Mountain Quadrangle, Juab County, Utah: U.S. Geological Survey Open-File Map, scale 1:24,000.

Morris, H. T., and Lovering, T. S., 1961, Stratigraphy of the East Tintic Mountains, Utah: U.S. Geological Survey Professional Paper 361, 145 p.

Morris, H. T., and Shepard, W. M., 1964, Evidence for a concealed tear fault with large displacement in the central East Tintic Mountains, Utah: U.S. Geological Survey Professional Paper 501-C, p. 19–21.

Oriel, S. S., and Armstrong, F. C., 1966, Times of thrusting in Idaho-Wyoming thrust belt: Reply: American Association of Petroleum Geologists Bulletin, v. 50, p. 2614–2621.

Oriel, S. S., and Platt, L. B., 1979, Younger-over-older thrust plates in southeastern Idaho [abs.]: Geological Society of America Abstracts with Programs, v. 11, no. 6, p. 298.

Price, R. A., 1981, The Cordilleran foreland thrust and fold belt in the southern Canadian Rocky Mountains, in McClay, K. R., and Price, N. J., eds., Thrust and nappe tectonics: Geological Society of London Special Publication 9, p. 427–447.

Rehrig, W. A., and Reynolds, S. J., 1980, Geologic and geochronologic reconnaissance of a northwest-trending zone of metamorphic core complexes in southern and western Arizona, in Crittenden, M. D., Jr., Coney, P. J., and Davis, G. H., eds., Cordilleran metamorphic core complexes: Geological Society of America Memoir 153, p. 131–157.

Roberts, R. J., Crittenden, M. D., Jr., Tooker, E. W., Morris, H. T., Hose, R. K., and Cheney, T. M., 1965, Pennsylvanian and Permian basins in northwestern Utah, northeastern Nevada and south-central Idaho: American Association of Petroleum Geologists Bulletin, v. 49, p. 1926–1956.

Royse, F., Jr., Warner, M. A., and Reese, D. L., 1975, Thrust belt structural geometry and related stratigraphic problems, Wyoming-Idaho-northern Utah, in Bolyard, D. W., ed., Deep drilling frontiers of the central Rocky Mountains: Rocky Mountain Association of Geologists, Symposium, p. 41–54.

Stewart, J. H., 1980, Regional tilt patterns of late Cenozoic basin-range fault blocks, western United States: Geological Society of America Bulletin, Part 1, v. 91, p. 460–464.

Stokes, W. L., compiler, 1963, Geologic map of northwestern Utah: Salt Lake City, University of Utah, College of Mines and Mineral Industries.

Stringham, B. F., 1942, Mineralization in the West Tintic mining district, Utah: Geological Society of America Bulletin, v. 53, p. 267–290.

Suppe, J., 1979, Fault-bend folding [abs.]: Geological Society of America Abstracts with Programs, v. 11, no. 7, p. 525.

Thomas, G. H., 1958, The geology of Indian Springs Quadrangle, Tooele and Juab Counties, Utah [M.S. thesis]: Provo, Brigham Young University, 35 p.

Wernicke, B., 1982, Mesozoic evolution, hinterland of the Sevier orogenic belt: Comment: Geology, v. 10, p. 3–5.

Winslow, M. A., 1981, Mechanisms for basement shortening in the Andean foreland fold belt of southern South America, in McClay, K. R., and Price, N. J., eds., Thrust and nappe tectonics: Geological Society of London Special Publication 9, p. 513–528.

MANUSCRIPT ACCEPTED BY THE SOCIETY AUGUST 20, 1982

Geological Society of America
Memoir 157
1983

Mesozoic and Cenozoic tectonic development of the Muddy, North Muddy, and northern Black Mountains, Clark County, Nevada

Robert G. Bohannon
U.S. Geological Survey
Mail Stop 917, Box 25046, Denver Federal Center
Denver, Colorado 80225

ABSTRACT

The Muddy and North Muddy Mountains were highly deformed during the Cretaceous and early Tertiary(?) Sevier orogeny. Two major thrust plates, the Muddy Mountain and Summit–Willow Tank plates, and a well-developed belt of recumbent folds are present there. A foreland basin developed east of the thrust plates as well. Many of the Paleozoic rock units exposed in these ranges vary in thickness and character between the thrust plates and the authochthon; the thicker and more shelflike units occur in the thrust plates, and the thinner, cratonic ones occur in the autochthon. This is especially true of the lower Paleozoic formations, whereas Mississippian through Permian formations vary much less in thickness. Pre-Sevier age Mesozoic formations exposed locally in the upper plate of the Summit–Willow Tank thrust and in the autochthon are consistent in thickness, and they are stratigraphically similar to equivalent rocks on the nearby Colorado Plateau. The oldest syntectonic Albian and Cenomanian(?) rocks of the foreland basin are cut by the Summit–Willow Tank thrust, but the youngest of them overlap it. The Muddy Mountain thrust overrode all of the foreland deposits.

New interpretations are presented herein regarding many of the Mesozoic structures of the Muddy and North Muddy Mountains. The Muddy Mountain, Glendale, and Arrowhead faults are interpreted to be parts of the same thrust and are collectively called the Muddy Mountain thrust. Three major faults, the Summit, the Willow Tank, and the North Buffington, form the older and structurally lower Summit–Willow Tank thrust. The North Muddy Mountain fold belt is interpreted to have developed synchronously with thrusting as it occurs between the two major thrusts. The monocline or drape fold described by early workers is reinterpreted as a truncation of one thrust plate by another. The North Muddy Mountains are now interpreted to underlie the Muddy Mountain plate and are thus autochthonous with respect to it.

After thrusting and foreland sedimentation, the next major tectonic and sedimentologic events occurred during Miocene time. Several complex, nonmarine, closed basins of the Horse Spring Formation formed in taphrogenic environments in conjunction with left slip on the Lake Mead fault system and simultaneous growth of the andesitic Hamblin-Cleopatra volcano. Basin and ranges of late Miocene age are superposed over all of these earlier-formed structures.

South of the Muddy Mountains, upper plate Paleozoic rocks dip southward at varying degrees and are overlapped by Tertiary rocks. The upper plate rocks are in apparent tectonic contact with autochthonous Mesozoic rocks in the Gale Hills,

Bowl of Fire, northern Bitter Spring Valley, and east of the Longwell Ridges. Unfortunately, this contact is almost completely covered and it is poorly understood. Available data suggest that the contact is a fault that formed prior to the deposition of the Thumb Member of the Horse Spring Formation (pre-15 m.y.), and it might have been (1) a north-dipping normal fault with a large amount of displacement; (2) the trace of the Las Vegas Valley shear zone with a large amount of right slip; or (3) the trace of the Muddy Mountain thrust (Mississippian over Cretaceous and Jurassic rocks at that point). The interpretation that it is the trace of the Las Vegas Valley shear zone is favored herein because neither of the other interpretations is as consistent with the known local and regional geology.

Post-15-m.y.-old deformation south of the Muddy Mountains includes a complex group of tectonic elements. A relatively minor amount of extensional tectonism occurred in the Gale Hills and was bounded to the north by the northwest-striking Gale Hills fault and to the south by the east-trending West Bowl of Fire fault. Uplift and relative southward displacement of the Bowl of Fire horst occurred on major oblique slip faults probably at the same time as the extension in the Gale Hills. North-south directed compression in the Gale Hills also apparently occurred at this time, as east-trending folds and reverse faults are present there. Normal faulting and graben formation occurred in White Basin in conjunction with oblique slip on the Rogers Spring fault. All of these local tectonic elements were probably active synchronously with left slip on the Lake Mead fault system and right slip on the Las Vegas Valley shear zone, but the details of interaction are poorly understood.

INTRODUCTION

The Muddy, North Muddy, and northern Black Mountains lie north of Lake Mead in southeastern Nevada, about 50 km east-northeast of Las Vegas (Fig. 1). The ranges are structurally and stratigraphically diverse because they occur both in the Cordilleran hingeline at the eastern margin of the Cretaceous thrust belt and in the highly deformed transition zone between the southeastern Great Basin and northern Sonoran Desert sections of the Basin and Range province. Cretaceous and early Tertiary(?) compressional tectonic events greatly affected the Muddy and North Muddy Mountains and resulted in thrust faults, overturned folds, and the development of a foreland basin. Along the most conspicuous and regionally extensive compressional structure, the Muddy Mountain thrust, Cambrian rocks were thrust eastward over a geologically complex terrane. Relations between the Muddy Mountain thrust and other nearby associated compression-related structures are complex and, as such, somewhat enigmatic. Nevertheless, near-perfect exposures, excellent three-dimensional control at the surface, and a broad spectrum of tectonic features related to thrusting make the Muddy and North Muddy Mountains ideal for study of thrust and foreland tectonism in southern Nevada.

Intense middle to late Tertiary tectonism, volcanism, and sedimentation overprinted most of the Cretaceous and early Tertiary features south of the Muddy Mountains and in the northern part of the Black Mountains. In the Gale Hills, White Basin, and Bitter Spring Valley, thick taphrogenic deposits formed in complexly evolving basins which developed in part sychronously with lateral faulting on faults of the Lake Mead fault system and with growth of a large andesitic to rhyodacitic stratovolcano. Although Cretaceous and Tertiary structures are spatially coincident and consequently hard to differentiate from one another throughout the Muddy Mountain area, the thick, widespread, and dateable synorogenic sedimentary and volcanic rocks south of the Muddy Mountains provide a good record of the extent of Tertiary deformation that is not available in most other places.

The first geologic excursions into southern Nevada were those of the Wheeler expeditions of 1871 and 1872, but detailed work on the structure and stratigraphy of the Muddy, North Muddy, and northern Black Mountains was not begun until about 1917 by C. R. Longwell. From then until his death in 1975, Longwell contributed considerably to knowledge of the local geologic framework with numerous publications including a particularly outstanding contribution in which he and his collaborators (Longwell and others, 1965) summarized much of the geology of Clark County. R. E. Anderson augmented Longwell's work by mapping extensional tectonic features south of Lake Mead (Anderson, 1971, 1977, 1978) and strike-slip faults in the northern Black Mountains (Anderson, 1973). Structural studies of relations along the Muddy Mountain thrust have been subsequently conducted by Brock and Engelder (1977), Temple (1977), and Arnold (1977).

This paper is an outgrowth of new geologic mapping (Bohannon, 1977a, 1977b, 1981, 1983a) in the Muddy and

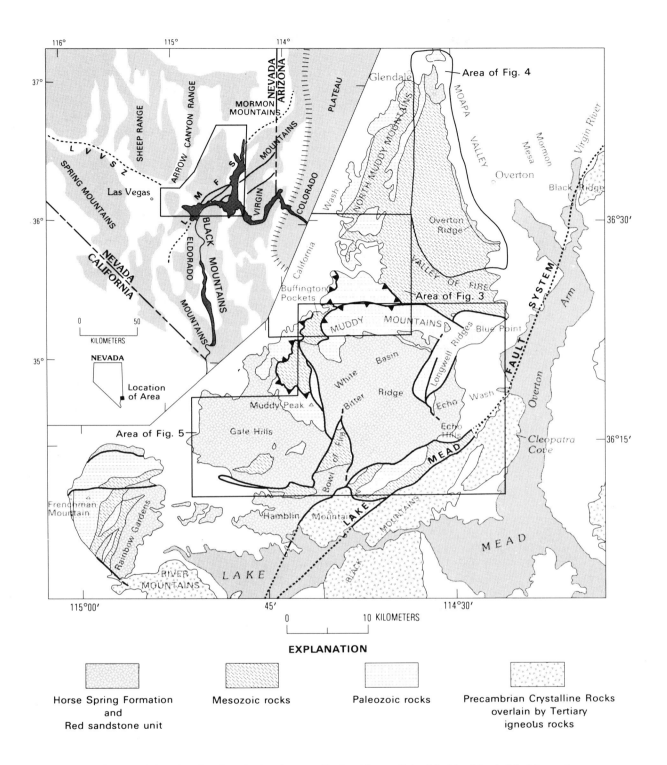

Figure 1. Map showing location and generalized geology of the Muddy, North Muddy, and northern Black Mountains in the southeastern Great Basin and northeastern Sonoran Desert. LMFS, Lake Mead fault system; LVVSZ, Las Vegas Valley shear zone.

North Muddy Mountains and is based on the geologic foundation built by C. R. Longwell. My mapping has been integrated with Anderson's (1973) geologic map of the northern part of the Black Mountains. Other related structural and stratigraphic studies (Bohannon, 1979, 1983b) also originated from the same mapping effort. The purpose of this paper is to provide up-to-date descriptions and interpretations of structures and tectonic features in the Muddy and North Muddy Mountains and to compare the geology there with that in the northern part of the Black Mountains and at Frenchman Mountain. Geologic maps of the northwestern part of the North Muddy Mountains by Longwell (1949), of Frenchman Mountain by C. R. Longwell (unpublished mapping on file in the library at the University of Nevada at Las Vegas) and by Bell and Smith (1980), and of the Virgin Mountains by Morgan (1964) provide a good information base in those areas.

STRATIGRAPHIC SETTING

Lower Paleozoic rocks in the Muddy, North Muddy, and northern Black Mountains were deposited in the hinge zone between the relatively stable cratonic platform to the east and the slowly subsiding continental shelf to the west. Large differences in thickness and facies exist between the rocks deposited on the platform and those deposited on the miogeoclinal shelf. Miogeoclinal Middle to Upper Cambrian and Ordovician rocks in the Muddy Mountain thrust allochthon include the Bonanza King, Nopah, and Monocline Valley Formations (Fig. 2) and are similar to equivalent rocks in the Keystone allochthon of the western part of the Spring Mountains (Burchfiel and others, 1974). In contrast to the thick allochthonous Middle and Upper Cambrian and Ordovician rocks, autochthonous stratigraphic sections exposed at Frenchman Mountain (C. R. Longwell, unpublished mapping of Frenchman Mountain) and in the Virgin Mountains (Morgan, 1964) include only thin Lower and Middle Cambrian units such as the Bright Angel Shale and Muav Formation and discontinuous Upper Cambrian and Ordovician rocks similar to those exposed at the western edge of the Colorado Plateau (Palmer, 1981; Palmer and Nelson, 1981). Similarly, the Devonian Sultan (Muddy Peak) Limestone in the allochthon of the Muddy Mountains (Fig. 2) is divisible into members equivalent to the Ironside, Valentine, and Crystal Pass described by Hewett (1931) in the Spring Mountains, whereas the authochthonous Devonian rocks at Frenchman Mountain, in the Virgin Mountains, and on the Colorado Plateau contain strata correlative with only the Valentine Member of the Sultan (McNair, 1951; Morgan, 1964; Moore, 1972).

The shelf-craton configuration was disrupted throughout the eastern Great Basin at about the same time as the Late Devonian to Mississippian Antler orogeny. The Mississippian Rogers Spring, Monte Cristo, and Redwall Limestones maintain consistent thicknesses and four- to five-member stratigraphies (from oldest to youngest in the Monte Cristo, they are the Dawn, Anchor, Bullion, Arrowhead, and Yellowpine Members as described by Hewett, 1931) from the Spring Mountains to the western Colorado Plateau. This evidence of remarkably stable conditions suggests that subsidence was minimal across the former continental shelf and cratonic platform transition zone during Mississippian time. The Bird Spring Formation, Callville Limestone, and Pakoon Formation of Pennsylvanian and Permian age are also relatively consistent in thickness throughout the Muddy Mountain area, but are considerably thicker than the Mississippian rocks. There are few discernable differences in the Bird Spring Formation between the allochthon and the authochthon in the Muddy Mountain area (Fig. 2), but a major facies change between carbonate rocks of the Bird Spring and sandstone of the cratonal lower Esplanade Sandstone of the Supai group occurs on the Colorado Plateau to the east (Longwell and Dunbar, 1936). The autochthonous Permian red beds [upper Esplanade(?)], Toroweap Formation, and Kaibab Limestone of the Muddy Mountain area are consistent in thickness and stratigraphy from the Spring Mountains to the Colorado Plateau and thus indicate widespread, moderately stable, environmental conditions regionally.

Throughout the Lake Mead region, Triassic and Jurassic rocks are lithologically similar to, and of about the same thickness as, equivalent rock units in southwestern Utah and northwestern Arizona, indicating widespread, relatively stable conditions during that time as well. The early and middle Mesozoic rocks are autochthonous in all the areas on Figure 1, except in parts of the North Muddy Mountains where they are in the Summit and Willow Tank thrust plates (Fig. 2). None are known in the allochthon of the Muddy Mountain thrust, although Burchfiel and others (1974) reported Lower and Middle(?) Triassic Moenkopi Formation in the equivalent Keystone allochthon of the Spring Mountains. Locally, the Moenkopi consists of six continental, marginal marine, and nearshore marine members that are, from oldest to youngest, the Timpoweap, the lower red, the Virgin Limestone, the middle red, the Shnabkaib, and the upper red (Reif and Slatt, 1979). The Upper Triassic Chinle Formation is mostly of fluvial origin and consists of two members, the basal Shinarump and overlying Petrified Forest (Stewart and others, 1972). Although the Upper Triassic Moenave and Kayenta Formations of southwest Utah are not differentiable from one another in the Muddy Mountain area, equivalent rocks do occur (Wilson and Stewart, 1967) and probably indicate deposition in a fluvial-deltaic or marginal marine environment. The Lower Jurassic Aztec Sandstone is part of a thick, widespread eolian quartz arenite deposit that includes the Navajo Sandstone of southwest Utah (Poole,

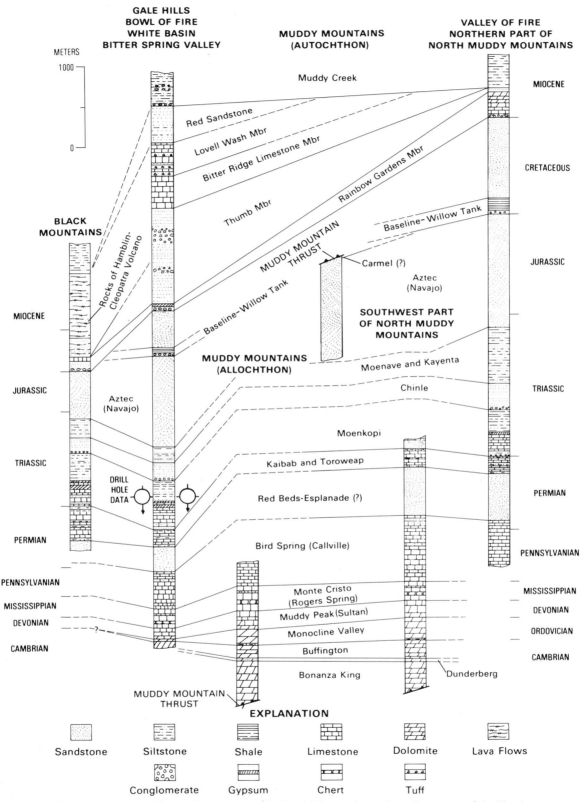

Figure 2. Stratigraphic relations between the Black Mountains and northern part of the North Muddy Mountains. Terms "allochthon" and "autochthon" used only with respect to Muddy Mountain thrust of Cretaceous age.

1964). At one locality south of Buffington Pockets, near-shore marine limestone crops out just beneath the Muddy Mountain thrust near the top of the Aztec Sandstone. This limestone possibly correlates with the Jurassic Carmel Formation of southwest Utah described by Cashion (1967), and it might rest depositionally on the Aztec. Alternatively, it might correlate with the Virgin Limestone Member of the Moenkopi and be caught in a thrust slice. If the limestone is indeed Carmel, is it the only known occurrence in southern Nevada.

In contrast to the preorogenic Triassic and Jurassic rocks, the synorogenic Cretaceous Willow Tank and Baseline Formations (Fig. 2) are foreland basin deposits and are probably time-equivalents to parts of the Iron Spring Formation in southwestern Utah. The foreland basin extends from north-central Utah to southern Nevada east of the Sevier thrust belt (Armstrong, 1968). The Lower Cretaceous Willow Tank is chiefly claystone deposited in a swamp environment (Ash and Read, 1976), although conglomerate at its base is probably a pediment gravel which formed above the Aztec Sandstone in an arid environment. The Upper and Lower Cretaceous Baseline consists of quartz arenite and conglomerate (Overton Conglomerate Member). Local source areas of Mesozoic rocks in the folded terrane of the North Muddy Mountains provided the quartz arenite and some of the conglomerate, but most of the Overton Conglomerate Member was derived from a source in the Muddy Mountain allochthon. The Willow Tank and Baseline, which are widespread in the Valley of Fire, are also present in the Gale Hills, the Bowl of Fire, and the Virgin Mountains. The fern *Tempskya*, several other ferns, ostracodes, fish teeth and bones, and *Chara* from the Willow Tank and the lower part of the Baseline indicate that these beds are late Early Cretaceous (Albian) in age (Ash and Read, 1976). The upper part of the Baseline is not well dated, and though a Cenomanian(?) age has been applied (Ash and Read, 1976), the upper part might be as young as early Tertiary. K/Ar age determinations of 96.4 and 98.4 m.y. on biotite from tuffaceous beds in the Willow Tank reported by Fleck (1970, table I) agree with the fossil ages.

During Miocene time, the structural configuration inherited from Cretaceous and early Tertiary(?) compressional events was altered considerably by the formation of several taphrogenic nonmarine basins with interior drainage and the growth of the Hamblin-Cleopatra stratovolcano. The following summary is abstracted from my description of these basins (Bohannon, 1983b) and from Anderson's (1973) description of the stratovolcano.

The four members of the Miocene Horse Spring Formation, which are, from oldest to youngest, the Rainbow Gardens, Thumb, Bitter Ridge Limestone, and Lovell Wash (Fig. 2), formed in three tectonically and temporally distinct basins. The Rainbow Gardens basin, a broad, shal-low depression in which carbonate and clastic rocks accumulated on a broad pediment surface and in lacustrine environments, formed about 20 to 16.5 m.y. ago above gently deformed autochthonous Mesozoic and upper Paleozoic beds between Frenchman Mountain and the Virgin Mountains (Fig. 1). The total extent of the basin and the structural and topographic conditions at its margin are uncertain, but relict high topography in the Muddy Mountain allochthon probably formed its northwestern margin. The Thumb basin developed about 16.5 to 14 or possibly 13.5 m.y. ago and covered roughly the same area as the Rainbow Gardens basin. However, unlike the earlier basin, the Thumb basin was a deep, northeast(?)-trending, fault-bounded trough in which a thick deposit of breccia, conglomerate, sandstone, gypsum, and tuffaceous rocks accumulated. The Lake Mead fault system, with about 65 km of left slip (Anderson, 1973; Bohannon, 1979), fragmented the Rainbow Gardens and Thumb basins and strongly controlled the geometry of the syntectonic Bitter Ridge–Lovell Wash basin which formed about 14 or possibly 13.5 to 12 m.y. ago. Lacustrine limestone, claystone, sandstone, and tuffaceous rocks of the Lovell Wash and Bitter Ridge Limestone Members are confined to the north side of the Lake Mead fault system between the southern part of the Muddy Mountains and Frenchman Mountain.

Andesite to rhyodacite flows and breccia beds accumulated to form the Hamblin-Cleopatra volcano in the northern Black Mountains during late Thumb and Bitter Ridge–Lovell Wash time. The volcano is part of a reginal volcanic field (Longwell, 1963) whose northern margin is disrupted by the Lake Mead fault system.

Early basin and range crustal extension resulted in the formation of White Basin (Fig. 1) in which red sandstone, tuff, conglomerate, and gypsum were deposited about 12 to 10.5 or possibly 10 m.y. ago. Several younger, larger basin and range basins surround the Muddy, North Muddy, and northern Black Mountains and contain thick deposits of sandstone, siltstone, conglomerate, and gypsum of the Miocene Muddy Creek Formation which accumulated between 10 and 5 m.y. ago.

MESOZOIC AND EARLY TERTIARY(?) STRUCTURES AND TECTONIC FEATURES

East-directed compressional deformation during the Cretaceous and early Tertiary(?) Sevier orogenic disturbance (Armstrong, 1968) created a complex structural configuration in the Muddy and North Muddy Mountains. Traces of thrust faults in the vicinity of Buffington Pockets (Fig. 1) and several remnants of a thrust plate scattered throughout the North Muddy Mountains were interpreted by Longwell (1922, 1949, 1962) as belonging to two major structures, the Muddy Mountain and Glendale thrusts (Fig. 3). He thought that the Muddy Mountain thrust was

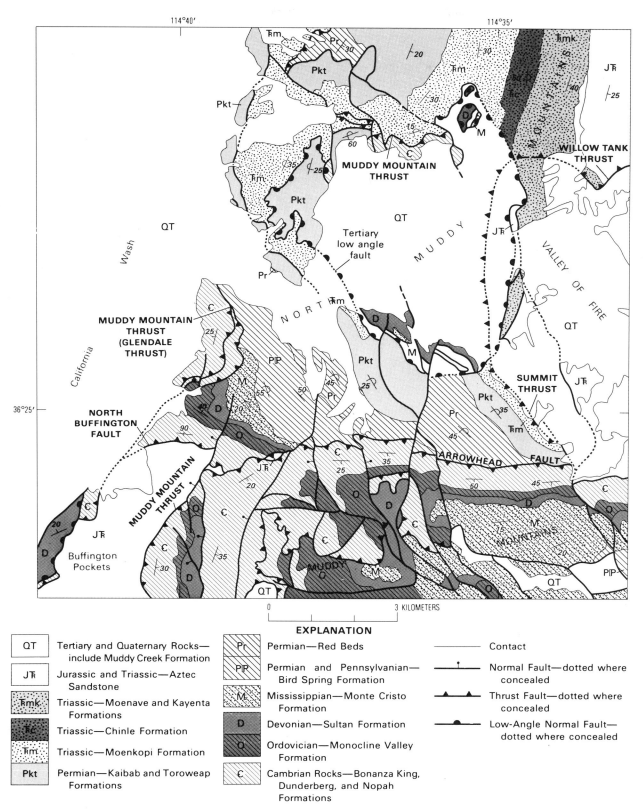

EXPLANATION

QT — Tertiary and Quaternary Rocks—
include Muddy Creek Formation

JŦ — Jurassic and Triassic—Aztec
Sandstone

Ŧmk — Triassic—Moenave and Kayenta
Formations

ŦC — Triassic—Chinle Formation

Ŧm — Triassic—Moenkopi Formation

Pkt — Permian—Kaibab and Toroweap
Formations

Pr — Permian—Red Beds

PłP — Permian and Pennsylvanian—
Bird Spring Formation

M — Mississippian—Monte Cristo
Formation

D — Devonian—Sultan Formation

O — Ordovician—Monocline Valley
Formation

Ϲ — Cambrian Rocks—Bonanza King,
Dunderberg, and Nopah
Formations

——— Contact

——⊥—— Normal Fault—dotted where
concealed

——▲▲—— Thrust Fault—dotted where
concealed

——●●—— Low-Angle Normal Fault—
dotted where concealed

Figure 3. Geologic map of the northwestern part of the Muddy Mountains and southwestern
part of the North Muddy Mountains.

the older structure, that it was stratigraphically controlled with Middle and Upper Cambrian Bonanza King Formation over Jurassic and Triassic(?) Aztec Sandstone regionally, and that it was deformed by the Glendale thrust. Cambrian to Mississippian rocks occur at the sole of the Glendale thrust of Longwell (1949), and Ordovician to Triassic rocks occur beneath it. Lesser thrusts reported by Longwell (1949) include the Summit thrust in the southwest part of the Valley of Fire, which places overturned Permian and Triassic rocks over Aztec Sandstone, and the Willow Tank thrust in the northern part of the Valley of Fire, which places Upper Triassic and Jurassic rocks over Cretaceous rocks. In addition to these thrusts, Longwell (1928) described the east-trending Arrowhead fault (Fig. 3) as a left-slip fault genetically related to thrusting. He later reinterpreted it (Longwell, 1949, 1962) as a post-thrusting left-reverse fault with about 10,000 feet of up-to-the-south displacement which decreased westward, ultimately dissipating in a monoclinal fold north of the Buffington Pockets area. Temple (1977) provided a similar interpretation of the geology at the west end of the Arrowhead fault and called upon drape folding of the Paleozoic section over the fault as the mechanism to produce the apparent monoclinal structure. North of the Arrowhead fault in the North Muddy Mountains, Longwell (1949) mapped overturned strata in several recumbent folds which he thought developed in the Muddy Mountain thrust plate beneath the Glendale thrust. East of the folds in the Valley of Fire area, Longwell (1949) described his Upper Cretaceous and Tertiary(?) Overton Fanglomerate as a wedge of alluvial material, which he thought was deposited on the then-inactive Muddy Mountain thrust plate east of the advancing Glendale thrust.

I favor the following alternate interpretations regarding the above features. (1) The Muddy Mountain, Glendale, and Arrowhead structures are all parts of one thrust fault, collectively called the Muddy Mountain thrust. If this interpretation is correct, the Muddy Mountain thrust is not everywhere stratigraphically controlled; a wide variety of rock units occur directly beneath it in the North Muddy Mountain region. (2) The Summit and Willow Tank thrusts are parts of another smaller and older thrust system which also includes a fault north of Buffington Pockets called the North Buffington fault. These faults are collectively called the Summit-Willow Tank thrust. (3) The apparent monocline or drape fold of Longwell (1949, 1962) and Temple (1977) is, instead, a truncation of one thrust plate by another. The steeply dipping Summit-Willow Tank plate is truncated beneath the low-dipping Muddy Mountain thrust plate. (4) The North Muddy Mountains are overlain by the Muddy Mountain thrust plate, as redefined, and are thus autochthonous with respect to it. (5) As a consequence of interpretation 4, the folds in the North Muddy Mountains developed beneath the Muddy Mountain thrust in the

upper plate of the Summit-Willow Tank thrust. (6) Foreland basin deposition began with the fine-grained Willow Tank Formation and became progressively coarser grained with the eventual deposition of the Overton Conglomerate Member of the Baseline Sandstone. This lithologic change formed in direct response to the eastward advance of the Summit-Willow Tank and Muddy Mountain thrust plates into the depositional basin.

Thrust Relations

West of Buffington Pockets the Muddy Mountain thrust places Bonanza King Formation over Aztec Sandstone and dips westward (Fig. 3). A younger, west-dipping, low-angle, normal fault locally juxtaposes Devonian rocks against the Bonanza King and the Aztec in that area, but the Devonian is dropped from a stratigraphic position high in the thrust plate (Fig. 3). East of Buffington Pockets, in the Muddy Mountain block, the Muddy Mountain thrust and the Cambrian through Devonian upper-plate rocks dip eastward about 30° into a younger, north-trending, high-angle normal fault. East of this normal fault, allochthonous Cambrian rocks can be traced eastward continuously (barring slight displacement by north-trending normal faults) on the south side of the Arrowhead fault of Longwell (1949). The Cambrian rocks south of the Arrowhead dip about 20° southeast in the western Muddy Mountain block and about 50° south near the Valley of Fire. The Arrowhead fault dips south more or less parallel to the dip of the upper-plate rocks in the Muddy Mountain block, and the fault plane is confined to the Bonanza King Formation. The Arrowhead is a relatively continuous structure between the southwestern Valley of Fire and the northeast part of Buffington Pockets where it is cut by the same north-trending, high-angle normal fault that cuts the Muddy Mountain thrust (Fig. 3). About 1.5 km east of the normal fault a large slice of Aztec Sandstone is caught between branches of the Arrowhead, and at its western termination at the normal fault, the Arrowhead juxtaposes Cambrian rock on the south over Aztec on the north. These relations strongly suggest that the Arrowhead fault is simply the steeply dipping trace of the Muddy Mountain thrust, the trend of which changes from northerly near Buffington Pockets to easterly near the Valley of Fire. The Cambrian Bonanza King forms the sole of the upper plate west of Buffington Pockets and throughout the Muddy Mountain block. Aztec Sandstone and locally post-Aztec clastic rocks (Brock and Engelder, 1977) are directly beneath the thrust in the Valley of Fire and in Buffington Pockets. Rocks in both the upper and lower plates dip roughly parallel to the thrust surface in the latter two areas. Between the Valley of Fire and Buffington Pockets, on the other hand, the lower plate consists of a terrane of steeply dipping and overturned

Cambrian through Triassic rocks that strike at a high angle to the fault trace. The steeply dipping, east-striking North Buffington fault bounds the southwest part of this terrane and separated vertical to overturned Bonanza King in the hanging wall from Aztec to the south, whereas the northwest-striking, southwest-dipping Summit thrust bounds the eastern side of the terrane and separates overturned upper plate Moenkopi Formation from autochthonous Aztec Sandstone to the northeast. Both the North Buffington fault and the Summit thrust are inferred to structurally underlie the thrust plate exposed in the Muddy Mountain block.

No evidence supports the claim of Longwell (1949, 1962) and Temple (1977) that the Cambrian rocks of the Muddy Mountain block (those south of the Arrowhead fault of Longwell, 1949) are continuous in an unbroken monoclinal fold with those adjacent to the North Buffington fault in the southwestern part of the North Muddy Mountains. On the contrary, the fault labeled as the Muddy Mountain thrust on Figure 3 (western part of the Arrowhead) is a conspicuous structure separating Middle Cambrian Papoose Lake(?) Member of the Bonanza King Formation in its hanging wall (Muddy Mountain block) from Middle and Upper Cambrian Banded Mountain Member through Triassic rocks in its footwall (terrane of steeply dipping to overturned rocks in the southwest part of the North Muddy Mountains). In this vicinity, both the Muddy Mountain thrust and the overlying Papoose Lake Member dip southeast at 20°, whereas the underlying Banded Mountain dips steeply and strikes at a high angle into the fault contact. No rocks like those thought to be Papoose Lake Member occur north of the fault. Longwell (1962, fig. 144.3, p. D84) and Temple (1977, pl. 1) both show similar relations on their geologic maps in spite of the contrary conclusions they draw in their texts. Stewart (1980, p. 80) provided an oblique air photograph of the area taken from the west by J. S. Shelton that clearly shows the dark, resistant dolomite thought to be the Papoose Lake Member faulted over the alternating light and dark Banded Mountain Member; no monoclinal fold is evident.

On the northwest side of Buffington Pockets, the Muddy Mountain thrust is covered by the Miocene Muddy Creek Formation and by Quaternary deposits of California Wash (Fig. 3). North of the covered zone both the North Buffington fault and the Glendale thrust of Longwell (1949) have Cambrian rocks in their upper plates similar to those above the Muddy Mountain thrust on the west side of Buffington Pockets. Longwell (1949, 1962) connected the thrust west of Buffington Pockets with the North Buffington fault, but because the North Buffington fault occurs *beneath* the Muddy Mountain thrust plate less than 2 km to the east on the east side of Buffington Pockets, it is probably also beneath it in the subsurface of California Wash. Thus, the Glendale thrust is probably the same as the

Muddy Mountain thrust and is not a structurally higher thrust as Longwell (1949) considered it to be.

The Muddy Mountain thrust, Glendale thrust, and Arrowhead fault are now mapped as one regionally continuous, low-angle structure collectively called the Muddy Mountain thrust (Bohannon, 1981, 1983a). Its upper plate is inferred to tectonically overlie the rocks of the North Muddy Mountains. Aztec Sandstone occurs beneath the thrust in Buffington Pockets and the Valley of Fire, but in the southwestern part of the North Muddy Mountains the lower plate is complex and includes the terrane of steeply dipping to overturned Cambrian through Triassic rocks between the North Buffington fault and the Summit thrust (Fig. 3). The lower and middle Paleozoic rocks in this fault-bounded stratigraphic section are very similar to those in the nearby equivalent section in the upper plate of the thrust in the Muddy Mountain block (Longwell, 1962; Longwell and Mound, 1967). Because the Paleozoic rocks in both plates were deposited in the transition zone between craton and shelf where lateral variability is common, their stratigraphic similarity suggests only minor transport on the Muddy Mountain thrust at that particular location. This contrasts with the large net displacement thought to have occurred on the Muddy Mountain thrust at most other places (Longwell and others, 1965).

The apparent discrepancy in displacement can be explained if the fault-bounded Paleozoic rocks in the lower plate are also tectonically detached from the autochthon by a lower structure which has a large net tectonic transport nearly equivalent to that described on the Muddy Mountain thrust in other areas. The North Buffington fault and possibly the Summit fault are the most likely surface traces of such a tectonically lower structure, but the attitude of the former is not entirely consistent with such an interpretation. At its west end, the North Buffington fault appears to have a nearly vertical attitude nearly parallel to that in the bedding in its hanging wall, but at its east end near where it passes beneath the Muddy Mountain thrust, the hanging-wall rocks dip to the northeast and possibly the fault does also. If the North Buffington fault does dip to the northeast, it seems prudent to interpret it as part of a major thrust, below the Muddy Mountain thrust, which extends in the subsurface to the northeast beneath the southwestern part of the North Muddy Mountains.

The geographic extent of the lower thrust plate in the North Muddy Mountains is uncertain. The nearest exposed thrust east of the North Buffington fault is the southwest-dipping Summit thrust, and the northwest- and west-dipping Willow Tank thrust is the nearest one to the northeast (Fig. 3). These two thrusts are probably continuous with one another in the subsurface beneath a Tertiary low-angle, younger-over-older fault (Fig. 3). The two thrusts collectively form a structure called the Summit–Willow Tank thrust. The rocks in the North Muddy Mountains above the subsurface projection of the Summit–Willow Tank thrust are highly folded, but they are not broken by other major thrusts except at a locality in the north-central part of Figure 3 where Cambrian rocks are structurally above overturned Triassic beds. The Triassic rocks there are part of the upper plate of the Summit–Willow Tank thrust, and the overlying thrust could be an outlier of the Muddy Mountain thrust. Thus, the allochthon above the North Buffington fault includes all of the rocks in the fold belt of the North Muddy Mountains and must extend at least as far to the east as the trace of the Summit–Willow Tank thrust and possibly beyond. If the North Buffington fault and the Summit–Willow Tank thrust are parts of the same fault, the fault surface is relatively flat beneath the southwestern part of the North Muddy Mountains and cuts up-section, truncating the steeply dipping to overturned upper-plate rocks eastward. If so, it is highly possible that Aztec Sandstone occurs directly beneath the fault throughout the subsurface of the southwestern North Muddy Mountains, because Aztec is exposed beneath both the North Buffington fault and the Summit thrust in Buffington Pockets and the Valley of Fire, respectively. The rocks in the Valley of Fire east of the thrust trace are autochthonous with respect to all of the thrusts in the latter case. Alternatively, if the major thrust extends beyond and below the trace of the Summit–Willow Tank thrust, its upper plate includes the rocks of the Valley of Fire and its eastern trace is probably buried beneath Tertiary rocks near Overton. Rocks east of Overton in the Virgin Mountains are thought to be autochthonous with respect to all of the Sevier thrust sheets.

The interpretation in which the Summit–Willow Tank thrust is the east margin of the allochthon and is continuous with the North Buffington fault is favored here. The rocks in the Valley of Fire are thought to be autochthonous because Cretaceous foreland basin deposits of the Willow Tank Formation and Baseline Sandstone are lithologically and stratigraphically similar to known autochthonous rocks in the Virgin Mountains. The alternative case is that the North Buffington fault is stratigraphically controlled in the Cambrian rocks and remains so beneath the North Muddy Mountains. However, the vertical, overturned and steep northeast dips in the Cambrian rocks north of the North Buffington fault relative to the gentle northwesterly dips in the Aztec Sandstone beneath the fault (Bohannon, 1981, 1983a) suggest that if that fault follows the Cambrian rocks, it must cut down-section in the lower-plate rocks toward the north and northeast. The latter, southwest-directed thrust geometry is not considered likely.

If the North Buffington fault is part of the Summit–Willow Tank thrust, as proposed herein, its displacement appears to decrease to the northeast. Although absolute displacement is uncertain, stratigraphic offset on the North Buffington fault must be large, whereas it appears to be intermediate on the Summit thrust and probably is minor on the Willow Tank thrust. The displacement discrepancy can easily be explained as a result of the intense folding described by Longwell (1949) in the North Muddy Mountains in the upper plate of the Summit–Willow Tank thrust. The large recumbent folds are present north and west of the area where the thrust appears to lose displacement, suggesting the transfer of east-directed crustal shortening from the fault to the folds in that area.

Foreland Basin

The Willow Tank Formation (around 98 to 96 m.y. old) and the basal white member of the Baseline Sandstone (Kwt and Kbw, respectively, Fig. 4), both of Albian age, are thick clastic units that record the progressive uplift and erosion of rocks in the upper plate of the Summit–Willow Tank thrust. These rocks provide evidence for thrusting synchronous with deposition. Pediment gravel with a diverse clast suite at the base of the Willow Tank Formation was deposited above a regional erosion surface cut into uppermost Aztec Sandstone and Carmel(?) Formation, but the age of the erosional episode is not well defined. The unconformity might herald early local uplift related to thrusting (Fleck, 1970); however, abundant quartzite clasts in the gravel do not have a local source in the fold belt of the North Muddy Mountains or from the upper plate of the Muddy Mountain thrust. Above the gravel, fine-grained swamp and pond deposits of the Willow Tank Formation give way upward to clean quartz arenite with minor amounts of conglomerate in the white member of the Baseline Sandstone. The voluminous, fluvial quartz arenite was probably locally derived from the Aztec to the west (Longwell, 1949), possibly during folding and uplifting of the Summit–Willow Tank thrust plate. The oldest quartz arenite beds are the first positive signs that thrusting had probably begun locally. The Summit–Willow Tank thrust, in turn, cuts both the Willow Tank Formation and the overlying white member of the Baseline Sandstone. Localized facies of coarse-grained conglomerate and breccia in the white member north of Tearfault Mesa (Fig. 4) of Longwell (1949) are adjacent to the trace of the Summit–Willow Tank thrust and indicate synchronous thrusting and deposition (Kbwc, Fig. 4). North of there the Summit–Willow Tank thrust tectonically overlies all of the white member. Thus, the Summit–Willow Tank plate, advancing from the west, probably acted as both a source area and as the western limit for sediment accumulation in the early foreland basin.

The red member and Overton Conglomerate Member of the Baseline Sandstone (Kbr and Kbo, respectively, Fig. 4), both of Albian to Cenomanian(?) age, record (1) the cessation of activity on the Summit–Willow Tank thrust; (2) the development, uplift, and erosion of the

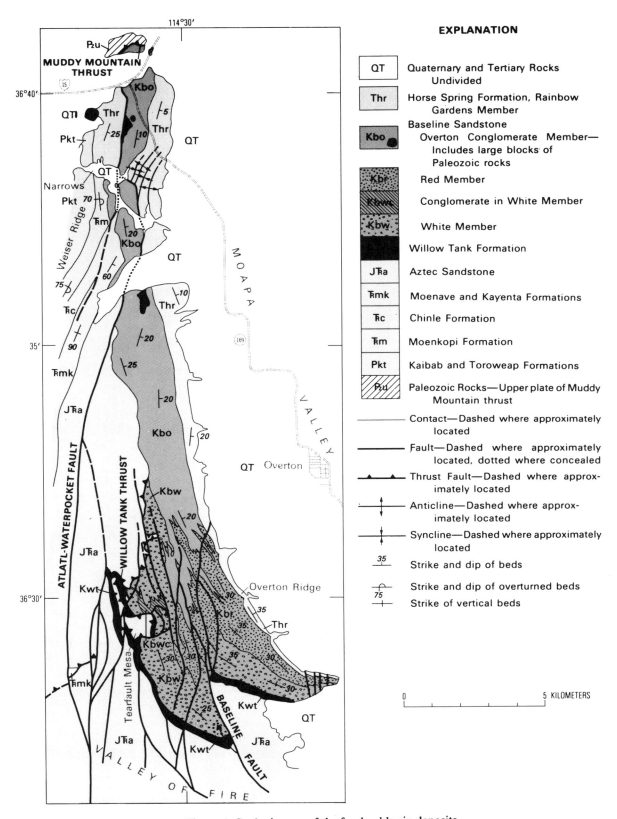

EXPLANATION

QT — Quaternary and Tertiary Rocks Undivided

Thr — Horse Spring Formation, Rainbow Gardens Member

Baseline Sandstone

Kbo — Overton Conglomerate Member— Includes large blocks of Paleozoic rocks

Kbr — Red Member

Kbwc — Conglomerate in White Member

Kbw — White Member

Willow Tank Formation

JŦa — Aztec Sandstone

Ŧmk — Moenave and Kayenta Formations

Ŧc — Chinle Formation

Ŧm — Moenkopi Formation

Pkt — Kaibab and Toroweap Formations

Pᵤ — Paleozoic Rocks—Upper plate of Muddy Mountain thrust

—— Contact—Dashed where approximately located

—— Fault—Dashed where approximately located, dotted where concealed

▲—▲ Thrust Fault—Dashed where approximately located

Anticline—Dashed where approximately located

Syncline—Dashed where approximately located

35 Strike and dip of beds

75 Strike and dip of overturned beds

Strike of vertical beds

0 — 5 KILOMETERS

Figure 4. Geologic map of the foreland basin deposits.

North Muddy Mountain fold belt (Longwell, 1949, 1952); and (3) the advance of the Muddy Mountain thrust plate from the west. The red member of the Baseline is conformable with the underlying white member and differs from the latter chiefly by its red color. In the northern part of the Valley of Fire, sandstone of the red member interfingers to the north with coarse-grained conglomerate of the Overton Conglomerate Member, which is also conformable with the underlying white member (Fig. 4). North of the Valley of Fire, the conglomerate overlaps the trace of the Summit–Willow Tank thrust and rests unconformably on Aztec Sandstone of the upper plate. This suggests that the Summit–Willow Tank thrust ceased activity by the time of deposition of the red member and Overton Conglomerate Member. Contrary to Longwell's (1949) interpretation, the unconformity at the base of the Overton Conglomerate Member is restricted to the region north of the trace of the Summit–Willow Tank thrust, and it does not extend into the basin to the south. Clasts in the Overton Conglomerate Member were derived chiefly from Paleozoic and Mesozoic carbonate rocks like those locally exposed in the North Muddy Mountain fold belt and in the upper plate of the Muddy Mountain thrust (Longwell, 1952). Several intact bodies of Paleozoic rock over 100 m in size (unlabeled masses shown in black on Fig. 4) occur within the conglomerate and document a source area with high relief (Longwell, 1949). The clast suite in the Overton Conglomerate Member indicates that upper Paleozoic and Triassic rocks were exposed in folds and were being eroded from the North Muddy Mountains and that middle Paleozoic rocks were being eroded from the advancing upper plate of the Muddy Mountain thrust during late Albian and Cenomanian(?) time. It appears that much of the folding in the North Muddy Mountains was synchronous with the eastward advance of the Muddy Mountain thrust and probably occurred in response to activity on that thrust. However, the folds are restricted to the Summit–Willow Tank thrust plate, and if they absorbed some of that fault's displacement, they must have started to form during the time of Summit–Willow Tank thrusting.

During its late stages of activity, the Muddy Mountain thrust overrode part of the Overton Conglomerate Member. Lower Paleozoic rocks north of the Narrows (Fig. 4), which are probably part of the Muddy Mountain thrust allochthon, tectonically overlie the Overton Conglomerate Member.

Displacement on the Summit–Willow Tank and Muddy Mountain thrusts resulted in little deformation in the autochthonous rocks of the Valley of Fire other than gentle folding and very local overturning of strata directly beneath the former thrust. In the Valley of Fire southeast of the trace of the Summit–Willow Tank thrust, autochthonous Mesozoic rocks are essentially disconformable with the overlying Miocene Rainbow Gardens Member of the Horse Spring Formation exposed at Overton Ridge (Fig. 4), indicating that the autochthon was nearly flat lying until at least the Miocene. Overturned and steeply dipping beds beneath and in front of the Summit–Willow Tank thrust at Tearfault Mesa of Longwell (1949) probably resulted from the ploughing of soft sediment during thrusting, but this type of deformation is limited in extent. In spite of the apparent lack of Mesozoic deformation in the autochthon, the folds of the North Muddy Mountain fold belt can be demonstrated to be Mesozoic in age because the Cretaceous Overton Conglomerate Member overlaps folded Triassic and Jurassic beds near the Narrows (Fig. 4, north end).

MIDDLE AND LATE TERTIARY STRUCTURES AND TECTONIC FEATURES

The structure beneath the unconformity at the base of the Rainbow Gardens Member of the Horse Spring Formation is regionally simple, indicating little pre–Horse Spring deformation throughout the autochthon of the Sevier thrust belt. Prior to deposition of the Tertiary sedimentary rocks at Frenchman Mountain, in the Muddy Mountains, and in the Virgin Mountains (Fig. 1), and prior to eruption of volcanic rocks south of Lake Mead and in the northern Black Mountains, a broad, north-trending, north-plunging arch cored by Precambrian crystalline rocks existed in the Lake Mead region between the Spring Mountains to the west and the Colorado Plateau to the east (Bohannon, 1983b). Gentle northeast dips, probably of 5° or less, were present in the Paleozoic and Mesozoic sedimentary rocks along the east flank of the arch, near the present western margin of the Colorado Plateau, prior to the eruption of the Peach Springs Tuff about 17 m.y. ago (Lucchitta, 1966, 1972, 1979; Young, 1966, 1970; Young and Brennan, 1974). Between the Virgin Mountains and Frenchman Mountain (Fig. 1), the Tertiary sedimentary rocks of the Horse Spring Formation were deposited on the nose of the arch where Mesozoic and Upper Paleozoic strata dipped northward or northeastward at about 5° to possibly as much as 15° (Bohannon, 1983b). Autochthonous Paleozoic and Mesozoic rocks in the Spring Mountains and McCullough Range apparently had a slight northwest dip prior to volcanism. These rocks formed part of the western flank of the arch. Precambrian rocks were exposed in the core of the arch between the McCullough Range and the Colorado Plateau prior to eruption of an ash flow tuff thought to be about 20 m.y. old (Anderson, 1969, 1971, 1973, 1977, 1978; Anderson and others, 1972). Thus, the arch was a pre–20-m.y. B.P. feature in the autochthon east of the thrust belt, but its temporal and genetic relations to the belt are not certain. The arch may be a Sevier, Laramide, or later feature.

The early Tertiary terrane, which was dominated by

the north-trending Cretaceous thrusts of the Sevier belt and the broad, gentle, basement-cored arch east of the thrusts, apparently was little deformed before Miocene time when strike-slip faults fragmented it, new basins were superposed on it, and the region was distorted by extensional tectonics of various local intensities (Bohannon, 1983b). The major Miocene structures and tectonic elements that evolved are the Las Vegas Valley shear zone, the Lake Mead fault system, the three basins of the Horse Spring Formation, several other basins that formed during a protracted period of basin and range deformation, and localized zones of intense extension where closely spaced normal faults are associated with stratal repetition. The Las Vegas Valley shear zone caused oroflexural bending of the ranges bordering Las Vegas Valley and resulted in the fault offset of several Sevier thrust traces and Paleozoic facies boundaries. Estimates of total net offset across the shear zone range from less than 20 km (Ketner, 1968) to greater than 65 km, but most authors agree that the combined effect of bending and faulting resulted in at least 40 and possibly as much as 70 km of Tertiary displacement (Longwell, 1960; Longwell and others, 1965; Ross and Longwell, 1964; Stewart, 1967; Stewart and others, 1968; Osmond, 1962; Poole and others, 1967; Burchfiel, 1965). The northeast-striking, left-slip Lake Mead fault system offsets rocks in the nose of the pre-Miocene arch apparently by as much as about 60 or 65 km (Anderson, 1973; Bohannon, 1979; Smith, 1981). Some of the basins of the Horse Spring Formation and those associated with early episodes of post–Horse Spring basin and range deformation formed around the nose of the arch while the Lake Mead fault system was active (Bohannon, 1983b). Anderson (1971) and Anderson and others (1972) detailed the development of one area of thin-skinned Miocene crustal extension south of Lake Mead, and Anderson (1973) and Bohannon (1979, 1980, 1983b) have examined the role of that extension with respect to strike-slip faulting. Other areas of similar thin-skinned crustal extension are possibly present in the Virginia Mountains and at Frenchman Mountain (Morgan, 1964; C. R. Longwell, unpub. mapping; Bell and Smith, 1980). Younger, north-trending basin and range valleys overprint many of the earlier formed structures.

Although the Muddy Mountains and North Muddy Mountains lack a significant Miocene sedimentary record, the stratigraphy and structure of White Basin, the Gale Hills, Bitter Spring Valley, and the northern part of the Black Mountains are dominated by Miocene features. The Horse Spring Formation is widespread throughout the latter areas and is present high in the mountain ranges as well as in the valleys. Post–Horse Spring rocks of Miocene age, on the other hand, are localized in valleys and basins, the oldest being the 10- to 12-m.y.-old red sandstone unit of White Basin and the youngest being the Muddy Creek Formation which partially fills the basins that surround the

Muddy Mountains. Deformation of the Muddy Creek Formation is nonexistent to slight in most areas, but the pre–Muddy Creek Miocene rocks are commonly deformed moderately to greatly.

Much of the local deformation in the Gale Hills can probably be attributed to activity on the Las Vegas Valley shear zone which projects into that area from Las Vegas Valley, a few tens of kilometers to the west-northwest. Although no surface ruptures in the Gale Hills can be positively correlated with the shear zone, I have identified two faults (the Gale Hills fault and another herein referred to as the western part of the West Bowl of Fire fault) that I interpret as extensions of the shear zone (Bohannon, 1979, 1983b). Likewise, much of the local deformation in the northern Black Mountains and in the southern part of Bitter Spring Valley is probably the result of activity on two major branches of the Lake Mead fault system—the Bitter Spring Valley fault and the Hamblin Bay fault. However, on a regional basis it is difficult to establish a clear-cut relationship between most of the Miocene deformation and major lateral-slip faults because some major structures appear to have been produced by extensional tectonism as well.

Gale Hills–Muddy Peak Area

The northwest-striking, southwest-dipping Gale Hills fault (Fig. 5) separates topographically high Paleozoic rocks that dip homoclinally to the south-southeast in the Muddy Peak block from the gently east-dipping Thumb and Bitter Ridge Limestone Members of the Horse Spring Formation in the northern part of the Gale Hills. The Paleozoic rocks at Muddy Peak (Fig. 5) are allochthonous above the Muddy Mountain thrust. Paleozoic rocks are also exposed directly southeast of the Gale Hills fault where they underlie the locally thin Thumb Member. The Thumb thickens considerably and changes facies from coarse-grained breccia and conglomerate into fine-grained sandstone and siltstone away from the fault to the southwest (Bohannon, 1983b). The lowest beds of the Bitter Ridge Limestone Member, which overlies the Thumb, are also faulted at several localities along the Gale Hills fault (Fig. 5), but the fault is overlapped by the younger beds of the Bitter Ridge to the west-northwest of Muddy Peak (Bohannon, 1981, 1983a, 1983b; not visible on Fig. 5). South of the Gale Hills fault, in the northern part of the Gale Hills, the Thumb is cut and tilted by a set of north-northeast-trending normal faults that dip between 60° and 85° (a large number of these to the west), and many of these terminate to the north at the Gale Hills fault. Only the largest normal faults are depicted on Figure 5.

The geologic relations associated with the Gale Hills fault suggest that it was active during the deposition of the Thumb Member of the Horse Spring Formation between

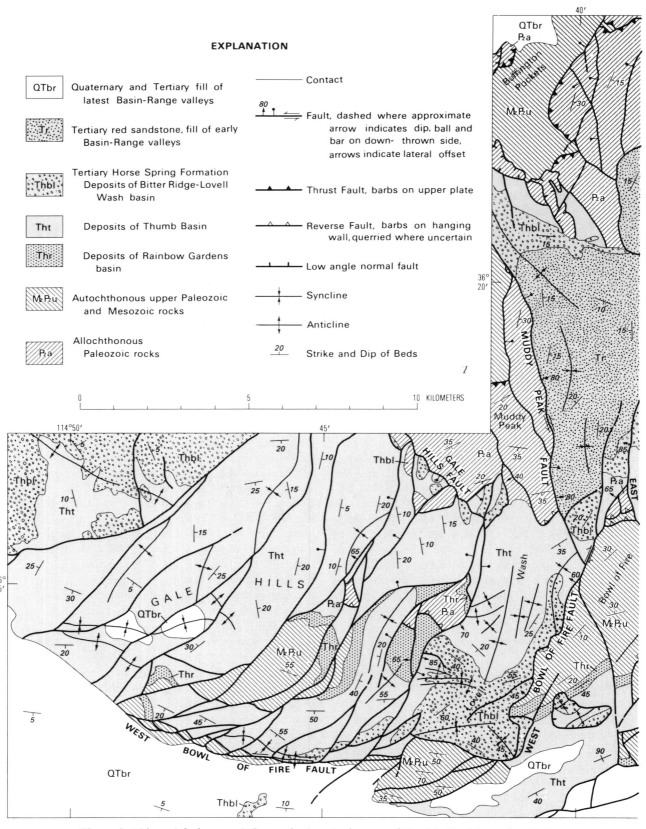

EXPLANATION

QTbr	Quaternary and Tertiary fill of latest Basin-Range valleys
Tr	Tertiary red sandstone, fill of early Basin-Range valleys
Thbl	Tertiary Horse Spring Formation Deposits of Bitter Ridge-Lovell Wash basin
Tht	Deposits of Thumb Basin
Thr	Deposits of Rainbow Gardens basin
M₂P₂u	Autochthonous upper Paleozoic and Mesozoic rocks
P₂a	Allochthonous Paleozoic rocks

—————— Contact

Fault, dashed where approximate arrow indicates dip, ball and bar on down-thrown side, arrows indicate lateral offset

▲▲ Thrust Fault, barbs on upper plate

△△ Reverse Fault, barbs on hanging wall, querried where uncertain

Low angle normal fault

Syncline

Anticline

Strike and Dip of Beds

0   5   10 KILOMETERS

Figure 5. (this and facing page) Generalized geologic map of the Muddy Mountains, Gale Hills, Bowl of Fire, Bitter Spring Valley, and northern part of the Black Mountains.

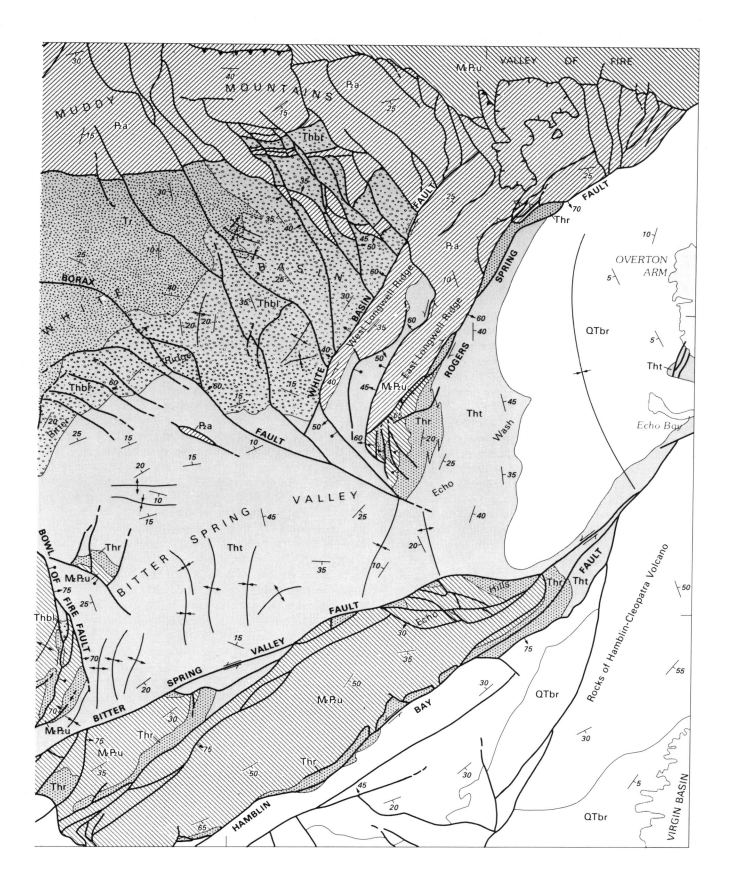

about 16.5 and 14 m.y. ago and that it probably acted as the northeastern margin of the Thumb basin in the Gale Hills area (Bohannon, 1983b). The Gale Hills fault remained active during the early stages of deposition of the Bitter Ridge Limestone Member because the oldest Bitter Ridge beds are cut by this fault. Fault activity must have ceased by approximately 13 m.y. ago because most of the Bitter Ridge Limestone Member overlaps the fault (Bohannon, 1981, 1983a). By the simplest interpretation, the Gale Hills fault is a southwest-dipping normal fault at the local northeastern margin of the Thumb basin. During the late stages of faulting it probably also accommodated the slight differential extension between the normally faulted area to its southwest and the Muddy Peak block to its northeast. Thus, a slight amount of right slip may have occurred on the Gale Hills fault during the episode of normal faulting if the extension south of the fault was west directed relative to the Muddy Peak block. Also, the possibility of a large amount of right slip on the Gale Hills fault cannot be entirely ruled out, because its geographic position and trend are aligned with the southeastern projection of the Las Vegas Valley shear zone (Bohannon, 1979). However, there is no direct evidence for large amounts of lateral slip on the Gale Hills fault.

South of the Gale Hills fault in the Gale Hills, displacement on the group of north-northeast-trending normal faults has created a complex horst and graben and tilted block pattern which appears to be better developed in the eastern hills than in the western ones (Fig. 5). In the western Gale Hills the Thumb Member is exposed in tilted blocks and horsts with small patches of the Muddy Creek Formation preserved above the Thumb in synclinal troughs within grabens. In the central Gale Hills, erosion in the horsts has exposed Aztec Sandstone, Willow Tank Formation, and Baseline Sandstone, all of which are unconformably overlain by the Rainbow Gardens Member. All three members of the Horse Spring Formation are preserved in the adjacent grabens in that area. A large graben, in which the Lovell Wash and Thumb Members are preserved, is present in the eastern part of the Gale Hills between the moderately west dipping, north-northeast-trending (eastern) part of the West Bowl of Fire fault and a steeply east dipping, north-northeast-trending fault west of Lovell Wash.

In two horsts in the central part of the Gale Hills, Mississippian and Pennsylvanian rocks are in reverse fault contact (open sawteeth, Fig. 5) with the Rainbow Gardens and Thumb Members (Fig. 5). The smaller and more westerly outcrop of Paleozoic rock is completely fault bounded, but the eastern one dips moderately to the northeast and is unconformably overlain by the Rainbow Gardens Member. In the graben between the two horsts there is no evidence of a reverse fault nor are any Paleozoic rocks present. Thus, the exposures of Paleozoic rocks and

the reverse faulting associated with them are localized in the horsts. Apparently a considerable amount of north-south shortening occurred in the horsts during their early uplift that apparently did not occur in the adjacent grabens. A Tertiary landslide or gravity glide origin might also be postulated for the Paleozoic outcrops, but the compressional hypothesis is favored because the outcrops are internally intact, contacts between major rock units in the outcrops are little deformed, and overturning of bedding occurs near the reverse fault contact in the eastern outcrop.

The set of north-northeast-trending normal faults terminates to the south at the east-trending (western) part of the West Bowl of Fire fault south of the Gale Hills (Fig. 5). At this locality the West Bowl of Fire fault changes from a complex zone of anastomosing faults in the west to a single break in the east. The normal faults peel off the north side of the east-west fault zone and abruptly change strike from easterly to northeasterly and ultimately to a north-northeasterly trend. The western part of the West Bowl of Fire fault separates Moenkopi Formation overlain by Muddy Creek Formation to its south from Thumb and Bitter Ridge Limestone Members to its north. Within fault branches of the anastomosing part are Moenkopi Formation, Kaibab Limestone, Toroweap Formation, and Permian red beds. A wide variety of dips are preserved on individual branches of the Bowl of Fire fault in this area, but the overall fault attitude appears to be vertical.

The West Bowl of Fire fault and the group of north-northeast-trending normal faults to its north appear to have been active synchronously, and they can probably be related kinematically in a manner similar to that proposed by Anderson (1973) in the Black Mountains. The West Bowl of Fire fault cuts the oldest beds of the Muddy Creek Formation, which ranges from about 10- to 5-m.y.-old locally. Some of the north-northeast-trending normal faults in the western Gale Hills cut the Muddy Creek also, but the displacements on all these faults are greater in older beds such as the Thumb and Bitter Ridge Limestone Members. These facts, coupled with the close geometric ties between the faults, suggest that fault movement occurred within a time period of about 13 to possibly 9 or 10 m.y. ago. The absolute slip on the West Bowl of Fire fault is not known, but several folds on either side of the fault, some large enough to show prominently on aerial photographs, exhibit what appears to be well-developed left drag. Also, R. E. Anderson (1980, written commun.) concluded, after a detailed study of the fault, that the direction and amount of plunge of slickensides, coupled with the known stratigraphic separation across the fault, are consistent with left slip. Left slip on the east-trending part of the West Bowl of Fire fault seems to be kinematically consistent with the inferred west-directed extension in the Gale Hills in the same manner as right slip, resulting from the same extension, can be inferred on the Gale Hills fault on the north side of the extensional area. Unfortunately, differential extension across the West Bowl of Fire fault implies a different extensional history for the area to its south, and little is known of the kinematic development of that area during the time period of fault activity because of cover by younger strata. Also, a left-slip interpretation is not consistent with the regional interpretation that the West Bowl of Fire fault may be the eastern extension of the right-slip Las Vegas Valley shear zone (Bohannon, 1979, 1983b).

The role that Mesozoic structure played in the development of the geology of the Gale hills is difficult to decipher. The Paleozoic rocks at Muddy Peak are known to be allochthonous above the Muddy Mountain thrust. Paleozoic rocks immediately south of the Gale Hills fault and in the two horsts of the central Gale Hills are assumed to be allochthonous also because of their proximity and similarity to the rocks at Muddy Peak. The Mesozoic rocks in the Gale Hills, on the other hand, are interpreted to be autochthonous with respect to the Muddy Mountain thrust. This interpretation seems prudent because (1) thick Cretaceous foreland deposits of the Willow Tank Formation and Baseline Sandstone in the Gale Hills are similar to foreland rocks in other parts of the autochthon, (2) thick Mesozoic sequences are unknown in the upper plate of the Muddy Mountain thrust anywhere else—in fact, the sedimentological history of the local foreland deposits strongly suggests that Mesozoic cover must have been almost completely eroded from the upper plate during thrusting because clasts of Upper Paleozoic rocks are common in foreland conglomerates derived from that plate, and (3) facies in the Mesozoic rocks and those nearby on the south side of the West Bowl of Fire fault are identical to facies in autochthonous Mesozoic rocks at Frenchman Mountain and in the northern part of the Black Mountains. Thus, there must be an important contact between autochthonous and allochthonous rocks buried beneath the Tertiary beds in the Gale Hills, but the nature of this contact is poorly understood.

I have presented one possible interpretation of the subsurface geology of the Gale Hills (Bohannon, 1983b, Fig. 44) in which I assumed that south-dipping allochthonous Paleozoic rocks could be traced continuously southward in the subsurface from the Muddy Peak area to a reverse fault contact involving Tertiary rocks in the central part of the Gale Hills. In this interpretation the contact between the allochthon and autochthon was thought to have had an early history of strike-slip faulting in addition to Tertiary reverse fault displacement. However, other solutions are possible and some of these are examined in the Discussion section after pertinent facts from other nearby areas have been described.

A poorly understood aspect of the geology of the Gale Hills is the role that folding played in its development. Numerous symmetrical to asymmetrical upright, broad to tight, open folds with wavelengths between 0.5 and 2 km occur throughout the hills (Fig. 5). One set of these folds trends north-northeast parallel to the trend of the normal faults, and the other trends about east, parallel to the trend of the western part of the West Bowl of Fire fault. North-northeast-trending folds are commonly found in the northern part of the Gale Hills adjacent to the normal faults. In many cases dips in the limbs of these folds steepen toward the faults, suggesting drag or sagging near the faults. The east-trending folds are concentrated in the southern part of the Gale Hills near the West Bowl of Fire fault. Some of these folds are strongly asymmetric and can be attributed to fault drag, but others appear unrelated to faulting. Some of these folds have amplitudes of several hundred meters, especially those in the southeastern Gale Hills and Lovell Wash area. They seem to indicate a strong component of north-south shortening throughout the southern part of the Gale Hills. The folds apparently started to develop after or during deposition of the Lovell Wash Member of the Horse Spring Formation (about 13 m.y. ago), and folding continued through the early stages of Muddy Creek deposition (probably about 9 or 10 m.y. ago) as evidenced by dips in the former unit being twice as steep as those in the latter unit (Fig. 5).

Bowl of Fire

The Bowl of Fire area is a triangular-shaped horst whose apex points north (Fig. 5). The bowl derives its name from the rim of topographically high rocks that surround it. Mesozoic strata (Moenkopi Formation through Baseline Sandstone) in the eastern part of the horst are folded into a broad, open, northeast-trending anticline. At the north end of the horst they are in the footwall of a reverse fault with Mississippian rocks in the hanging wall; to the south they are unconformably overlain by the Rainbow Gardens Member of the Horse Spring Formation. A north-striking, high-angle fault transects the Bowl of Fire and separates the anticline to the east from a southeast-dipping homocline consisting of Aztec Sandstone and Rainbow Gardens Member to the west. The moderately west-dipping, north-northeast-striking West Bowl of Fire fault bounds the horst on the west, and the steeply east-dipping, north-striking East Bowl of Fire fault bounds it to the east. To the southeast, the horst is truncated by the Bitter Spring Valley fault of the Lake Mead fault system. A low-angle fault that places Bitter Ridge Limestone and Thumb Members over the Mesozoic and Tertiary rocks in the southern part of the Bowl of Fire may be an outlier of the West Bowl of Fire fault.

The northerly striking segment of the West Bowl of Fire fault dips between 45° and 65° to the west, but near the southwest corner of the bowl, the fault trace abruptly changes trend to easterly in the Lovell Wash area. The trend change takes place without disruption, and no major splays continue to the east or the south of the bend. At the bend, the Lovell Wash and Bitter Ridge Limestone Members are conglomeratic next to the fault (Bohannon, 1983b). The facies change from conglomerate to fine-grained lithologies northward away from the fault is abrupt and is interpreted to have resulted from deposition concurrent with faulting. The timing of this faulting is consistent with that presented above for the western part of the West Bowl of Fire fault.

The East Bowl of Fire fault dips about 70° to 75° to the east and splays southward. The eastern splay dies to the south in a tight, open syncline (Fig. 5). The western splay cuts Mesozoic rocks and ultimately offsets part of the Bitter Spring Valley fault south of the Bowl of Fire.

The West and East Bowl of Fire faults converge at the apex of the Bowl of Fire horst, and a north-trending, steeply west-dipping fault extends northward from the junction (Fig. 5). Displacement on this northward extension is minimal compared with the stratigraphic separation on the faults that bound the Bowl of Fire. A line marking the greatest displacement follows the East Bowl of Fire fault northward to its convergence with the West Bowl of Fire fault where it abruptly bends southwest to conform to the latter fault trace. Ultimately it trends east, following the trace of the western part of the West Bowl of Fire fault. Thus, the shape of the Bowl of Fire horst resembles an inverted ship's prow with a steeper dipping eastern than western side. Kinematically, the relative uplift of the horst could have taken place by left-normal slip on the East Bowl of Fire fault and right-normal slip on the north-northeast-trending segment of the West Bowl of Fire fault. Thus, the bowl has seemingly been displaced to the south and upward relative to the rocks to the west, north, and east of it. Left-slip on the east-west segment of the West Bowl of Fire fault due to west-directed extension in the Gale Hills seems compatible with such relative movement.

The Mesozoic rocks in the bowl are interpreted to be autochthonous with respect to the Muddy Mountain thrust for the same reasons that were outlined for those in the Gale Hills. In addition to those reasons, drilling information from the Shell No. 1 Bowl of Fire drill hole (Bohannon and Bachhuber, 1979) suggests that Devonian rocks in the subsurface of the bowl are of the same thickness as, and similar in lithology to, autochthonous Devonian rocks in the Virgin Mountains and at Frenchman Mountain. Likewise, Ordovician rocks are reported to be thin in the drill hole, unlike the thick Ordovician sections known in the allochthon.

Mississippian rocks, which probably are part of the allochthon, are in reverse fault contact with the autochthonous Mesozoic beds at the north end of the bowl. Although the fault appears to be a reverse fault because it dips to the north beneath the older rocks, it could, instead, be a normal fault upon which structurally higher older rocks from the upper plate of the thrust have been dropped against younger rocks of the lower plate. Various possibilities concerning this contact and its relation to the inferred contact between the allochthon and autochthon in the Gale Hills, plus its relations to other similar contacts in Bitter Spring Valley and in the Longwell Ridge area, are evaluated in the Discussion section.

White Basin

White Basin is a physiographic depression bordered on the north, west, and east by Paleozoic rocks in the upper plate of the Muddy Mountain thrust and on the south by Bitter Ridge, a prominent north-to northwest-dipping hogback in the Bitter Ridge Limestone Member of the Horse Spring Formation. White Basin is a broad, north-trending graben about 12 km across, formed between the Muddy Peak fault on the west and the White Basin fault on the east. Displacements on both faults apparently diminish northward because the graben is not as well developed in the Muddy Mountains. Structurally the graben continues to the south beyond Bitter Ridge and is present in northern Bitter Spring Valley (Fig. 5).

The Muddy Peak fault is a north- to northwest-striking, left-normal slip fault that dips between 75° and 80° to the east. Just north of its southern termination at the West Bowl of Fire fault, the Muddy Peak fault separates a northeast-trending syncline in the Thumb to its west from a faulted, northwest-dipping homocline in Tertiary beds to its east (Fig. 5). Farther north, south-dipping Paleozoic rocks in the upper plate of the Muddy Mountain thrust near Muddy Peak are in reverse fault contact with the Thumb which is conglomeratic and dips steeply beneath that contact. East of the Muddy Peak fault in western White Basin, conglomerate with clasts of upper-plate rocks derived from Muddy Peak forms a fault-front facies in the red sandstone unit. This facies continues northward to the structural pinch-out of the upper-plate rocks north of Muddy Peak where it interfingers with deposits of the Lovell Wash basin which underlie the aforementioned red sandstone. The conglomerate is not present north of the structural pinch-out. I have interpreted these relations as indicating that there has been little lateral offset on the Muddy Peak fault and that it was active during deposition of the Lovell Wash and the red sandstone, probably between about 13.5 and 10 m.y. ago (Bohannon, 1979, 1983b). Slickensides on the fault surface that have an average pitch of 55° to the north indicate a slight component of left slip in addition to the normal slip.

The White Basin fault is a northwest- and northeast-striking, left-normal slip fault that dips between 40° and 60° (Fig. 5). South of the Longwell Ridges the fault trends northwest, dips 50° to the southwest, and lies entirely within the Thumb. It bends sharply to the northeast at the south end of West Longwell Ridge and north of the bend dips 40° to 60° northwestward where it separates Bitter Ridge Limestone and Lovell Wash in White Basin from northeast-striking and southeast-dipping, upper-plate Paleozoic rocks on West Longwell Ridge. Slickensides pitch 67° south on the fault surface. Because of its similar trend, dip, and offset, the normal fault bounding the northwest side of East Longwell Ridge is assumed to have developed synchronously with the White Basin fault, and both faults are inferred to share the motion history of the Muddy Peak fault.

A set of northwest-trending normal faults (some possibly having oblique slip) is present in White Basin, at Bitter Ridge, and in northern Bitter Spring Valley (Fig. 5). Some displace the Muddy Mountain thrust (Arrowhead fault of Longwell, 1949) and continue to the north beyond it, whereas others apparently terminate at, or join with, the eastern projection of the Arrowhead fault south of the Valley of Fire, indicating that the Muddy Mountain thrust surface was used by the Tertiary faults. Some of the faults in eastern White Basin dip between 35° and 50° to the southwest, whereas others dip at about the same amount to the northeast, creating a complex horst and graben terrane in which upper-plate Paleozoic rocks are exposed in the horsts and Tertiary rocks are preserved in the grabens. This horst-graben terrane trends at a high angle to the White Basin graben and is probably secondary to it. In central White Basin, three of these faults cut the red sandstone unit, but displace it less than they do older beds such as the Lovell Wash and Bitter Ridge Limestone Member. Most of the faults in the Bitter Spring Valley and Bitter Ridge areas dip about 60° to the northeast and have well-developed right separation. Strikes of bedding along Bitter Ridge are at a high angle to the strike of the faults, and the dips do not appear to be the result of stratal rotation due to faulting. One of these northwest-striking faults, the Borax fault, extends to the southeast beyond the southern extent of the Longwell Ridges where its displacement apparently dies in the Thumb. In northwestern White Basin the fault abruptly bends into a northerly trend and then back to an east-west trend where it bounds the northwest and north sides of the basin. Possibly this fault has a spoon shape in three dimensions. One small outcrop of Paleozoic rocks (probably allochthonous above the Muddy Mountain thrust) is present in northern Bitter Spring Valley adjacent to one of the northwest-trending faults.

The internal structure of the White Basin graben is essentially that of a large, doubly plunging syncline, the axis of which lies in the western part of the basin. Several smaller folds are present in the western part of the basin as well. The Bitter Ridge Limestone and Lovell Wash Members of the Horse Spring Formation dip to the northwest beneath the red sandstone unit at Bitter Ridge,

and they emerge on the northwest side of White Basin north of Muddy Peak where their average dip is to the southeast. Most of the smaller folds in White Basin are near faults of the northwest-trending set and probably relate to those faults as second-order folds.

I have drawn interpretive structure sections in White Basin (Bohannon, 1981, 1983a) in which I assumed that Paleozoic carbonate rocks above the Muddy Mountain thrust underlie all of the basin. This assumption is consistent with the fact that Paleozoic rocks crop out in northern Bitter Spring Valley, and if valid, it implies that there is a contact between the allochthon and autochthon buried beneath the Thumb Member in the northern part of Bitter Spring Valley between the latter outcrop and the Mesozoic rocks exposed in western Bitter Spring Valley. The nature of this contact is uncertain, and several possibilities regarding it are developed in the Discussion section where it is compared to other similar contacts.

Longwell Ridges–Bitter Spring Valley Area

The northeast-striking Rogers Spring fault, which bounds the southeast side of the Longwell Ridges and the Muddy Mountains, dips between 60° and 70° to the southeast (Fig. 5). Near its southern truncation at the normal fault that bounds the northwest side of East Longwell Ridge, the Rogers Spring fault separates resistant upper-plate middle Paleozoic rocks to the northwest from a fault-repeated section of Aztec Sandstone and Rainbow Gardens Member of the Horse Spring Formation to the southeast. The repeating faults strike north and dip west, and that strata adjacent to them dip moderately eastward. Along much of its length the Rogers Spring fault separates allochthonous Paleozoic rocks as old as Ordovician from Thumb and Rainbow Gardens Members, but at its northern end, near the southeast part of the Valley of Fire, it displaces the Muddy Creek Formation.

The nature of the Rogers Spring fault is poorly understood. Superficially, it appears to be a simple normal fault that bounds the west side of the Overton Arm valley, especially when the displacement indicated by Tertiary beds is examined alone. Between the two Longwell Ridges the Thumb has been dropped on the fault bounding the northwest side of East Longwell Ridge. The fault between the ridges is younger than the Rogers Spring fault because it cuts the latter structure. The Thumb rests directly on the Paleozoic rocks of West Longwell Ridge on a steep buttress unconformity (Bohannon, 1983b, Fig. 16b), and the Rainbow Gardens probably is absent in the subsurface between the ridges. Palinspastic restoration of normal faulting between the ridges leaves the Thumb and its basal buttress unconformity structurally higher than the Rainbow Gardens and basal Thumb Members across the Rogers Spring fault to the southeast. Also, the Tertiary beds southeast of the Rogers Spring fault bend into it in a manner suggesting right-normal drag. Thus, short of the facts that the Tertiary rocks were deposited on Paleozoic rocks on one side of the fault and on Mesozoic rocks on the other, and that a thick section of Rainbow Gardens is present southeast of the fault whereas none occurs northwest of it, the relations in the Tertiary rocks appear to be satisfied by a simple normal displacement model.

Structural relations in the Paleozoic and Mesozoic rocks, on the other hand, are not satisfied by a simple normal fault model. The Paleozoic rocks exposed on the Longwell Ridges are allochthonous above the Muddy Mountain thrust which presumably underlies the ridges and all of White Basin. In contrast, the Mesozoic rocks southeast of the Rogers Spring fault are probably autochthonous for the same reasons that those in the Gale Hills and Bowl of Fire are. Thus, if the Muddy Mountain thrust ever existed southeast of the Rogers Spring fault above the autochthonous Aztec Sandstone it must have been eroded prior to the deposition of the Rainbow Gardens Member which now overlies the Aztec. If that is the case, pre-Rainbow Gardens reverse separation is suggested, considering the present dip of the fault.

Possibly, the Rogers Spring fault has had a multistage history with a Miocene normal fault event and a pre-Miocene event that resulted in the apparent reverse separation. Another possibility is that one episode of post- or late-Thumb-age lateral slip of major proportions took place. If that is the case, the juxtaposition of the unlike terranes on either side of the fault is easily explained. However, the truncation of the Rogers Spring fault by younger faults to the south and the lack of an offset counterpart on the south side of those faults in Bitter Spring Valley do not seem to be consistent with interpretations invoking more than about a kilometer of pre–Rainbow Gardens or Thumb lateral slip. For example, the Thumb Member in Bitter Spring Valley is not cut by any major faults, and there are no folds or areas of crustal extension that are sufficiently large enough to absorb a large amount of lateral slip in that area. Although pre-Thumb lateral slip is possible, it does little to avoid the apparent two-stage history on the Rogers Spring fault. Because relations along the Rogers Spring fault bear heavily on the nature of the contact between the allochthon and autochthon described in other areas, further analysis is attempted in the Discussion section.

DISCUSSION

The probable pre-Miocene configuration of the terrane described in the report is shown on Figure 6. The Lake Mead fault system is shown on the diagram, as is Lake Mead, but only for reference as neither is pre-Miocene. The trace of the Muddy Mountain thrust, which apparently trended northerly on a regional scale, is shown west of the

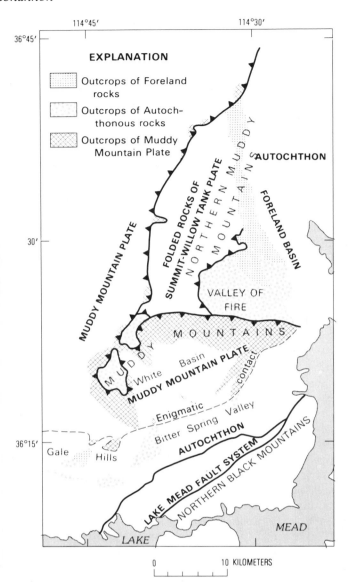

Figure 6. Simplified reconstruction of relations between thrusts, folded belt associated with thrusts, and the autochthon in the Muddy, North Muddy, and northern Black Mountains.

North Muddy Mountains. The upper plate probably once covered the North Muddy Mountains and was later eroded, but the extent of cover and time period of erosion are not known, so the thrust trace is shown in its present position on the diagram. The thrust appears to have been folded into a broad anticline whose northern end was open in the Buffington area. An east-trending segment of the thrust was probably also present south of the Valley of Fire. South of there, the limit of the Muddy Mountain plate is located along the enigmatic contact between the allochthon and the autochthon, which has been described in preceding sections. The Summit–Willow Tank thrust and the fold belt in its upper plate were present beneath and to the northeast of the Muddy Mountain plate. Autochthonous rocks, including foreland basin deposits, to the east of both of the thrusts were gently deformed and probably dipped slightly to the northeast. Displacement has not been restored on the Lake Mead system. The autochthonous rocks of the Muddy Mountain area were probably adjacent to those of the northern part of the Virgin Mountains prior to Miocene time (Bohannon, 1979, 1983b).

The enigmatic contact between the allochthon and the autochthon south and east of the Muddy Mountains was a fault contact that may have been pre-Miocene and was certainly pre-Thumb in age. However, the character, precise age, and significance of the structure are debatable. Geologic relations between the Gale Hills and Bitter Spring Valley suggest that the contact might have been either (1) a large north- to northeast-dipping, east-trending normal fault that dropped upper-plate rocks against lower-plate rocks; (2) a large lateral fault, possibly related to the Las Vegas Valley shear zone; or (3) simply the trace of the Muddy Mountain thrust.

If the enigmatic contact was a normal fault, no sedimentation was associated with it, it apparently predated all normal fault activity in the area, and it developed at a high angle to the trends of later normal faults. The southerly dip of the Paleozoic rocks near Muddy Peak might be explained by rotation due to normal faulting, but the attitude of the Rogers Spring fault seems contrary to the northwest-dipping fault interpretation.

If the contact was a large lateral fault associated with the Las Vegas Valley shear zone, the zone must have been active prior to at least 15 m.y. ago because many radiometric age determinations in the overlapping Thumb are that old (Bohannon, 1983b). Although R. J. Fleck (1981, written commun.) demonstrated that rocks as young as 13 m.y. north of the shear zone are deformed by it, and concluded from this that the shear zone is that young or younger, there are no data constraining large amounts of pre-15-m.y. right slip on the zone.

If the enigmatic fault is simply the buried and slightly modified trace of the Muddy Mountain thrust, then the thrust must climb up-section abruptly from its position in the Cambrian Bonanza King Formation north of Muddy Peak to the Mississippian Monte Cristo Limestone at the enigmatic contact. Mississippian rocks are exposed all along the enigmatic contact in direct and near juxtaposition to the autochthonous rocks. If this is the case, the thrust beneath Muddy Peak would probably be nearly flat and would cut the south-dipping homocline to emerge in the Mississippian near the north end of the Bowl of Fire. Aztec Sandstone and Baseline Sandstone in the lower plate would remain relatively undeformed beneath the thrust. This interpretation is not the one that I proposed on the structure sections with the geologic map of the region (Bohannon, 1981, 1983a), but those interpretations assumed that the Muddy Mountain thrust was strictly controlled stratigraphically within the Bonanza King. That assumption may not be valid regionally, but if it is, either the normal fault or the lateral-slip interpretations probably fit the data best. If it is not, the Muddy Mountain thrust interpretation seems plausible.

The contrasting Tertiary history on either side of the Rogers Spring fault, coupled with its limited extent and the constraints on large amount of Miocene lateral slip thus imposed, suggests that the faulted allochthon-autochthon relation occurred in nearly its present configuration in the Longwell Ridge area prior to deposition of the Thumb and possibly the Rainbow Gardens Members. The Miocene and younger structural geometry in the Longwell Ridge area suggests that a simple picture of uplift and relative northeasterly displacement of East Longwell Ridge by left normal slip on the fault to its northwest and right normal slip on the Rogers Spring fault represents the most likely post-Thumb kinematic history in the area of the ridges. However, these interpretations do not constrain the possibility that pre-Miocene lateral slip generated the original allochthon-autochthon relations. If the Rogers Spring fault was the Las Vegas Valley shear zone prior to deposition of the Thumb Member of the Horse Spring Formation, its northeast trend requires a large bend in the normally east-west trending zone. Certainly, 14- to 10-m.y.-old deformation associated with the Lake Mead fault system could have caused such bending, but the apparent post-13-m.y.-old activity on the shear zone north of Las Vegas is left unaccounted for in the Lake Mead area by this shear zone model alone. Possibly another fault besides the enigmatic contact, such as the western part of the West Bowl of Fire fault, might have been the eastern extension of the Las Vegas Valley shear zone about 13 m.y. ago in the Lake Mead area.

If the enigmatic contact was not the Las Vegas Valley shear zone and the allochthon-autochthon contact was simply the pre-Miocene trace of the Muddy Mountain thrust, several other relations along the Rogers Spring fault are left unexplained. Tertiary rocks there were apparently deposited across a high relief contact that placed Mississippian and Pennsylvanian rocks to the northwest against the Aztec Sandstone to the southeast. The thrust trace must have been buried northwest of the Rogers Spring fault at this time, as it is today, because palinspastic restoration of any Miocene normal faulting on the Rogers Spring fault (which relatively uplifted East Longwell Ridge) only buries the thrust deeper than it is now. Also, because Ordovician rocks are exposed on East Longwell Ridge near the area where it joins West Longwell Ridge, the Muddy Mountain thrust is not as high as the Mississippian at that locality, although it could be climbing southward toward younger rocks.

A large part of the deformation south of the Muddy Mountains took place between about 14 and 10 m.y. ago, and several seemingly contrasting styles appear to have developed synchronously during this period. East-west extension on normal faults in the Gale Hills and White Basin was synchronous with north-south compression which resulted in reverse faults and folds in the horsts, grabens, and tilted blocks associated with the normal faults. At the same time, the proposed local kinematic uplift history for the Bowl of Fire horst implies that north-south extension was occurring there adjacent to the compression in the southern Gale Hills. Lastly, the left slip suggested by kinematic analysis and minor structure evaluation on the western part of the West Bowl of Fire fault seems incompatible with it being the post-13-m.y.-old eastern extension of the right-slip Las Vegas Valley shear zone. However, the latter fault is the only one in the Lake Mead area with the necessary trend and age to correspond to R. J. Fleck's (1981, written commun.) 13-m.y.-old and younger zone.

CONCLUSION

No clear-cut solutions are apparent to me regarding the enigmatic contact, the origin of the West Bowl of Fire fault, or the kinematics of the apparently contrasting extension and compression south of the Muddy Mountains. Unfortunately, the genesis of these features bears heavily on the age and origin of the Las Vegas Valley shear zone, the extent and character of the Muddy Mountain thrust, and the general evolution of Muddy Mountain geology. At present, the hypothesis that the enigmatic contact was the pre–Rainbow Gardens Member–Las Vegas Valley shear zone seems the most plausible. If this is the case, probably the greatest part of the total slip on the shear zone took place during this early stage, which was prior to activity on the Lake Mead fault system. Thus, the simple kinematic models that I presented in which nearly all of the displacement on both fault systems developed synchronously (Bohannon, 1979, 1983b) lack credence.

During the time period between about 14 and 10 m.y. ago, both major fault systems might have been active if the West Bowl of Fire fault was part of the Las Vegas Valley shear zone. However, the details concerning their interaction and the respective roles of east-west extension in the Gale Hills, north-south compression in the southern Gale Hills, north-south extension in the Bowl of Fire, and basin and range faulting are not well understood.

The precise time of initiation of basin and range deformation in the area is difficult to pinpoint. Possibly some of the Horse Spring units were deposited during such deformation, but if so, basin geometry and location has changed considerably since then. Local basin and range valleys such as California Wash, the Virgin River Valley, and Overton Arm valley all contain Muddy Creek Formation at the surface. Bitter Ridge Limestone Member is thought to be present in the subsurface of California Wash

(Bohannon, 1983b) which suggests that basin and range deformation might have begun as early as about 13 m.y. ago locally, but even this could be a minimum figure.

ACKNOWLEDGMENTS

This paper was reviewed by E. D. DeWitt, E. B. Ekren, and G. A. Davis. Their comments helped greatly. R. E. Anderson provided many valuable discussions and much pertinant criticism throughout the course of the study.

REFERENCES CITED

Anderson, R. E., 1969, Notes on the geology and paleohydrology of the Boulder City pluton, southern Nevada, in Geological Survey research 1969: U.S. Geological Survey Professional Paper 650-B, p. B35–B40.
—— 1971, Thin-skinned distension in Tertiary rocks of southeastern Nevada: Geological Society of America Bulletin, v. 82, p. 43–58.
—— 1973, Large magnitude late Tertiary strike-slip faulting north of Lake Mead, Nevada: U.S. Geological Survey Professional Paper 794, 18 p.
—— 1977, Geologic map of the Boulder City 15-minute quadrangle, Clark County, Nevada: U.S. Geological Survey Geologic Quadrangle Map GQ-1395.
—— 1978, Geologic map of the Black Canyon 15-minute quadrangle, Mohave County, Arizona, and Clark County, Nevada: U.S. Geological Survey Geologic Quadrangle Map GQ-1394.
Anderson, R. E., Longwell, C. R., Armstrong, R. L., and Marvin, R. F., 1972, Significance of K-Ar ages of Tertiary rocks from the Lake Mead region, Nevada-Arizona: Geological Society of America Bulletin, v. 83, p. 273–288.
Armstrong, R. L., 1968, Sevier orogenic belt in Nevada and Utah: Geological Society of America Bulletin, v. 79, p. 429–458.
Arnold, H. B., 1977, Geology of part of the Muddy Mountains, Clark County, Nevada [M.S. thesis]: Cheeny, Washington, Eastern Washington State University, 60 p. 2 pls.
Ash, S. R., and Read, C. B., 1976, North American Species of Temskya and their stratigraphic significance, *with a section* Stratigraphy and age of the *Tempskya*-bearing rocks of southern Hidalgo County New Mexico by R. A. Zeller, Jr.: U.S. Geological Survey Professional Paper 874, 42 p.
Bell, J. W., and Smith, E. I., 1980, Geologic map of the Henderson quadrangle, Nevada: Nevada Bureau of Mines and Geology Map 67, scale 1:24,000.
Bohannon, R. G., 1977a, Geologic map of the Valley of Fire region, North Muddy Mountains, Clark County, Nevada: U.S. Geological Survey Miscellaneous Field Studies Map MF-849, scale 1:25,000.
—— 1977b [1978], Preliminary geologic map and sections of White Basin and Bitter Spring Valley, Muddy Mountains, Clark County, Nevada: U.S. Geological Survey Miscellaneous Field Studies Map MF-922, scale 1:25,000.
—— 1979, Strike-slip faults of the Lake Mead region of southern Nevada, in Armentrout, J. M., Cole, M. R., and TerBest, Harry, eds., Cenozoic paleogeography of the Western United States—Pacific Coast Paleogeography Symposium 3: Los Angeles, Pacific Section, Society of Economic Paleontologists and Mineralogists, p. 129–139.
—— 1980, Middle and late Tertiary tectonics of a part of the Basin and Range province in the vicinity of Lake Mead, Nevada and Arizona, [abs.], in Howard, K. A., Carr, M. D., and Miller, D. M., Tectonic framework of the Mojave and Sonoran Deserts, California and Arizona: U.S. Geological Survey Open-File Report 81-503, p. 7–9.
—— 1981, Geologic map, tectonic map, and structure sections of the Muddy and northern Black Mountains, Clark County, Nevada: U.S.

Geological Survey Open-File Map 81-796.

—— 1983a, Geologic map, tectonic map, and structure sections of the Muddy and northern Black Mountains, Clark County, Nevada: U.S. Geological Survey Miscellaneous Geologic Investigations Series Map I-1406, scale 1:62,500 (in press).

—— 1983b, Nonmarine sedimentary rocks of Tertiary age in the Lake Mead region, southeastern Nevada and northwestern Arizona: U.S. Geological Survey Professional Paper 1259 (in press).

Bohannon, R. G., and Bachhuber, Fred, 1979, Road log from Las Vegas to Keystone thrust area and Valley of Fire via Frenchman Mountain, *in* Newman, G. W., and Goode, H. D., eds., Basin and Range symposium and Great Basin field conference: Denver, Rocky Mountain Association of Geologists Guidebook, p. 579–596.

Brock, W. G., and Engelder, Terry, 1977, Deformation associated with the movement of the Muddy Mountain overthrust in the Buffington window, southeastern Nevada: Geological Society of America Bulletin, v. 88, p. 1667–1677.

Burchfiel, B. C., 1965, Structural geology of the Specter Range quadrangle, Nevada, and its regional significance: Geological Society of America Bulletin, v. 48, no. 1, p. 40–65.

Burchfiel, B. C., Fleck, R. J., Secor, D. J., Vincelette, R. R., and Davis, G. A., 1974, Geology of the Spring Mountains, Nevada: Geological Society of America Bulletin, v. 85, p. 103–122.

Cashion, W. B., 1967, Carmel Formation of the Zion Park region, southwestern Utah—A review: U.S. Geological Survey Bulletin 1244-J, 9 p.

Fleck, R. J., 1970, Tectonic style, magnitude, and age of deformation in the Sevier orogenic belt in southern Nevada and eastern California: Geological Society of America Bulletin, v. 81, no. 6, p. 1705–1720.

Hewett, D. F., 1931, Geology and ore deposits of the Goodsprings quadrangle, Nevada: U.S. Geological Survey Professional Paper 162, 172 p.

Ketner, K. B., 1968, Origin of Ordovician quartzite in the Cordilleran miogeosyncline: U.S. Geological Survey Professional Paper 600-B, p. B169–B177.

Longwell, C. R., 1922, The Muddy Mountain overthrust in southeastern Nevada: Journal of Geology, v. 30, no. 1, p. 63–72.

—— 1928, Geology of the Muddy Mountains, Nevada, with a section through the Virgin Range to the Grand Wash Cliffs, Arizona: U.S. Geological Survey Bulletin 798, 152 p.

—— 1949, Structure of the northern Muddy Mountain area, Nevada: Geological Society of America Bulletin, v. 60, p. 923–967.

—— 1952, Structure of the Muddy Mountains, Nevada, *in* Thune, H. W., ed., Cedar City, Utah, to Las Vegas, Nevada: Utah Geological Society, Guidebook to the Geology of Utah, no. 7, p. 109–114.

—— 1960, Possible explanation of diverse structural patterns in southern Nevada: American Journal of Science, v. 258–A (Bradley Volume), p. 192–203.

—— 1962, Restudy of the Arrowhead fault, Muddy Mountains, Nevada: U.S. Geological Survey Professional Paper 450-D, p. D82–D85.

—— 1963, Reconnaissance geology between Lake Mead and Davis Dam, Arizona-Nevada: U.S. Geological Survey Professional Paper 374-E, 51 p.

Longwell, C. R., and Dunbar, C. O., 1936, Problems of Pennsylvanian Permian boundary in southern Nevada: American Association of Petroleum Geologists Bulletin, v. 20, no. 9, p. 1198–1207.

Longwell, C. R., and Mound, M. C., 1967, A new Ordovician formation in Nevada dated by conodonts: Geological Society of America Bulletin, v. 78, no. 3, p. 405–412.

Longwell, C. R., Pampeyan, E. H., Bower, Ben, and Roberts, R. J., 1965, Geology and mineral deposits of Clark County, Nevada: Nevada Bureau of Mines Bulletin 62, 218 p.

Lucchitta, Ivo, 1966, Cenozoic geology of the upper Lake Mead area adjacent to the Grand Wash Cliffs, Arizona [Ph.D. thesis]: University

Park, Pennsylvania State University, Ph.D. thesis, 218 p.

—— 1972, Early history of the Colorado River in the Basin and Range province: Geological Society of America Bulletin, v. 83, p. 1933–1948.

—— 1979, Late Cenozoic uplift of the southwestern Colorado Plateau and adjacent lower Colorado River region: Tectonophysics, v. 61, p. 63–95.

McNair, A. H., 1951, Paleozoic stratigraphy of northwestern Arizona: American Association of Petroleum Geologists Bulletin, v. 35, no. 3, p. 524–525.

Moore, R. T., 1972, Geology of the Virgin and Beaverdam Mountains, Arizona: Arizona Bureau of Mines Bulletin 186, 65 p.

Morgan, J. R., 1964, Structure and stratigraphy of the northern part of the south Virgin Mountains, Clark County, Nevada [M.S. thesis]: Albuquerque, University of New Mexico, 103 p.

Osmond, J. C., 1962, Stratigraphy of Devonian Sevy Dolomite in Utah and Nevada: American Association of Petroleum Geologists Bulletin, v. 46, p. 2033–2056.

Palmer, A. R., 1981, Lower and Middle Cambrian stratigraphy of Frenchman Mountain, Nevada, *in* Taylor, M. E., and Palmer, A. R., eds., Cambrian stratigraphy and paleontology of the Great Basin and vicinity, Western United States: 2nd International Symposium on the Cambrian System, Guidebook for Field Trip 1, p. 11–13.

Palmer, A. R., and Nelson, C. A., 1981, Lower and Middle Cambrian stratigraphy and paleontology of the southern Great Basin, California and Nevada, *in* Taylor, M. E., and Palmer, A. R., eds., Cambrian stratigraphy and paleontology of the Great Basin and vicinity, Western United States: 2nd International Symposium on the Cambrian System, Guidebook for Field Trip 1, p. 1–10.

Poole, F. G., 1964, Paleowinds in the Western United States, *in* Nairn, A.E.M., Problems in paleoclimatology: London, John Wiley, p. 394–406.

Poole, F. G., Baars, D. L., Drewes, Harald, Hayes, P. T., Ketner, K. B., McKee, E. D., Tiechert, Curt, and Williams, J. S., 1967 [1968], Devonian of the southwestern United States, *in* Oswald, D. H., ed., International Symposium on the Devonian System, Calgary, Alberta, September 1967, Vol. 1: Calgary, Alberta, Alberta Society of Petroleum Geologists, p. 879–912.

Reif, D. M., and Slatt, R. M., 1979, Red bed members of the Triassic Moenkopi Formation, southern Nevada: Sedimentology and paleogeography of a muddy tidal flat deposit: Journal of Sedimentary Petrology, v. 49, no. 3, p. 869–890.

Ross, R. J., Jr., and Longwell, C. R., 1964, Paleotectonic significance of Ordovician sections south of the Las Vegas shear zone, *in* Ross, R. J., Jr., Middle and Lower Ordovician formations in southernmost Nevada and adjacent California: U.S. Geological Survey Bulletin 1180-C, p. C88–C93.

Smith, E. I., 1981, Contemporaneous volcanism, strike-slip faulting and exotic block emplacement in the River Mountains, Clark County, southern Nevada [abs.]: Geological Society of America Abstracts with Programs, v. 13, no. 2, p. 107.

Stewart, J. H., 1967, Possible large right-lateral displacement along fault and shear zones in the Death Valley–Las Vegas area, California and Nevada: Geological Society of America Bulletin, v. 78, p. 131–142.

—— 1980, Geology of Nevada: Nevada Bureau of Mines and Geology Special Publication 4, 136 p.

Stewart, J. H., Albers, J. P., and Poole, F. G., 1968, Summary of regional evidence for right-lateral displacement in the western Great Basin: Geological Society of America Bulletin, v. 78, p. 131–142.

—— 1980, Geology of Nevada: Nevada Bureau of Mines and Geology Special Publication 4, 136 p.

Stewart, J. H., Albers, J. P., and Poole, F. G., 1968, Summary of regional evidence for right-lateral displacement in the western Great Basin: Geological Society of America Bulletin, v. 79, p. 1407–1414.

Stewart, J. H., Poole, F. G., and Wilson, R. F., 1972, Stratigraphy and origin of the Chinle Formation and related Upper Triassic strata in the Colorado Plateau region: U.S. Geological Survey Professional Paper 690, 336 p.

Temple, V. J., 1977, Structural relations along the western end of the Arrowhead fault, Muddy Mountains, Nevada [M.S. thesis]: Texas A & M University, 92 p.

Wilson, R. F., and Stewart, J. H., 1967, Correlation of Upper Triassic and Triassic(?) Formations between southwestern Utah and southern Nevada: U.S. Geological Survey Bulletin 1244-D, 20 p.

Young, R. A., 1966, Cenozoic geology along the edge of the Colorado Plateau in northwestern Arizona [Ph.D. thesis]: St. Louis, Missouri, Washington University, 155 p.

—— 1970, Geomorphological implications of pre-Colorado and Colorado tributary drainage in the western Grand Canyon region: Plateau, v. 42, no 3, 107–117.

Young, R. A., and Brennan, W. J., 1974, The Peach Strings Tuff—its bearing on structural evolution of the Colorado Plateau and development of Cenozoic drainage in Mohave County, Arizona: Geological Society of America Bulletin, v. 85, p. 83–90.

Manuscript Accepted by the Society August 20, 1982

Geological Society of America
Memoir 157
1983

Stratigraphic variation and low-angle faulting in the North Hansel Mountains and Samaria Mountain, southern Idaho

Richard W. Allmendinger*
Department of Geological Sciences
Cornell University
Ithaca, New York 14853

Lucian B. Platt
Department of Geology
Bryn Mawr College
Bryn Mawr, Pennsylvania 19010

ABSTRACT

Sandstone and limestone of the Pennsylvanian and Lower Permian Oquirrh Formation in a gently dipping homocline on Samaria Mountain are about 850 m thick. In the North Hansel Mountains, several kilometers to the west and across a fault inferred to be a thrust, rocks of the same age range are four times thicker. The North Hansel section is deformed into open folds with wavelengths of a few kilometers, except where basal Oquirrh units are overturned eastward on a large scale near a low-angle fault between the Oquirrh and the Mississippian and Pennsylvanian Manning Canyon Shale. On the west side of Samaria Mountain, Oquirrh units are more tightly folded and overturned eastward in the upper plate of the Samaria Mountain thrust. This steep fault places older deformed parts of the Oquirrh on undeformed younger parts. We infer that the Samaria Mountain thrust is laterally continuous with the Manning Canyon detachment in the North Hansel Mountains.

The large change in thickness of the Oquirrh across the fault and the severe disharmonic deformation in upper and lower plates in both ranges suggest that initial movement on the fault was eastward-thrusting, probably during the Mesozoic. A structurally higher allochthon of Oquirrh limestones in the North Hansel Mountains crosscuts these older structures and may be Tertiary in age. Reactivation of the Samaria Mountain thrust and Manning Canyon detachment as a normal fault during the Tertiary is also possible.

INTRODUCTION

The North Hansel Mountains and Samaria Mountain are located in southernmost Idaho, along the border with neighboring Utah (Fig. 1). They are located at the eastern margin of an area of extensive exposures of Pennsylvanian and Permian strata that cover many of the mountains in the northeastern Great Basin. Until recently, this part of the Cordillera was poorly known, particularly from a structural standpoint, but recent mapping by the U.S. Geological Survey, under the direction of Max D. Crittenden, Jr. in

northern Utah and of Steven S. Oriel in southern Idaho, has begun to document some of the fascinating and important relationships in the region. These relationships have a bearing on the nature of late Paleozoic, Mesozoic, and Cenozoic tectonics that have affected the region.

A long-standing problem in this part of the Cordillera is the origin and genesis of faults that are oriented at a low angle to bedding but that place younger rocks over older (Armstrong, 1972; Hose and Danes, 1973; Roberts and Crittenden, 1973; Oriel and Platt, 1979; Allmendinger and Jordan, 1981). Such faults are more common than older-

*Also at: U.S. Geological Survey, Box 25046, Federal Center, Denver, Colorado 80225

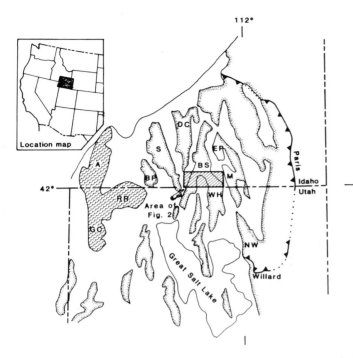

Figure 1. Map of northern Utah and southern Idaho, showing major physiographic features and the location of the study area (diagonal-ruled box). Geographic names mentioned in text: A—Albion Range; BP—Blackpine Mountains; BS—Blue Springs Hills; DC—Deep Creek Mountains; EP—Elkhorn peak; GC—Grouse Creek Mountains; M—Malad Range; NW—northern Wasatch Range; RR—Raft River Mountains; WH—West Hills. Line with sawteeth—Paris and Willard thrust faults; wavy line pattern—metamorphic "core" complex.

over-younger faults in the region referred to as the hinterland of the Mesozoic-early Cenozoic Idaho-Wyoming-Utah (or "Sevier") thrust belt. However, it is not clear whether the younger-over-older, low-angle faults are related to thrust belt deformation, or if they are a product of middle and late Cenozoic deformation of a different character (Armstrong, 1972). Furthermore, we do not know whether these faults formed in a regional compressive or extensional stress regime.

This paper describes low-angle faults and related structures in the North Hansel Mountains and Samaria Mountain. Most of these structures we interpret to have formed *initially* by west-to-east thrusting during the Mesozoic. Some of these faults may have been reactivated and others probably formed initially during horizontal extension in the Cenozoic. Therefore, we use the term "low-angle fault" as a descriptive term for any fault, oriented at a low angle to bedding. This definition is strictly geometrical and may encompass both thrust faults and low-angle normal faults. With respect to the above controversy, we regard "thrust fault" and "normal fault" as genetic terms and hence use them only in interpretations. There, "thrust fault" is used as follows: "A fault that produces lateral translation and re-

gional shortening." Likewise, "low-angle normal fault" is used as: "A fault that produces lateral translation and regional lengthening." These faults are inferred to be produced, respectively, in a compressive or tensional* stress regime.

A main point of the paper is that the significance of the low-angle faults cannot be determined without also considering late Paleozoic paleogeography. Moreover, the structural geometry of one phase of deformation influences that of succeeding phases.

We are fortunate in not being the first to deal with geologic problems in Samaria Mountain and the North Hansel Mountains. Earlier general strratigraphic frameworks for the two areas were provided by Beus (1968) and Ballou (unpub.), respectively, and our work complements theirs. Regional syntheses, directly pertinent to this study, have been made by Oriel and Platt (1979), Jordan and Douglass (1980), and Allmendinger and Jordan (1981).

REGIONAL GEOLOGIC SETTING

The rocks in the North Hansel Mountains and Samaria Mountain were deposited in the Cordilleran miogeocline (F. C. Armstrong and Oriel, 1965; R. L. Armstrong, 1968a), which was initiated during Late Precambrian rifting of the western North American continental margin (Stewart, 1972). Facies of Ordovician through Mississippian rocks similar to those exposed on the east side of Samaria Mountain can be found throughout the eastern and central Great Basin. During the Pennsylvanian and Permian in southern Idaho and northern Utah, sandstones and limestones of the Oquirrh Formation were deposited in a relatively deep, northwest-trending, fault-bounded basin, interpreted to have been related to the Ancestral Rocky Mountains (Jordan and Douglass, 1980). This basin cut obliquely across the grain of the miogeocline. Permian and Lower Triassic strata overlie Oquirrh basin sediments in mountain ranges surrounding those described here. Upper Triassic and Jurassic sediments are well known farther east, but their original western extension into the region of the North Hansel Mountains and Samaria Mountain is unknown. The next youngest deposits are Neogene conglomerates, ash flow and air fall tuffs, and basalts.

The study area lies in the region commonly referred to as the hinterland of the Cordilleran fold-thrust belt (Armstrong, 1972; Roberts and Crittenden, 1973; Oriel and Platt, 1979; Allmendinger and Jordan, 1981). Metamorphic rocks of the Albion, Raft River, and Grouse Creek Mountains lie west of the North Hansel Mountains and surrounding ranges (Fig. 1). This metamorphic terrane exhibits a complex structural and metamorphic history, spanning Jurassic through Miocene (Armstrong, 1976; Compton and

---

*"Compression" and "tension" as used in this paper refer to *deviatoric* stresses.

others, 1977; Davis and Coney, 1979; Crittenden, 1980; Miller, 1980; Todd, 1980). The relations between metamorphic complex and foreland thrust belt have been a subject of debate for many years (e.g., Crittenden, 1979; Compton and Todd, 1979), and part of the reason for that debate has been the lack of information on the structural evolution of the intervening region. To the east of the study area lie the imbricated thrusts and folds of the Idaho-Wyoming-northern Utah segment of the Cordilleran fold-thrust belt. These structures formed as an eastward-propagating, deformational front during latest Jurassic to early Eocene times (F. C. Armstrong and Oriel, 1965; Oriel and Armstrong, 1966; R. L. Armstrong, 1968b; Royse, Warner, and Reese, 1975; Oriel and Platt, 1980).

Oriel and Platt (1979) first described tectono-stratigraphic relations within the hinterland east of the metamorphic complexes in the Albion and Raft River Mountains (Fig. 1). They recognized three allochthonous units of possible regional extent, separated by low-angle faults that place younger rocks over older rocks. Here, we present a structural interpretation of one locality with exposure of their upper and middle allochthons. Jordan and Douglass (1980) and Crittenden (1982) have suggested that the allochthons defined by Oriel and Platt may extend into northern Utah. The upper allochthon, composed primarily of Pennsylvanian through Lower Triassic strata, was informally named the Hansel plate by Allmendinger and Jordan (1981) who inferred that initial eastward translation of the plate might have been in the Middle Jurassic.

The most recent and continuing deformation in the region is Basin and Range extension. The North Hansel Mountains and Samaria Mountain lie in the intermountain seismic belt (Smith, 1978). Faults cutting Quaternary lake terraces on the west side of the North Hansel Mountains are *en echelon* and trend northeasterly (Allmendinger, in press), and the Pocatello Valley earthquake of 1975 was located beneath the valley separating Samaria Mountain and the North Hansel Mountains (Platt, 1975). Older, presumably at least partly Tertiary, high-angle faults have several different trends and are very common in the North Hansel Mountains (Allmendinger, in press) and to the north in the Blue Springs Hills but are mostly lacking in Samaria Mountain (Platt, 1977).

STRATIGRAPHIC RELATIONS

The lithologies and facies of upper Paleozoic strata in southern Idaho and northern Utah have been well described elsewhere (Jordan and Douglass, 1980; Skipp and Hall, 1980; Yancey and others, 1980). A discussion of the stratigraphy of the North Hansel Mountains and Samaria Mountain may be found in Ballou (1982), Allmendinger (in press), Beus (1968), and Platt (1977), respectively. For the sake of brevity, we summarize here only those general aspects of the stratigraphy pertinent to the major subject of

this paper. (Figure 2 shows stratigraphic and structural symbols used in all succeeding figures.)

Unnamed sandstone

The highest pre-Cenozoic stratigraphic unit in this region is a Permian (Leonardian?) dolomitic sandstone with thin interbeds of dolomite and limestone. In the North Hansel Mountains, this unnamed unit is exposed in the core of the syncline and has a minimum thickness of 500 m. Rocks of similar lithology also overlie the Oquirrh Formation in nearby mountain ranges and have been named Diamond Creek Sandstone (Doelling, 1981), Hudspeth Cutoff Sandstone (Cramer, 1971), or informally, the unnamed formation (Doelling, 1981). Because of continuing confusion over facies relations and nomenclature, this unit in the North Hansel Mountains is informally referred to here as the unnamed sandstone.

Oquirrh Formation

The monotonous sequence of sandstone, sandy limestone, and limestone that comprises the Oquirrh Formation (Lower Permian-Pennsylvanian) crops out over much of the study area. Approximately 95 percent of the North Hansel Mountains and 50 percent of Samaria Mountain are underlain by Oquirrh strata (Fig. 3). To the north, the Blue Springs Hills are almost totally underlain by Oquirrh Formation, except for a tectonic window exposing Mississippian rocks (Coward, 1979).

At Samaria Mountain, Platt (1977) divided Lower Permian to Lower Pennsylvanian (Wolfcampian to Morrowan) strata of the Oquirrh into three units. His upper unit (Fig. 4) comprises about 250 m of dark brownish-gray calcareous sandstone with subordinate ledges of bioclastic limestone, all yielding Wolfcampian fusulinids (R. Douglass, personal communicaton, 1975, 1976). This unit has an eroded top and occurs only west of the Samaria Mountain fault (see below). East of the Samaria Mountain fault, Platt's sandy (middle) member and limestone (lower) member (Fig. 4) constitute about the top 850 to 900 m of a structurally uninterrupted sequence of Ordovician to Permian strata (Fig. 3). Diagnostic fossils from the limestone member suggest Atokan and Morrowan ages (Beus, 1968, p. 800), and fusulinids from the sandy member indicate Wolfcampian ages (G. Steele in Beus, 1968, p. 801 and 806). Beus (1968) suggested the presence of a major unconformity omitting all Upper Pennsylvanian rocks in the Oquirrh Formation east of the North Canyon fault. This may be the westernmost exposure of a widely documented Upper Pennsylvanian unconformity in Idaho, Wyoming, and northern Utah east of the region described here (Mallory, 1972). Beus (1968, p. 801) reported a collection of Guadalupian fauna from the Phosphoria(?) Formation, approximately 880 m above the top of the Manning Canyon Shale, east of the North Canyon fault.

STRATIGRAPHIC SYMBOLS

GENERAL UNITS

Q	Quaternary
Ts	Tertiary Salt Lake Fm.
PℙPo	Oquirrh Fm.(Permian & Pennsylvanian)
M	Mississippian
D	Devonian
S	Silurian
O	Ordovician

PERMIAN & PENNSYLVANIAN STAGES

L	Leonardian
W	Wolfcampian
V	Virgilian
Mo	Missourian
D	Desmoinesian
A	Atokan
Mw	Morrowan

NORTH HANSEL MTNS.

Ps	Unnamed sandstone
Poe	Oquirrh Fm., unit e
Pod	Oquirrh Fm., unit d
ℙoc	Oquirrh Fm., unit c
ℙob	Oquirrh Fm., unit b
ℙoa	Oquirrh Fm., unit a
ℙox	Oquirrh Fm., unit x
ℙMmc	Manning Canyon Shale

SAMARIA MTN.

Pou	Oquirrh Fm., upper member
ℙPos	Oquirrh Fm., sandy member
ℙol	Oquirrh Fm., limestone member
ℙMmc	Manning Canyon Shale
Mgb	Great Blue Limestone
Mh&Ml	Humbug Fm. & Lodgepole Limestone
Db	Beirdneau Fm.
Dh	Hyrum Dolomite
Dwc	Water Canyon Fm.
Sl	Laketown Dolomite

STRUCTURAL SYMBOLS

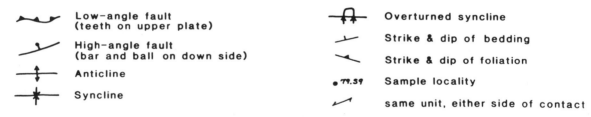

Low-angle fault (teeth on upper plate)	Overturned syncline
High-angle fault (bar and ball on down side)	Strike & dip of bedding
Anticline	Strike & dip of foliation
Syncline	Sample locality
	same unit, either side of contact

Figure 2. Explanation for stratigraphic and structural symbols used in Figures 3-7. Units from Platt (1977) and Allmendinger (in press).

In the North Hansel Mountains, there are six different units of the Oquirrh (Fig. 4), ranging from Middle Pennsylvanian (Atokan) through Lower Permian (Leonardian?). This section, well exposed in an east-dipping homocline, is remarkably complete though broken by high-angle faults. The top of the formation—the contact with the overlying unnamed Permian sandstone—is present on the northeast side of the range, and late Wolfcampian fusulinids (R. Douglass, personal communication, 1980) have been collected from about 200 m below that contact. The base of the Oquirrh is faulted and the lowest unit is found only in a structurally complex area on the west side of the range. The lithologies of the lowest unit, which has yielded Atokan fusulinids (R. Douglass, personal communication, 1980), clearly identify it as representative of the regionally extensive basal limestone of the Oquirrh Formation (the West Canyon limestone member of Beus, 1968). Cross sections

indicate that the lowest unit in the North Hansel Mountains has a minimum thickness of 400 m though its original thickness is unknown. This thickness is similar to that of the basal member at Samaria Mountain (Platt, 1977). The similar thicknesses of the Atokan facies of the Oquirrh Formation over most of southern Idaho and northern Utah suggest that this part of the section exposed in the North Hansel Mountains is probably close to its full thickness. Thus, the thickness of the entire Oquirrh Formation in the North Hansel Mountains is more than 3500 m, with the overlying unnamed sandstone adding another 500 m or more to the Permian and Pennsylvanian section (Fig. 4).

In summary, at Samaria Mountain approximately 500 m of Oquirrh underlie the lowest Wolfcampian fusulinid in the section, but in the North Hansel Mountains at least 2200 m of Oquirrh occur beneath the lowest Wolfcampian fusulinids (Fig. 4). Thus, the Pennsylvanian section thickens

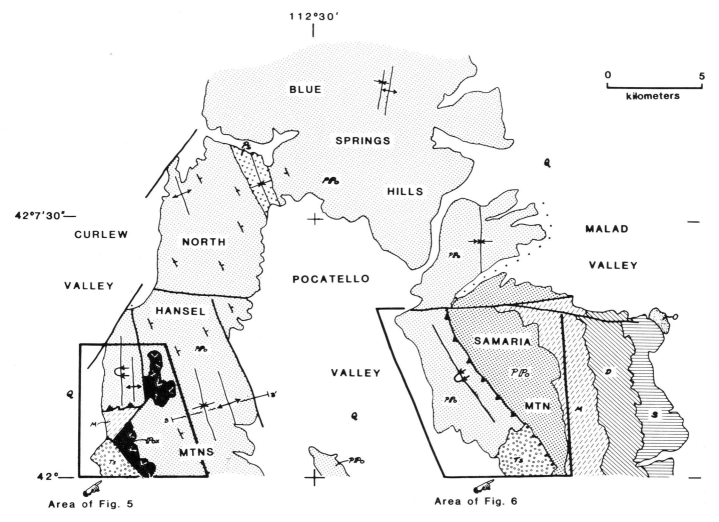

Figure 3. Generalized geologic and tectonic map of the field area, based on Platt (1977) and Allmendinger (in press). Location of section BB′ in Figure 5A shown. See Figure 2 for explanation of symbols.

more than four times across a distance of 15 to 20 km. Thickness changes in the Permian rocks are difficult to determine. If Beus' Guadalupian fauna is in place, then the entire Pennsylvanian and Permian section thickens westward by more than a factor of four. The rapid change in thickness would not be surprising if the Oquirrh Formation was deposited in a northwest-trending trough bounded at this locality by high-angle faults. However, coarse conglomerates and megabreccias, which suggest a steep, fault-bounded margin on the west side of the Oquirrh base (Jordan and Douglass, 1980), are completely lacking here. Furthermore, clast sizes in thin lenses of chert-pebble conglomerate increase *westward* across this area, suggesting derivation from the west rather than a fault-bounded, eastern margin. Therefore, we favor the alternative explanation that the rapid thickness changes result from telescoping of thicker and thinner facies due to west-to-east, low-angle

faulting. In either explanation, though, the geometric margin of the Oquirrh basin at this latitude probably lay between Samaria Mountain and the North Hansel Mountains.

Manning Canyon Shale

The argillite, siltstone, and scattered, poorly sorted sandstones of the Manning Canyon Shale (Pennsylvanian and Mississippian) are exposed on the east side of Samaria Mountain where the shale is in a complete, little deformed, and gently west-dipping sequence ranging from Ordovician Swan Peak Quartzite through the middle sandy member of the Oquirrh Formation (Figs. 3 and 5). The Manning Canyon there is 80 m thick. Manning Canyon Shale, lithologically similar to the exposures on Samaria Mountain, is also exposed in the North Hansel Mountains and Blue Springs Hills (the latter is described by Coward, 1979). In

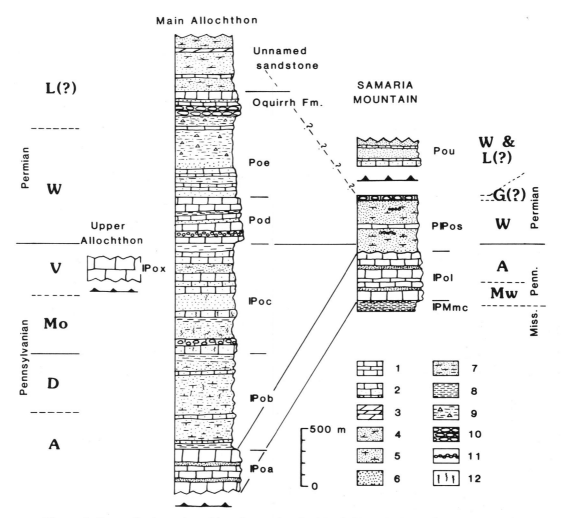

Figure 4. Generalized stratigraphic columns for the North Hansel Mountains and Samaria Mountain. Ages shown to left and right of columns based on Beus (1968) and fusulinid determinations by R. C. Douglass (personal communication, 1980). The upper member of the Oquirrh occurs only west of the Samaria Mountain fault, while sections for the sandstone and limestone members and the Manning Canyon Shale are from east of the fault. G(?)-Guadalupian(?). See Figure 2 for explanation of other symbols. Horizontal tie-line at first occurrence of Wolfcampian fusulinids in each section. North Hansel column shows overlying unnamed Permian sandstone. Lithology symbols: 1—limestone; 2—sandy limestone; 3—dolomite; 4—dolomitic sandstone; 5—calcareous sandstone; 6—sandstone; 7—siltstone and siltly sandstone; 8—shale; 9—silicified siltstone; 10—bedded chert; 11—conglomeratic lens; 12—trace fossils and bioturbation.

each area, the formation occurs in a window separated from the overlying Oquirrh by a low-angle fault. In the window on the west side of the North Hansel Mountains, most of the contacts with the Oquirrh are presently formed by young, high-angle faults, but pieces of the older, low-angle fault are still preserved (Fig. 3). The thickness of the Manning Canyon Shale in the North Hansel Mountains is unknown.

Older Paleozoic Units

Units older than the Pennsylvanian and Mississippian Manning Canyon Shale are exposed on the east side of Samaria Mountain (Fig. 3) (Beus, 1968; Platt, 1977). The Ordovician through Mississippian section there occurs in a structurally coherent block and is typical of the Cordilleran miogeocline (Armstrong, 1968a). On a more regional scale,

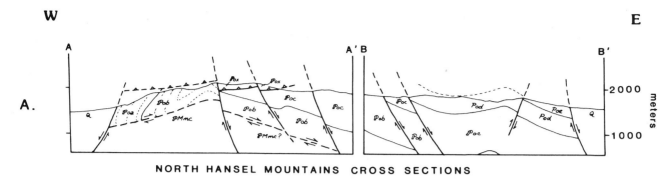

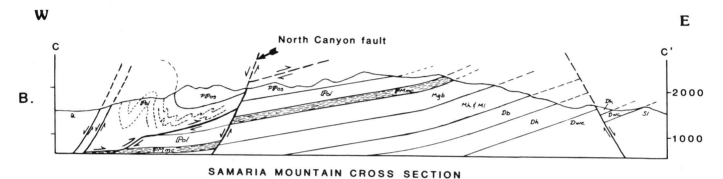

Figure 5. Geologic cross sections of A) the North Hansel Mountains, and B) Samaria Mountain, after Allmendinger (in press) and Platt (1977), respectively. No vertical exaggeration. Location of section AA' shown in Figure 6; BB' in Figure 3; and CC' in Figure 7. See Figure 2 for explanation of symbols.

the Samaria Mountain section can be considered part of a belt of lower and middle Paleozoic sedimentary rocks that extend from Wellsville Mountain in the northern Wasatch Range at least as far north as Elkhorn Peak (Fig. 1). At Wellsville Mountain, this section lies in the upper plate of the Willard thrust (Crittenden, 1972).

STRUCTURAL GEOMETRY

Folding

Three domains of folding can be identified in a west-to-east section across the North Hansel Mountains and Samaria Mountain. These are: (1) long wavelength, medium to large amplitude folds in the North Hansel Mountains; (2) shorter wavelength, small to medium amplitude folds on the west side of Samaria Mountain and in the Blue Springs Hills; and (3) the gently dipping, unfolded section on the east side of Samaria Mountain. The transition from 1 to 2 is gradational but from 2 to 3 is sharp, occurring across the North Canyon fault (see below). The homoclinal dip in the North Hansel Mountains and possibly Samaria Mountain may largely be due to rotation on young Basin-and-Range normal faults rather than the folding described here.

North Hansel Mountains. Several major folds have been identified by mapping in the North Hansel Mountains. The most interesting of these is an overturned syncline that trends north-south along the western edge of the range where it is cut by a later normal fault (section A-A', Fig. 5A). The overturned limb dips to the west at between 45° and 70°, indicating an east-verging fold and west-over-east shear. The lowest units involved in the fold are the basal Oquirrh limestone (Atokan), and the overturning is spatially associated with the low-angle fault between the Oquirrh and Manning Canyon (Fig. 6).

Most of the other folds in the North Hansel Mountains are broad, open structures with steeply west-dipping axial surfaces (section B-B', Fig. 5A). Though variable, the folds have an average wavelength (anticlinal crest to anticlinal crest) of about 5 to 6 km and an average amplitude (crest to trough along a single stratigraphic horizon) of 0.25 to 1.5 km. These folds involve stratigraphically higher units in the Oquirrh and show that fold intensity decreases away from the low-angle fault separating Oquirrh from Manning Canyon.

West Samaria Mountain-Blue Springs Hills. A complete, structural cross section from the North Hansel Mountains to Samaria Mountain must include a dogleg north to the southern Blue Springs Hills because of the

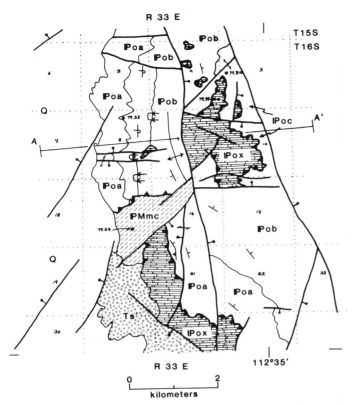

Figure 6. Detailed geologic map of part of the North Hansel Mountains, showing structural relations of upper allochthon (limestone pattern), main allochthon (no pattern), and lower allochthon (diagonal dashed lines). Location of cross section AA' in Figure 5A shown. Grid of dotted lines—sections in T.16 S., R.33 E. Selected section numbers shown. See Figure 2 for explanation of symbols.

intervening Pocatello Valley (Fig. 3). Unfortunately, identification of individual folds in the Blue Springs Hills is difficult due to poor exposure and monotonous Quirrh stratigraphy. Consistent variations of strikes and dips suggests at least local folding with wavelengths of about 1.0 to 1.5 km and poorly known, but probably relatively small amplitudes (based on gentle dips). These folds, occurring in all Oquirrh units mapped by Platt (1977), apparently have slight to moderate eastward vergence and plunge southward.

Folds on the west side of Samaria Mountain involve lower and middle members of the Oquirrh Formation and are much tighter (Fig. 5B). They have wavelengths of about 1.0 km or less and steeply dipping or overturned limbs that suggest eastward vergence. This vergence is confirmed by cleavage that dips 45° to 60° west. The folds trend north-northwest, paralleling the surface trace of the North Canyon fault (see below).

East Samaria Mountain. There is almost no evidence of folding on the east side of Samaria Mountain (Fig. 5B). The strike and dip of all units at most localities are N. 0° to

10° E., 15° to 25° W. On the north side of Samaria Mountain, approaching the North Canyon fault from the east, there is a suggestion of a very gentle, north-northeast trending, south-plunging, anticline-syncline pair, but the relation of this structure to the north-northwest trending fault is unknown. Clearly, the North Canyon fault separates a domain of moderate to intensive folding on the west (upper plate) from one of almost no folding at all on the east (Figs. 5B and 7).

Faulting

Manning Canyon Fault. The low-angle fault between the Manning Canyon and the Oquirrh in the North Hansel Mountains is exposed for about 2 km on the west side of the range (Fig. 6). The remainder of the tectonic boundaries of the window are formed by younger, high-angle faults. The fault surface is poorly exposed, but deformation along it is characterized by a zone of moderate brecciation that extends 10 to 50 m into the upper plate. The Manning Canyon Shale of the lower plate is intensely cleaved, forming pencil structures that trend northward, parallel to minor fold axes. There is no continuity of folding between upper and lower plates. The small exposed part of the fault dips gently westward, less steeply than the west-dipping overturned bedding in the basal Oquirrh units in the upper plate (Figs. 5 and 6). Thus, the fault locally cuts across bedding in the upper plate at an angle of 40° to 50° and places the lowest units of the Oquirrh tectonically over the Manning Canyon Shale. Cross sections suggest that a few hundred metres of stratigraphic section may be omitted along the fault at this locality (section A-A', Fig. 5A). However, the decrease in intensity of folding in higher stratigraphic levels of the Oquirrh Formation on the east side of the mountains (section B-B', Fig. 5A) indicates that this low-angle fault, if regionally extensive, remains at or near the base of the Oquirrh Formation throughout the North Hansel Mountains.

Upper low-angle fault of the North Hansel Mountains. One of the most enigmatic and intriguing structures of the entire region is located along the central crest and on the southwest side of the North Hansel Mountains. This upper low-angle fault emplaces on allochthon of Oquirrh carbonates quite unlike the exposed Oquirrh rocks in the main part of the mountains (Fig. 6). The rocks within this allochthon (unit x, Fig. 3) are dominantly light and dark colored, thick to massively bedded limestones that are relatively free of quartz sand or silt. Locally, thick (0.5 to 2 m) calcite veins are present and recrystallization is more obvious than in the rest of the Oquirrh Formation in the main part of the mountains. Deformed crinoid ossicles and coral can be seen on the outcrops, and Ballou (unpub.) reported fusulinids of Virgilian age at two localities within the allochthon. Virgilian facies in the rest of the North Hansel Mountains

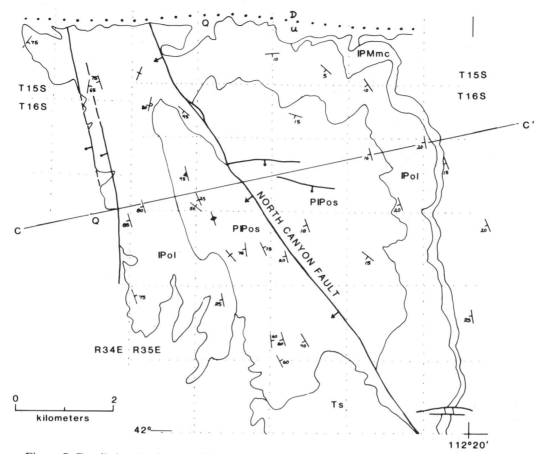

Figure 7. Detailed geologic map of the west side of Samaria Mountain, showing contrast in structural styles across the North Canyon fault. Location of cross section CC' in Figure 5B shown. Grid of dotted lines—sections in the townships and ranges shown. See Figure 2 for explanation of symbols.

are thin- to medium-bedded, bioturbated silty micrite and bioclastic silty limestone, interbedded with quartzose to subarkosic sandstone (Allmendinger, in press). This contrast in facies (Fig. 4) implies substantial transport of the upper allochthon, but we do not know from which direction.

The allochthon is presently exposed in a number of scattered, small klippen (Fig. 6), lying on Oquirrh strata ranging in age from Atokan to Virgilian. In general, the fault surface dips west or southwest, and at one locality cleavage in lower-plate limestone dips 50° west and parallels the fault (cleavage in limestone is not common in the North Hansel Mountains). Elsewhere, the fault surface apparently has a more gentle dip. At one locality, the upper allochthon rests directly on Manning Canyon Shale and appears to have been emplaced across the low-angle fault between Manning Canyon and the Oquirrh sequence in the main part of the range (Fig. 6). Also, the upper low-angle fault cuts across and truncates the major overturned fold on the west side of the mountains. These relations suggest

that the upper plate was emplaced after the major folding and low-angle faulting in the North Hansel Mountains, though the fault may have been gently folded after emplacement as well.

North Canyon Fault and Samaria Mountain Thrust. A steep, north-northwest trending fault is exposed for about 10 km along the west side of Samaria Mountain (Fig. 7). As stated above, this fault separates a nearly undeformed sequence of Ordovician through Permian-Pennsylvanian strata on the east from moderately to highly folded Oquirrh Formation on the west. At its north end in Idaho, stratigraphic throw on the fault is apparently small, with the lower member of the Oquirrh on the west against the middle member on the east (Platt, 1977). The fault has a narrow zone of brecciation and at least one fault sliver. It is truncated on the north side of Samaria Mountain by a younger, east-striking normal fault with a downthrown north side.

Tracing the North Canyon fault south across the border into Utah, it cuts progressively downsection in the footwall. At its southernmost exposure, Permian-Pennsylvanian rocks in the hanging wall are against Devonian rocks in the footwall (Beus, 1968). We interpret the North Canyon fault as a high-angle normal fault that drops the west side down.

However, the striking contrasts in structural style and stratigraphic thicknesses and facies between upper Paleozoic rocks west and east of the fault strongly suggest that those to the west are allochthonous with respect to those on the east. Thus, the North Canyon fault must cut and displace an older low-angle fault, which we interpret to be a thrust (the "Samaria Mountain thrust") because equivalent age rocks are present in both upper and lower plates. Displacements on the younger normal fault were small at its north end; older rocks in the upper plate of the low-angle fault were dropped down against younger rocks in the lower plate of the thrust. The "older-over-younger" relations of the earlier thrust fault were preserved there, but farther southeast with increasing displacement on the North Canyon fault, stratigraphic relations more typical of a normal fault are observed. At the southeast end, rocks west of the normal fault remain allochthonous relative to those to the east, but the thrust origin has been obscured by superposed deformations.

Because unfaulted Manning Canyon is present on the east side of Samaria Mountain and similar age rocks belonging to different plates are present on either side of the North Canyon fault, there must be a ramp in the subsurface where the older thrust fault cut up section across the Permian-Pennsylvanian section. Although we do not know its exact position, given the small amount of displacement on the North Canyon normal fault and the tight folding in the upper plate of the thrust at the west side of Samaria Mountain, the ramp is probably located not far west of the normal fault (Fig. 5b).

MICROSTRUCTURE ANALYSIS OF ALLOCHTHONOUS ROCKS

Analysis of microstructures in quartz and carbonate minerals can provide additional insight into the nature and mechanical significance of deformation. In rocks that have been deformed under low-grade or nonmetamorphic conditions (e.g., rocks in the North Hansel Mountains), such studies provide information on strain and, by inference, on paleostress orientation that cannot be determined by macroscopic observation alone. Unfortunately, the techniques do not provide timing, and thus interpretation of the results of microstructure analysis in terms of regional geology may be ambiguous.

In this section, we discuss some of the results of microstructure analyses of four samples collected from various tectonic units in the North Hansel Mountains. Classical analysis techniques for quartz sub-basal I deformation la-

mellae and calcite *e*-lamellae were used (Turner and Weiss, 1963; Ave'Lallement and Carter, 1971). Because young high-angle faulting has caused rigid body rotations of the strata, bedding planes are used as the principal frame of reference for interpretation of the microstructures. The four oriented samples in the North Hansel Mountains were collected from: (1) a quartzite in the Manning Canyon (loc. 79.S4); (2) a limestone from the homoclinal sequence of Oquirrh Formation in the main part of the range (loc. 79.S10); (3) a sandy limestone from the lower part of the Oquirrh on the overturned limb of the syncline on the west side of the mountains (loc. 79.S3); and (4) a bioclastic limestone from the upper Oquirrh allochthon along the crest of the range (loc. 79.S9). Sample localities were shown in Figure 6, and results of the analyses are shown in Figure 8.

Dynamic analyses of quartz deformation lamellae in the Manning Canyon quartzite show that the paleocompression axis presently plunges 15° to 20° due east and lies within 15° to the plane of bedding (loc. 79.S4; Fig. 8A). The tension axis is oriented near vertical and perpendicular to bedding. Nearly as prominent as the tension axis, the intermediate direction is parallel to the strike direction of bedding in the sample. The sample from the homoclinal sequence of Oquirrh in the main part of the range (loc.

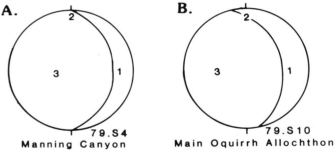

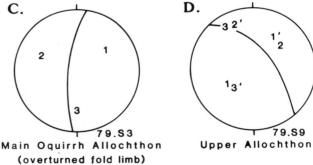

Figure 8. Summary of microfabric analysis in various tectonic units of the North Hansel Mountains. Equal area, lower hemisphere projections; 1—compression; 2—intermediate axis; 3—extension; great circle shows bedding. A) 50 quartz grains measured; B) 25 calcite grains; C) 25 calcite, 10 quartz grains; D) 32 calcite grains—primed numbers were determined from 7 calcite grains in matrix; unprimed numbers, from 25 calcite grains in a vein. See text for discussion.

79.S10; Fig. 8B) exhibits a similar pattern using calcite *e*-lamellae analysis. The other sample from the main allochthon of Oquirrh in the North Hansel Mountains (loc. 79.S3; Fig. 8C), collected from the overturned limb of the syncline, yielded somewhat different results. Compression was at a high angle to bedding and is presently oriented plunging 50° to the northeast; tension plunges gently south-southeast, about parallel to the fold axis.

The sample collected from the upper allochthon is more difficult to interpret (loc. 79.S9; Fig. 8D). Most of the measurements were made on sparry calcite in a vein, and only a few could be made in the matrix due to the overall micritic texture. *E*-lamellae in the vein indicate a strong tension maximum oriented parallel to bedding and N 20°-30° W (Fig. 8D). The associated compression fabric is more diffuse but shows a compression maximum perpendicular to bedding. The number of grains measured in the matrix (seven) is so small as to be nearly meaningless. However, those grains have a relatively tight cluster and indicate a possible compression direction that may be oriented plunging northeast, within 30° to bedding and nearly perpendicular to the compression maximum from the vein calcite. The associated tension direction may plunge steeply southwest, at a high angle to bedding. These apparent complexities in the fabric can be tentatively interpreted as evidence of multiple deformation, which would be in accord with field observations indicating emplacement of the upper allochthon at a later time than folding and the movement along the low-angle fault between Manning Canyon and Oquirrh.

INTERPRETATIONS

Relative Timing of Faulting and Folding

The short map traces of the low-angle faults and the long-wavelength, open nature of many of the folds prevent a definitive determination of the relations between these two types of structures. Where observed, the faults are not folded significantly but separate different domains of folding. However, a correlation of large, tight, overturned folds in the main Oquirrh allochthon in the North Hansel Mountains and the west side of Samaria Mountain with the Manning Canyon low-angle fault and the thrust cut by the North Canyon, fault, respectively, is evident. In contrast, the upper low-angle fault in the North Hansel Mountains probably is unrelated to any of the folds there.

Thus, we infer that folding, low-angle faulting, and initial movement on the Samaria Mountain thrust are generally the products of the same tectonic event. In detail, faulting may be slightly younger than or possibly synchronous with the fold development, accounting for the observed cross-cutting relations. Emplacement of the upper allochthon in the North Hansel Mountains postdates all significant folding but predates most high-angle faulting.

Translation Direction of the Allochthons

The consistent eastward vergence of folds throughout the region and the paleogeography of the Oquirrh basin show that the main Oquirrh allochthon was transported from west to east. The rapid change in thickness of the Oquirrh Formation between Samaria Mountain and the North Hansel Mountains can be interpreted as a result of west-to-east translation of a central part of the Oquirrh basin over a marginal part. The amount of translation cannot be determined because we do not know in detail the original configuration of the Oquirrh basin.

The translation direction of the upper allochthon of the North Hansel Mountains is unknown. A large amount of translation is implied by the contrast in facies between upper and main allochthons. However, more information is needed about facies relations of the Oquirrh Formation in southern Idaho.

Mechanics of the Deformation

The orientations of microscopic strains, with respect to bedding and present geographic coordinates, suggest that most of the intracrystalline deformation probably occurred during a single (though perhaps progressive in time and space) tectonic event. This deformation probably occurred by nearly horizontal, east-west compression and vertical tension when the strata were approximately flat-lying. The nearly identical orientations in both Manning Canyon and the overlying allochthon of Oquirrh indicate that both tectonic units were subjected to the same or similar deformations.

The inferred compression at a high angle to bedding and tension subparallel to fold axis noted in the sample from the overturned limb (loc. 79.S3) are similar to that found on overturned fold limbs in the Meade plate of the Idaho-Wyoming thrust belt (Allmendinger, 1982). This observation and the fact that the inferred regional compression directions are essentially perpendicular to the mapped fold axes suggest that the intracrystalline strain was approximately synchronous with folding. If folding and low-angle faulting were synchronous, then the faulting was produced by east-west, horizontal compression.

This interpretation raises the paradox of horizontal shortening in a region characterized by younger-over-older, low-angle faulting, which implies geometric lengthening of the section. The strata omitted along the low-angle fault between Manning Canyon and Oquirrh constitute only a small proportion of the entire Oquirrh Formation. Thus, this lower fault occurs at approximately the normal stratigraphic contact between the Oquirrh and Manning Canyon, based on exposures and on inferences from fold geometry. If so, the fault may be a stratigraphically controlled detachment horizon in the Manning Canyon Shale. The microstructure analysis suggests that the fault is best interpreted as a bedding-plane thrust (compression parallel to bedding). The omissions of section may be due to reactivation during later extensional deformation or perhaps to local vagaries in folding and irregularities in the fault plane during thrusting.

Correlation of Low-angle Faults

Though cut by late high-angle faults, the main Oquirrh allochthon in the North Hansel Mountains is structurally and stratigraphically continuous with the upper plate of the Samaria Mountain thrust west of the North Canyon fault. At the deepest level of exposure, the thrust must cut well into the basal limestone member of the Oquirrh Formation at about the same stratigraphic level as the detachment between Oquirrh and Manning Canyon in the North Hansel Mountains. Even though it cannot be directly shown on

the surface that the two faults are, or were, connected, we suggest that the Samaria Mountain thrust originated as a frontal ramp that was laterally continuous westward with the Manning Canyon detachment. The correlation of these two faults and their geometries with respect to stratigraphic horizons suggest that that thrust was parallel to bedding under the North Hansel Mountains and ramped upward at the west side of Samaria Mountain (Fig. 9). Folds are tighter over the ramp and more open over the flat.

Oquirrh Thickness Changes and Ramping

At least half, and perhaps all the Oquirrh Formation thickens by more than four times between Samaria Mountain and the North Hansel Mountains. This change in thickness generally correlates with the inferred position at which the thrust changes from a bedding plane fault to where it begins to ramp upward in the upper Paleozoic

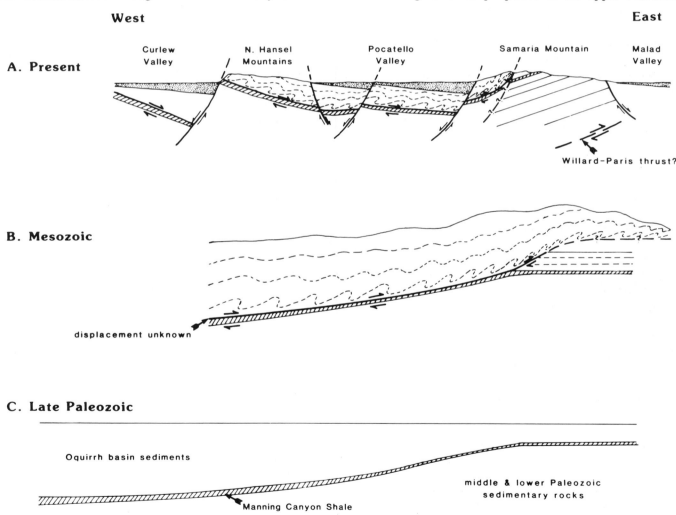

Figure 9. Schematic cross-sections showing our suggested evolution of the structures (particularly low-angle faults and folds) in the North Hansel Mountains and Samaria Mountain. Not to scale. See text for discussion.

section. Folds in the upper plate appear to "pile up" (i.e., become more closely or tightly spaced at the lower corner of this ramp—see Fig. 9). Our interpretation of these relations is that the original geometry of the Oquirrh basin at this location, combined with the mechanical properties of the Manning Canyon (i.e., a weak zone favoring detachment), controlled the geometry of the Samaria Mountain thrust and its westward extension in the North Hansel Mountains. Perhaps many of the enigmatic structures of the greater region represent the interaction of compressional and subsequent tensional deformation with the geometry of this major Paleozoic sedimentary basin.

Age of Deformation

Folding of the unnamed sandstone and overlap of the Salt Lake Formation suggest that the age of deformation is post-Early Permian and pre-Miocene and Pliocene. Considering regional relations, this range can be narrowed somewhat to post-Early Triassic because of the involvement of Lower Triassic Dinwoody and Thaynes Formations in similar structures at several localities in southern Idaho and northern Utah.

Within these wide constraints, the relation of this faulting and folding to a specific regional tectonic episode is indeterminate. The structural geometry of the Samaria Mountain thrust-Manning Canyon detachment and the associated folds, in combination with scattered microstructure analyses in the region (Allmendinger, 1980), suggest that the deformation was a result of regional, east-west horizontal compression. Deformation of this style is known to have occurred to the east, in the Idaho-Wyoming-northern Utah thrust belt during latest Jurassic, Cretaceous, and early Tertiary (Armstrong and Oriel, 1965; Oriel and Armstrong, 1966); palinspastic reconstructions of that belt (Royse, Warner, and Reese, 1975) indicate that the basal decollement for the thrust belt probably originally extended westward beneath the area described here. In addition, Mesozoic low-angle faulting, folding, and metamorphism have been documented or suggested to the west (Armstrong, 1976; Compton and Todd, 1979; Miller, 1980; Allmendinger and Jordan, 1981; Miller and Hoggatt, 1981), although not all authors ascribe the strain to layer-parallel compression. No similar compressional deformation has been documented during the middle and late Cenozoic. Eocene block faults (Armstrong and Oriel, 1965) had a significantly different geometry than those faults described here. Thus we suggest that folding and thrusting described here occurred at some time during the Mesozoic, following the deposition of the Lower Triassic Dinwoody and Thaynes Formations. Allmendinger and Jordan (1981) have proposed that the thrust in the Manning Canyon Shale might have first moved during the Middle Jurassic, on the basis of metamorphic and igneous relations much

farther west. While not contradicting their hypothesis, our data do not constrain the time of thrusting any more closely than discussed above.

Widespread, Tertiary, low-angle normal faulting has been well documented both to the west and south (Compton et al., 1977; Todd, 1980; McDonald, 1976; Davis and Coney, 1979). Though age data are lacking for the upper allochthon in the North Hansel Mountains, the cross-cutting relations and microstructure analysis indicate that it may have been emplaced during the regional Tertiary extension. Furthermore, just as the Manning Canyon horizon was apparently favored for faulting during the Mesozoic, it may have been equally favored during the Cenozoic, and probably some or all of the stratigraphic omission noted occurred at this later time.

CONCLUSIONS

The low-angle faulting and associated folding in the North Hansel Mountains and Samaria Mountain was probably a result of east-west horizontal compression that affected the entire region during the Mesozoic. We interpret the Manning Canyon detachment and Samaria Mountain thrust as the bedding plane flat and ramp, respectively, of a single thrust. The position of the ramp in this thrust may have been determined by the original geometry of the Pennsylvanian-Permain Oquirrh basin (Fig. 9). The regional significance of this fault is not well established; however, its local importance is best demonstrated by the fact that the Pennsylvanian facies and maybe the Permian facies of the Oquirrh are more than four times thicker above and west of the thrust than east of it. Tertiary extension probably has also affected the region, resulting in a reactivation of older low-angle faults and formation of new ones.

Remaining problems and unknowns raised by this study indicate two avenues of research that would be valuable. First, a more complete understanding of the nature and regional distribution of Oquirrh facies will help determine the amount of translation along the Samaria Mountain fault-Manning Canyon detachment because hanging wall cutoffs have been mostly eroded and the footwall is seldom exposed. Second, dating of these structures is critical if we are to establish their regional tectonic significance. Understanding the structural development of this region during the Mesozoic is important for determining what relations may exist between Mesozoic plutonism and metamorphism to the west and the Idaho-Wyoming-Utah thrust belt to the east.

ACKNOWLEDGMENTS

We have benefited from discussions with many geologists working in the region, including R. L. Armstrong, R. Coward, M. D. Crittenden, Jr., R. C. Douglass, M. H.

Hait, Jr., R. Hoggan, T. E. Jordan, P. Link, S. S. Oriel, and J. F. Smith. R. C. Douglass dated many of the fusilinids that document temporal correlation among different units. Thanks are due to R. L. Armstrong, W. Hobbs, T. E. Jordan, D. M. Miller, and S. S. Oriel for their helpful reviews of the manuscript. Max D. Crittenden, Jr. has continually provided encouragement toward understanding the regional problems. We are most indebted to S. S. Oriel who directed this research, and whose own mapping in the region has helped establish the tectonic framework on which our work is based. The U.S. Geological Survey supported the field work for this study. The microfabric work and preparation of the manuscript have been supported by National Science Foundation grant EAR 80-18758. Cornell contribution no. 715.

REFERENCES CITED

Allmendinger, R. W., 1980, Tectonic significance of microstructures in Idaho-Wyoming thrust belt and hinterland [abs.]: American Association of Petroleum Geologists Bulletin, v. 64, p. 669.

—— 1982, Analysis of microstructures in the Meade plate of the Idaho-Wyoming foreland thrust belt, U.S.A.: Tectonophysics, v. 85, p. 221–251.

——, (in press), Geologic map of the North Hansel Mountains, Idaho and Utah: U.S. Geological Survey MF series, scale 1:24,000.

Allmendinger, R. W., and Jordan, T. E., 1981, Mesozoic evolution, hinterland of the Sevier orogenic belt: Geology, v. 9, p. 308–313.

Armstrong, F. C., and Oriel, S. S., 1965, Tectonic development of Idaho-Wyoming thrust belt: American Association of Petroleum Geologists Bulletin, v. 49, p. 1847–1866.

Armstrong, R. L., 1968a, The Cordilleran miogeosyncline in Nevada and Utah: Utah Geological and Mineral Survey Bulletin 58, 58 p.

——, 1968b, Sevier orogenic belt in Nevada and Utah: Geological Society of America Bulletin, v. 79, p. 429–458.

——, 1972, Low-angle (denudation) faults, hinterland of the Sevier orogenic belt, eastern Nevada and western Utah: Geological Society of America Bulletin, v. 83, p. 1729–1754.

——, 1976, Albion Range: Isochron/West, no. 15, p. 12–23.

Ave 'Lallemant, H. G., and Carter, N. L., 1971, Pressure dependence of quartz deformation lamellae orientations: American Journal of Science, v. 270, p. 218–235.

Ballou, R., (unpub. man.), Geology of the North Hansel Mountains, Idaho and Utah: Brigham Young University.

Beus, S. S., 1968, Paleozoic stratigraphy of Samaria Mountain, Idaho-Utah: American Association of Petroleum Geologists Bulletin, v. 52, p. 782–808.

Compton, R. R., and Todd, V. R., 1979, Oligocene and Miocene metamorphism, folding, and low-angle faulting in northwestern Utah; Reply: Geological Society of America Bulletin, Part I, v. 90, p. 307–309.

Compton, R. R., Todd, V. R., Zartman, R. E., and Naeser, C. W., 1977, Oligocene and Miocene metamorphism, folding, and low-angle faulting in northwestern Utah: Geological Society of America Bulletin, v. 88, p. 1237–1250.

Coward, R., 1979, Geology of the Buist quadrangle, southern Idaho [M.S. thesis]: Bryn Mawr, Pennsylvania, Bryn Mawr College.

Cramer, H. R., 1971, Permian rocks from the Sublette Range, southern Idaho: American Association of Petroleum Geologists Bulletin, v. 55, p. 1787–1801.

Crittenden, M. D., Jr., 1972, Willard thrust and the Cache allochthon,

Utah: Geological Society of America Bulletin, v. 83, p. 2871–2880.

——, 1979, Oligocene and Miocene metamorphism, folding, and low-angle faulting in northwest Utah; Discussion: Geological Society of America Bulletin, Part I, v. 90, p. 305–306.

——, 1980, Metamorphic core complexes of the North American Cordillera; Summary, in Crittenden, M. D., Jr., and others, eds., Cordilleran metamorphic core complexes: Geological Society of America, Memoir 153, p. 485–490.

——, 1982, Younger-over-older low-angle faults of the Sevier Orogenic Belt in the Promontory range, northern Utah [abs.]: Geological Society of America Abstracts with Programs (1982 Rocky Mountain Section), v. 14, p. 308.

Davis, G. H., and Coney, P. J., 1979, Geological development of the Cordilleran metamorphic core complexes: Geology, v. 7, p. 120–124.

Doelling, H., 1981, Geology and mineral resources of Box Elder County, Utah: Utah Geological and Mineral Survey Bulletin 115, 251 p.

Hose, R. K., and Danes, Z. F., 1973, Development of the late Mesozoic to early Cenozoic structures in the eastern Great Basin, in de Jong, K. A., and Scholten, R. eds., Gravity and tectonics: New York, Interscience, p. 429–441.

Jordan, T. E., and Douglass, R. C., 1980, Paleogeography and structural development of the Late Pennsylvanian to Early Permian Oquirrh basin, northwestern Utah, in Fouch, T. D., and Magathan, E. R., eds., Paleozoic paleogeography of west-central United States; west-central United States paleogeography symposium 1: Denver, Rocky Mountain Section, Society of Economic Paleontologists and Mineralogists, p. 217–238.

Mallory, W. M., (compiler), 1972, Regional synthesis of the Pennsylvanian System, in Geologic atlas of the Rocky Mountain region, United States of America: Denver, Rocky Mountain Association of Geologists, p. 111–132.

McDonald, R. E., 1976, Tertiary tectonics and sedimentary rocks along the transition, Basin and Range province to plateau and thrust belt province, Utah: Rocky Mountain Association of Geologists symposium, p. 281–317.

Miller, D. M., 1980, Structural geology of the northern Albion Mountains, south-central Idaho, in Crittenden, M. D., Jr., and others, eds., Cordilleran metamorphic core complexes: Geological Society of America, Memoir 153, p. 399–423.

Miller, D. M., and Hoggatt, W. C., 1981, Mesozoic metamorphism, low-angle faulting, and folding in the Pilot Range, Nevada and Utah [abs.]: Geological Society of America Abstracts with Programs, v. 13, p. 97.

Oriel, S. S., and Armstrong, F. C., 1966, Times of thrusting in Idaho-Wyoming thrust belt; Reply: American Association of Petroleum Geologists Bulletin, v. 50, p. 2614–2621.

Oriel, S. S., and Platt, L. B., 1979, Younger-over-older thrust plates in southeastern Idaho [abs.]: Geological Society of America Abstracts with Programs, v. 11, p. 298.

——, 1980, Geologic map of the Preston 1° x 2° Quadrangle, southeastern Idaho and western Wyoming: U.S. Geological Survey Miscellaneous Investigations Map I-1127, scale 1:250,000.

Platt, L. B., 1975, Recent faulting at Samaria Mountain, southeastern Idaho [abs.]: Geological Society of America Abstracts with Programs, v. 7, p. 1229–1230.

——, 1977, Geologic map of the Ireland Springs-Samaria area, southeastern Idaho and northern Utah: U.S. Geological Survey Miscellaneous Field Studies Map MF-890, scale 1:48,000.

Roberts, R. J., and Crittenden, M. D., Jr., 1973, Orogenic mechanism, Sevier Orogenic Belt, Nevada and Utah, in de Jong, K. A., and Scholten, R., eds., Gravity and tectonics: New York, Interscience, p. 409–428.

Royse, F., Warner, M. A., and Reese, D. L., 1975, Thrust belt structural geometry and related stratigraphic problems, Wyoming-Idaho-northern Utah: Rocky Mountain Association of Geologists Symposium (1975), p. 41–54.

Skipp, B., and Hall, W. E., 1980, Upper Paleozoic paleotectonics and paleogeography of Idaho, *in* Fouch, T. D., and Magathan, E. R., eds., Paleozoic paleogeography of west-central United States; west-central United States paleogeography symposium 1: Denver, Rocky Mountain Section Society of Economic Paleontologists and Mineralogists, p. 387–422.

Smith, R. B., 1978, Seismicity, crustal structure, and intraplate tectonics of the interior of the western Cordillera, *in* Smith, R. B., and Eaton, G. P., eds., Cenozoic tectonics and regional geophysics of the western Cordillera: Geological Society of America, Memoir 152, p. 111–144.

Stewart, J. H., 1972, Initial deposits in the Cordilleran geosyncline: evidence of a Late Precambrian (850 m.y.) continental separation: Geological Society of America Bulletin, v. 83, p. 1345–1360.

Todd, V. R., 1980, Structure and petrology of a Tertiary gneiss complex in northwestern Utah, *in* Crittenden, M. D., Jr., and others, eds., Cordilleran metamorphic core complexes: Geological Society of America, Memoir 153, p. 349–383.

Turner, F. J., and Weiss, L. J., 1963, Structural analysis of metamorphic tectonites: New York, McGraw-Hill, 545 p.

Yancey, T. E., Ishibashi, G. D., and Bingman, P. T., 1980, Carboniferous and Permian stratigraphy of the Sublette Range, south-central Idaho, *in* Fouch, T. D., and Magathan, E. R., eds., Paleozoic paleogeography of west-central United States; west-central United States paleogeography symposium 1: Denver, Rocky Mountain Section Society of Economic Paleontologists and Mineralogists, p. 259–270.

MANUSCRIPT ACCEPTED BY THE SOCIETY AUGUST 20, 1982

Geological Society of America
Memoir 157
1983

Glacial and tectonically influenced sedimentation in the Upper Proterozoic Pocatello Formation, southeastern Idaho

Paul Karl Link
Department of Geology
Idaho State University
Pocatello, Idaho 83209

ABSTRACT

The Upper Proterozoic Pocatello Formation, southeastern Idaho, contains heterogeneous clastic and carbonate strata (Scout Mountain Member), volcanic and volcaniclastic rocks (Bannock Volcanic Member), and laminated argillite (upper member). These rocks are interpreted to have been deposited during a period of regional glaciation and local volcanism during initial formation of the Cordilleran miogeocline. They are presently exposed in two bands of outcrop, as allochthons, in the northern Bannock Range south and east of Pocatello and the southern Bannock Range north and west of Preston.

Strata of the Scout Mountain Member are divided into lithofacies interpreted to have been deposited in a variety of shallow to deep marine sedimentary environments. Massive diamictite containing rare striated clasts is interpreted as subaqueous lodgement tillite or flow tillite. Interbedded graded sandstones and diamictites are interpreted as turbidites and deep water mass flow deposits. Thick-bedded cobble conglomerate is interpreted as a shallow to deep marine channel-fill deposit. Clast assemblages in these coarse strata indicate that both extrabasinal quartzitic and granitic terranes and intrabasinal volcanic and siltstone terranes contributed sediment. Sandy and silty facies, locally with thin carbonate interbeds, are interpreted as intertidal to subtidal shallow marine deposits.

Syn-sedimentary tectonism is suggested by local abrupt facies changes, the presence of interbedded volcanics, and the abundance of clasts in diamictite interpreted to have been derived from exposed and uplifted portions of the Pocatello Formation. Strata exposed in the southern Bannock Range are interpreted to have been deposited in a northward-deepening sub-basin of the early Cordilleran miogeocline. A separate, northward-deepening sub-basin, containing glaciated volcanic islands at the north end, is inferred for deposition of strata now exposed in the northern Bannock Range. An alternate explanation for the disparity in facies between the northern and southern Bannock Range involves differential tectonic transport along previously undetected tear faults between the two areas during the Mesozoic Sevier Orogeny.

INTRODUCTION

Sedimentary sequences that lie conformably below fossiliferous Cambrian strata are strikingly similar along much of the length of the North American Cordillera (Crittenden and others, 1971; Stewart, 1972; Stewart and Suczek, 1977; Christie-Blick and others, 1980). Poorly sorted clastic and intercalated mafic volcanic rocks occur at the base of many of these sequences. Limited radiometric dating and lithologic correlation summarized by Stewart (1972) suggest that these basal rocks were deposited during the Late Proterozoic (<900 m.y.) as defined by Harrison and Peterman (1980). A distinctive lithology in these basal rocks is diamicitite (Flint and others, 1960) or mixtite (Schermerhorn, 1966), a very poorly sorted rock in which sand and larger particles float in a muddy matrix. The Upper Proterozoic diamictites in Utah and southeastern Idaho have long been interpreted as tillite, or strata deposited by or in proximity to glacial ice (Blackwelder, 1932; Ludlum, 1942).

The present study documents the stratigraphy and depositional environment of the diamictite-bearing Pocatello Formation (Crittenden and others, 1971; Trimble, 1976; Link, 1981, 1982) in the Bannock Range of southeastern Idaho (Fig. 1). The Pocatello Formation is the oldest rock unit exposed in the Idaho-Wyoming thrust belt (Armstrong and Oriel, 1965; Royse and others, 1975) and is allochthonous, having been transported eastward 50 to 100 km during the late Mesozoic Sevier Orogeny (Armstrong, 1968b; Allmendinger and Jordan, 1981).

The Pocatello Formation contains three members: the Scout Mountain Member, the Bannock Volcanic Member, and the upper member (Fig. 1). Crittenden and others (1971) and Trimble (1976) originally described a "lower member," which is exposed in only one locality, north of Portneuf Narrows (Section 7, Fig. 2). This area has been shown to be structurally overturned, and the originally defined "lower member" is now recognized to be the upper member (Link and others, 1980).

The base of the Pocatello Formation is not exposed. Its upper contact is gradational with the Blackrock Canyon Limestone which is overlain by the Papoose Creek Formation and several thousand meters of unfossiliferous siltstone and quartzite of the Brigham Group. The first Cambrian fossils occur in carbonate rocks and shales above the Brigham Group quartzite.

The Scout Mountain Member contains quartzite, sandstone, siltstone, conglomerate, diamictite, limestone, and dolomite and is up to 900 m thick. The Bannock Volcanic Member consists of up to 450 m of mafic and intermediate volcanic and volcaniclastic rocks that form a lenticular body within the Scout Mountain Member. The Bannock Volcanic rocks include lava (locally pillowed), volcanic breccia, intrusive rocks, and large exposures of foliated greenstone of probable volcaniclastic origin. The upper member of the Pocatello Formation contains more than 600 m of laminated argillite, siltstone, and quartzite (Crittenden and others, 1971; Trimble, 1976).

STRATIGRAPHY OF THE POCATELLO FORMATION

Exposures of the Pocatello Formation occur in the northern Bannock Range near Pocatello (Sections 4 to 9, Fig. 3) and in the southern Bannock Range, west of Preston (Sections 1 to 3, Fig. 3). The locations of measured sections

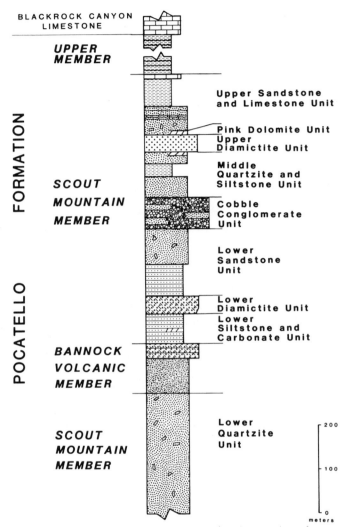

Figure 1. Stratigraphic units, Upper Proterozoic Pocatello Formation, Northern Bannock Range.

are listed in Table 1. In the northern Bannock Range, the stratigraphy proposed in this paper is, except for the omission of the "lower member," similar to that of Crittenden and others (1971) and Trimble (1976). The Pocatello Formation as exposed in the northern Bannock Range is divided into informal stratigraphic units in Figure 1 and 3. Exposures in the southern Bannock Range contain similar rocks but cannot be correlated directly to those in the northern part of the range.

Northern Bannock Range

Portneuf Narrows to Garden Creek Gap. The most

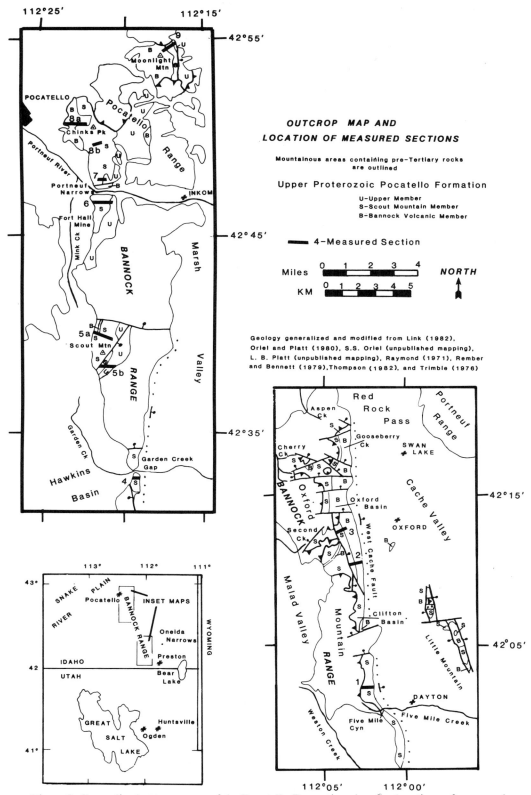

Figure 2. Generalized outcrop map of the Pocatello Formation showing locations of measured sections. Geology generalized and modified from Link (1982), Oriel and Platt (1980), S. S. Oriel (unpublished mapping), L. B. Platt (unpublished mapping), Raymond (1971), Rember and Bennett (1979), Thompson (1982), and Trimble (1976).

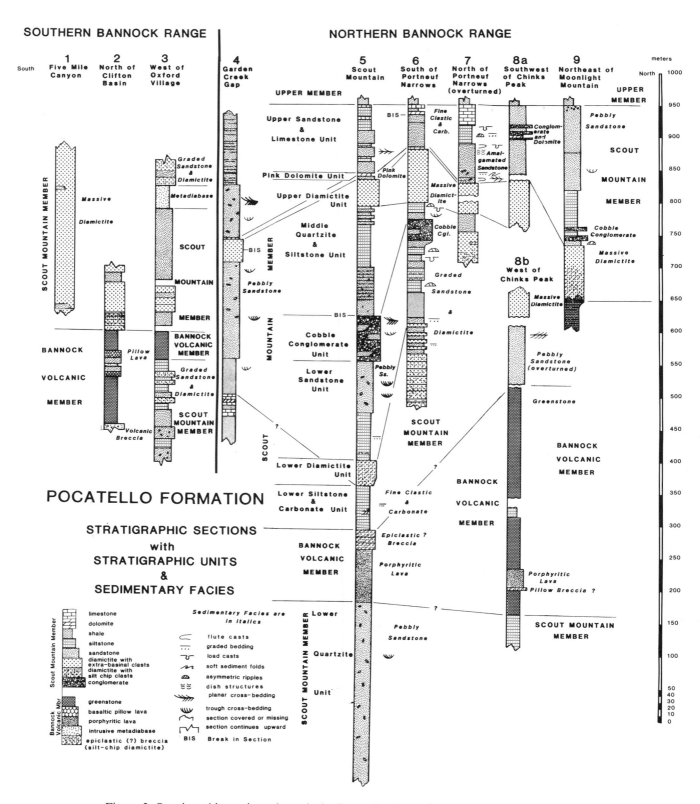

Figure 3. Stratigraphic sections through the Pocatello Formation. Localities are plotted on Figure 2.

TABLE 1: LOCATIONS OF MEASURED SECTIONS SHOWN IN FIGURES 2 AND 3

Section 1: Five Mile Canyon, measured in the south wall, unnamed canyon north of Five Mile Canyon, north ½ Section 17, T. 15 S., R. 38 E., Clifton and Weston Canyon 7½' quadrangles. (Section 1 of Plate 3)

Section 2: North of Clifton Basin, measured on north side of creek draining east from Clifton Basin, NW¼ Section 29 and NE¼ Section 30, T. 14 S., R. 38 E., 5800 to 6800 ft. elevation, Clifton 7½' quadrangle. (Section 3 of Plate 3)

Section 3: South of Oxford Basin, measured on the west side of large bowl-shaped canyon, southwest of Oxford Village, south ½, Section 6, T. 14 S., R. 38 E., from 6100 to 8380 ft., Clifton 7½' quadrangle. (Section 5 of Plate 3)

Section 4: Composite section measured at Garden Creek Gap by Thompson (1982). 0-300 m is Thompson's Section C-C', 300-500 m is Thompson's Section B-B'. Both sections are measured on the west flank of ridge south of Garden Creek Gap, NE¼, Section 12, T. 10 S., R. 35 E., Hawkins 7½' quadrangle. Section B-B' is approximately 350 m north of Section C-C'. These are reproduced as Sections 9 and 10 of Plate 4.

Section 5: Composite section on Scout Mountain. 0-700 m is Section 14 of Plate 4, measured on the west side of the north end of the Scout Mountain Ridge, along intermittent stream that drains west towards Scout Mountain Campground, north ¼, Section 34, T. 8 S., R. 35 E., from 6800 to 8300 ft. elevation, Scout Mountain 7½' quadrangle. 700 to 100 m (above the cobble conglomerate) is from 152 m to 479 m of Section 12 of Plate 4, measured west of the north fork of Goodenough Canyon, west slope, south end of Scout Mountain Ridge, NE¼ Section 10, T. 9 S., R. 35 E., elevation 7400 to 8000 ft., Scout Mountain 7½' quadrangle. This latter section continues upward into Section G-H, of Crittenden and others (1971, p. 584-585) through the upper member, Pocatello Formation.

Section 6: South of Portneuf Narrows and above Fort Hall Mine, measured east of Bannock County Dump on the west slope of the north-trending ridge of the Bannock Range, north ½, southwest ¼, Section 27, T. 7 S., R. 35 E., 5000 to 6000 ft. elevation, Inkom 7½'° quadrangle (Section 16 of Plate 4). Details are included from above the highest sandstone bed of Section 15 of Plate 4, which is measured above the Fort Hall Mine, north ½ Section 34, T. 7 S., R. 35 E., Inkom 7½' quadrangle. (This is Section C'-C" of Crittenden and others, 1971, p. 583-584.)

Section 7: North of Portneuf Narrows, measured in structurally overturned beds on the south end of the Chinks Peak ridge, SE¼ Section 16 and SW¼ Section 15, T. 7 S., R. 35 E., Inkom 7½' quadrangle. The section starts on the west slope of the ridge and at the crest jogs about 300 m to the south and continues down the east side of the ridge to the large mine dump at 5240 ft. elevation. This section is C-C' and A-B of Crittenden and others (1971, p. 583-584), but the rocks were not recognized to be overturned. This section is reproduced from Section 18 of Plate 5.

Section 8A: Chinks Peak, southwest ridge, measured in canyon with small grove of trees in middle that drains southwest from saddle south of peak 6361, from 5600 to 6200 ft. elevation, in NE¼, Section 9, and NW¼ Section 10, T. 7 S., R. 35 E., Inkom 7½' quadrangle. (Section 19 of Plate 5)

Section 8B: Chinks Peak, west side, measured in canyon south of road up Chinks Peak and along the south side of that road in the south ½ Section 32 and SW¼ Section 33, T. 6 S., R. 35 E., from elevation 5060 to 6260 ft., Pocatello South and Inkom 7½' quadrangles. This is Section 20 of Plate 5 and is approximately the same as Section E-F of Crittenden and others (1971, p. 584).

Section 9: Northeast slope of Moonlight Mountain, measured from jeep road to east over peak 6521 and down hill to old mine building on east slope of Moonlight Mountain, elevation 6521 to 6080 ft., NW¼ Section 17, T. 6 S., R. 36, Moonlight Mountain 7½' quadrangle (Section 21 of Plate 5)

complete and structurally uncomplicated exposures of the Scout Mountain Member occur on both sides of the north-trending ridge of the Bannock Range, from Portneuf Narrows south to Garden Creek Gap (Sections 4 to 6, Fig. 3). The thickest measured section occurs on Scout Mountain (Section 5).

The lowest exposed beds (lower quartzite unit, Fig. 3) consist of white to green, vitreous, locally pebbly and trough cross-bedded quartzite and lie at the base of the slope on the west side of Scout Mountain (Section 5). If unfaulted, this lower quartzite may be as much as 280 m thick.

A poorly exposed and lenticular volcanic body 110 m thick overlies the lower quartzite west of Scout Mountain

and is assigned to the Bannock Volcanic Member. It contains green siliceous metatuff, porphyritic felsic lava, and greenschist containing chips (clasts?) up to 10 cm in diameter that are recrystallized to fine-grained chlorite.

These volcanic rocks are overlain, in a very poorly exposed interval, by perhaps 70 m of brown thin-bedded sandstone, siltstone, and silty dolomite (lower siltstone and carbonate unit). A similar sequence containing both gray limestone and brown dolomite occurs at the base of the section at Garden Creek Gap (Section 4, Fig. 3) (Thompson, 1982).

The lower diamictite unit overlies the lower siltstone and carbonate unit on the west side of Scout Mountain (Section 5) and is the stratigraphically lowest unit exposed

Table 2: Clast percentages in various units and localities of the Scout Mountain Member

Count Number	General Locality	Stratigraphic Unit	Extrabasinal (%)		Intrabasinal (%)		Total Number of Clasts Counted
			Quartzite	Basement	Siltstone	Volcanic	
	Southern Bannock Range						
A	Five Mile Canyon	Massive Diamictite	31.1	11.6	44.7	12.5	720
B	Clifton Basin	Bedded Diamictite	32.8	9.2	48.6	9.2	649
C	Oxford and Davis Basins	Bedded Diamictite	17.3	8.4	58.2	15.9	438
	Northern Bannock Range						
D	Garden Creek Gap	Upper (Massive) Diamictite	48.7	39.5	10.4	1.2	162
E	Scout Mountain	Lower (Bedded) Diamictite	43.0	0	52.0	4.9	325
F	Scout Mountain	Cobble Conglomerate	85.4	11.2	1.6	1.6	62
G	Scout Mountain	Upper (Massive) Diamictite	35.7	42.5	21.2	.4	221
H	South of Portneuf Narrows	Lower (Bedded) Diamictite	10.3	2.0	78.7	3.0	165
I	South of Portneuf Narrows	Cobble Conglomerate	91.1	6.3	1.2	1.2	79
J	South of Portneuf Narrows	Upper (Massive) Diamictite	34.1	4.8	43.9	21.9	41
K	North of Portneuf Narrows	Massive Diamictite	13.4	48.3	11.2	26.9	89
L	Southwest of Chinks Peak	Massive Diamictite	30.6	3.6	47.2	18.4	163
M	Southwest of Chinks Peak	Cobble and Boulder Conglomerate	61.7	3.1	35.0*	0	157
N	Northeast of Moonlight Mountain	Massive Diamictite	31.1	5.3	58.0	5.3	186
O	Northeast of Moonlight Mountain	Cobble Conglomerate	90.0	6.0	2.0	2.0	50

* Includes mostly locally derived dolomite clasts.

in Section 6. The partly covered contact below the diamictite is probably gradational. The diamictite is crudely bedded and as much as 100 m thick.

As a means of comparing the provenance of the coarse strata of the Scout Mountain Member, clast counts were made at several localities and are detailed in Table 2. Ninety-five percent of the clasts in the lower diamictite unit on Scout Mountain (Count 5) consist of quartzite and siltstone. A much different clast assemblage occurs in the lower diamictite unit south of Portneuf Narrows (Count 8), as 80 percent of the clasts are volcanic rocks.

The quartzite pebble category contains diverse, moderately well-rounded, pink, gray, white, red, purple, and brown orthoquartzites. These rocks are inferred to have been eroded from a quartzitic terrane outside of the depositional basin and are labeled extrabasinal in Table 2. Basement rocks, consisting of granite, gneiss, vein quartz, and schist compose the other extrabasinal category in Table 2.

The siltstone category in Table 2 contains flattened and recrystallized gray, green, and white pelitic rocks. These siltstones are inferred to have been derived within the depositional basin, from exposed strata of the Scout Mountain Member, and are classed as intrabasinal on Table 2. Other intrabasinal clasts present are green porphyritic mafic volcanic rocks inferred to have been derived from areas of exposure of the Bannock Volcanic Member.

In Section 6, south of Portneuf Narrows, the lower diamictite unit passes upward into the lower sandstone unit through about 50 m of interbedded diamictite and graded sandstone. The sandstone is brown, medium- to fine-grained lithic wacke and contains graded beds several cm thick overlain by siltstone (A-E graded bedding of the Bouma Sequence; Bouma, 1962).

The base of the lower sandstone unit is placed at the top of the stratigraphically highest diamictite bed. The lower sandstone unit consists of medium to coarse, magnetite-bearing trough cross-bedded slightly pebbly sandstone at Scout Mountain (Section 5) and thin-bedded to laminated, rippled fine sandstone south of Portneuf Narrows (Section 6).

The upper part of the lower sandstone in Section 6 contains a coarsening- and thickening-upward sequence of interbedded siltstone and medium to coarse sandstone. The basal 2 cm of some sandstone beds are inversely graded, while the main parts of the beds are normally graded. The upper portions of some beds contain purple siltstone rip-up clasts up to 6 cm in diameter. Other structures in these sandstones include trough cross-beds up to 30 cm wide, scour and fill, load structures, and lenticular bedding. On Scout Mountain, the lower sandstone unit coarsens and becomes more pebbly upward. Pebble lags occur along the bases of trough cross-beds.

The cobble conglomerate unit overlies the lower sandstone unit with an abrupt, and locally erosional, contact. South of Portneuf Narrows it is massively bedded, whereas it is interbedded with sandstone on Scout Mountain. The cobble conglomerate unit is correlated with trough cross-bedded slightly pebbly sandstone at Garden Creek Gap. Clasts in the cobble conglomerate unit are up to 60 cm in diameter and are predominantly quartzite (Table 2, Counts 6 and 9). Basement clasts are subordinate, and intrabasinal siltstone and volcanic clasts are very sparse.

The middle quartzite and siltstone unit overlies the cobble conglomerate and contains medium-grained quartzite and brown siltstone with local thin limestone interbeds. The contact is gradational over about 10 m. South of Portneuf Narrows, the middle quartzite and siltstone unit contains siltstone that is cross-laminated and interbedded with fine sandstone in beds up to 50 cm thick. The sandstone contains flame structures, load casts, and trough crossbeds. Above Fort Hall Mine, this sandy unit is replaced by conglomerate overlain by 1 m of gray silty limestone below the upper diamictite unit.

The upper diamictite unit overlies the middle quartzite and siltstone with a sharp contact and is about 50 m thick in Sections 4, 5, and 6. The upper diamictite is the stratigraphically lowest unit exposed in overturned beds north of Portneuf Narrows (Section 7). It is generally thick-bedded to massive and contains clasts up to 150 cm long in a black, brown, purple, or gray muddy matrix. A striated quartzite clast found in this unit south of Portneuf Narrows is shown in Figure 4. The striations are interpreted to be glacial because they occur in subparallel sets, which cut across bedding in the clast. They occur on a flattened and possibly faceted face of the cobble. The clast-to-matrix ratio in the upper diamictite unit is highly variable and, locally, clasts are imbricated. The clast composition is also variable (Table 2, Counts 4, 7, 10, and 11), changing from predominantly intrabasinal clasts at Garden Creek Gap and Scout Mountain (Counts 4 and 7) to a more equal ratio between intrabasinal and extrabasinal clasts near Portneuf Narrows (Counts 10 and 11). There is a significant basement fraction at Garden Creek Gap, which is not present south of Portneuf Narrows, but which occurs again north of the Narrows.

A distinctive porphyritic volcanic clast found in the upper diamictite north of Portneuf Narrows is shown in Figure 5. It is identical to plagioclase-porphyritic basalt that occurs extensively in the Bannock Volcanic Member west of Chinks Peak.

The lenticular pink dolomite unit overlies the upper diamictite. The dolomite is wavy laminated and silty, and about 1 m thick. It rests on the upper diamictite with abrupt contact in Sections 4 and 6. In Section 7, north of Portneuf Narrows, pink dolomite occurs as clasts in a dolomite chip breccia with a coarse sandy matrix directly above the upper diamictite. Identical dolomite appears in talus on Scout Mountain (Section 5) and probably occupies

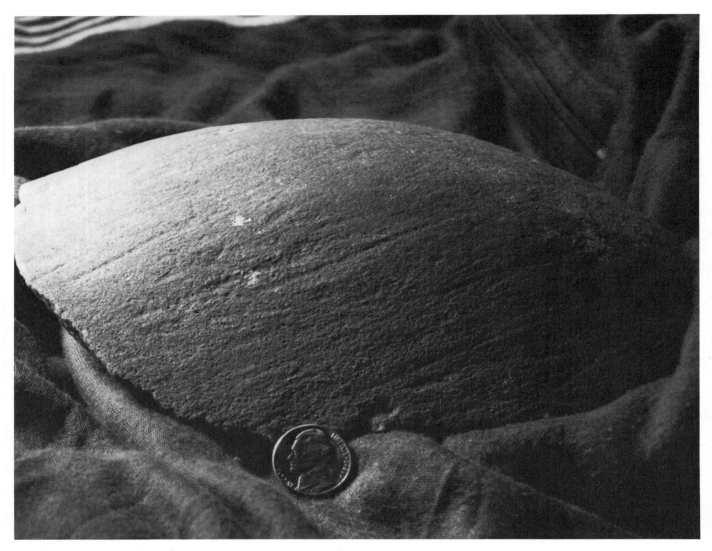

Figure 4. Striated purple quartzite cobble from upper diamictite south of Portneuf Narrows. The striations occur in subparallel sets and are interpreted to be glacial.

the same stratigraphic position. Outcrops of the dolomite are laterally continuous for tens of meters south of Garden Creek Gap but are discontinuous elsewhere.

The upper sandstone and limestone unit fines upward from coarse sandstone to argillite and limestone through 40 m of section. Generally, the basal sandstone of this unit is medium- to coarse-grained subarkosic wacke and quartz wacke. The sandstone is locally pebbly and contains flute casts, dish structures, and large planar and trough cross-beds up to several meters in wavelength. North of Portneuf Narrows, it contains slump structures up to 1 m in diameter. The sandstone fines upward and grades into siltstone that contains flaser bedding, asymmetrical ripples, and load structures. A persistent bed of gray or white, locally recrystallized, silty limestone or marble forms the top of the

upper sandstone and limestone unit, and is the uppermost bed in the Scout Mountain Member.

Thinly laminated, gray to black, silver-weathering, locally pyritic shale and argillite of the upper member, Pocatello Formation, overlie the Scout Mountain Member with a gradational contact. The upper member is about 600 m thick southeast of Scout Mountain (Trimble, 1976), and contains interbedded brown quartzite and argillite. The Blackrock Canyon Limestone gradationally overlies the upper member.

Chinks Peak and Moonlight Mountain. Structurally complicated and dismembered exposures of the Pocatello Formation occur on Chinks Peak and Moonlight Mountain in the Pocatello Range (northernmost Bannock Range, Sections 8 and 9, Figs. 1 and 3). Because of thrust-

Figure 5. Clast of porphyritic basalt derived from the Bannock Volcanic Member and found in diamictite of the Scout Mountain Member north of Portneuf Narrows.

ing and recumbent folding, the present location of these rocks may not preserve their original depositional relation to those south of Portneuf Narrows or to each other.

The isolated exposures northeast of Moonlight Mountain (Section 9) contain foliated greenstone of the Bannock Volcanic Member at the base. This greenstone grades upward into silt-chip-bearing, green, foliated (volcanigenic?) diamictite, and then to sandy diamictite that contains extrabasinal clasts (Count 14, Table 2).

The diamictite interval is overlain by coarse-grained, cross-bedded sandstone with pods and thin beds of cobble conglomerate containing dominantly quartzite clasts (Count 15). Above the sandstone is dark argillite, more coarse sandstone, and granule conglomerate, all below argillite of the upper member, Pocatello Formation. The

fining-upward sequence of the upper sandstone and limestone unit present near Portneuf Narrows does not occur on Moonlight Mountain.

The exposures on the crest and west side of Chinks Peak (Sections 8a and 8b) contain both the Bannock Volcanic Member and the Scout Mountain Member. The Bannock Volcanic Member contains plagioclase-porphyritic basalt (Fig. 5) and abundant variable greenstone. Some greenstone consists of monolithologic breccia containing green chlorite clasts that may have originated as basaltic pillows or parts of pillows (G. D. Harper, personal communication, 1982). Such rock is known as pillow breccia (deWit and Stern, 1978). Schistose greenstone containing stretched dark siltstone and volcanic clasts as well as isolated extrabasinal quartzite clasts is a common, and puzzling, lithology. Greenschist facies metamorphism is pervasive in this area; chlorite, albite, epidote, and biotite are common in thin section.

Rocks on the south end of the Chinks Peak ridge, north of Portneuf Narrows, are structurally overturned (Link and others, 1980), and the extent of overturned strata northward towards Chinks Peak is uncertain. Dark siltstone of the Scout Mountain Member occurs topographically beneath the Bannock Volcanic Member at the base of the west slope of Chinks Peak (bottom of Section 8b). A structurally overturned gray and brown quartzite containing large planar cross-beds occurs on the summit and west side of Chinks Peak (near top of Section 8b).

A variable sequence occurs stratigraphically above what is probably the upper diamictite unit on the southwest slopes of the Chinks Peak ridge (Section 8a). The diamictite here is thickly bedded and contains both extra- and intrabasinal clasts (Count 12, Table 2). Maximum clast size is 1.5 m, and striated quartzites occur. This diamictite is interbedded with coarse lithic sandstone and pink laminated dolomite. Brown lithic sandstone rests upon the diamictite and is overlain in turn by interbedded laminated dolomite and quartzitic cobble and boulder conglomerate. Locally, the conglomerate lies with erosive contact on the dolomite and contains a significant proportion of angular dolomite clasts up to 1.3 m in diameter (Count 13, Table 2). More brown sandstone occurs above the conglomerate and below argillite of the upper member, Pocatello Formation. The carbonate horizon present immediately below the upper member in Sections 4 to 7 does not occur in the Chinks Peak area.

Southern Bannock Range, Oxford Mountain

Rocks of the Pocatello Formation, metamorphosed to greenschist facies, are the oldest and structurally lowest strata exposed on Oxford Mountain west of Preston, Idaho. Extensive exposures of the Bannock Volcanic Member and Scout Mountain Member occur in the upper Cherry Creek drainage on the north end of Oxford Mountain, along its east face, and on the southwest side of the mountain, in the upper part of Second Creek (see Fig. 2). There are isolated exposures in Cache Valley, notably on Little Mountain. Three partial stratigraphic sections are shown in Figure 3 (Sections 1 to 3).

Extrusive rocks of the Bannock Volcanic Member are lenticular within the Scout Mountain Member on Oxford Mountain where they consist of at least 200 m of greenstone, agglomerate, pillow breccia, and pillow lava in Section 2. A body of metadiabase or metagabbro assigned to the Bannock Volcanic Member intrudes the overlying Scout Mountain Member west of Oxford Village (Section 3).

Brown sandstone, thick-bedded diamictite, and lenticular conglomerate of the Scout Mountain Member are interbedded with and overlie pillow lava and greenstone of the Bannock Volcanic Member in Section 2. The sandstone is lithic wacke and contains some graded beds. Clasts in the diamictite are heterogeneous, consisting chiefly of siltstone chips with minor volcanic, granitic, and quartzite clasts (Table 2, No. 2, 3). Clast-to-matrix ratio varies, and some parts of the diamictite are crowded with stones. Lenses of conglomerate a few meters thick and tens of meters long are interbedded with both sandstone and diamictite.

The southernmost exposures of the Pocatello Formation are at Five Mile Canyon (Section 1) and are assigned to the Scout Mountain Member. They consist of massive cobble and boulder-bearing diamictite with indistinct thick bedding, rare clast imbrication, and minor interbeds of brown, medium-grained sandstone. The diamictite matrix is a brown, hematitic, lithic wacke. Siltstone and quartzite clasts predominate here in the diamictite (Table 1, Count 1).

INTERPRETATION OF THE DEPOSITIONAL ENVIRONMENT OF THE SCOUT MOUNTAIN MEMBER

Tectonic and Climatological Framework

A tectonic environment of continental rifting has been proposed for the diamictite- and volcanic-bearing strata which occur near the base of Upper Proterozoic sequences of the North American Cordillera (Stewart and Suczek, 1977). The glaciogenic nature of the diamictites in northern Utah has been documented most recently by Varney (1976), Blick (1979), Ojakangas and Match (1980), and Crittenden, Christie-Blick, and Link (in press). A glacial origin for diamictites in southeastern Idaho has been proposed by Ludlum (1942, 1943), Crittenden and others (1971), Trimble (1976), and Link (1981).

The importance of the regional depositional setting of diamictite-bearing strata to their environmental interpretation has been emphasized by Crowell (1978), Edwards (1978), Frakes (1979), Link and Gostin (1981) and Boulton and Deynoux (1981). Because diamictites can be deposited both by glaciers and nonglacial mass flows (Schermerhorn, 1974), the sedimentary facies framework can be a critical factor in determining if a glacial depositional model is feasible.

In the Scout Mountain Member, massive diamictites contain diverse, well-rounded, and locally glacially striated quartzite and granitic clasts up to 1.5 m in diameter (Fig. 4). In addition, the diamictites have lateral continuity over tens of kilometers in both the northern and southern Bannock Range and are correlative with proven glaciogenic strata in northern Utah. A glacial depositional framework is therefore strongly suggested.

Tectonically influenced sedimentation is also suggested in the Scout Mountain Member. A large percentage of clasts in bedded diamictites are volcanic rocks and siltstones, which are inferred to be intrabasinal, eroded from areas in which previously deposited parts of the Pocatello Formation were exposed.

Sedimentary Facies

The strata of the Scout Mountain Member are divided into eight lithofacies which are italicized in the text and summarized in Table 3. The rationale for these lithofacies is discussed below.

Diamictites within the Scout Mountain Member are divided into two types: massive and bedded. The *massive diamictite* facies composes the diamictite at Five Mile Canyon (Section 1, Fig. 3) and the upper diamictite unit in the northern Bannock Range (Sections 4 to 9).

The *massive diamictite* facies includes the only strata of the Scout Mountain Member which are interpreted to have been deposited from glacial ice. The facies contains the most diverse extrabasinal clast assemblage, as well as sparse striated stones, and forms beds up to 50 m thick that are laterally continuous for several kilometers. The *massive diamictite* facies is interpreted as either till that was deposited below a partially buoyant ice sheet (subaqueous lodgement tillite) or till that has flowed downslope after initial deposition (flow tillite). Terminology is that of Boulton and Deynoux (1981, p. 399).

Several alternate interpretations of the *massive diamictite* are possible. Similar poorly sorted rocks have been interpreted as originating as glaciomarine diamictite (Boulton and Deynoux, 1981, p. 407) consisting of debris dropped through water underneath a floating ice sheet, or as terrestrial lodgement tillite deposited beneath a subaerial ice sheet. A totally nonglacial origin as amalgamated subaqueous debris flows is also possible. The first interpretation is rejected because of the lack of stratified interbeds which would be expected if any bottom currents were present to sort the debris falling out of the ice. The second hypothesis seems unlikely because of the great thickness of diamictite at Five Mile Canyon (250 m) and the association in the northern Bannock Range with carbonate and sandstone interpreted to be marine. The totally nonglacial interpretation is doubtful because of the occurrence of striated stones and the regional Late Proterozoic glacial framework. An origin for some of the *massive diamictite* as flow till, as proposed in this paper, acknowledges the possibility of postdepositional flow of till initially deposited by glacial ice.

Bedded diamictites are assigned to the *graded sandstone and diamictite* facies, with an interpreted origin as sediment gravity flows on a submarine slope. Rocks assigned to this facies include the exposures of Scout Mountain Member near Clifton and Oxford Basins in the southern Bannock Range (Sections 2 and 3, Fig. 3) and the lower diamictite and parts of the lower sandstone unit in the northern Bannock Range (Sections 4-9). Intrabasinal siltstone and volcanic clasts generally predominante in diamictites of this facies; input of extrabasinal clasts that may have been glacially transported is subordinate (Counts 2, 3, 5, and 8, Table 2).

Sandstones within the *graded sandstone and diamictite* facies are fine- and medium-grained lithic wackes and contain both inversely and normally graded beds, ripple marks, and load structures. They are interpreted as deposited in deep water by turbidity and contour currents. Contour currents are relatively slow-moving fluid flows that produce thin-bedded, locally graded, rippled or cross-laminated, fine clastic deposits on the present-day continental rise (Rupke, 1978, p. 389).

The *cobble conglomerate* facies composes the cobble conglomerate unit in Sections 5 and 6 and thin conglomerates in the upper sandstone and limestone unit in Sections 8 and 9 in the northern Bannock Range. This facies consists of thick-bedded to massive conglomerate containing moderately well-rounded, predominantly quartzite clasts up to 80 cm in diameter. South of Portneuf Narrows (Section 6), the conglomerate occurs in disorganized, thick beds and contains minor, coarse, sandy interbeds. On Scout Mountain (Section 5), a significant amount of interbedded sandstone is present. The cobble conglomerate unit is interpreted as a deep water channel-fill deposit south of Portneuf Narrows, but as a shallow water or fluvial deposit to the south on Scout Mountain and to the north on Moonlight Mountain and Chinks Peak. In a study of stone shapes and sedimentary structures in this conglomerate, Lilley (1973) was unable to find definitive evidence for either a fluvial or a shallow marine depositional environment.

The *pebbly sandstone* facies contains medium to coarse, lithic and arkosic sandstone with pebble lags and isolated pebbles. It composes a number of the sandy and quartzitic stratigraphic units in the northern Bannock Range including the lower quartzite unit, lower sandstone unit, the sandstone interbeds in the cobble conglomerate unit, the sandy part of the middle quartzite and siltstone unit, and the bottom of the upper sandstone and limestone unit at Garden Creek Gap. Both planar and trough cross-beds occur in this facies. Heavy minerals (magnetite and ilmenite) locally mark the cross laminae. Planar laminations also occur. The *pebbly sandstone* facies is interpreted as as shallow marine deposit. At Garden Creek Gap, the middle quartzite and siltstone unit contains bidirectional trough cross-beds (Thompson, 1982), suggesting a tidally influenced depositional environment.

The *amalgamated sandstone* facies composes the base of the upper sandstone and limestone unit north of Portneuf Narrows (Section 7, Fig. 3). It contains abundant sed-

TABLE 3: SEDIMENTARY FACIES, SCOUT MOUNTAIN MEMBER

Facies	Lithology	Sedimentary Structures	Interpreted Genesis
Massive Diamictite	Extremely poorly sorted diamictite with dominantly extrabasinal pebble to boulder size clasts floating in a muddy or sandy matrix.	Diamictite contains indistinct beds 20-50 m thick, local clots of stones, rare clast imbrication, and rare stratified non-pebbly interbeds.	Subaqueous lodgement tillite or flow tillite.
Graded Sandstone and Diamictite	Interbedded diamictite and fine- to medium-grained lithic wacke. Diamictite contains dominantly intrabasinal clasts.	Sandstone is medium- to thick-bedded (.5 to 2 m). Sandstone contains A-E graded beds, inferred rip-up siltstone chips, microcross lamination and small current ripples.	Proximal turbidites and subaqueous mass flows.
Cobble Conglomerate	Cobble- to boulder-bearing conglomerate with coarse sandy matrix.	May be massive and disorganized or thickly interbedded with cross-bedded pebbly sandstone.	Where massive south of Portneuf Narrows, the conglomerate is interpreted as a deep water channel deposit. Elsewhere conglomerate is interbedded with the pebbly sandstone facies and is interpretated as a fluvial or shallow marine high energy deposit.
Pebbly Sandstone	Medium- to coarse-grained, moderately sorted, lithic and arkosic arenite and wacke with floating pebbles and pebble lags on bedding planes. Locally the sandstone contains magnetite and ilmenite laminations.	Thin- to medium-bedded (3-40 cm). Sandstone contains planar cross beds up to 3 m long and 1 m high, bidirectional trough cross beds up to 50 cm wide and 20 cm high, and planar lamination.	Nearshore, shoreface, and/or tidal channel, shallow marine deposit.
Amalgamated Sandstone	Medium- to coarse-grained lithic wacke, disconformably overlies pink dolomite and contains dolomite chips north of Portneuf Narrows.	Amalgamated beds, contains dish structures, slumprolls, flute marks, load structures, local parallel lamination and large trough cross beds.	Liquified sediment flow or gain flow deposit formed in shallow marine tidal channels.
Fine Clastic and Carbonate	Fine-grained quartz wacke, siltstone, limestone and dolomite.	Thin bedded to laminated, contains rhythmic lamination, local flaser bedding, load structures, ripples, and thin graded beds. Carbonate is wavy laminated.	Low energy offshore shallow marine deposit.
Pink Dolomite	Thinly laminated silty dolomite.	Medium- to thin-bedded, contains wavy laminations.	Low energy intertidal mud flat deposit. Wavy laminations interpreted to be stromatolites.

imentary structures indicative of sediment gravity flow, including dish structures, floating soft-sediment clasts, flute marks, load structures, and contorted beds.

The *fine clastic and carbonate* facies composes the lower siltstone and carbonate unit, and parts of the middle quartzite and siltstone and upper sandstone and limestone units in the northern Bannock Range. This facies contains siltstone, fine sandstone, limestone and dolomite. Sedimentary structures in the clastic rocks include rhythmic laminations, local flaser-bedding, ripples, and load structures. The carbonates are generally wavy laminated and vary from

thin lenses less than 1 m thick to laterally persistent beds up to 10 m thick. The *fine clastic and carbonate* facies is interpreted to be a low-energy, shallow marine deposit.

The *pink dolomite* facies composes the pink dolomite unit, which conformably overlies the upper diamictite unit in much of the northern Bannock Range. It consists of wavy laminated, silty dolomite in discontinuous beds and lenses less than 1 m thick. It is interpreted to have been deposited as stromatolitic limestone on algal mats in an intertidal or subtidal environment and to have subsequently recrystallized to dolomite. The work of Logan and others (1964) suggests that wavy laminated, laterally persistent stromatolites form in a low-current energy environment. The presence of dolomite-chip breccia in the upper sandstone and limestone unit (immediately above the pink dolomite unit) north of Portneuf Narrows suggests tidal channels between intertidal carbonate mud flats.

Another possible origin for the pink dolomite unit is suggested by the recent discovery by Parker and others (1981) of stromatolites growing in ice-covered, glacier-fed lakes in Antarctica. Similar lakes may have developed on the irregular top of the subaerially exposed till sheet represented by the upper diamictite unit. Storms or a sea-level rise may have resulted in ripping up of the stromatolitic mats and deposition of what is now the dolomite-chip breccia.

Similar pink dolomites occur in the Kelley Canyon Formation which overlies Upper Proterozoic diamictite near Huntsville, Utah (Crittenden and others, 1971; Sorensen and Crittenden, 1980). The presence of these apparently warm-water dolomites above many Upper Proterozoic diamictites worldwide led Schermerhorn (1974) to question whether the diamictites could have been deposited under cold glacial conditions. However, recent studies by Bjørlykke and others (1978) have documented Pleistocene, cold-water, postglacial carbonates, and the work of Leonard and others (1981) stresses that clastic supply rather than water temperature is the main limiting factor in carbonate formation. The finely crystalline *pink dolomite* of the Scout Mountain Member is probably diagenetic, and a warm depositional environment is therefore not required.

DEPOSITIONAL MODEL

The Pocatello Formation is exposed in two isolated outcrop bands, and these will be considered separately. The favored interpretation, supported below, is that these two areas, the northern and southern Bannock Range, were separate sub-basins during initial continental rifting and formation of the Cordilleran miogeocline. Figure 6 shows tentative reconstructions of the environments of deposition of the Pocatello Formation.

Northern Bannock Range

A shallow marine environment (Fig. 6A) is envisaged for the lowest exposures of the Scout Mountain Member, which contain the lower quartzite unit on the west slope of Scout Mountain (Section 5, Fig. 3). The topographically lowest exposures west of Chinks Peak (Section 8b) are dark siltstones of uncertain stratigraphic position or depositional environment. Mafic lavas and inferred epiclastic rocks or pillow breccias of the Bannock Volcanic Member overlie these siliceous clastic rocks in both localities but are much thicker on Chinks Peak, which is interpreted to be near the volcanic center.

The appearance of extrabasinal clasts in the inferred deep water deposits of the lower diamictite unit may represent a glacial sediment source (Fig. 6B). Basin-margin tectonism could also be a cause of the diverse clast assemblage in these bedded diamictites. The inferred turbidites of the *graded sandstone and diamictite* facies (the lower diamictite unit and overlying lower sandstone unit) south of Portneuf Narrows (Section 6) are much thicker than they are to the south on Scout Mountain, and a northward-deepening basin, shown in Figure 6B, is inferred. Correlative strata on Scout Mountain and at Garden Creek Gap contain shallow marine facies.

Faulting associated with extrusion of the Bannock Volcanics is shown in the area between Chinks Peak and Portneuf Narrows to account for the dissimilarity of exposures in the two localities. This dissimilarity also could have been caused by Mesozoic or Tertiary faulting.

The thickness variation and sedimentary structures of the cobble conglomerate unit are also interpreted as supporting the concept of a northward-deepening basin. The conglomerate south of Portneuf Narrows is thick and abruptly overlies thin-bedded, graded sandstones. The sequence of *graded sandstone and diamictite* and *cobble conglomerate* facies south of Portneuf Narrows resembles the deep water slope and inner fan associations of Mutti and Ricci-Lucchi (cited by Rupke, 1978, p. 411). The cobble conglomerate at Portneuf Narrows is thus interpreted as a deep water, disorganized conglomerate (Walker, 1975), deposited by submarine mass flow.

On Scout Mountain the cobble conglomerate unit is interbedded with sandstone of the shallow marine *pebbly sandstone* facies, and at Garden Creek Gap only sparsely pebbly trough cross-bedded sandstone of the *pebbly sandstone* facies is present and is correlated with the cobble conglomerate unit.

Alternate interpretations of the conglomerate as totally shallow marine or fluvial are also possible. A disconformity below the conglomerate south of Portneuf Narrows could be invoked to explain the abrupt contact from inferred turbidites of the *graded sandstone and diamictite* facies to a fluvial conglomerate.

The quartzitic clast make-up of the cobble conglomerate could represent either a glacio-fluvial or a nonglacial source. If glaciers are invoked to supply the extrabasinal clasts of the lower diamictite, they are also available for the cobble conglomerate. Again, however, a tectonically related source cannot be eliminated.

In either case, the supply of large clasts dwindled after deposition of the cobble conglomerate, and inferred shallow marine deposits of the *pebbly sandstone* and *fine clastic and carbonate* facies are present above the conglomerate in all sections south of Portneuf Narrows. *Massive diamictite* containing glacially striated clasts and interpreted as subaqueous lodgement tillite or flow tillite was deposited over

DEPOSITIONAL MODELS FOR THE POCATELLO FORMATION

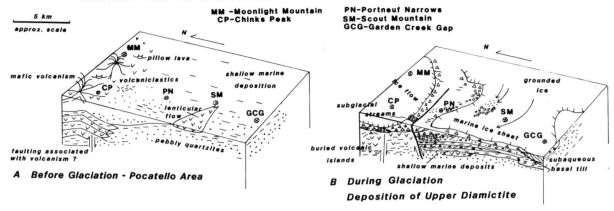

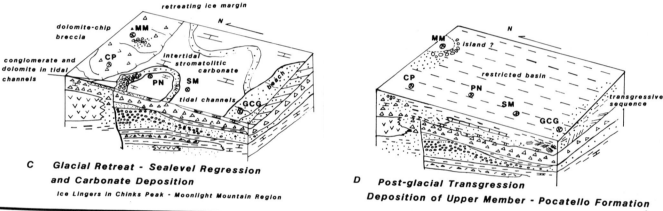

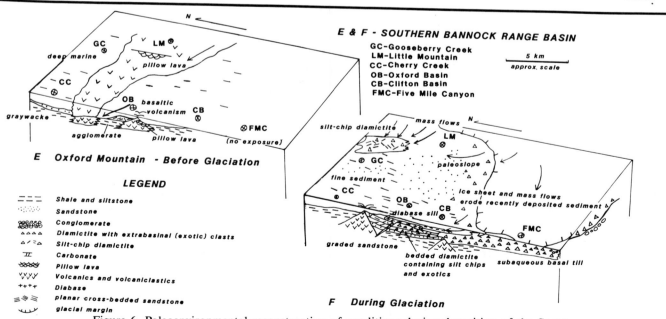

Figure 6. Paleoenvironmental reconstruction of conditions during deposition of the Scout Mountain Member.

all of the northern Bannock Range (Fig. 6B). The thickest deposits and the largest clasts occur southwest of Chinks Peak (Section 8a, Fig. 3) and near Moonlight Mountain (Section 9) and are interpreted to belong to the proximal glacial facies association of Edwards (1978), deposited closest to the glacial margin. Thinner diamictite from Portneuf Narrows south to Garden Creek Gap is interpreted to belong to the marginal glacial facies association described by Edwards. Tillite in this facies association was deposited only during maximum ice advance.

The number of glaciers and their flow directions are poorly constrained. Figure 6B shows two glaciers in order to explain regional variation in clast composition. *Massive diamictite* on Chinks Peak and Moonlight Mountain contains a higher percentage of intrabasinal clasts than *massive diamictite* further south (Portneuf Narrows to Garden Creek Gap).

Figure 6C shows interpreted conditions during the inferred postglacial regression which allowed deposition of the *pink dolomite* facies as stromatolitic carbonate on intertidal mud flats. Local conglomerate interbedded with dolomite in Section 8a and parts of the *amalgamated sandstone* facies are interpreted as representing tidal channel deposits.

This local sea-level regression may have been caused by isostatic rebound following deglaciation of the Bannock Range. It was followed by regional postglacial transgression (Fig. 6D), caused by large-scale melting of ice sheets. The upper sandstone and limestone unit was deposited during this transgression. This unit belongs to the *fine clastic and carbonate* facies, fines upward, and is overlain by dark siltstone and argillite of the upper member, Pocatello Formation, interpreted to have been deposited in a deep, quiet basin.

Southern Bannock Range

The composite stratigraphic section of the Pocatello Formation on Oxford Mountain, southern Bannock Range (Sections 1 to 3, Fig. 3), contains at least 200 m of mafic volcanic and volcaniclastic strata, assigned to the Bannock Volcanic Member, and at least 450 m of diamictite-bearing strata assigned to the Scout Mountain Member. Rocks of the *massive diamictite* and *graded sandstone and diamictite* facies are present and are interpreted to have been deposited in a northward-deepening basin. Figure 6E shows the inferred depositional environment of the lowest exposed strata of the Pocatello Formation on the east face of Oxford Ridge. This sequence contains volcanigenic greenstone overlain by interbedded graded sandstones, pillow lavas, and thin diamictites. Diamictite becomes thicker and forms a greater portion of the outcrop upwards. The upper part of this sequence is interpreted to represent a turbidite

assemblage, coarsening upward due to the input of glacially transported debris.

Thick, *massive diamictite* at Five Mile Canyon (Section 1) is interpreted as subaqueous lodgment tillite (Fig. 6E) deposited during maximum glacial advance. It is correlated with bedded diamictite interpreted as flow till and turbidite deposited in deeper water to the north near Clifton Basin (Section 2).

REGIONAL PALEOENVIRONMENTAL RECONSTRUCTION

Assuming no significant tectonic dislocation between the northern and southern Bannock Ranges during the Mesozoic Sevier Orogeny, a paleoenvironmental model involving two northward-deepening sub-basins is inferred for deposition of the Scout Mountain Member, Pocatello Formation. This interpretation is favored because the northernmost exposures on Oxford Mountain are inferred to contain deep water strata, and the southernmost exposures in the northern Bannock Range (at Garden Creek Gap) are inferred to have been deposited in shallow water. The northern sub-basin is interpreted to have deepened northwards towards volcanic centers in the Chinks Peak region, and the southern sub-basin is inferred to have deepened northward from Five Mile Canyon toward Oxford Basin. The basins are inferred to have become integrated by the time of deposition of the upper member, Pocatello Formation, which can be correlated from the northern Bannock Range to central Utah (Christie-Blick and others, 1980). The Kelley Canyon Formation in Utah contains identical, thinly laminated strata to that in the upper member, Pocatello Formation. Small sub-basins are to be expected during initial rifting and foundering of a new ocean basin, the tectonic environment in which the Pocatello Formation is inferred to have been deposited. Lacustrine sedimentation was possible during initial development of these sub-basins.

An alternate interpretation is that only one basin existed, and the two bands of outcrop now present cannot be correlated because of a) lateral facies change or b) later tear faulting. A single basin could have simply been persistently shallower in the Garden Creek Gap and Scout Mountain areas, deepening both to the north toward Portneuf Narrows and to the south toward Oxford Mountain. Alternatively, thrusting and tear faulting during the Sevier Orogeny could have moved the generally deep water strata now present on Oxford Mountain farther eastward than the generally shallow water Garden Creek Gap rocks.

CONCLUSIONS AND SPECULATION ABOUT SOURCE AREAS

The Pocatello Formation contains diverse clastic and

volcanic strata deposited during initial sedimentation in the Cordilleran miogeocline. A glacial sediment input is suggested by the presence of diamictites containing diverse and locally striated clasts. Clasts in the Scout Mountain Member are of four general types:

1) Extrabasinal Quartzites. A variety of vitreous, fine- to medium-grained quartzites occur as rounded and subrounded cobbles and boulders. No obvious nearby source of these rocks is exposed although they are grossly similar to Middle Proterozoic (Precambrian Y) quartzites of the Lemhi Group of north central Idaho (Ruppel, 1975), and to quartzites of uncertain stratigraphic affinity which overlie and are faulted against Archean basement rocks in the Albion-Raft River Ranges, 200 km west of the Bannock Range in south central Idaho (Compton and others, 1977; Crittenden, 1979; Compton and Todd, 1979; Miller, 1980; Crittenden and Sorensen, 1980). Middle Proterozoic quartzites of the Big Cottonwood Formation and the Uinta Mountain Group in northern Utah (Crittenden and Wallace, 1973) are less vitreous and less well sorted than quartzite clasts in the Scout Mountain Member. These rock units are inferred not to have been a source.

2) Extrabasinal Basement Rocks. Granitic and gneissic felsic plutonic and metamorphic rocks make up a small, but locally dominant percentage of clasts. These rocks are similar to basement terranes of Archean age in northwestern Wyoming, the Green Creek Complex of the Albion Range (Armstrong, 1968a), and the Lower Proterozoic to Archean Farmington Canyon Complex of the Wasatch Range of Utah (Hedge and others, this volume).

3) Intrabasinal Volcanic Rocks. Heterogeneous vesicular and porphyritic volcanic clasts occur in the Scout Mountain Member, some of them identical to parts of the Bannock Volcanic Member. No other volcanic source terranes of Proterozoic or older age are exposed in the Utah-Idaho-Wyoming area.

4) Intrabasinal Siltstones. Flat, angular, micaceous chips up to 40 cm long are common in bedded diamictite throughout the study area. These are inferred to have been soft siltstone clasts ripped up from marginal portions of the basin and redeposited in diamictite. In some diamictites these are the only clasts present. Another possible origin could be as chips of volcanic rock from pyroclastic deposits of the Bannock Volcanic Member.

In summary, the large size and variable composition of clasts in the Scout Mountain Member are interpreted to be caused by sediment input both from nearby block-faulted highlands or volcanic edifices and from distant terranes through erosion and transportation by glaciers.

ACKNOWLEDGEMENTS

Max D. Crittenden, Jr. has been a teacher and mentor since our first meeting on a field trip examining Proterozoic diamictites in South Australia in 1976. I feel lucky to have been able to call Max and his wife Mabel my friends. I am also particularly indebted to John C. Crowell and Steven S. Oriel who supervised this study. I have benefited from the help and encouragement of Stanley M. Awramik, Robert C. Bright, Nicholas Christie-Blick, Eugene Domack, Richard V. Fisher, Victor A. Gostin, Michael Green, G. D. Harper, Dennis Kurtz, George B. LeFebre, Julia M. G. Miller, H. Thomas Ore, Lucian B. Platt, Braden J. Thompson, Donald E. Trimble, and Marjorie Ulin. National Science Foundation Grant EAR77-06008, (J. C. Crowell, principal investigator) funded much of the research for a Ph.D. thesis at the University of California, Santa Barbara. Dennis Dunn, Alex Fischer, Sandy Pitman, Shirley M. Sargent, and Vicky Zimmerman provided invaluable aid in drafting and typing. Gary J. Axen, Lucian B. Platt, John H. Stewart, and Victoria R. Todd offered valuable comments on an early draft of the manuscript.

REFERENCES CITED

Allmendinger, R. W., and Jordan, T. E., 1981, Mesozoic evolution, hinterland of the Sevier orogenic belt: Geology, v. 9, p. 308–313.

Armstrong, F. C., and Oriel, S. S., Tectonic development of Idaho-Wyoming thrust belt: American Association of Petroleum Geologists Bulletin, v. 49, p. 1847–1866.

Armstrong, R. L., 1968a, Sevier orogenic belt in Nevada and Utah: Geological Society of America Bulletin, v. 79, p. 429–458.

——1968b, Mantled gneiss domes in the Albion Range, southern Idaho: Geological Society of America Bulletin, v. 79, p. 1295–1314.

Bjørlykke, K., Bue, B., and Elverhøi, A., 1978, Quaternary sediments in the northwestern part of the Barents Sea and their relation to the underlying Mesozoic bed rock: Sedimentology, v. 25, p. 227–246.

Blackstone, D. L., Jr., 1977, Tectonic map of the Overthrust Belt, western Wyoming, southeastern Idaho and northeastern Utah: Wyoming Geological Association 28th Annual Field Conference Guidebook (in pocket).

Blackwelder, E., 1932, An ancient glacial formation in Utah: Journal of Geology, v. 40, p. 289–304.

Blick, N. H., 1979, Stratigraphic, structural and paleogeographic interpretation of Upper Proterozoic glaciogenic rocks in the Sevier orogenic belt (Ph.D. thesis): Santa Barbara, University of California, 633 p.

Boulton, G. S., and Deynoux, M., 1981, Sedimentation in glacial environments and the identification of tills and tillites in ancient sedimentary sequences: Precambrian Research, v. 15, p. 397–422.

Bouma, A. H., 1962, Sedimentology of some flysch deposits: a graphic approach to facies interpretation: Amsterdam, Elsevier, 168 p.

Christie-Blick, N., Link, P. K., Miller, J.M.G., Young, G. M., and Crowell, J. C., 1980, Regional geologic events inferred from Upper Proterozoic rocks of the North American Cordillera: Geological Society of America Abstracts with Programs, v. 12, p. 402.

Compton, R. R., and Todd, V. R., 1979, Oligocene and Miocene metamorphism, folding, and low-angle faulting in northwestern Utah: Reply: Geological Society of America Bulletin, Part I, v. 90, p. 307–309.

Compton, R. R., Todd, V. R., Zartman, R. E., and Naeser, C. W., 1977, Oligocene and Miocene metamorphism, folding, and low-angle faulting in northwestern Utah: Geological Society of America Bulletin, v. 88, p. 1237–1250.

Crittenden, M. D., Jr., 1979, Oligocene and Miocene metamorphism, folding, and low-angle faulting in northwestern Utah, discussion: Geological Society of America Bulletin, Part I, v. 90, p. 305–306.

Crittenden, M. D., Jr., Christie-Blick, N., and Link, P. K., in press, Evidence for two pulses of glaciation during the late Proterozoic in northern Utah: Geological Society of America Bulletin.

Crittenden, M. D., Jr., Schaeffer, F. E., Trimble, D. E., and Woodward, L. A., 1971, Nomenclature and correlation of some upper Precambrian and basal Cambrian sequences in western Utah and southeastern Idaho: Geological Society of America Bulletin, v. 82, p. 581–602.

Crittenden, M. D., Jr., and Sorensen, M. L., 1980, The Facer Formation, a new Early Proterozoic unit in northern Utah: U.S. Geological Survey Bulletin 1482-F, 28 p.

Crittenden, M. D., Jr., and Wallace, C. A., 1973, Possible equivalents of the Belt Supergroup in Utah: Belt Symposium, v. 1, Moscow, Idaho, Idaho Bureau of Mines and Geology, p. 116–138.

Crowell, J. C., 1978, Gondwanan glaciation, cyclothems, continental positioning, and climate change: American Journal of Science, v. 278, p. 1345–1372.

deWit, M. J., and Stern, C., 1978, Pillow Talk: Journal of Volcanology and Geothermal Research, v. 4, p. 55–80.

Edwards, R. F., Sanders, J. E., and Rodgers, J., 1960, Diamictite: a substitute name for symmictite: Geological Society of American Bulletin, v. 71, p. 1809–1810.

Frakes, L. A., 1979, Climates throughout geologic time: New York, Elsevier, 310 p.

Harrison, J. E., and Peterman, Z. E., 1980, North American Commission on Stratigraphic Nomenclature Note 52—A preliminary proposal for a chronometric time scale for the Precambrian of the United States and Mexico: Geological Society of America Bulletin, Part I, v. 91, p. 377–380.

Hedge, C. E., Stacey, J. S., and Bryant, B., 1983, Geochronology of the Farmington Canyon Complex, Wasatch Mountains, Utah, *in* Miller, D. M., Todd, V. R., and Howard, K. A., eds., Tectonic and stratigraphic studies in the eastern Great Basin: Geological Society of America Memoir 157 (this volume).

Leonard, J. E., Cameron, B., Pildey, O. H., and Friedman, G. M., 1981, Evaluation of cold water carbonates as a possible paleoclimatic indicator: Sedimentary Geology, v. 28, p. 1–28.

Lilley, W. W., 1973, Stratigraphy and possible paleoenvironments of the upper Scout Mountain Member of the Pocatello Formation (late Precambrian), Bannock County, Idaho: Northwest Geology, v. 2, p. 59–64.

Link, P. K., 1981, Upper Proterozoic diamictites in southeastern Idaho, *in* Hambrey, M. J., and Harland, W. B., eds., Earth's pre-Pleistocene glacial record: Cambridge, Cambridge University Press, p. 736–739.

——, 1982, Geology of the Pocatello Formation (Upper Proterozoic) and geologic mapping in the Bannock Range, southeastern Idaho [Ph.D. Thesis]: Santa Barbara, University of California, 131 p.

Link, P. K., Bright, R. C., and Trimble, D. E., 1980, New discoveries in the upper Proterozoic stratigraphy of southeastern Idaho: Geological Society of America Abstracts with Programs, v. 12, p. 278.

Link, P. K., and Gostin, V. A., 1981, Facies and paleogeography of Sturtian glacial strata (late Precambrian), South Australia: American Journal of Science, v. 281, p. 353–374.

Logan, B. W., Rezak, R., and Ginsburg, R. N., 1964, Classification and environmental significance of algal stromatolites: Journal of Geology, v. 72, p. 68–83.

Ludlum, J. C., 1942, Pre-Cambrian formations at Pocatello, Idaho: Journal of Geology, v. 50, p. 85–95.

——, 1943, Structure and stratigraphy of part of the Bannock Range, Idaho: Geological Society of America Bulletin, v. 54, p. 973–986.

Miller, D. M., 1980, Structural geology of the northern Albion Mountains, south-central Idaho, *in* Crittenden, M. D., Jr., Coney, P. J., and Davis, G. H., eds., Cordilleran Metamorphic Core Complexes: Geological Society of America Memoir 153, p. 399–423.

Ojakangas, R. W., and Matsch, C. L., 1980, Upper Precambrian (Eocambrian) Mineral Fork Tillite of Utah: a continental glacial and glaciomarine sequence: Geological Society of America Bulletin, Part I, v. 91, p. 495–501.

Oriel, S. S., and Platt, L. B., 1980, Geologic map of the Preston 1° x 2° quadrangle, southeastern Idaho and western Wyoming: U.S. Geological Survey Map I-1127, Scale 1:250,000.

Parker, B. C., Simmons, G. M., Jr., Love, F. G., Wharton, R. A., Jr., and Seaburg, K. G., 1981, Modern Stromatolites in Antarctic Dry Valley Lakes: Bioscience, v. 31, p. 656–661.

Raymond, L. C., 1971, Structural geology of the Oxford Peak area, Bannock Range Idaho [M.S. thesis]: Logan, Utah, Utah State University, 48 p.

Rember, W. C., and Bennett, E. H., 1979, Geologic map of the Pocatello Quadrangle, Idaho: Idaho Bureau of Mines and Geology, Scale 1:250,000.

Royse, F. C., Warner, M. A., and Reese, D. L., 1975, Thrust belt structural geometry and related stratigraphic problems, Wyoming-Idaho-northern Utah, *in* Bolyard, D. W., ed., Deep drilling frontiers of the central Rocky Mountains: Rocky Mountain Association of Geologists, Symposium, p. 41–54.

Rupke, N. A., 1978, Deep clastic seas, *in* Reading, H. G., ed., Sedimentary environments and facies: New York, Elsevier, p. 372–415.

Ruppel, E. T., 1975, Precambrian Y sedimentary rocks in east-central Idaho: U.S. Geological Survey Professional Paper 889-A, 23 p.

Schermerhorn, L.J.G., 1966, Terminology of mixed coarse-fine sediments: Journal of Sedimentary Petrology, v. 36, p. 831–835.

——, 1974, Late Precambrian mixtites: glacial and/or non-glacial?: American Journal of Science, v. 274, p. 673–824.

Sorensen, M. L., and Crittenden, M. D., Jr., 1980, Geologic map of the Huntsville Quadrangle, Weber and Cache Counties, Utah: U.S. Geological Survey Map GQ-1503, Scale 1:24,000.

Stewart, J. H., 1972, Initial deposits in the Cordilleran geosyncline: evidence of a late Precambrian (<850 m.y.) continental separation: Geological Society of America Bulletin, v. 83, p. 1345–1360.

Stewart, J. H., and Surzek, C. A., 1977, Cambrian and latest Precambrian paleogeography and tectonics in the western United States, *in* Stewart, J. H., Stevens, C. H., and Fritsche, A. E., eds., Paleozoic paleogeography of the western United States: Society of Economic Paleontologists and Mineralogists, Pacific Section, Pacific Coast Paleogeography Symposium I, p. 1–18.

Thompson, B. J., 1982, Geology of Garden Creek Gap, Bannock Range, southeastern Idaho [M.S. thesis]: Pocatello, Idaho State University, 41 p.

Thompson, B. J., and Link, P. K., 1981, Stratigraphy of Upper Proterozoic Pocatello Formation, Garden Creek Gap, Bannock Range, southeast Idaho: Geological Society of America Abstracts with Programs, v. 13, p. 228.

Trimble, D. E., 1976, Geology of the Michaud and Pocatello Quadrangles, Bannock and Power Counties, Idaho: U.S. Geological Survey Bulletin 1400, 88 p.

Varney, P. J., 1976, Depositional environment of the Mineral Fork Formation (Precambrian), Wasatch Mountains, Utah, *in* Hill, J. G., ed., Geology of the Cordilleran hingeline: Rocky Mountain Association of Geologists, Symposium, p. 91–102.

Walker, R. G., 1975, Generalized facies models for resedimented conglomerates of turbidite association: Geological Society of America Bulletin, v. 86, p. 737–748.

Manuscript Accepted by the Society August 20, 1982

Geological Society of America
Memoir 157
1983

Phanerozoic magmatism and mineralization in the Tooele 1° × 2° quadrangle, Utah

William J. Moore
Edwin H. McKee
U.S. Geological Survey
345 Middlefield Road
Menlo Park, CA 94025

ABSTRACT

A comprehensive suite of radiometric age determination clearly defines three periods of igneous activity in the Tooele 1° × 2° quadrangle, Utah. These periods and the associated igneous activity are (1) Jurassic calcalkaline plutonism, (2) late Eocene to early Miocene calc-alkaline plutonism and volcanism, and (3) middle and late Miocene bimodal basaltic and rhyolitic volcanism. These periods are recognized throughout the entire Great Basin. A characteristic metal or suite of metals is associated with each period. Tungsten skarns are commonly related to Jurassic plutonism; base-metal and precious-metal deposits are characteristic of the late Eocene to early Miocene igneous activity; and beryllium, fluorine, and uranium are elements associated with silicic rocks of the middle and late Miocene period of igneous activity.

INTRODUCTION

Recent compilation of a geologic map for the Tooele, Utah, 1° × 2° quadrangle (Moore and Sorensen, 1979) provided the opportunity to assemble and evaluate all radiometric age determinations for igneous rocks in this eastern part of the Great Basin. Radiometric age data in several areas (particularly the Oquirrh Mountains-Bingham mining district and Deep Creek Mountains-Gold Hill mining district) was reasonably complete, but elsewhere ages were suspect, anomalous, or lacking. A supplemental program of K-Ar dating was undertaking to augment available data and to establish an internally consistent framework for evaluating the chronology of Phanerozoic magmatism and associated hydrothermal mineralization. Certain igneous bodies were redated, and others were dated for the first time. In all, 20 new determinations were made. These, combined with about 45 ages that had been reported previously, represent most of the major intrusive bodies and larger volcanic fields in the quadrangle.

The "Uinta trend," a belt of middle Tertiary igneous centers 275 km long in the eastern Great Basin, is a central element in the geology of the Tooele quadrangle. Our geologic understanding of this "trend" and, in particular, the segment that includes the phaneritic plutons in the central Wasatch Mountains to the east, owes much to the detailed mapping done by Max Crittenden over a period of 40 years. His pioneering application of the lead-alpha method (Crittenden and others, 1952) in dating the Wasatch plutons established a baseline for subsequent radiometric studies.

We have been challenged and stimulated by Max's knowledge of the Great Basin, and we have also had the good fortune to join lively discussions in several of his Wasatch field camps. It is our privilege to add to the geologic framework that Max has shaped.

ANALYTICAL TECHNIQUES

Materials dated included biotite, muscovite, hornblende, and acid-treated, whole-rock samples of basalt, dacite, rhyolite. Mineral separates were used wherever unaltered phenocrysts were available. The whole-rock samples seem to provide reliable age determinations that, in most cases, were close to those expected on the basis of previous dates, correlation of lithologic types, or regional chronologic patterns.

Potassium-argon dating for this study was done in the laboratories of the U.S. Geological Survey, Menlo Park, California, using standard isotope-dilution procedures as described by Dalrymple and Lanphere (1969). Pure-mineral concentrates were prepared using heavy liquid, magnetic, electrostatic, and hand-picking procedures. Whole-rock samples were crushed, sieved to 60 to 100 mesh size, washed, and treated for 1 minute in 5-percent HF and 30 minutes in 14-percent HNO_3 solutions. This treatment has proved advantageous in eliminating ex-traneous argon from whole-rock samples. Potassium anal-yses were performed using a lithium metaborate flux fusion-flame photometry technique, the lithium serving as an internal standard (Ingamells, 1970). Argon analyses were performed using a 60° sector, 15.2 cm radius, Nier-type mass spectrometer, operated in the static mode for mass measurement. The precision of the data, shown as the ± value, is the estimated analytical uncertainty at one standard deviation. It represents uncertainty in the meas-urement of radiogenic ^{40}Ar and K in the sample and is based on experience with replicated analyses in the Menlo Park laboratories. The decay constants used for ^{40}K are those adopted by the International Union of Geological Sciences Subcommission on Geochronology (Steiger and Jager, 1977). Dates and analytical data determined for this study are in Table 1.

MESOZOIC MAGMATISM

The first suggestion that Mesozoic plutonism had oc-curred in the eastern Great Basin was a lead-alpha age of 193 m.y. (Early Jurassic) reported by Whelan (1970) for zircon from the granitic rock of Notch Peak in the west-central House Range south of the Tooele quadrangle. This date—which has possible inherent errors due to the uncer-tain Th/U ratio of the zircon, presence of primary lead, and possible loss or gain of parent or daughter isotopes—is considered only a general indication of the rock age. Later studies by Armstrong and Suppe (1973), using the more precise and more accurate K-Ar method, determined a 143-m.y. age (latest Jurassic) for biotite from the same pluton and a 140-m.y. age for the quartz monzonite of Crater Island in the northern Silver Island Mountains in the northwest corner of the Tooele quadrangle. Subsequently, Caroon (1977) reported concordant K-Ar ages of 150 m.y. (biotite) and 144 m.y. (hornblende) for the Newfoundland stock about 40 km northeast of Crater Island. Stacey and Zartman (1978) established by Rb-Sr dating that the main mass of a quartz monzonite pluton in the Gold Hill mining district, long considered to be post-Eocene in age, was Late Jurassic (152 m.y.). No other Jurassic plutons were known

in the Tooele 1° × 2° quadrangle or in the entire state of Utah.

Because of this demonstrated existence of Jurassic plutonism in the northern Silver Island Mountains, ages were determined for a biotite and hornblende from pre-viously undated granodiorites south of Crater Island (loc. 1, Fig. 1). These samples yielded ages of 160 m.y. and 174 m.y., respectively, and reaffirm a Mesozoic magmatic event in this region.

When this study was begun in the mid-1970s, the only radiometric age for the Gold Hill stock was a biotite K-Ar age of 42 m.y. reported by Armstrong (1970); this date was compatible with the late Eocene or post-Eocene age pre-viously inferred by Nolan (1935). A sample of fresh quartz monzonite from the abandoned railroad grade east of Gold Hill—the approximate location of Armstrong's sample (Fig. 1, locs. 6, 7)—was dated for this study and produced a similar age of 44 m.y. After the earlier results were verified, no additional work seemed needed until Stacey and Zart-man (1978) reported Jurassic ages of 151 m.y. and 153 m.y. from the southern part of the stock (locs. 10, 11, Fig. 1). It now seems probable that the major part of the Gold Hill stock is Late Jurassic in age, and that the Eocene ages determined by Armstrong (1970) and ourselves are the result of thermal resetting by a smaller, more leucocratic pluton emplaced at its northern margin (Fig. 1). The con-tact between the two granitic bodies is concealed by allu-vium north of the Gold Hill townsite. Thermal effects can be detected at least 3 km south of the concealed contact (loc. 9, Fig. 1), where we obtained an age of 135 m.y., slightly younger than that of Stacey and Zartman (1978).

Younger Mesozoic magmatism has not yet been rec-ognized in the Tooele 1° × 2° quadrangle although the dating results in the Gold Hill area urge caution in generali-zations. Cretaceous magmatism and, in particular, latest Cretaceous to early Tertiary "Laramide" magmatism—significant in the metallogenesis of the Rocky Mountains and the Basin and Range provinces of Arizona and New Mexico—have not been firmly established anywhere in Utah. The sole exception may be in the Big Cottonwood mining district of the central Wasatch Mountains, just east of the Tooele quadrangle. Here James (1979) reports a hornblende K-Ar age of 72.4 m.y. for a diorite pluton in the Argenta area. Two subsequent hornblende dates (a repli-cate sample from the same body and a single determination for another smaller diorite sill in the Argenta intrusive complex) are 60.9 m.y. and 67.3 m.y., respectively (E. H. McKee, unpublished data, 1982).

The geologic significance of these data is clouded not only by scatter among the late Cretaceous ages but by four biotite K-Ar ages ranging from 36.7 m.y. to 32.5 m.y. for small dikelike intrusions mapped as part of the Argenta complex by Crittenden (1965). These small intrusions are considered by James (written communication, 1982) to be

TABLE 1. NEW POTASSIUM-ARGON AGE DETERMINATIONS FOR IGNEOUS ROCKS IN THE TOOELE, UTAH $1° \times 2°$ QUADRANGLE

Sample location (Fig. 2)	Field number	Latitude (N.)	Longitude (W.)	Rock type	Mineral	K_2O percent	$^{40}Ar_{rad}$ mole/g	$^{40}Ar_{rad}$ percent	Age $\times 10^6$ yr
1	74-KA-11	40°51'40"	113°55'00"	Biotite-hornblende granodiorite	Biotite	9.64	1.8347×10^{-9}	49.7	159.6±5
	74-KA-12			Hornblende diorite	Hornblende	0.596, 0.652	1.6392×10^{-10}	64.3	173.8±3
2	74-KA-13	40°50'10"	113°57'30"	Hornblende-biotite andesite	Biotite	8.53	5.0784×10^{-10}	75.7	40.9±0.6
5	74-KA-15	40°11'45"	113°58'00"	Latite	do.	8.45	4.8238×10^{-10}	75.2	39.2±0.6
7	74-KA-16	40°10'00"	113°49'30"	Hornblende biotite	do.	6.60	4.2241×10^{-10}	67.3	43.9±0.8
9	75-KA-9	40°08'40"	113°48'50"	Quartz monzonite	Hornblende	0.413, 0.414	8.3399×10^{-11}	52.7	134.9±4
13	74-KA-18	40°08'30"	113°44'20"	Muscovite granite	Muscovite	10.27	2.1103×10^{-10}	33.8	14.2±0.6
15	74-KA-9	40°42'10"	112°55'15"	Augite basalt	Basalt	1.316	2.6257×10^{-11}	48.1	13.8±0.4
16	74-KA-4	40°37'35"	112°38'50"	Alkali olivine basalt	Basalt(?)	4.320	7.5326×10^{-11}	51.2	12.1±0.3
17	74-KA-3	40°37'35"	112°35'15"	do.	do.	5.050	9.2993×10^{-11}	72.0	12.7±0.2
18	74-KA-6	40°33'25"	112°35'25"	Hornblende-biotite latite	Biotite	8.58	4.9153×10^{-10}	39.4	39.4±0.5
19	74-KA-7	40°33'15"	112°41'50"	Hornblende-augite dacite	Hornblende	1.173	6.9351×10^{-11}	40.9	40.6±1.7
20	70-KA-1	40°31'40"	112°36'00"	Biotite-augite granodiorite	Biotite	8.60	4.8855×10^{-10}	68.6	39.0±0.6
21	70-KA-2A	40°31'00"	112°35'45"	Hornblende latite	do.	8.35	4.8963×10^{-10}	75.9	40.3±0.5
22	74-KA-5	40°30'05"	112°34'45"	Biotite-hornblende latite	do.	8.31	5.0589×10^{-10}	72.2	41.8±0.5
33	69-SM-2	40°27'55"	112°26'30"	Nephaline basalt	Basalt	2.195	1.2823×10^{-10}	72.5	40.1±0.5
35	69-TS-32	40°25'40"	112°05'45"	do.	do.	2.355	1.3181×10^{-10}	90.7	38.5±0.3
36	WT-41	40°20'45"	112°13'20"	Biotite granodiorite porphyry	Biotite	8.91	4.7503×10^{-10}	66.3	36.7±0.5
39	74-KA-2	40°18'15"	112°12'00"	Biotite-hornblende rhyolite	do.	7.48	4.1390×10^{-10}	71.0	38.0±0.5
40	74-KA-1	40°09'15"	112°00'35"	Olivine basalt	Basalt	2.061	6.3869×10^{-11}	24.8	21.4±2.5

$^{40}K/K_{total} = 1.167 \times 10^{-4} mol/mol$; $\lambda_\epsilon = 0.572 \times 10^{-10} yr^{-1}$; $\lambda_\beta = 4.963 \times 10^{-10} yr^{-1}$

offshoots of the Alta stock. In our estimation, the results are equivocal and will require additional study in order to be adequately interpreted. If the latest Cretaceous ages are not spurious, they record an extremely localized Laramide magmatic event in the eastern Great Basin.

CENOZOIC MAGMATISM

Eocene and (or) Oligocene

Middle Tertiary age determination reported by Armstrong (1963) for igneous rocks in the Bingham mining district, east-central Oquirrh Mountains (Fig. 1) were among the first in the eastern Great Basin. Prior to sampling for this study, the only published Tertiary age for igneous rocks in other parts of the Tooele quadrangle were 30 m.y. for muscovite from a beryl-bearing pegmatite on Granite Peak Mountain (loc. 14, Fig. 1) and the 42-m.y.-age for the Gold Hill stock (Armstrong, 1970). Intrusive equivalents of the Laguna Springs Latite of Morris and Lovering (1961), exposed in the southeastern corner of the quadrangle, had been dated in the adjoining Delta 1° × 2° quadrangle, where they yielded biotite K-Ar ages of about 32 m.y. (Laughlin and others, 1969). Numerous later studies have focused on the chronology of Cenozoic magmatic events in the Great Basin in relation to the tectonic evolution of this region (see Hintze, 1973; Best and others, 1980).

Armstrong's (1963) determinations in the Oquirrh

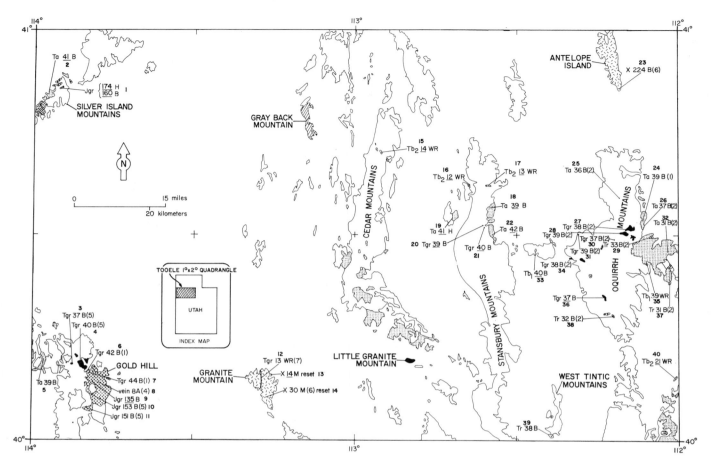

Figure 1. Occurrence of igneous rocks in the Tooele 1° x 2° quadrangle, Utah, and location of samples dated by radiometric methods. Outcrop areas of sedimentary rocks are unpatterned Geology after Moore and Sorensen (1979).

EXPLANATION

		Pliocene or Miocene	
Tb₂			
Tgr	Ta Tr	Tb₁	Oligocene or Eocene
	Jgr		JURASSIC
	X		PRECAMBRIAN(?)

Tb₂   BASALT AND BASALTIC ANDESITE FLOWS

Tgr   GRANODIORTIE, QUARTZ MONZONITE, AND QUARTZ LATITE

Ta   ANDESITE, DACITE, AND QUARTZ LATITE FLOWS AND BRECCIAS

Tr   RHYOLITE FLOWS AND BRECCIAS

Tb₁   BASALT AND BASALTIC ANDESITE FLOWS AND SILLS

Jgr   DIORITE, GRANODIORITE, AND QUARTZ MONZONITE

X   CRYSTALLINE METAMORPHIC ROCKS

Notation

Map unit	Age in m.y.	References	Sample material	Locality
Granodiorite, quartz, monzonite, and quartz latite	40	(1) Armstrong (1970)	B, biotite	33
		(2) Moore (1973)	H, hornblende	
		(3) Stacey and Zartman (1978)	WR, whole rock	
		(4) Whelan (1970)	M, muscovite	
		(5) C. E. Hedge (oral commun, 1982)	A, adularia	
		(6) Warnaars and others (1978)		
		(7) This report (Table 1)		

Mountains have been supplemented and largely superseded by results of several comprehensive investigations (Moore, 1973; Warnaars and others, 1978). The later studies, totaling over 40 analyses from a suite of epizonal calc-alkaline plutons and coeval volcanic rocks, provide the most complete documentation of middle Tertiary magmatism in the Tooele 1° × 2° quadrangle. To these data, we add new ages of 39 m.y. and 40 m.y. for a nepheline basalt flow in the eastern Oquirrh Mountains (loc. 35, Fig. 1) and for a nepheline basalt sill exposed on the crest of the South Mountain about 30 km to the west (loc. 33, Fig. 1), respectively. Granodiorite from the Pophyry Knob in the southwestern Oquirrh Mountains (Fig. 1, loc. 36) yielded a biotite age of 37 m.y.

Compared to those of Oquirrh Mountains, exposures of intermediate flows and laharic breccias in the Stansbury Mountains to the west are not as extensive, and intrusive bodies are smaller and fewer. A date of 39 m.y. was obtained for a biotite-augite monzonite plug in North Willow Canyon (loc. 20, Fig. 1). Petrographically, this rock resembles the Last Chance stock in the Bingham mining district. A north-trending latite dike in the North Willow Canyon area (loc. 21, Fig. 1) yielded a 40 m.y. age, and three age determinations fall between 39 m.y. and 42 m.y. for presumably coeval latites in eruptive sequences exposed on the western and eastern flanks of the range (locs. 18, 19, 22, Fig. 1). Thus, the igneous rocks of the Stansbury Mountains are the same age and are similar lithologically to those of the Oquirrh Mountains.

Other newly dated Eocene and Oligocene igneous rocks in the Tooele 1° × 2° quadrangle include a 38-m.y.-old crystal-vitric rhyolite flow near Bennion Ranch in the southern West Tintic Mountains (loc. 39, Fig. 1). Previously dated rhyolites in the area (35 to 50 km northeast in the Oquirrh Mountains) are approximately 6 m.y. younger (locs. 37 and 38, Fig. 1).

In the Cedar Mountains, attempts to collect samples of the volcanic rocks and subvolcanic intrusions suitable for radiometric dating have proven unsuccessful; pervasive oxidation of mafic minerals precludes the use of these minerals. The intermediate lavas, breccias, and their inferred feeders—the Tabbys Peak and Little Granite Mountain plugs—are provisionally assigned a middle Tertiary age on the basis of lithologic similarity to dated rocks in the Stansbury and Oquirrh Mountains.

Eocene and Oligocene ages determined for andesitic eruptive rocks in the Silver Island and northern Deep Creek Mountains are consistent with the predominance of middle Tertiary volcanism throughout the Tooele 1° × 2° quadrangle. These determination necessitate revisions in the reconstructed magmatic histories in both areas because younger ages had been inferred previously. A 41-m.y.-old biotite K-Ar age was obtained for Schaeffer's (1960) "andesite porphyry #3" (loc. 2, Fig. 1), interpreted by him to be the youngest flow in the Silver Island volcanic sequence. Accordingly, a minimum age is established for the remainder of the sequence, which, in addition to andesitic lavas and breccias, includes several thick rhyolite porphyry flows in the lower part.

About 12 km northwest of the Gold Hill townsite, a 39-m.y. age was obtained for biotite from a clast in a latite breccia overlying the Eocene(?) White Sage Formation (loc. 5, Fig. 1). The Eocene K-Ar age for the latite breccia, which corroborates gastropod ages for the White Sage Formation reported by Nolan (1935, p. 43), suggests that calc-alkaline volcanism was contemporaneous with emplacement of the smaller leucocratic quartz monzonite north of Gold Hill.

Miocene

Basalt and basaltic andesite flows scattered through the quadrangle are part of a widespread Great Basin basaltic eruptive event that began generally about 17 m.y. ago (McKee and others, 1970). The basalts commonly are olivine bearing and quartz normative.

The oldest of the four Miocene basalts dated here was collected near Goshen Pass in the southeastern corner of the Tooele quadrangle (loc. 40, Fig. 1), and it is about 21 m.y. old. Nearly concordant ages of 12.1 ± 3 m.y. and 12.7 ± 0.2 m.y. were obtained for vesicular basalts on the western flanks and near the crest, respectively, of the northern Stansbury Mountains (locs. 16, 17, Fig. 1). A basalt flow from Hastings Canyon in the northern Cedar Mountains (loc. 15, Fig. 1) yielded a whole-rock K-Ar age of 14 m.y. These ages are slightly older than the median for mafic volcanism suggested by the age histogram of Best and others (1980, Fig. 3).

We include the 14-m.y. age determined for coarse muscovite from the two-mica leucogranite at the northern end of Granite Peak Mountain (loc. 13, Fig. 1) with our Miocene determinations. This age is younger than the 30-m.y. age reported previously by Whelan (1970) for muscovite from a pegmatite that intrudes granite gneiss near the southern tip of Granite Peak Mountain; the difference in ages may not be significant because preliminary Rb-Sr isotopic data suggest that these leucocratic rocks are Precambrian in age (C. E. Hedge, oral communication, 1979). Apparently, radiogenic argon concentrations were substantially, but erratically modified by a later Tertiary thermal event.

SOME PETROLOGIC AND TECTONIC IMPLICATIONS OF THE RADIOMETRIC AGE DETERMINATIONS

We consider the radiometric geochronology of igneous rocks in the Tooele 1° × 2° quadrangle to be comprehensive enough to justify generalization. Certain apparent time-space patterns clarify or extend tectonic and magmatic patterns recognized elsewhere in the Great Basin, and characteristic metals are associated with the major magmatic and tectonic events.

Mesozoic Magmatism

Two factors appear significant regarding the distribution of Jurassic plutons in the Tooele quadrangle and adjoining areas: (1) the absence of granitic rocks of this age east of longitude 113° W and (2) a general east-southeasterly alignment of plutons extending westward from Gold Hill. It is possible, however, that these patterns result from sampling bias, as outcrop distribution is controlled by late Tertiary Basin and Range faulting and alluvial cover. The eastern limit of Jurassic magmatism apparently marks the edge of a magmatic arc or superposed arcs active during most of Mesozoic time, when an oceanic plate was subducted eastward beneath the North American plate (see Burchfiel, 1980). The apparent alignment of plutons is interpreted to be a manifestation of east-trending zones of structural weakness; a similar alignment has been suggested based on geophysical anomalies and mineral occurrences (Stewart and others, 1977; Rowley and others, 1978). The Jurassic plutons lying west of the Gold Hill stock are along an extension of the Deep Creek-Tintic mineral belt of Hilpert and Roberts (1964), which may represent a transverse flaw in the overriding continental plate above the subduction zone.

There is a strong relationship between skarn-type tungsten mineralization and Mesozoic plutonism in the Great Basin of western Nevada (H. L. Stager, oral communication, 1981). This relationship is repeated in western Utah and the Tooele quadrangle. Occurrences of contact-metamorphic tungsten (scheelite) mineralization are associated with each of the four known areas of Jurassic plutonism in Utah. Lemmon (1964) described these in the Newfoundland Range, Deep Creek Mountains (Gold Hill), and House Range; the fourth, at the contact between the Crater Island pluton and Permian carbonate rocks in the northern Silver Island Mountains, was being actively prospected in 1978. Whereas minor occurrences of tungsten mineralization are locally associated with Tertiary plutons in Nevada and Utah, it is our impression that most production, although limited, has come from deposit related to Mesozoic plutonism. Newberry and Einaudi (1981) attribute this association to deeper levels of erosion in the Mesozoic plutons compared to those in Tertiary plutons. The deeper parts of the pluton system are more favorable for tungsten mineralization.

Cenozoic magmatism

Eocene and Oligocene. An alignment of plutons and associated mineral deposits trends southwestward from the axis of the Uinta arch through the central Wasatch Mountains and across the Tooele 1° × 2° quadrangle. This feature, here termed the "Uinta trend," was first noted over a century ago and has been discussed many times since. Erickson (1976) summarized the various lines of geological and geophysical evidence suggesting that the Uinta trend has been an intermittently active structural zone since at least late Precambrian time. It is possible that this alignment marks a zone of deep crustal weakness, which guided emplacement of the plutons in a manner similar to that previously invoked for Jurassic plutons along an extension of the Deep Creek-Tintic mineral belt. The juncture of the two zones may have localized the Gold Hill plutons.

Crittenden and others (1952), in an early application of the lead-alpha method, were the first to attempt radiometric dating of the plutons in the Cottonwood area of the Wasatch Mountains. A subsequent study (Crittenden and others, 1973) incorporated results of K-Ar and fission-track dating methods. These two studies were followed by radi-

ometric age determinations of igneous activity and mineralization in both the Park City district east of the Cottonwood area (Bromfield and others, 1977) and in the Bingham mining district (Moore, 1973; Warnaars and others, 1978) to the west. In each area plutonism, coeval volcanism, and spatially associated mineralization took place within a period of roughly 5 to 15 m.y. in the latest Eocene to latest Oligocene time (40 to 25 m.y.). Lead-silver mineralization associated with sills of the Porphyry Knob-Lion Hill area on the western slope of the Oquirrh Mountains is another example of this middle Tertiary event.

Results of this study establish Eocene and Oligocene igneous centers in the Stansbury and northern Deep Creek Mountains. The ages of plutonic and volcanic rocks in these centers are clustered between 44 m.y. and 37 m.y. It is probable that the lavas and subvolcanic intrusions of the Cedar Range will also prove to be the upper parts of a middle Tertiary system. All but a small part of the metallic mineral production in the Tooele quadrangle and Wasatch Mountains is from deposits associated with this regional Eocene and Oligocene magmatic event. Base and precious metals dominate these deposits, and varied assemblages of copper, argentiferous lead, and zinc sulfides are characteristic of many mineralized areas. The igneous centers of the Uinta trend form a belt about 25 km wide and at least 275 km long; they define, as Tooker (1971) emphasizes, a major metallogenetic element in the eastern Great Basin.

Miocene. Integrated stratigraphic, structural, and radiometric data from many localities in the Great Basin show that extensional tectonics (Basin and Range block faulting) began at many places about 17 m.y. ago and continue to the present day. Like other large rift systems throughout the world, basalt and its differentiates were erupted along with the rifting. No one area in the Great Basin has a complete 17-m.y. sequence of lavas, and most areas contain flows spanning only a few million years. The lava flows are commonly a useful geologic indicator of the age of faulting in a given area (for example, Hamblin, 1970; Best and Hamblin, 1978). Olivine basalt flows in the northern Stansbury Mountains probably give a maximum age for Basin and Range faulting in this range. The nearly concordant ages for petrographically indistinguishable olivine basalt flows near the crest and in the western foothills indicate that major block faulting began no more than 12 to 13 m.y. ago.

Late Cenozoic lavas in the Great Basin, although dominantly basalt, locally include silicic lavas in bimodal associations. One example of silicic volcanism is the sequence of topaz-bearing rhyolites in the Thomas Range-Spor Mountain area immediately south of the Tooele 1° x 2° quadrangle. The small outcrop of alkali rhyolite lava at the southern tip of Granite Peak Mountain (Fig. 1) is probably an erosional outlier of the Spor Mountain eruptives, which range in age from 21 to 6 m.y. old (Lindsey, 1977). A Miocene age of about 13 m.y. is also reported for the alkali rhyolite porphyry dike (loc. 12, Fig. 1) that intrudes metamorphic crystalline rocks of Granite Peak Mountain (C. E. Hedge, oral communication, 1979).

Although the evidence is fragmentary, these data indicate that parts of the Tooele quadrangle, including Granite Peak Mountain, were affected by a Miocene silicic volcanic event. Steepened thermal gradients, resulting from emplacement of a pluton at a shallow depth, may have caused loss of argon resetting the ages of muscovite in the leucogranite and granite gneiss of Granite Peak Mountain. This age resetting might account for the otherwise enigmatic K-Ar ages of 14 m.y. and 30 m.y. in rocks that were previously correlated by Moore and Sorensen (1978) with lithologically similar parts of the Farmington Canyon Complex of Archaen and Proterozoic age. Alternatively, these data may reflect the thermal effects of a process involving anatectic remobilization of old sialic basement rocks (Best and others, 1974), whereby Granite Peak Mountain might represent the deeply eroded central part of a metamorphic core complex stripped of its metasedimentary carapace.

Occurrences of beryllium minerals, fluorite, and (or) uranium minerals are characteristic of Miocene rhyolites in western Utah (Shawe, 1966; Lindsey, 1977) and characteristic of bimodal rhyolite throughout the world; all of these minerals have been produced from rocks in the Spor Mountain area. In the Tooele quadrangle, fluorite has been produced at Wildcat Mountain (Thurston and others, 1954), bertrandite occurs in 8-m.y.-old (Whelan, 1970) quartz-calcite-adularia veins that cut the Jurassic pluton at Gold Hill (Griffitts, 1965), and beryl occurs in pegmatite dike swarms that intrude biotite granite gneiss on Granite Peak Mountain (Hanley and others, 1950). All these occurrences in western Utah, except for the Wildcat Mountain fluorite deposit, lie within the limits of the Deep Creek-Tintic mineral belt, as delineated by Hilpert and Roberts (1964). It appears that this belt and possible zone of crustal weakness, which transects the southern part of the Tooele quadrangle, had a recurring and long-lived role in localizing Phanerozoic plutons and eruptive rocks.

ACKNOWLEDGEMENTS

R. A. Armin, D. A. John, and M. L. Lanphere of the U.S. Geological Survey carefully reviewed the manuscript and offered useful suggestions for its improvement. W. J. Tafuri of Getty Oil Company kindly provided a sample of the granodiorite porphyry from the Porphyry Knob area in the Oquirrh Mountains for radiometric dating.

REFERENCES CITED

Armstrong, R. L., 1963, Geochronology and geology of the eastern Great Basin in Nevada and Utah [Ph.D. thesis]: New Haven, Conn., Yale University, 202 p.

——, 1970, Geochronology of Tertiary igneous rocks, eastern Basin and Range province, western Utah, eastern Nevada, and vicinity, U.S.A.: Geochimica et Cosmochimica Acta, v. 34, p. 203–232.

Armstrong, R. L., and Suppe, J., 1973, Potassium-argon geochronometry of Mesozoic igneous rocks in Nevada, Utah, and southern California: Geological Society of America Bulletin, v. 84, p. 1375–1392.

Best, M. G., Armstrong, R. L., Graustein, W. C., Embree, G. F., and Ahlborn, R. C., 1974, Mica granites of the Kern Mountains pluton, eastern White Pine County, Nevada: Remobilized basement of the Cordilleran miogeosyncline?: Geological Society of America Bulletin, v. 85, p. 1277–1286.

Best, M. G., and Hamblin, W. K., 1978, Origin of the northern Basin and Range Province: Implications from the geology of its eastern boundary: Geological Society of America Memoir 152, p. 313–340.

Best, M. G., McKee, E. H., and Damon, P. E., 1980, Space-time-composition patterns of late Cenozoic mafic volcanism, southwestern Utah and adjoining areas: American Journal of Science, v. 280, p. 1035–1050.

Bromfield, C. S., Erickson, A. J., Jr., Haddadin, M. A., and Mehnert, H. H., 1977, Potassium-argon ages of intrusion, extrusion, and associated ore deposits, Park City mining district, Utah: Economic Geology, v. 72, p. 837–848.

Burchfiel, B. C., 1980, Tectonics of noncollisional regimes—the modern Andes and the Mesozoic Cordilleran orogen of the western United States in continental tectonics, Studies in geophysics: Washington, D. C., National Academy of Sciences, ch. 6, p. 65–72.

Caroon, C., 1977, Petrochemistry and petrography of the Newfoundland stock, northwestern Utah [Ph.D. thesis]: Palo Alto, Calif., Stanford University, 241 p.

Crittenden, M. D., Jr., 1965, Geology of the Mount Aire quadrangle, Salt Lake County, Utah: U.S. Geological Survey Map Geologic Quadrangle Map GQ–379, scale 1:24,000.

Crittenden, M. D., Jr., Sharp, B. J., and Calkins, F. G., 1952, Geology of the Wasatch Mountains east of Salt Lake City, Parleys Canyon, and Traverse Range, in Guidebook to the geology of Utah: Utah Geological Society, no. 8, p. 1–37.

Crittenden, M. D., Jr., Stuckless, J. S., Kistler, R. N., and Stern, T. W., 1973, Radiometric dating of intrusive rocks in the Cottonwood area, Utah: U.S. Geological Survey Journal of Research, v. 1, p. 173–178.

Dalrymple, G. R., and Lanphere, M. A., 1969, Potassium-argon dating: San Francisco, W. H. Freeman, 258 p.

Erickson, A. J., Jr., 1976, The Uinta-Gold Hill trend: an economically important lineament, in Hogdson, R. R., ed., Proceedings of the First International Conference on the New Basement Tectonics: Utah Geological Association, p. 126–138.

Griffitts, W. R., 1965, Recently discovered beryllium deposits near Gold Hill, Utah: Economic Geology, v. 60, p. 1298–1305.

Hamblin, W. K., 1970, Structure of the western Grand Canyon region, in Hamblin, W. K., and Best, M. G., eds., Guidebook to the geology of Utah: Utah Geological Society, no. 23, 156 p.

Hanley, J. B., Heinrich, E. W., and Page, C. R., 1950, Pegmatite investigations in Colorado, Wyoming, and Utah: U.S. Geological Survey Professional Paper 227, 125 p.

Hilpert, L. S., and Roberts, R. J., 1964, Economic Geology, in Mineral and water resources of Utah: Washington, D.C., U.S. Congress, 88th, 2nd Session, Committee Print, p. 28–34.

Hintze, L. F., 1973, Geologic history of Utah: Brigham Young University, Geology Studies, v. 20, pt. 3, 181 p.

Ingamells, C. O., 1970, Lithium metaborate flux in silicate analysis: Analytica Chimica Acta, v. 52 (z), p. 323–334.

James, L. P., 1979, Geology, ore deposits and history of the Big Cottonwood mining district: Utah Geological and Mineralogical Survey Bulletin 114, 98 p.

Laughlin, A. W., Lovering, T. S., and Mauger, R. L., 1969, Age of some Tertiary igneous rocks from the East Tintic district, Utah: Economic Geology, v. 64, p. 915–918.

Lemmon, D. M., 1964, Tungsten, *in* Mineral and water resources of Utah: Washington, D. C., U.S. Congress, 88th, 2nd Session, Committee Print, p. 28–34.

Lindsey, D. A., 1977, Epithermal beryllium deposits in water-laid tuff, western Utah: Economic Geology, v. 72, p. 219–232.

McKee, E. H., Noble, D. C., and Silberman, M. L., 1970, Middle Miocene hiatus in volcanic activity in the Great Basin area of the western United States: Earth and Planetary Science Letters, v. 8, p. 93–96.

Moore, W. J., 1973, A summary of radiometric ages of igneous rocks in the Oquirrh Mountains, north-central Utah: Economic Geology, v. 68, p. 97–101.

Moore, W. J., and Sorensen, M. L., 1978, Metamorphic rocks of the Granite Peak area, Tooele County, Utah [abs.]: Geological Society of America Abstract with Programs, v. 10, p. 303.

——, 1979, Geologic map of the Tooele 1° x 2° quadrangle, Utah: U.S. Geological Survey Miscellaneous Investigation Series Map I–1132, scale 1:250,000.

Morris, H. T., and Lovering, T. S., 1961, Stratigraphy of the East Tintic Mountains, Utah: U.S. Geological Survey Professional Paper 361, 145 p.

Newberry, R. J., and Einaudi, M. T., 1981, Tectonic and geochemical setting of tungsten skarn mineralization in the Cordillera, *in* Dickinson, W. R., and Payne, W. D., eds., Relations of tectonics to ore deposits in the southern Cordillera: Arizona Geological Society Digest, no. 14, p. 99–111.

Nolan, T. B., 1935, The Gold Hill mining district, Utah: U.S. Geological Survey Professional Paper 177, 172 p.

Rowley, P. D., Lipman, P. W., Mehnert, H. H., Lindsey, D. A., and Anderson, J. J., 1978, Blue Ribbon lineament, an east-trending structural zone within the Pioche mineral belt of southwestern Utah and eastern Nevada: U.S. Geological Survey Journal of Research, v. 6, no. 2, p. 175–192.

Schaeffer, F. E., 1960, Igneous rocks of the central and southern Cedar Mountains, *in* Guidebook to the geology of Utah: Utah Geological Society, no. 15, 192 p.

Shawe, D. R., 1966, Arizona-New Mexico and Nevada-Utah beryllium belts: U.S. Geological Survey Professional Paper 550–C, p. C206–C213.

Stacey, J. S., and Zartman, R. E., 1978, A lead and strontium isotope study of igneous rocks and ores from the Gold Hill mining district, Utah: Utah Geology, v. 5, p. 1–15.

Steiger, R. H., and Jager, E., 1977, Convention on the use of decay constants in geo- and cosmochronology: Earth and Planetary Science Letters, v. 36, p. 359–362.

Thurston, W. R., Staatz, M. H., Cox, D. C., and others, 1954, Fluorspar deposits of Utah: U.S. Geological Survey Bulletin 1005, 53 p.

Tooker, E. W., 1971, Regional structural controls of the deposits, Bingham mining district, Utah, U.S.A.: Society of Mining Geologists, Japan Special Issue 3, p. 76–81.

Warnaars, F. W., Smith, W. H., Bray, R. E., Lanier, G., and Shafiqullah, M., 1978, Geochronology of igneous intrusions and porphyry copper mineralization at Bingham, Utah: Economic Geology, v. 73, no. 7, p. 1242–1249.

Whelan, J. A., 1970, Radioactive and isotopic age determinations of Utah rocks: Utah Geological and Mineralogical Survey Bulletin 81, 75 p.

MANUSCRIPT ACCEPTED BY THE SOCIETY AUGUST 20, 1982

Geological Society of America
Memoir 157
1983

Allochthonous quartzite sequence in the Albion Mountains, Idaho, and proposed Proterozoic Z and Cambrian correlatives in the Pilot Range, Utah and Nevada

David M. Miller

U.S. Geological Survey
345 Middlefield Road
Menlo Park, California 94025

ABSTRACT

A thick, allochthonous sequence of quartzite and schist exposed on Mount Harrison in the Albion Mountains, Idaho, is proposed to stratigraphically correlate with the upper part of the Proterozoic Z McCoy Creek Group and the Proterozoic Z and Lower Cambrian Prospect Mountain Quartzite in the Pilot Range, Utah and Nevada. The presence of these miogeoclinal strata in the Albion Mountains suggests that anomalous stratigraphic units elsewhere in the range are not of Cambrian age.

The correlation of units in the Albion Mountains and the Pilot Range is based on lithology, thickness, and sedimentary structures. Ambiguous low-angle fault relations in the Albion Mountains and moderate metamorphism of the section in that range preclude a positive correlation with units in the Pilot Range. Correlation of strata in the Albion Mountains with middle Proterozoic or early Paleozoic strata exposed in central Idaho is shown to be less probable.

The strata on Mount Harrison identified as Proterozoic Z and Lower Cambrian in this study are part of an overturned, structurally complex sequence of metasedimentary rocks that lie tectonically on overturned, metamorphosed Cambrian(?) shale and/or Mississippian shale and metamorphosed Ordovician carbonate strata; if the intervening rocks are Cambrian rather than Mississippian, these relations suggest that a largely miogeoclinal sequence (Proterozoic Z to Ordovician) similar to sequences in nearby mountains may once have been present near the Albion Mountains area. Elsewhere in the Albion Mountains and the adjoining Raft River and Grouse Creek Mountains, however, metamorphosed Ordovician carbonate rocks appear to stratigraphically overlie metamorphosed clastic rocks of uncertain age that differ from miogeoclinal rocks of the region. The section on Mount Harrison may be a relict of a more complete section that is tectonically thinned elsewhere in the metamorphic terrane, or it may be a far-traveled allochthonous slice of miogeocline juxtaposed with nonmiogeoclinal-facies strata of similar age.

INTRODUCTION

Proterozoic Z strata conformably underlie Cambrian strata throughout much of the Cordilleran miogeocline (Stewart and Poole, 1974). Largely as a result of Max Crittenden's painstaking studies of Proterozoic Z sequences, these thick clastic sequences have been well documented near Huntsville in northern Utah and in southeastern Idaho

(Crittenden and others, 1971). The broadly lithologically similar Proterozoic Z McCoy Creek Group of Misch and Hazzard (1962) occurs beneath Cambrian rocks in eastern Nevada (Misch and Hazzard, 1962; Woodward, 1963, 1965, 1967), but in no place is the base of the section exposed (Fig. 1). In contrast, metasedimentary rocks of uncertain

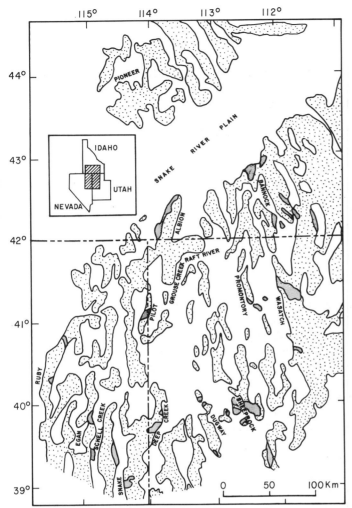

Figure 1. Sketch map of northeastern Great Basin area, showing mountain ranges largely composed of pre-Tertiary sedimentary rocks (stippling). Areas of Proterozoic Z rocks are shaded. Data from Stokes (1963), Hope and Coats (1976), Bond (1978), and Blick (1979).

age depositionally overlie Archean basement and structurally and/or depositionally underlie miogeoclinal Ordovician strata in the Albion, Raft River, and Grouse Creek Mountains of southern Idaho and northeastern Utah (Fig. 1). These strata of uncertain age constitute an exceptionally thin clastic sequence that differs lithologically from typical miogeoclinal rocks exposed elsewhere in southern Idaho, northern Utah, and northeastern Nevada. Either the common assemblage of thick Proterozoic Z strata has been tectonically removed in the Raft River, Grouse Creek, and Albion Mountains, or pronounced facies changes render the assemblage unidentifiable (Compton and Todd, 1979; Crittenden, 1979). Another group of metamorphosed strata, allochthonous with respect to the thin clastic sequences and also of uncertain age, occurs along the western flank of the Albion Mountains. This al-

lochthonous sequence contains thicker units of clastic rocks than in the autochthon.

Based on recent mapping in the Pilot Range, Nevada and Utah (Miller and Lush, 1981; Miller and others, 1982), and in the northern Albion Mountains (Miller, 1980) (Fig. 1), I consider the upper part of the McCoy Creek Group and overlying strata exposed in the Pilot Range correlative with lithologically similar allochthonous strata in the northwestern Albion Mountains. Details in support of this stratigraphic correlation are given in this report, along with some of the tectonic implications.

REGIONAL SETTING

The northeastern Great Basin (Fig. 1) is characterized by generally north-trending, fault-controlled ranges separated by basins that are filled with Cenozoic deposits across which the correlation of some stratigraphic units and structural elements is difficult. These basin-and-range structures are superposed on Mesozoic and Tertiary high- and low-angle faults of local and regional extent. Despite the resulting structural complexity, stratigraphic correlations of Paleozoic and younger strata have been carried out confidently over much of the region. Structurally complex areas, many of which contain metamorphosed strata, still pose problems despite extensive study in recent years. Some of these problems exist because correlations with unmetamorphosed strata are difficult, as in the metamorphic terrane in the Albion, Raft River, and Grouse Creek Mountains region (Fig. 1), where anomalous strata of largely undetermined age occur. Ranges east and west of this terrane expose only upper Paleozoic strata, and north of the terrane, sedimentary rocks are covered by volcanic rocks of the Snake River Plain. South of the Albion-Raft River-Grouse Creek metamorphic terrane, the Pilot Range contains a relatively well known and extensive Proterozoic Z and Paleozoic stratigraphic section that is slightly to moderately metamorphosed; it is, therefore, a logical area in which to look for units possibly correlative with the highly metamorphosed and deformed strata in the adjacent metamorphic terrane.

Although the Albion-Raft River-Grouse Creek metamorphic terrane is characterized by stratigraphic attenuation due to low-angle faulting and plastic flow, the work of Armstrong (1968), Compton (1972, 1975), Todd (1975, 1980), Compton and others (1977), Armstrong and others (1978), and Miller (1980) has established a laterally consistent sequence of thin metasedimentary units throughout most of this terrane. This sequence, here termed the "Raft River Mountains sequence" from Compton's studies (1972, 1975) in the Raft River Mountains, is composed of two parts. The lower part encompasses six units of uncertain age (Compton and others, 1977; Compton and Todd, 1979; Crittenden, 1979), including the Precambrian(?) Elba

Quartzite, schist of the Upper Narrows, quartzite of Yost, and schist of Stevens Spring, and the Cambrian(?) quartzite of Clarks Basin and schist of Mahogany Peaks. These lower units are overlain by strata, also assigned to the Raft River Mountains sequence, representing the metamorphosed Ordovician Pogonip Group, Eureka Quartzite, and Fish Haven Dolomite, and the Mississippian Chainman-Diamond Peak Formations, undivided. The Raft River Mountains sequence as described in this paper rests unconformably on Archean (Precambrian W) granite, gneiss, and schist and is in many places tectonically overlain by upper Paleozoic sedimentary and metamorphic rocks. Major low-angle faults, common within the Raft River Mountains sequence, contribute to the difficulty of assigning ages to the lower units. Southward in the Grouse Creek Mountains, thin tectonic slices of metamorphosed Devonian strata appear within the sequence in their proper stratigraphic position above Ordovician units and below Mississippian rocks (Miller and others, 1980).

Strata underlying a major part of the Albion Mountains differ from the relatively simple sequence of units found elsewhere in the metamorphic terrane. Along the west side of the mountain range, including Middle Mountain, the west flank of Mount Harrison, and the northward continuation of the range west and north of the town of Albion (Fig. 2), is a subterrane composed largely of metaquartzite, minor schist and marble, and varying amounts of layered gneissic granite. This subterrane, here termed the "quartzite assemblage," is of uncertain age and appears to be a high-level composite tectonic sheet that lies above all rocks except the sheet of Pennsylvanian strata in the northern part of the range; the field relations, however, are unclear in many exposures. Part of the quartzite assemblage is well exposed on Mount Harrison, where an overturned sequence of quartzite and schist identified by Armstrong (1968, 1970) is here termed the "Mount Harrison sequence."

The Mount Harrison sequence tectonically underlies another part of the quartzite assemblage that lies along the west side of the Albion Mountains in the Mount Harrison area; this higher part of the assemblage is here termed the "sequence of Robinson Creek" (Fig. 3). Although the sequence of Robinson Creek is lithologically similar to the Mount Harrison sequence, I interpret that the two sequences are separated by a low-angle fault and, therefore, are not definitely correlative. R. L. Armstrong (unpublished data, 1977) considered the Mount Harrison sequence and the sequence of Robinson Creek to be partial stratigraphic equivalents. The Mount Harrison sequence is faulted above overturned, metamorphosed Paleozoic strata belonging to the upper part of the Raft River Mountains sequence (Miller, 1980).

The proposed correlation of the Mount Harrison sequence with Precambrian and Cambrian rocks of the mi-

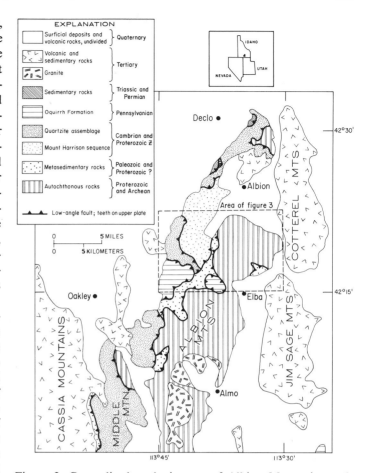

Figure 2. Generalized geologic map of Albion Mountains and vicinity, modified from Armstrong and others (1978) and D. M. Miller (unpublished data, 1979). Autochthon consists of the Green Creek Complex (Archean), and the Elba Quartzite and schist of the Upper Narrows (Proterozoic X or Z). The Proterozoic(?) and Paleozoic metasedimentary rocks unit consists of the quartzite of Yost, schist of Stevens Spring, quartzite of Clarks Basin, schist of Mahogany Peaks, the Pogonip Group, the Eureka Quartzite, and the Chainman-Diamond Peak Formations, undivided. Mount Harrison sequence consists of the Dayley Creek Quartzite of Armstrong (1968), schist of Willow Creek, and the Harrison Summit Quartzite of Armstrong (1968). Quartzite sequence consists of all rocks in quartzite assemblage as defined in text, exclusive of units in Mount Harrison sequence.

ogeocline, coupled with the tectonic setting of the Mount Harrison sequence, has important implications for the age of the lower part of the Raft River Mountains sequence. Armstrong (1968, 1970) and Compton and Todd (1979) had already noted that strata within the Raft River Mountains sequence, which differ dramatically from those of the Mount Harrison sequence, may be Cambrian. If the correlation proposed herein is correct, two different sequences of Cambrian age—juxtaposed by low-angle faults—are present, or earlier suggestions that part of the Raft River sequence is Cambrian are in error.

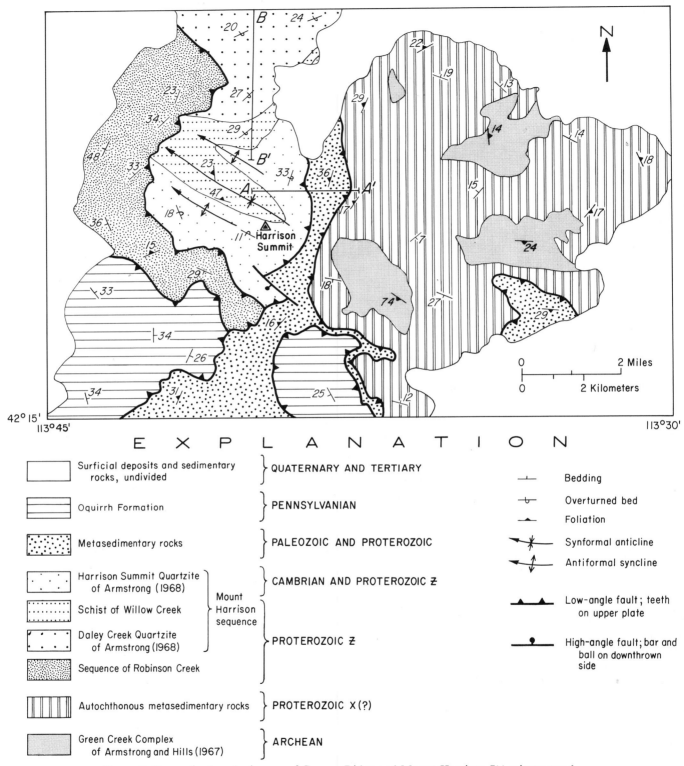

E X P L A N A T I O N

Surficial deposits and sedimentary rocks, undivided	} QUATERNARY AND TERTIARY	
Oquirrh Formation	} PENNSYLVANIAN	
Metasedimentary rocks	} PALEOZOIC AND PROTEROZOIC	
Harrison Summit Quartzite of Armstrong (1968)	} CAMBRIAN AND PROTEROZOIC Ƶ	Mount Harrison sequence
Schist of Willow Creek		
Daley Creek Quartzite of Armstrong (1968)	} PROTEROZOIC Ƶ	
Sequence of Robinson Creek		
Autochthonous metasedimentary rocks	} PROTEROZOIC X (?)	
Green Creek Complex of Armstrong and Hills (1967)	} ARCHEAN	

Bedding ⊥
Overturned bed
Foliation
Synformal anticline
Antiformal syncline
Low-angle fault; teeth on upper plate
High-angle fault; bar and ball on downthrown side

Figure 3. Generalized geologic map of Connor Ridge and Mount Harrison 7½-minute quadrangles (Miller, 1980; unpublished data, 1979), showing locations of cross sections *A-A'* and *B-B'* in Figure 4. The Proterozoic and Paleozoic metasedimentary rocks unit is upper part of the Raft River Mountains sequence, and the autochthonous metasedimentary rocks unit is lower part of the sequence. Sequences of Robinson Creek and Mount Harrison comprise the quartzite assemblage.

MOUNT HARRISON SEQUENCE

Although several premetamorphic low-angle faults occur in the Raft River Mountains sequence on Mount Harrison, their number and location are commonly obscure, so that tectonic relations among the metasedimentary units within the sequence are still uncertain. However, on Mount Harrison, inversion of metamorphosed Ordovician(?) strata in the upper part of the Raft River Mountains sequence (Miller, 1980) suggests that a major recumbent fold is present within the Raft River Mountains sequence. Overlying the inverted metamorphosed Ordovician units of the Raft River Mountains sequence, and separated from them by a low-angle fault zone containing broken dark schist and graphitic phyllite, is a sequence of two quartzites and an intervening schist that is also overturned, on the basis of cross beds (Armstrong, 1970; Miller, 1980). Despite the folds and the possibility of hidden faults, I consider the Mount Harrison sequence to be stratigraphically intact because all well-exposed contacts are conformable and thickness and facies variations are minimal. It should be noted that lenticular outcrop patterns of thin quartzites enclosed by schist are present; they may have either a tectonic or a primary depositional origin. In fact, R. L. Armstrong (unpublished data, 1977) interpreted local elimination of one of these lenticular quartzites as being due to low-angle faulting, whereas I consider its elimination to be a large-scale boudinage feature simply accentuating primary channel forms.

The Mount Harrison sequence is structurally truncated by the overlying sequence of Robinson Creek (Fig. 3), which, on Mount Harrison, consists largely of quartzite and schist that generally resembles the rock types of the Mount Harrison sequence. Lesser amounts of clean to schistose calcitic marble and rare dolomitic marble, also present in the sequence of Robinson Creek, do not occur in the Mount Harrison sequence.

Stratigraphic Description

The Mount Harrison sequence consists of three distinctive units: the Dayley Creek Quartzite of Armstrong (1968), the schist of Willow Creek, and the Harrison Summit Quartzite of Armstrong (1968) (Figs. 3 and 4). Armstrong (1968) originally defined the two quartzites as well as an intervening unit named the "Land Creek Formation"; part of his Land Creek Formation is now included in the tectonically separate sequence of Robinson Creek that overlies the Mount Harrison sequence. Consequently, I herein refer to the part of his Land Creek Formation in the Mount Harrison sequence as the "schist of Willow Creek" to avoid confusion (Fig. 5).

Dayley Creek Quartzite of Armstrong (1968). The Dayley Creek Quartzite of Armstrong (1968) is best exposed along the north-trending ridge north of Mount Harrison, in sections 29 and 32, T. 13 S., R. 24 E.; here the unit is entirely overturned. The base of the section is cut by a high-angle fault, below which is more schist and quartzite, subsequently considered by R. L. Armstrong (unpublished data, 1977) as also belonging to the Dayley Creek. The section described here includes (1) a lower part composed of an approximately 800 to 1200-m-thick sequence of well-bedded and crosslaminated micaceous quartzite that crops out north of the area of Figure 3, and (2) a stratigraphically overlying, approximately 1,435-m-thick sequence of similar crossbedded quartzite that includes one schist unit. Thicknesses were estimated from structure sections.

The lowest part of the Dayley Creek consists of about 1100 to 1500 m of gray-weathering, light-gray to white, micaceous quartzite, with thin interbeds of biotite and white-mica schist. The rocks are fully recrystallized, except for granular and larger quartz clasts, and are generally medium to coarse grained. Bedding is pronounced and ranges from thin to very thick but is generally medium to thick. Crosslaminations occur in planar sets and are ubiquitous; they are generally tabular, wedge shaped, or form gentle troughs and lie at a low angle to bedding. Granular quartz is common on the bottoms of troughs; however, only one conglomerate bed, about 1 m thick, was observed.

A 50-m-thick mica schist unit abruptly overlies the lowest-occurring quartzite. The schist is mainly composed of biotite, white mica, feldspar, and quartz but also contains some garnet and large crystals of staurolite; it weathers brown to tan and is tan on fresh surfaces.

Gradationally overlying the mica schist unit is schistose quartzite that contains abundant biotite and white-mica-schist partings, as well as scattered coarse and fine mica crystals. Granular quartz beds lie 10 m above the contact with the mica schist unit; highly micaceous dark-gray quartzite occurs for approximately 100 m above the contact. The dark-gray quartzite contains green, fuchsite(?)-bearing horizons; it grades upward into gray quartzite similar to that underlying the mica schist unit.

Approximately 350 m above the mica schist unit, the gray quartzite rapidly grades into clean white thin- to medium-bedded crosslaminated quartzite. This white quartzite unit, about 385 m thick, contains rare granular layers at the base of troughs and no schist partings; it is overlain by about 300 m of darker gray quartzite that contains about 5 percent biotite and white mica and some iron oxide minerals. The white quartzite resembles the lower white micaceous quartzite in most respects. In its uppermost 50 meters, the darker quartzite grades with increasing mica content and decreasing bed thickness into schistose quartzite. Feldspar fragments are an important constituent in many beds near the top of the Dayley Creek. Thin interbeds of tan garnet schist are common near the top of the

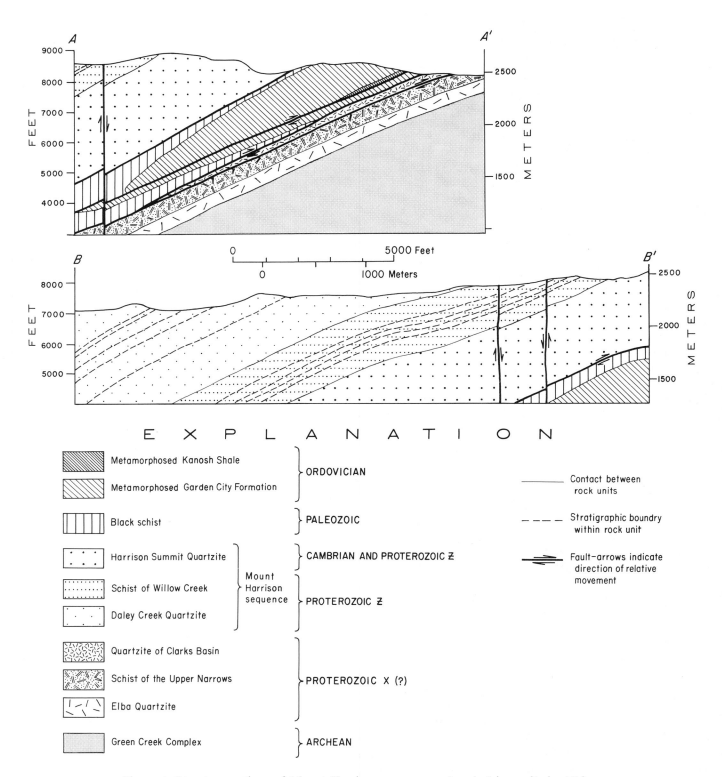

EXPLANATION

Metamorphosed Kanosh Shale	⎱ ORDOVICIAN
Metamorphosed Garden City Formation	
Black schist	⎱ PALEOZOIC
Harrison Summit Quartzite	⎱ CAMBRIAN AND PROTEROZOIC Z
Schist of Willow Creek	Mount Harrison sequence ⎱ PROTEROZOIC Z
Daley Creek Quartzite	
Quartzite of Clarks Basin	
Schist of the Upper Narrows	⎱ PROTEROZOIC X (?)
Elba Quartzite	
Green Creek Complex	⎱ ARCHEAN

——————— Contact between rock units

– – – – – Stratigraphic boundry within rock unit

⇒ Fault–arrows indicate direction of relative movement

Figure 4. Structure sections of Mount Harrison sequence and underlying units in Albion Mountains, showing inferred tectonic relations. Recumbently folded Ordovician strata are enclosed by the black schist unit, which is a composite of strata possibly of Cambrian and Mississippian age. Units are described more completely in text. The autochthonous metasedimentary rocks unit of Figure 3 consists of the Elba Quartzite and schist of the Upper Narrows; the metasedimentary rocks unit of Figure 3 consists of the metamorphosed Kanosh Shale, the metamorphosed Garden City Formation, the black schist unit, and quartzite of Clarks Basin.

Northern Albion Mountains			Pilot Range
Armstrong (1968,1970)	This report		This report
Conner Creek Formation	Black schist	Paleozoic / Cambrian / ? / Proterozoic Z	Phyllite of Killian Springs
LOW-ANGLE FAULT	*LOW-ANGLE FAULT*		*MINOR LOW-ANGLE FAULT*
Harrison Summit Quartzite	Harrison Summit Quartzite		Prospect Mountain Quartzite of Misch and Hazzard (1962)
Land Creek Formation	Schist of Willow Creek: Upper schist member / Middle quartzite member / Middle schist member / Lower quartzite member / Lower schist member		McCoy Creek Group: Unit G
Dayley Creek Quartzite	Dayley Creek Quartzite		Unit F
(faulted base)	*(faulted base)*		Unit E

Figure 5. Correlation chart showing nomenclature for Mount Harrison sequence and correlation with rock units in Pilot Range proposed in this report. Structural and stratigraphic interpretations for Mount Harrison sequence of R. L. Armstrong (written communication, 1980) and D. M. Miller (this report) are indicated. Armstrong (1968) considered the sequence of quartzite and schist on Mount Harrison to young from the Conner Creek Formation of Armstrong and Hills (1967) to his Dayley Creek Quartzite but later reversed the younging direction on the basis of overturned crossbeds in much of the sequence (Armstrong, 1970). His present interpretation (R. L. Armstrong, written communication, 1980) makes no assumptions about relative ages of the Harrison Summit Quartzite and Dayley Creek Quartzite.

unit, which is drawn at the base of the first schist bed thicker than 4 m.

Schist of Willow Creek. Armstrong (1968) defined the Land Creek Formation as a dominantly schistose unit occurring between two thick quartzite units, the Dayley Creek and Harrison Summit Quartzites of Armstrong (1968). He incorrectly considered the schist, quartzite, and marble exposed south of Mount Harrison to be correlative with the schist and quartzite exposed north of the mountain (the sequence described in this report) and thus included several rock types in the Land Creek Formation that do not occur in the sequence described here. To avoid confusion with this earlier terminology, I here refer to the rocks within the Mount Harrison sequence termed by Armstrong (1968) the Land Creek Formation as the "schist of Willow Creek." This relation is illustrated in Figure 5 of this report.

The schist of Willow Creek crops out between essentially continuous quartzite beds of the Dayley Creek and Harrison Summit Quartzites. Much of the unit is well exposed along the ridge north of Mount Harrison (Fig. 3) in sec. 32, T. 12 S., R. 24 E. although the uppermost member is best exposed in the upper reaches of Willow Creek (sec. 5, T. 13 S., R. 24 E.), for which the unit is named. The unit

is composed of five members of alternating schist and quartzite, here termed the lower schist and lower quartzite, the middle schist and middle quartzite, and upper schist members. These members are broadly lenticular, and some are locally missing. The entire unit is approximately 510 m thick, as determined from structure sections (Fig. 4).

The lower schist member, about 105 m thick, consists of garnet- and staurolite-bearing tan pelitic and quartzose schist and interbedded lenses of light-gray, coarse-grained to granular, micaceous quartzite. Schistose quartzite beds near the base, as much as 3 m thick, are coarse grained to conglomeratic. The top of the member is placed at the sharp contact with the overlying, thick-bedded, lower quartzite member.

The lower quartzite member consists of about 60 m of impure, thick-bedded quartzite. The basal part is flaggy micaceous and feldspathic gray quartzite, ranging in grain size from coarse sand to granular. The remainder of the member is heterogeneous and consists of thin beds of brown quartzose garnet schist, feldspathic gray quartzite, clean white quartzite, and distinctive boulder conglomerate containing ripup clasts of schist and boulders of quartzite and rare granite (Fig. 6A); muscovite becomes increasingly common near the top. The top of the member is drawn at the base of the first thick schist bed of the overlying middle schist member.

The middle schist member, about 130 m thick, consists of tan to brown homogeneous garnetiferous and quartzose schist. A few thin beds of tan quartzite occur near the base; staurolite is locally present in the schist. The top of the member is drawn at the base of the first thick quartzite bed of the overlying middle quartzite member.

The middle quartzite member, approximately 90 m thick, has interbedded contacts with the underlying and overlying schist members. Interbedded tan quartzose schist and gray coarse-grained to pebbly quartzite occur within the basal part. The central part of the member consists of gray coarse-grained to pebbly micaceous poorly bedded quartzite; thin interbeds of calcareous schist occur in the upper part. The top of the member is drawn at the top of the uppermost quartzite of the sequence.

The upper schist member, about 125 m thick, consists of a lower part of pelitic schist, a distinctive middle part of interlayered marble and schist, and an upper part of quartzose schist. The lower part is tan to brown pelitic to quartzose schist, that is thin bedded (1 to 8 cm) and laminated; regular color alternations caused by varying quartz content give a banded appearance similar to a rhythmite. The middle part is medium- to dark-gray and brown, very thinly interbedded calcite marble and pelitic schist (Fig. 6B). The upper part, 10 to 30 m thick, consists of dark-gray to dark-brown spotted calcareous schist, quartzose schist, and schistose quartzite. Quartz content increases progressively upward within the top 20 m. The top of the member is

Figure 6. Characteristic rock types in schist of Willow Creek. A: Ripup clasts in coarse quartzite matrix from lower quartzite member. B: Thinly interbedded marble and phyllite with pronounced crosscutting foliation from upper schist member. Pencil is 15 cm long.

drawn at the base of the first white thick-bedded quartzite layer of the Harrison Summit Quartzite.

Harrison Summit Quartzite of Armstrong (1968). Armstrong (1968) named the Harrison Summit Quartzite from its exposure on Mount Harrison. The entire unit is overturned (Armstrong, 1970; Miller, 1980). The best exposure of the upper part of the unit is in the cliffs west of Lake Cleveland in sec. 4, T. 13 S., R. 24 E., where the unit is estimated to be about 740 m thick. The stratigraphic top of the unit is poorly exposed in a low-angle fault zone about 3 km south of the summit region, and contact relations between the Harrison Summit and the stratigraphically lower schist of Willow Creek are best exposed in the upper part of Willow Creek.

The Harrison Summit Quartzite is thick bedded and prominently crosslaminated throughout. The crossbeds are generally tabular or wedge shaped and form planar bedding units; festoons are uncommon. The Harrison Summit is white to light gray or creamy white, generally medium grained, with a few granular layers at the bases of beds and rare conglomerate beds containing pebbles less than 2 cm in diameter. Thin, 1 to 5 cm-thick, dark-brown schist interbeds commonly occur near the top of the unit, as well as a few ripup clasts of similar lithology. Feldspathic interbeds are uncommon. Large cubes of hematite occur locally. The top of the unit is not exposed near Lake Cleveland and is only poorly exposed south of Mount Harrison, where the unit is juxtaposed structurally with underlying schist and phyllite by a low-angle fault.

Overturned Rock Units that Structurally Underlie the Mount Harrison Sequence. Underlying the faulted structural base of the Harrison Summit Quartzite is a fault-bounded unit of dark schist, graphitic phyllite, and quartzite (black schist unit, Fig. 4) that locally is highly jumbled and broken. The black schist unit contains rock types similar to both the schist of Mahogany Peaks and the Chainman Shale-Diamond Peak Formation, undivided (Miller, 1980). The rock types in the schist unit are also generally similar to dark graphitic phyllite and schist of probable Early and Middle Cambrian age in the Pilot Range, although fossils collected by Armstrong (1968) from a distinctive conglomerate in the black schist unit several kilometers south of Mount Harrison indicate that part of the unit there is post-Middle Cambrian. Between the exposures of overturned metamorphosed Ordovician rocks and overturned Harrison Summit Quartzite, the black schist unit contains none of this distinctive conglomerate and, therefore, may represent metamorphosed Cambrian rocks similar to those in the Pilot Range. The base of the black schist unit is bounded by a low-angle fault, below which overturned metamorphosed Ordovician units thought to represent the Garden City Formation and the Kanosh Shale (of the Pogonip Group) successively occur (Miller, 1980).

PILOT RANGE STRATA

Slightly to moderately metamorphosed Proterozoic Z and Cambrian strata occur beneath a gently arched decollement in the Pilot Range. This decollement, herein named the "Pilot Peak decollement" for its exposures south of Pilot Peak, is a dominant structure in the range that separates metamorphosed rocks below from unmetamorphosed rocks above; part of the Cambrian section is generally removed by the decollment but in places there may be duplication of unlike Upper Cambrian strata across the structure. The metamorphosed strata include three major rock units (Fig. 7): the Proterozoic Z McCoy Creek Group of Misch and Hazzard (1962); the Proterozoic Z and Lower Cambrian Prospect Mountain Quartzite of Misch and Hazard (1962); and previously unnamed overlying phyllite, marble, and limestone of probable Cambrian age. Strata above the decollement range in age from Upper Cambrian to Permian and are unmetamorphosed (Blue, 1960; O'Neill, 1968; D. M. Miller, unpub. data, 1981; Miller and Lush, 1981; Miller and others, 1982).

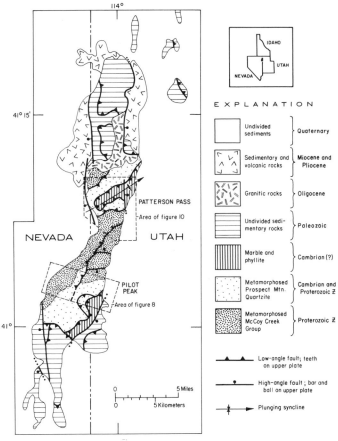

Figure 7. Geologic sketch map of Pilot Range. Geology of central part of range after Miller and Lush (1981) and Miller and others (1982), and of northern and southern parts after Blue (1960), Stokes (1963), O'Neill (1968), Hope and Coates (1976), and D. M. Miller (unpublished data, 1981).

A window in the Pilot Peak decollement exposes the metamorphosed McCoy Creek Group near its center; the metamorphosed Prospect Mountain Quartzite, overlain by Cambrian phyllite and carbonate rock, occurs near the south and north limits of the window. Two composite stratigraphic sections, ranging from the upper part of the McCoy Creek Group to Cambrian phyllite, are here described, one from each end of the window because marked facies changes occur in some units. Although the southern exposures, near Pilot Peak, contain covered areas and are in part low-angle faulted, the structure is relatively simple in many places so that the resulting composite stratigraphic section is probably accurate. The northern exposures of the Prospect Mountain Quartzite and adjacent strata, near Patterson Pass, are complicated by locally complex low-angle faults, major tight folds, and rapid facies changes; thus, the composite section for that area can be treated with less confidence.

Pilot Peak Area

The Pilot Peak area (Fig. 8) is underlain by a gently to steeply east-dipping sequence of Proterozoic Z and Cambrian strata. A low-angle fault lies near the top of the Prospect Mountain Quartzite or within metamorphosed Cambrian phyllite in most localities. Low-angle faults with separations of less than 400 m are mapped within unit G of the McCoy Creek Group where marker beds are present. Similar low-angle faults possibly occur but are largely unrecognized in the more lithologically homogeneous parts of the section. In addition, minor low-angle faults are recognized near the base of the Prospect Mountain Quartzite in other parts of the Pilot Range; an analogous fault in the Pilot Peak area cannot be ruled out because exposures are meager. The quartzite units generally are broadly folded, and tight folds commonly occur in rocks adjacent to low-angle faults. The schists in unit G, particularly those poor in quartz or calcite, are generally moderately to tightly folded. The rocks were metamorphosed to the middle greenschist facies and retrogressed to the chlorite zone.

Woodward's (1967) correlation of units underlying the Prospect Mountain Quartzite with the McCoy Creek Group is followed here. The stratigraphic section described by Woodward (1967) was measured in the same general area as that of this report; however, because his section included complex structure, covered areas, and areas of exceedingly poor exposure, I have remeasured the section in several localities. Section thicknesses were determined from structure sections because the deformation of the rocks renders any exact thickness of little value.

McCoy Creek Group, Units A(?), B(?), C, D, and E. In the Pilot Range, heterogeneous quartzite overlain by pelitic and amphibole schist is provisionally assigned to unit A, overlying marble is provisionally assigned to unit B, tan metasiltstone is assigned to unit C (~ 70 m), massive pebbly and granular quartzite is assigned to unit D (~ 215 m), and brown quartzose schist and metasiltstone are assigned to unit E (~ 150 m). Units A(?) and B(?) are structurally juxtaposed with unit F and cannot be correlated with certainty with the lower units of the McCoy Creek Group in the Schell Creek Range (O'Neill, 1968). These units are not considered in the regional correlation described here and are not discussed further in this report.

McCoy Creek Group, Unit F. A 300-m-thick quartzite was designated "unit F" by Woodward (1967). Near Pilot Peak (sec. 4, Fig. 8), unit F overlies platy brown schist of unit E with sharp contact. The lower part of the quartzite is gray- to brown-weathering (light gray to white on fresh surfaces), poorly sorted medium-sand-size to granular quartzite, and is generally moderately to poorly bedded in the lower 40 m. The central part is uniformly cross-laminated and contains tabular, wedge, and trough shapes common in sets that form generally medium to thick planar beds. There is a general trend from high-angle tabular cross sets to festoons upward in the quartzite. Pebbles, less than 2 cm in diameter, commonly form layers in channels at the bases of the cross sets; these pebbles are well rounded and consist of white vein quartz and quartzite. Clasts in conglomerate beds are matrix supported, and rare microcline crystals, as large as 2 cm in diameter, occur with quartz pebbles. Feldspar is a minor constituent of most of the fine-grained beds. Higher in the central part of the section, some zones are slightly micaceous, and quartzite beds thicker than 1 m are locally present. The upper 80 m contain a few beds of dark mica schist, as thick as 15 cm, and dark fine-grained quartzite beds, 10 to 20 cm thick. The upper part typically is less distinctly bedded than the central part, and crosslaminations are less pronounced. The top 30 m of the unit vary laterally and contain broad channels filled with pebbles. In a few localities, the top 20 m consist of thick beds of micaceous quartzite, with thin interbeds of dark-gray slate and slaty quartzite. Typically, a conglomeratic subunit, as thick as 30 m, is present at the top; this subunit consists of massive to very thick-bedded, matrix-supported conglomerate that is poorly sorted. Clasts are 1 to 6 cm in diameter and consist dominantly of white vein quartz; in one locality, red jasper or chert and purple quartzite clasts were noted. In most areas, ripup wedges of phyllite also are common (Fig. 9A). The top of unit F is drawn at the base of the first thick phyllite bed.

McCoy Creek Group, Unit G. Unit G, approximately 480 m thick as determined from section 1 (Fig. 8), can be divided into a lower part of quartzose phyllite, a middle part of interbedded phyllite and calcitic marble, and an upper part of phyllite, quartzose phyllite, and phyllitic quartzite.

The lower part is typically brown, laminated quartzose phyllite that contains lenses or interbeds of brown con-

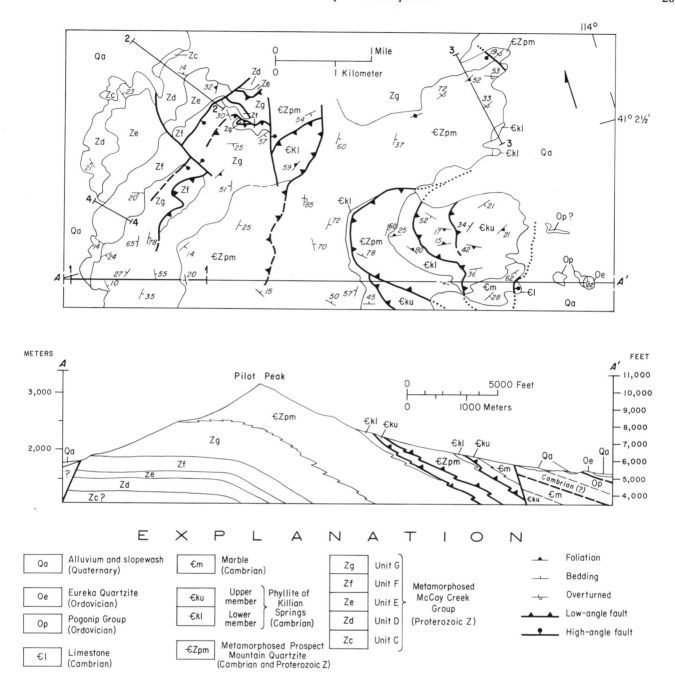

Figure 8. Generalized geologic map and cross section of Pilot Peak area, Pilot Range (see Fig. 7 for location). Measured sections are numbered as described in text.

glomerate similar to that in the top of unit F; and slate gray, micaceous metasiltstone that is in part rhythmically bedded and contains interbeds of phyllite, quartzose phyllite, quartzite, and conglomerate. Amphibole schist is present locally. In exposures of unit G near section 1 (Fig. 8), as many as three 10- to 20-m-thick conglomerate layers are present in a zone beginning about 80 m above the base of the unit. The conglomerate, which crops out more than

4 km along strike, is poorly sorted, medium sand size to pebbly, and white. The matrix consists of biotite, feldspar, white mica, and quartz; the clasts are generally of quartzite although rounded and wedge-shaped ripup clasts of phyllite are common. Above the conglomerate is light- to medium-gray fine-grained quartzite separated by 10 to 15 m of green or gray slate and phyllite containing quartzite and calcareous pods and stringers that may represent len-

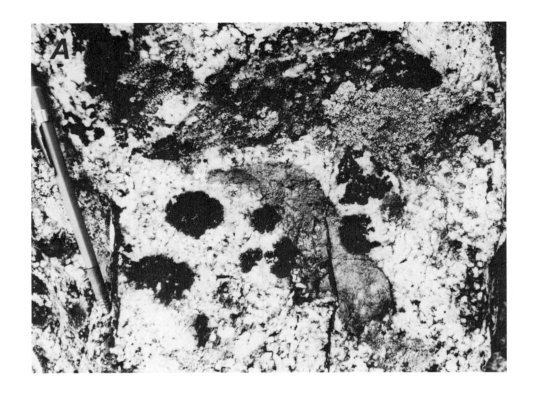

Figure 9. Distinctive lithologies in the upper part of the McCoy Creek Group of Misch and Hazzard (1962). A: Phyllite ripup wedges in pebbly quartzite at top of unit F. Pencil is 15 cm long. B: Interbedded marble and phyllite in central part of Unit G.

ticular or flaser bedding. Several kilometers north of the Pilot Peak area, interbedded conglomerate and phyllite make up the lower 100 m of unit G; the lower part there is informally designated the "conglomerate subunit."

The middle part, about 60 m thick, is characterized by interbedded calcitic marble, calcareous phyllite, and phyllite or metasiltstone in layers 1 to 7 cm thick (Fig. 9B). The marble is white, tan, dark gray, or blue-gray, and slightly micaceous. Intervening beds are dark gray, green, or brown metasiltstone and phyllite, and brown, gray, or tan calcareous phyllite and micaceous marble. This middle part of unit G weathers more resistantly than the remainder of the unit and generally stands out on the gentle slopes formed by unit G.

Overlying the middle part of unit G is about 200 m of medium-gray, thin-bedded, fine-grained quartzite and metasiltstone containing biotite, feldspar, and quartz. This upper part of the unit grades upward into thin- and medium-bedded, medium-gray, fine- to medium-grained quartzite that weathers brown. The thin bedding and color variations between beds gradually disappear upward as the dominant rock type becomes uniformly medium bedded, gray, faintly crossbedded quartzite. This quartzite changes abruptly to light-gray, thick-bedded, prominently cross-laminated, coarse-grained quartzite of the Prospect Mountain Quartzite.

Prospect Mountain Quartzite of Misch and Hazzard (1962). Approximately 955 m of the Prospect Mountain Quartzite was determined from section 3 (Fig. 8). The unit is conspicuously bedded, crosslaminated, and remarkably homogeneous. The lower 70 to 100 m of the Prospect Mountain Quartzite is white fine-grained porcelaneous quartzite, overlain by 5 to 15 m of poorly sorted bedded conglomerate and grit containing abundant microcline clasts. The overlying 700 m or so is all thick-bedded, crosslaminated, white to light-gray quartzite. Grain size ranges from medium sand to pebble (2 cm in diameter), but the rock is generally coarse sand size. Pebbly and granular beds are uncommon, and some rare distinctive units form conglomeratic intervals, several meters thick, that are feldspathic. Generally, the granules occur on the bottoms of cross sets and at the bases of graded beds. These cross sets are commonly tabular or wedge shaped although festoons are noted in several localities. The quartzite ranges from clean in about 40 percent of the section to slightly or moderately micaceous elsewhere; feldspar is not abundant except in rare beds, and metamorphic hornblende is rare. The uppermost 70 to 80 m of the unit commonly contains white mica on parting surfaces, 1- to 10-cm-thick beds of metagraywacke or biotite-white mica feldspar quartz schist, and rare medium to thick beds of dark-rusty-brown, iron-rich, micaceous quartzite interlayered with the generally light-gray thick-bedded quartzite. The topmost 10 m of the Prospect Mountain consist of white thick-bedded quartzite,

with medium-thick interbeds of dark-brown to blue-gray quartzite that is overlain by medium- and dark-gray, medium-bedded, crosslaminated, micaceous quartzite near the sharp contact with the overlying metamorphosed phyllites of Cambrian age.

Metamorphosed Cambrian Rocks. The metamorphosed clastic rocks overlying the Prospect Mountain Quartzite in the Pilot Range are here informally named the "phyllite of Killian Springs" to emphasize the lithologic differences between these dark graphitic rocks and the typical greenish-tan shale and siltstone of this interval (the Pioche Formation as defined by Hintze and Robison, 1975) found in many mountain ranges of the central and eastern Great Basin. The phyllite is named for excellent exposures south of Patterson Pass at Killian Springs (Sec. 16, T. 5 N., R. 19 W.). In the Pilot Peak area, the more metamorphosed equivalent of the Killian Springs is dark-brown quartzose graphitic phyllite or schist; it appears to grade upward into an upper, heterogeneous member consisting of brown quartzose schist and calcareous metasiltstone, brown schistose hornblende-bearing quartz-rich marble, and tan pelitic schist. Work in progress indicates that this upper member may not belong to the type Killian but that it may be part of an overlying calcareous section. Overlying the phyllite of Killian Springs is unnamed clean white marble and white schistose marble here regarded as Middle to Upper Cambrian(?) on the basis of lithology and stratigraphic position. Middle Cambrian fossils have been recovered from similar, but less metamorphosed, strata in the adjacent Silver Island Mountains and Toano Range (R. A. Robison, oral communication, 1981).

Patterson Pass Area

Immediately south of Patterson Pass (Fig. 7), the Prospect Mountain Quartzite is folded into an overturned southeast-vergent major fold with a northeast-trending axis. Low-angle faults that eliminate stratigraphic section occur near the upper and lower boundaries of the unit and are folded along with the Prospect Mountain and adjacent units (Fig. 10). Tectonically underlying the Prospect Mountain Quartzite on the southern limb of the fold is a thick sequence of phyllite and conglomeratic quartzite that correlates with the conglomerate subunit of McCoy Creek Group unit G. A thin slice of calcareous phyllite occurs in the fault zone separating unit G and the Prospect Mountain Quartzite; it is generally similar to unit G of the Pilot Peak area. On the northern limb of the fold, the conglomerate subunit underlies a thick section of unit G, but it is highly folded and faulted and, therefore, difficult to compare with the southern exposures. Overlying the Prospect Mountain Quartzite are the phyllite of Killian Springs, locally truncated by a low-angle fault, and an overlying sequence of slightly recrystallized platy limestone of probable Middle

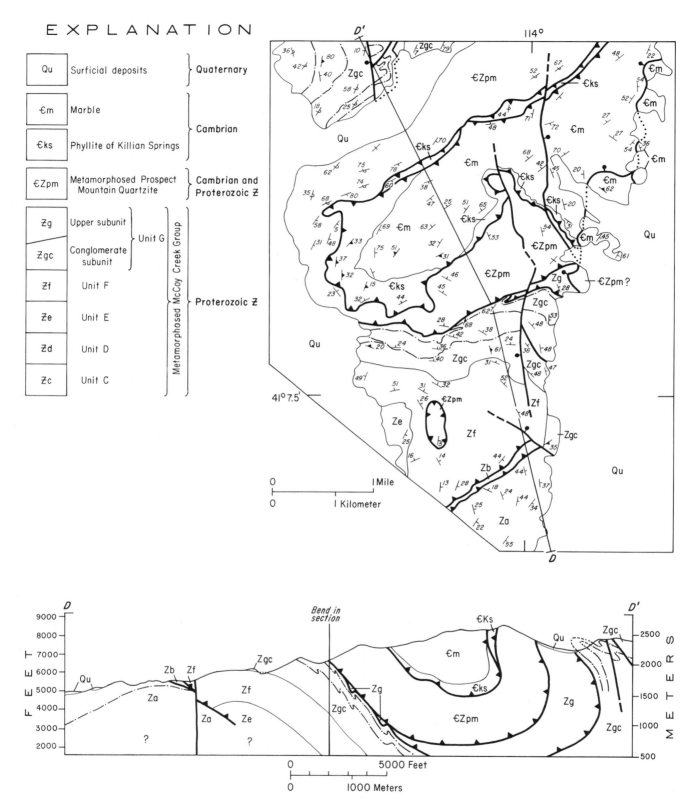

Figure 10. Generalized geologic map and cross section of Patterson Pass area, Pilot Range (see Fig. 7 for location). Stratigraphic markers in conglomerate subunit of unit G of the McCoy Creek Group of Misch and Hazzard (1962) are indicated by dot-dashed lines. Structure symbols are identified in Figure 8.

Cambrian age. The units of the McCoy Creek Group and the Prospect Mountain Quartzite are described below; thicknesses were determined from the structure sections shown in Figure 10.

McCoy Creek Group Unit F. An approximate thickness of 430 m is indicated for unit F of the McCoy Creek Group where measured on section 1 (Fig. 10). Exposures of the unit in the Patterson Pass area are similar in all respects to those in the Pilot Peak area. The top of unit F is marked by a conglomeratic zone, about 20 m thick, containing ripup clasts of phyllite, boulders of quartzite, and, locally, chert or jasperoid clasts; interbeds of dark phyllite similar to that in overlying unit G are common as well. The top of unit F is placed at the top of the uppermost thick conglomerate bed and is generally marked by a pronounced break in slope from the cliffy exposures of unit F to the moderate slopes of unit G.

McCoy Creek Group Unit G. Unit G of the McCoy Creek Group in this area is divided into the conglomerate subunit and an upper subunit. The conglomerate subunit is further subdivided into four intervals, designated (in ascending order) 1 through 4, on the basis of lithology; it is well exposed on the southern limb of the fold (Fig. 10), where the sequence of alternating phyllite and conglomerate is about 595 m thick. The uppermost conglomerate bed is folded and truncated by a low-angle fault. The northern limb of the fold contains a thick section of massive conglomerate, with interbedded phyllite that in part is recumbently folded. I have been unable to correlate these exposures in detail with the section on the southern limb, owing to structural complexities, but the conglomerate subunit appears to be much more conglomeratic in the northern exposures.

Interval 1 of the conglomerate subunit, about 380 m thick, is dominantly dark quartzose phyllite and interbedded coarse quartzite and conglomerate. About 12 m of blue-gray silvery-weathering schist containing 20 to 30 percent magnetite are present at the base, succeeded by interbedded zones of phyllite and quartzite, 10 to 25 m thick. Phyllite-rich zones are dark brown, dark green, and dark gray weathering, brown, green, and gray phyllitic, slaty, or fissile rocks that are thin bedded to laminated. Medium-grained micaceous quartzite interbeds are common. Quartzite-rich zones are dark-brown- and gray-brown-weathering, brown to gray and light-gray strata, chiefly medium-grained to conglomeratic micaceous quartzite; interbeds of dark slate and metasiltstone are common. Coarser quartzite beds are generally light-gray weathering. Conglomerate clasts and granules are commonly white vein quartz, dark and light quartzite, feldspar cleavage fragments (as much as 20 percent of the clasts in some beds), and plates of phyllite (as much as 30 percent of the clasts in some beds). Quartzite layers are indistinctly bedded and rarely crosslaminated. Near the top of the interval, blue-gray slate weathers white and rarely contains a small amount of calcite.

Interval 2 is mostly light-gray, coarse-grained to conglomeratic quartzite, about 145 m thick. The base is marked by a sharp contact between light-gray quartzite and the underlying fissile green slate. The lower 40 m or so is medium- to thick-bedded light-gray-, medium-gray-, and brown-weathering micaceous quartzite, crosslaminated with tabular and wedge-shaped sets. Pebbles occur in thin layers at the bases of beds and as sparse, randomly distributed clasts in the finer grained parts. The quartzite is generally coarse, poorly sorted, and impure. The upper 105 m or so of the interval is massive to very poorly bedded polymict conglomerate, with rare pebble-poor zones that contain as much as 20 percent feldspar. The quartzite ranges in color from light to dark gray and rarely is black; mica and feldspar are common matrix constituents. Clasts vary considerably in size, shape, roundness, and color. Although most clasts are of quartzite, plates and rounded cobbles of phyllite are common above rare phyllite interbeds in the conglomerate.

The lowermost part of interval 3 is characterized by phyllite interbedded with conglomerate belonging to underlying interval 2. Most of interval 3 is rhythmically bedded phyllite and metasiltstone, about 50 m thick. Dark-brown colors and gentle slopes serve to distinguish this interval from the underlying cliffy white and gray conglomerate. Interval 3 is overlain by 10 to 20 m of dark conglomerate, here assigned to interval 4, that resembles the conglomeratic quartzite in interval 2. Where the calcareous upper subunit of unit G overlies the conglomerate subunit, the structure is complex, and the contact is probably a low-angle fault.

Neither the base nor the top of the upper subunit of unit G of the McCoy Creek Group is definitely exposed in the Patterson Pass area. The base is probably defined by dark phyllite overlying conglomerate at the top of the conglomerate subunit in the northern fold limb. At the north border of the area of Figure 10, the upper subunit is continuously exposed from a fault(?) contact with underlying conglomerate to a point about 20 m below the base of the Prospect Mountain Quartzite. The complex structure in that area consists of several low-angle faults and tight folds, and so the measured thickness is of unknown value, and the lithologic sequence could be seriously distorted. The sequence does conform well with the section of unit G in the Pilot Peak area, both in thickness and lithology, and so I consider this section to be a good representation of the upper subunit. The estimated thickness of the upper subunit is 525 m.

The lower part of the upper subunit of unit G is dark phyllite and metasiltstone, with one or two lenses of coarse quartzite that may be fault slices of neighboring units. The phyllite grades upward into medium-interbedded to thickly interbedded marble and clastic rocks; the interbedded clastic rocks are metasiltstone and dark, fine-grained quartzite. The calcareous part of the section grades rapidly upward into dark, medium-bedded, crosslaminated, fine-grained quartzite, with 1- to 10-cm-thick interbeds of calcareous phyllite. Micaceous thin- to medium-bedded, faintly crosslaminated quartzite continues to near the contact with the Prospect Mountain Quartzite, where slope wash covers the rocks. In general, the upper subunit lithologically resembles unit G as exposed in the Pilot Peak area above the zone of conglomerate (about 110 m above the base), except that quartz is more common throughout most of the Patterson Pass section.

Prospect Mountain Quartzite. The Prospect Mountain Quartzite is 700 m thick on the northern limb of the syncline and 865 m thick on the southern limb. In both areas a low-angle fault cuts out the upper part of the unit, and on the southern limb a fault locally cuts out the lower part as well; therefore, these thicknesses are minimums. The quartzite is folded on a small scale in a few places, but generally it is unfolded except on a scale of several kilometers as a thick slab sandwiched between the overlying and underlying phyllitic units.

The lithology, sedimentary structures, and thickness of the Prospect Mountain Quartzite in the Patterson Pass area are generally similar to equivalent exposures in the Pilot Peak area; differences, however, are apparent near the base and top of the unit. Exposures in the lower part of the Prospect Mountain consist of the dark mica-rich quartzite and light-gray porcelaneous pure quartzite similar to rocks exposed near Pilot Peak; granular layers and some feldspathic beds containing 20 to 30 percent feldspar fragments are locally present. The upper part of the unit typically contains dark- to medium- or blue-gray quartzite that is locally conglomeratic; the coloring appears to be caused by iron oxides because the mica content is generally low. Small rounded shale pebbles are rare constituents of the conglomerate. Locally, the conglomerate dominates the upper part of the section; in these areas, it is as much as 4 m thick, contains thin interbeds of phyllite, and is characterized by heterogeneous coarse clasts that include phyllite ripup wedges.

Metamorphosed Cambrian(?) Rocks. The phyllite of Killian Springs is named for exposures in the hinge of the large syncline, where it is about 295 m maximum thickness. Homogeneous dark-gray graphitic and micaceous phyllite and slate comprise most of the unit, with metasiltstone and calcareous phyllite beds locally present in the upper part. The graphite content of the phyllite of Killian Springs contrasts sharply with that of green oxidized shale and siltstone of the lower member of the Pioche Formation. The phyllite of Killian Springs is succeeded by micaceous and silty thin-bedded to laminated marble and limestone that roughly correlates stratigraphically with Cambrian(?) marble in the

Pilot Peak area (Fig. 8) and platy, silty limestone in the nearby Toano Range and Silver Island Mountains, from which Middle Cambrian fossils have been recovered (R. A. Robison, oral communication, 1981).

CORRELATION OF ALBION MOUNTAINS AND PILOT RANGE STRATA

Stratigraphic correlation of the rock sequences in the Pilot Range and Albion Mountains is difficult because of the metamorphism and deformation in both areas. Moderate to extensive recrystallization in nearly all the rocks described above has caused changes in apparent grain size (except for granules and pebbles) and in primary clay mineralogy, and partial or complete elimination of fine sedimentary structures, fossils, and such trace fossils as worm burrows and tracks. In general, primary features in quartzite are preserved better than those in argillaceous rocks. Complex folds and numerous low-angle faults also introduce many uncertainties in thickness and continuity in the measured sections. For example, R. L. Armstrong (written communication, 1980) has interpreted the section he studied on Mount Harrison to be interrupted by a low-angle fault, whereas I consider the sequences to be intact.

I consider the following criteria to be most significant for correlating metamorphosed strata in the study area: (1) stratigraphic sequence, (2) general lithology and thickness, and (3) distinctive rock types and sedimentary structures.

Stratigraphic Position

In the Pilot Range, the sequence of quartzite and phyllite described here underlies marble and phyllite; this relation indicates that the sequences exposed in the Pilot Peak and Patterson Pass areas are equivalent. Low-angle faults within both sequences generally truncate less than 200 m of section and so are considered to be relatively unimportant in judging the equivalence of the two sequences. This conclusion is supported by the other correlation criteria summarized below.

In the Albion Mountains, however, the sequence of strata in question is bounded by probably major faults, and so its position in the stratigraphic sequence is difficult or impossible to determine.

General Lithology and Thickness

In the Pilot Range, the Prospect Mountain Quartzite is an easily distinguishable unit because it is so much thicker than other quartzites, is homogeneous, and is pervasively thick bedded and crosslaminated. Unit G of the McCoy Creek Group is readily identified by its calcareous interbeds and its moderately to conspicuously conglomer-

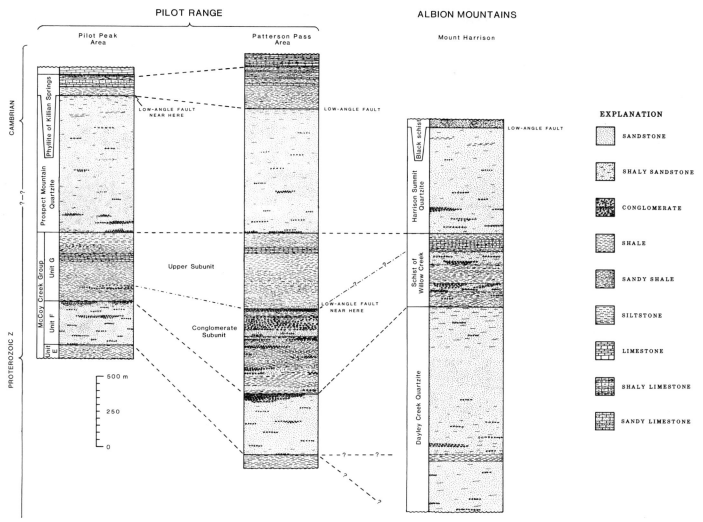

Figure 11. Tentative correlation of stratigraphic sections in Pilot Range and Albion Mountains. Bases and tops of sections not shown.

atic lower subunit. Unit F is characterized by medium to thick bedding and prominent crosslamination, in contrast to the underlying, more massive quartzite (unit D); it differs from the Prospect Mountain Quartzite in being much thinner and less well bedded near the base and top. These general lithologic characteristics allow the various rock units to be correlated with confidence throughout the Pilot Range (Fig. 11).

The Mount Harrison sequence contains a distinctive thick-bedded, conspicuously crosslaminated, homogeneous quartzite, 740 m thick, that is underlain by a unit with calcareous beds near the top and much conglomerate and quartzite in the middle and lower parts, which in turn is underlain by more than 1,400 m of quartzite characterized by pervasive crosslaminations and 5 to 10 percent mica. The sequence is, therefore, similar to the Pilot Range sequence on the basis of general features, except that the

lower quartzite is much thicker and more micaceous in the Albion Mountains (Fig. 11).

Strata within the Cordilleran miogeocline other than those of late Precambrian and early Paleozoic age cannot be correlated with the Mount Harrison sequence because the Paleozoic miogeoclinal section is dominated by abundant carbonate, and the Precambrian strata, other than those close to the Precambrian-Phanerozoic boundary, do not contain any thick crosslaminated quartzite units.

Distinctive Rock Types and Sedimentary Structures

Thinly interbedded calcareous marble and dark phyllite are a distinctive lithologic assemblage within unit G of the McCoy Creek Group in the Pilot Range and the schist of Willow Creek in the Albion Mountains. These rock types are strikingly similar in appearance in the two ranges despite the metamorphism and deformation of the rocks (compare Figs. 6 and 9). Elsewhere within the Cordilleran miogeocline, other sequences of interbedded phyllite and marble, or their shale and limestone protolithic equivalents, are present in unit A of the McCoy Creek Group and in the Pioche Formation (Misch and Hazzard, 1962). In both these intervals, however, the limestone units are thicker bedded, and quartzite or calcareous quartzite is a common associate. Therefore, the phyllite and marble on Mount Harrison most probably correlate with unit G of the McCoy Creek Group.

Laminated limestone, silty limestone, and siltstone were noted by Stewart (1974) as distinctive lithologies comprising the Osceola Argillite of Misch and Hazzard (1962), the equivalent of unit G of the McCoy Creek Group, in the Snake Range of east-central Nevada. The fact that the Rainstorm Member of the Johnnie Formation in the Death Valley area contains the same distinctive laminated limestone led Stewart (1974) to correlate the Rainstorm with the Osceola Argillite and unit G. Stewart (1974) also described lenticular quartzite and conglomerate in unit G and the Rainstorm; these rock types resemble the conglomerate channel fills(?) in unit G in the Pilot Range. The presence of these distinctive rock types in correlative units over such a wide area supports the correlation of unit G of the McCoy Creek Group with the schist of Willow Creek, as proposed here.

Local coarse conglomerate containing phyllite ripup wedges is common at several stratigraphic horizons in the sequences in both the Pilot Range and the Albion Mountains. Such conglomerate is absent elsewhere in the Precambrian and Paleozoic section in the Pilot Range and, to my knowledge, has been recognized only in the McCoy Creek Group and equivalent strata in other ranges within the northeastern Great Basin, such as in the Promontory Range (M. D. Crittenden, oral communication, 1979), the Deep Creek Range (Woodward, 1965), and the Sheeprock Mountains (Blick, 1979).

Other Possible Correlations of the Mount Harrison Sequence

Armstrong (1968) proposed that the Mount Harrison sequence and structurally overlying rocks of the quartzite assemblage are stratigraphically correlative with the Middle Ordovician Kinnikinic Quartzite exposed in central Idaho. The Kinnikinic Quartzite was first named by L. P. Ross (in Wilmarth, 1932) from exposures of quartzite and subordinate carbonate rocks in the Bayhorse area of central Idaho and was later more fully described by Ross (1934, 1937). It has since been redefined by Hobbs and others (1968) to include only the clean quartzite at the top of Ross's Kinnikinic. Two distinct sections were described by Hobbs and others (1968). The first section consists of the Kinnikinic Quartzite, underlain by massive and sandy dolomite, which overlies thick impure heterogeneous quartzite; the section probably is entirely Ordovician. The second section underlies Middle Cambrian shale and is composed of quartzite containing pebbles, shale, siltstone, and feldspar-bearing beds; this quartzite is underlain by silty and shaly dolomite and limestone, as well as siltstone, which in turn is underlain by thick-bedded crosslaminated locally dolomitic shaly and silty quartzite. Although some of these strata resemble some of those on Mount Harrison, the sequences do not match; in particular, the calcareous section of central Idaho contains a small percentage of clastic material, whereas the schist of Willow Creek contains only a minor amount of carbonate rocks within a thick pelitic and clastic sequence.

In the Pioneer Mountains of central Idaho (Fig. 1) is a lower, metamorphosed sequence of quartzite, schist, and banded calc-silicate rock that was assigned to the Hyndman and East Fork Formations by Dover (1969). Members G, F, and E of the East Fork Formation are now recognized by Dover (1980; oral communication, 1980) as metamorphosed Ordovician strata correlative with those described by Hobbs and others (1968). The underlying Hyndman Formation consists of thick-bedded, crosslaminated, locally pebbly quartzite (member D); distinctively banded green calc-silicate rock (member C); feldspathic and pebbly quartzite (member B); and pelitic schist (member A). The calc-silicate rock (member C) is distinguished from the schist of Willow Creek on Mount Harrison by a distinctive uniform banding throughout the entire unit and the absence of quartzite. The East Fork Formation and Mount Harrison Sequence, therefore, are probably not stratigraphically correlative.

Conclusion

The lithology, thickness, and sedimentary structures of the Mount Harrison sequence resemble those of the Proterozoic Z McCoy Creek Group and the Proterozoic Z and Lower Cambrian Prospect Mountain Quartzite exposed in the Pilot Range. Such distinctive strata as thinly interbedded phyllite and marble, and coarse conglomerate containing ripup wedges of phyllite, occur in both sequences. I conclude that the Mount Harrison sequence is probably stratigraphically correlative with the Proterozoic Z and Lower Cambrian rocks in the Pilot Range. In view of the uncertainties due to structural complexities in both mountain ranges, this correlation must remain tentative.

DISCUSSION

Proterozoic Z and Early Cambrian Depositional Setting

Facies changes within the McCoy Creek Group in the

Pilot Range and within correlative(?) rocks in the Albion Mountains reveal some environments of deposition in the Cordilleran miogeocline during Proterozoic Z time. Our understanding of the paleogeography of the miogeocline is poor owing to structural complexities, particularly thrust faulting, in the region; therefore, construction of palin-spastic maps must await further tectonic reconstructions of the region. Correlation of the McCoy Creek Group with sequences in similar stratigraphic position farther east in Utah and southeastern Idaho is not yet established in detail, and so discussion of the Proterozoic Z depositional setting is here restricted to exposures of the McCoy Creek Group.

Assuming that the present positions of the McCoy Creek Group and the Prospect Mountain Quartzite in the Pilot Range approximate their original spatial relations, the following northward facies changes are apparent: (a) unit F thickens, but the lithology and sedimentary structures remain essentially the same; and (b) unit G thickens considerably, and coarse clastic and pelitic materials increase near its base. The upper, calcareous part of unit G and the Prospect Mountain Quartzite are virtually unchanged northward.

On a regional scale, the upper units of the McCoy Creek Group change moderately in thickness from the Schell Creek Range (Misch and Hazzard, 1962) to the Deep Creek Range (Woodward, 1965), the Pilot Range, and the Albion Mountains. The upper part of unit G contains laminated limestone in the Snake Range (Stewart, 1974), rare limestone or dolomite in the Deep Creek Range (Woodward, 1965, p. 316), no carbonate rocks in the Egan Range, relatively abundant calcareous rocks in the Pilot Range, and common carbonate rock in the possibly stratigraphically correlative strata in the Albion Mountains.

Unit H of the McCoy Creek Group, as defined by Misch and Hazzard (1962), is clean white well-bedded crosslaminated quartzite that underlies the Prospect Mountain Quartzite in the Schell Creek Range, but the unit is problematical in other ranges. Woodward (1965, 1967) considered unit H to be missing in the Pilot Range and to be thinned in the Deep Creek Range of Utah, and inferred that a regional disconformity is present at the base of the Prospect Mountain Quartzite. Unit H probably cannot be distinguished from the Prospect Mountain Quartzite as defined by Misch and Hazzard (1962) and Wodward (1967) in many ranges, owing to northward facies changes. Unit H in the type area is even difficult to distinguish from the Prospect Mountain Quartzite; I, therefore, follow Stewart (1974) and include unit H in the lower part of the Prospect Mountain Quartzite.

The shelf on which the McCoy Creek Group was deposited was apparently unstable. Stable conditions began only during deposition of the upper part of unit G, as indicated by the regionally persistent facies in that unit and

in the younger Prospect Mountain Quartzite. Pervasive crosslamination in the well-sorted quartzite of unit F indicates that the strata were probably deposited in a shallow sea with strong currents. The general change from tabular to festoon cross sets and then to conglomerate upward in unit F suggests regressive conditions. The local conglomerate in unit F and moderate thickness changes in the unit may reflect variations in the depth of deposition. Lenses of conglomerate in unit G are most reasonably interpreted as channel fillings. Cobbles and shale ripups in the lenses suggest that deposition occurred after local erosion by a high-velocity flow or current. Rapid northward thickening of the conglomerate in unit G in the Pilot Range indicates that the influx of detritus varied both in source and volume, perhaps owing to varying degrees of subsidence. Probable flaser bedding in unit G suggests tidally dominated deposition. The upper, calcareous part of unit G must have been deposited during a period of low detritus influx, perhaps as the result of a barrier between the major detrital source and the area of deposition. The Prospect Mountain Quartzite was deposited in a remarkably stable, regionally continuous environment. Pervasive cross stratification in the unit suggests that deposition occurred near shore or on a broad shallow shelf with strong currents; its great thickness indicates a gradual and uniform downwarping of the shelf. However, Stewart and Poole (1974) suggested that the Proterozoic Z and Lower Cambrian strata of the mogeocline were deposited in shallow seas distant from a shoreline because current directions are unidirectional, rather than bidirectional as in a tidal regime.

Tectonic Implications of Proterozoic Z and Lower Cambrian Strata in the Albion Mountains

The tectonic implications of the proposed lithostratigraphic correlation between the Pilot Range and the Albion Mountains may provide new insights into the tectonic history of the northern Great Basin. Because of the complexity and partly unresolved problems of the stratigraphy and structure of the Albion-Raft River-Grouse Creek metamorphic terrane, only the aspects of this subject relevant to the proposed correlation are summarized here. Some implications for the tectonics of the metamorphic terrane that result from the presence of Proterozoic Z and Lower Cambrian strata on Mount Harrison are given below.

The central issue in the interpretation of this region is to what degree of confidence highly strained and faulted metamorphosed strata can be correlated with their unmetamorphosed counterparts. In lieu of any direct age determination of equivalence, a lithostratigraphic reconstruction of the metamorphic terrane has been attempted. The data reported here, which indicate that the strata on Mount Harrison probably belong to Proterozoic Z and Lower Cambrian sequences, represent the first identification of strata of this age in the terrane. It has been suggested (Armstrong, 1968; Compton, 1972, 1975; Compton and Todd, 1979) that some of the lower part of the Raft River Mountains sequence is Proterozoic Z or Cambrian and belongs to a nonmiogeoclinal facies, but the data are ambiguous (Crittenden, 1979).

One of the critical questions that emerges from the correlation made here is whether the Mount Harrison sequence is closely associated with the Raft River Mountains sequence or whether it has been tectonically juxtaposed as a result of large-scale low-angle faulting. If the two sequences are closely associated, the lower part of the Raft River Mountains sequence must be older than Proterozoic Z because it differs from the typical Proterozoic Z and lower Paleozoic miogeoclinal section of the region. If the Mount Harrison sequence was carried in with a far-traveled thrust plate, facies changes in Proterozoic Z and Lower Cambrian strata may have been telescoped, and the Raft River Mountains sequence may be partly or wholly coeval with the Mount Harrison sequence.

The tectonic relation of the Mount Harrison sequence to the structurally underlying Raft River Mountains sequence and the overlying sequence of Robinson Creek is complicated by low-angle faults of uncertain separation; similar faults occur within the Raft River Mountains sequence. Because these faults were metamorphosed concurrent with or after movement along them, any characteristics within the fault zones that might indicate magnitude of separation are generally absent.

The sequence of Robinson Creek is lithologically similar to the Mount Harrison sequence in its content of impure coarse quartzite, and lesser schist and limestone marble. However, the two sequences are not directly correlative because several distinctive rock types of the Harrison Summit Quartzite and the schist of Willow Creek are absent in the sequence of Robinson Creek, and dolomitic marble in the Robinson Creek does not have a counterpart in the Mount Harrison. My structural investigations on Mount Harrison verify Armstrong's (1970) findings that the Mount Harrison sequence is overturned, whereas most cross stratification I measured in the sequence of Robinson Creek indicates that the beds are upright (Miller, 1980). The sequence of Robinson Creek is locally tightly folded, and, in a few places, overturned crossbeds occur. A reasonable interpretation is that the sequence of Robinson Creek represents the upright limb of a major recumbent fold that has undergone a few kilometers of separation along a fault near its axial plane. According to this interpretation, the sequence of Robinson Creek represents Proterozoic Z strata older than unit G of the McCoy Creek Group, on the basis of the generally quartzitic composition of the Robinson Creek, which resembles that of the lower units of the McCoy Creek Group. In addition, dolomite, common in the lower part of the McCoy Creek (Misch and Hazzard,

1962), is also found in the Robinson Creek. An alternative interpretation that the sequence of Robinson Creek is separated from the Mount Harrison sequence by a major tectonic boundary and, therefore, is unrelated, is also possible but considered unlikely because the lithologies are generally similar. Approximately 35 km south of Mount Harrison, on Middle Mountain (Fig. 2), mapping by R. L. Armstrong has shown that the quartzite assemblage tectonically overlies the lower part of the Raft River Mountains sequence and is tectonically overlain by, but closely associated with, Ordovician miogeoclinal strata (Miller and others, 1980); these relations suggest that the quartzites on Middle Mountain are Cambrian and (or) Ordovician but do not constitute the typical miogeoclinal carbonate section. An alternate interpretation, that the Ordovician strata are separated from the quartzite assemblage by a major tectonic boundary, is possible in view of the obscuring of the contact by high-grade metamorphism and intense deformation of the rocks.

Structurally underlying the Mount Harrison sequence is a unit of dark schist, graphitic phyllite, and quartzite (black schist unit, Fig. 4), which includes rocks similar to the schist of Mahogany Peaks and the Chainman Shale-Diamond Peak Formation, undivided. These metamorphosed argillaceous rocks locally are tectonically jumbled by postmetamorphic (Tertiary?) faults that produce a highly sheared, melange-like aspect. The graphitic phyllite resembles the phyllite of Killian Springs in the Pilot Range. This black schist unit in the Albion Mountains is underlain by laminated marble representing the metamorphosed Garden City Formation of Early and Middle Ordovician age, which in turn structurally overlies the metamorphosed Kanosh(?) Shale of the Pogonip Group (Miller, 1980). This overturned Ordovician section is separated by a low-angle fault from underlying strata belonging to the Ordovician Eureka Quartzite and Pogonip Group, and the tectonically underlying units in the lower part of the Raft River Mountains sequence (Fig. 4), all upright.

The sequence of overturned strata, from oldest to youngest, on Mount Harrison consists of: the Proterozoic Z(?) Dayley Creek Quartzite and schist of Willow Creek, the Proterozoic Z(?) and Lower Cambrian(?) Harrison Summit Quartzite, the black schist unit possibly correlative with the Lower to Middle(?) Cambrian phyllite of Killian Springs, and Lower and Middle Ordovician strata. As in the Pilot Range, Middle(?) and Upper Cambrian carbonate rocks are missing at a low-angle fault between Ordovician carbonate rocks typical of the miogeocline and dark Cambrian phyllite of deep water, offshore facies. The sequence on Mount Harrison therefore may represent an overturned equivalent of the entire upper part of the Proterozoic Z and lower Paleozoic section as seen in the Pilot Range, with the Pilot Peak decollement present at the fault boundary that removes section and juxtaposes strata of different depositional facies. Southward from the ridge exposing the fault relations east of Mount Harrison, the black schist unit thickens considerably and locally includes rocks similar to metamorphosed Mississippian Chainman-Diamond Peak Formations (Manning Canyon Shale of Armstrong, 1968). Based on Paleozoic fossils recovered from part of the black schist unit south of Mount Harrison, Armstrong (1968) correlated the schists with the Mississippian Manning Canyon Shale. Additional lithologies in the unit are similar to the schist of Mahogany Peaks (Miller, 1980) and the phyllite of Killian Springs. If this black schist unit once contained only the Killian Springs below the Pilot Peak decollement, subsequent low-angle faulting and major folding are required to explain the present structural and stratigraphic relations.

The apparent contrast in depositional facies across a fault in the Albion Mountains that occurs at the same stratigraphic and structural position as the Pilot Peak decollement indicates that the Pilot Peak decollement possibly was present in the Albion Mountains, and by extension, throughout the Albion-Raft River-Grouse Creek metamorphic terrane. The stratigraphic position of the Mount Harrison sequence, in this view, was originally between the lower and upper parts of the Raft River Mountains sequence. This interpretation requires that the Mount Harrison sequence was tectonically removed from near the base of metamorphosed Ordovician strata, at approximately the position of the inferred Pilot Peak decollement, over most of the Albion-Raft River-Grouse Creek metamorphic terrane. In this interpretation, the Proterozoic Z(?) sequence represented by the quartzite assemblage along the west side of the Albion Mountains represents some of the strata missing from the more highly thinned and bedding-plane-faulted central part of the terrane. A perhaps analogous tectonic situation occurs in the southern Grouse Creek Mountains, where Silurian and Devonian strata at the south margin of the terrane occur as thin tectonic slices in their normal stratigraphic positions, but elsewhere in the terrane they are consistently removed by low-angle faults (Compton and others, 1977). However, the interpretation that the Mount Harrison sequence was tectonically removed throughout most of the terrane is problematic because no fault has been identified near the base of the Ordovician strata or adjacent to the schist of Mahogany Peaks in much of the terrane (Compton and Todd, 1979). Less highly metamorphosed rocks involved in thrusting in fold and thrust belts, such as in the North American Cordillera or the Appalachians, commonly contain large flat faults that remain within a consistent stratigraphic horizon and lack exotic slices over a large area. Perhaps such faults once broke the strata in the Raft River Mountains area and are now difficult to identify owing to subsequent metamorphism and ductile deformation.

An alternative view, that a major tectonic boundary unrelated to the Pilot Peak decollement occurs between the Mount Harrison sequence and the overturned Ordovician rocks, is possible because such a tectonic boundary south of Mount Harrison truncates several units in both the Raft River Mountains sequence and the quartzite assemblage, and because slices of less highly metamorphosed rock are locally present at this boundary (R. L. Armstrong, written communication, 1980). On Mount Harrison the black schist unit is composed of medium-grade schist and of graphitic phyllite that appears to be of lower grade than the schist, perhaps owing to the inhibiting effects of graphite on metamorphic reactions.

On the basis of the structural relations in the Mount Harrison area, I consider both interpretations of the tectonic contact at the base of the Mount Harrison sequence to be partly supported. I briefly discuss below some consequences of the view that the Mount Harrison sequence is separated from the underlying rocks by a fault analogous to the Pilot Peak decollement, resulting in a structural sequence similar to that in the Pilot Range. This sequence will be referred to as the miogeoclinal sequence, even though it is recognized that deep water Cambrian facies may be represented; the rocks in the Pilot Range are generally miogeoclinal compared with the clastic rocks of the lower Raft River sequence.

The hypothesis of a close relation between the Mount Harrison sequence and the miogeoclinal Ordovician section within the Raft River Mountains sequence allows two interpretations of the relation between the miogeoclinal sequence and the remaining five or six (depending on whether the schist of Mahogany Peaks is included) units of the lower part of the Raft River Mountains sequence: (1) the miogeoclinal section was originally separate from other units in the lower part of the Raft River Mountains sequence and was juxtaposed by considerable movement on low-angle faults; or (2) all the strata in thrust sheets in the metamorphic terrane moved only a few tens of kilometers on low-angle faults, and so nonmiogeoclinal facies of Proterozoic Z and Paleozoic strata cannot be present in the Raft River sequence.

The first interpretation implies that the relations between the lower part of the Raft River Mountains sequence and overlying units of the Cordilleran miogeocline are indeterminate. Therefore, part of the Raft River Mountains sequence could represent Proterozoic Z or Cambrian facies remarkably different from the miogeoclinal section seen in the Mount Harrison sequence. This interpretation is supported by a poorly defined 570-m.y. Rb-Sr isochron determined from samples of several different metasedimentary units in the Raft River Mountains sequence (Armstrong, 1976), and by the lack of evidence supporting an early metamorphic event affecting the lower units of the Raft River Mountains sequence similar to that observed in lower Proterozoic rocks elsewhere in the region.

The second interpretation implies that the lower part of the Raft River Mountains sequence is not Proterozoic Z or Cambrian and, therefore, is most probably older. Data supporting a possible Proterozoic X age for these units, summarized by Crittenden (1979), include the presence of distinctive rock types both in the lower part of the Raft River Mountains sequence and in a Proterozoic X sequence (Facer Formation) in the Wasatch Mountains. In addition, where faults are definitely recognized within the lower part of the Raft River Mountains sequence, they cannot displace strata significantly because units in both autochthonous and allochthonous positions are virtually identical. This evidence suggests that the lower Raft River sequence and the miogeoclinal section on Mount Harrison are not separated by major faults.

The recognition of Proterozoic Z and Lower Cambrian strata in the Albion-Raft River-Grouse Creek metamorphic terrane supports the view that a thick miogeoclinal section similar to those exposed in nearby mountain ranges exists here also. The possibility that the Pilot Peak decollement, which is present in several ranges to the southwest (Miller and others, 1982), may also occur within the metamorphic terrane, is suggestive of widespread low-angle faulting and crustal shortening of miogeoclinal rocks throughout this region. A complete understanding of the tectonic and stratigraphic relations within and between the Raft River Mountains sequence and the Mount Harrison sequence must still await resolution, but it now appears that two or more major low-angle faults are widely exposed in the northern Great Basin.

ACKNOWLEDGMENTS

I thank Max Crittenden, Robert Compton, Michael McCollum, Richard Armstrong, Nicholas Christie-Blick, and James Dover for substantially improving my understanding of concepts discussed in this report. Reviews by R. L. Armstrong, N. H. Christie-Blick, M. B. McCollum, H. T. Morris, and J. H. Stewart substantially improved earlier versions of this paper and are gratefully acknowledged. Andrew P. Lush, Martha A. Pernokas, and Joel D. Schneyer provided capable assistance during field mapping of the Pilot Range.

REFERENCES CITED

Armstrong, R. L., 1968, Mantled gneiss domes in the Albion Range, southern Idaho: Geological Society of America Bulletin, v. 79, p. 1295–1314.
——, 1970, Mantled gneiss domes in the Albion Range, southern Idaho: a revision: Geological Society of America Bulletin, v. 81, p. 909–910.
——, 1976, The geochronometry of Idaho (pt. 2): Isochron/West, no. 15, p. 1–33.
Armstrong, R. L., and Hills, F. A., 1967, Rb-Sr and K-Ar geochronologic studies of mantled gneiss domes, Albion Range, southern Idaho, U.S.A.: Earth and Planetary Science Letters, v. 3, no. 2, p. 114–124.
Armstrong, R. L., Smith, J. F., Jr., Covington, H. R., and Williams, P. L., 1978, Preliminary geologic map of the west half of the Pocatello 1° x 2° quadrangle, Idaho: U.S. Geological Survey Open-File Report 78–533, scale 1:250,000.
Blick, N. H., 1979, Stratigraphic, structural and paleogeographic interpretation of upper Proterozoic glaciogenic rocks in the Sevier Orogenic belt, northwestern Utah [Ph.D. dissertation]: Santa Barbara, University of California, 633 p.
Blue, D. M., 1960, Geology and ore deposits of the Lucin Mining District, Box Elder County, Utah, and Elko County, Nevada [M.S. thesis]: Salt Lake City, University of Utah, 122 p.
Bond, J. G., compiler, 1978, Geologic map of Idaho: Idaho Department of Lands, Bureau of Mines and Geology, scale 1:500,000.
Compton, R. R., 1972, Geologic map of the Yost quadrangle, Box Elder County, Utah, and Cassia County, Idaho: U.S. Geological Survey Miscellaneous Geologic Investigations Map I-672, 7 p., scale 1:31,680, 2 sheets.
——, 1975, Geologic map of the Park Valley quadrangle, Box Elder County, Utah, and Cassia County, Idaho: U.S. Geological Survey Miscellaneous Geologic Investigations Map I-873, 6 p., scale 1:31,680.
Compton, R. R., and Todd, V. R., 1979, Oligocene and Miocene metamorphism, folding, and low-angle faulting in northwestern Utah: Reply: Geological Society of America Bulletin, v. 90, p. 307–309.
Compton, R. R., Todd, V. R., Zartman, R. E., and Naeser, C. W., 1977, Oligocene and Miocene metamorphism, folding, and low-angle faulting in northwestern Utah: Geological Society of America Bulletin, v. 88, p. 1237–1250.
Crittenden, M. D., Jr., 1979, Oligocene and Miocene metamorphism, folding, and low-angle faulting in northwestern Utah: Discussion: Geological Society of America Bulletin, v. 90, p. 305–306.
Crittenden, M. D., Jr., Schaeffer, F. E., Trimble, D. E., and Woodward, L. E., 1971, Nomenclature and correlation of some upper Precambrian and basal Cambrian sequences in western Utah and southeastern Idaho: Geological Society of America Bulletin, v. 82, p. 581–602.
Dover, J. H., 1969, Bedrock geology of the Pioneer Mountains, Blaine and Custer Counties, central Idaho: Idaho Bureau of Mines and Geology Pamphlet 142, 66 p.
——, 1980, Status of the Antler orogeny in central Idaho—clarifications and constraints from the Pioneer Mountains: in Fouch, T. D., and Magatham, E. R., eds., Paleozoic paleogeography of west-central United States: Society of Economic Paleontologists and Mineralogists west-central U.S. Paleogeographic Symposium 1, p. 371–386.
Hintze, L. F., and Robison, R. A., 1975, Middle Cambrian stratigraphy of the House, Wah Wah, and adjacent ranges in western Utah: Geological Society of America Bulletin, v. 86, p. 881–891.
Hobbs, S. W., Hays, W. H., and Ross, R. J., Jr., 1968, The Kinnikinic Quartzite of central Idaho—redefinition and subdivision: U.S. Geological Survey Bulletin 1254-J, p. J1–J22.
Hope, R. A., and Coats, R. R., 1976, Preliminary geologic map of Elko County, Nevada: U.S. Geological Survey Open-File Map 76–779, scale 1:100,000.
Miller, D. M., 1980, Structural geology of the northern Albion Mountains, south-central Idaho, in Crittenden, M. D., Jr., and others, eds., Cordilleran metamorphic core complexes: Geological Society of America Memoir 153, p. 399–423.
Miller, D. M., and Lush, A. P., 1981, Preliminary geologic map of the Pilot Peak and adjacent quadrangles, Elko County, Nevada, and Box Elder County, Utah: U.S. Geological Survey Open-File Report 81–658, 18 p., scale 1:24,000, 2 sheets.
Miller, D. M., Lush, A. P., and Schneyer, J. D., 1982, Preliminary geologic map of Patterson Pass and Crater Island NW quadrangles, Box Elder County, Utah, and Elko County, Nevada: U.S. Geological Survey Open-File Report 82-834, 20 p., scale 1:24,000, 2 sheets.
Miller, D. M., Todd, V. R., Armstrong, R. L., and Compton, R. R., 1980,

Geology of the Albion-Raft River-Grouse Creek Mountains area, northwestern Utah and southern Idaho: Geological Society of America, Rocky Mountain Section Field Trip Guidebook, 32 p.

Misch, P., and Hazzard, J. C., 1962, Stratigraphy and metamorphism of late Precambrian rocks in central northwestern Nevada and adjacent Utah: American Association of Petroleum Geologists Bulletin, v. 46, p. 289–343.

O'Neill, J. M., 1968, Geology of the southern Pilot Range, Elko County, Nevada, and Box Elder County, Utah [M.S. thesis]: Albuquerque, University of New Mexico, 112 p.

Roberts, R. J., and Thomasson, M. R., 1964, Comparison of Late Paleozoic depositional history of northern Nevada and central Idaho, *in* Geological Survey research 1963: U.S. Geological Survey Professional Paper 475-D, p. D1–D6.

Ross, C. P., 1934, Correlation and interpretation of Paleozoic stratigraphy in south-central Idaho: Geological Society of America Bulletin, v. 45, p. 947–1000.

——, 1937, Geology and ore deposits of the Bayhorse region, Custer County, Idaho: U.S. Geological Survey Bulletin 877, 161 p.

Stewart, J. H., 1974, Correlation of uppermost Precambrian and Lower Cambrian strata from southern to east-central Nevada: U.S. Geological Survey Journal of Research, v. 2, p. 609–618.

Stewart, J. H., and Poole, F. G., 1974, Lower Paleozoic and uppermost Precambrian Cordilleran miogeocline, Great Basin, western United States, *in* Dickinson, W. R., ed., Tectonics and sedimentation: Society of Economic Paleontologists and Mineralogists Special Publication 22, p. 28–57.

Stokes, W. L., compiler, 1963, Geologic map of northwestern Utah: Salt Lake City, University of Utah, College of Mines and Mineral Industries, scale 1:250,000.

Todd, V. R., 1975, Late Tertiary low-angle faulting and folding in Matlin Mountains, northwestern Utah: Geological Society of America Abstracts with Programs, v. 7, p. 381–382.

——, 1980, Structure and petrology of a Tertiary gneiss complex in northwestern Utah, *in* Crittenden, M. D., Jr., and others, eds., Cordilleran metamorphic core complexes: Geological Society of America Memoir 153, p. 349–383.

Wilmarth, M. G., compiler, 1932, Tentative correlation of the named geologic units of Idaho: U.S. Geological Survey Correlation Chart No. 31, 2 sheets.

Woodward, L. A., 1963, Late Precambrian metasedimentary rocks of Egan Range, Nevada: American Association of Petroleum Geologists Bulletin, v. 47, p. 814–822.

——, 1965, Late Precambrian stratigraphy of northern Deep Creek Range, Utah: American Association of Petroleum Geologists Bulletin, v. 49, p. 310–316.

——, 1967, Stratigraphy and correlation of Late Precambrian rocks of Pilot Range, Elko County, Nevada, and Box Elder County, Utah: American Association of Petroleum Geologists Bulletin, v. 51, p. 235–243.

Manuscript Accepted by the Society August 20, 1982

Geological Society of America
Memoir 157
1983

Structural geometry and sequence, Bovine Mountain, northwestern Utah

Teresa E. Jordan
Department of Geological Sciences
Cornell University
Ithaca, New York 14853

ABSTRACT

Bovine Mountain, at the south end of the Raft River-Grouse Creek-Albion Mountains metamorphic core complex, exposes Ordovician and Silurian carbonates and quartzites, six members in the Pennsylvanian and Permian Oquirrh Group, upper Eocene intrusive rocks, and upper Tertiary volcanic and sedimentary rocks. Low-grade regional and contact metamorphism affected Oquirrh Group rocks, although Ordovician to Silurian units apparently are not metamorphosed.

A north-trending anticline, the Bovine fold, caused widespread overturning of units. A younger northeast-trending fold set is superimposed. Three sets of small-scale folds are recognized in the lowest member of the Oquirrh Group. The younger two small-scale fold sets probably formed coevally with the two known map-scale fold sets, but the earliest east-trending isoclinal recumbent small-scale fold set has no known map-scale equivalent at Bovine Mountain. The Bovine fold pre-dates a 38 m.y. old pluton.

Two faults at low-angles to bedding separate three structural plates. Devonian and Mississippian strata are structurally omitted along fault I, whereas fault II primarily follows bedding but also cuts out strata locally. Fault I is mapped at Bovine Mountain only on the overturned limb of the Bovine fold; fault II is inferred to exist in both overturned and upright limbs of that fold.

INTRODUCTION

Bovine Mountain is the southernmost exposure of the Raft River-Grouse Creek-Albion Mountains metamorphic core complex of northwestern Utah and southern Idaho (Fig. 1). However, the stratigraphic section is more complete and less metamorphosed at Bovine Mountain than in those areas to the north. Data presented in this paper suggest that rocks exposed at Bovine Mountain may be transitional between those typical of the metamorphic terrane and essentially unmetamorphosed neighboring regions of the eastern Basin and Range. The structural geology of Bovine Mountain will be described and compared to the complex deformation history of the Grouse Creek, Raft River, and Albion Mountains previously described by Compton and Todd (Compton and others, 1977; Compton and Todd, 1979; Todd, 1980) and Miller (1980).

Bovine Mountain forms the southern end of the Grouse Creek Mountain in northwestern Utah (Fig. 1). It lies in the northeastern Basin and Range province and within the hinterland of the Mesozoic to early Cenozoic Cordilleran thrust belt. Bovine Mountain trends east-west and is about 8 km long. Total relief is about 2700 ft (800 m) in the map area, and local relief is about 1200 ft (350 m) between valley bottoms and ridges within the mountain block. The area described in this report includes the southern two-thirds of Bovine Mountain and flanking terraces to the south and east formed by Pleistocene Lake Bonneville (Fig. 2). Baker (1959) mapped the Bovine Mountain area, and Compton and others (1977) incorporated preliminary mapping of Bovine Mountain by Stanford University field classes into a synthesis of the Grouse Creek and Raft River Mountains. The strati-

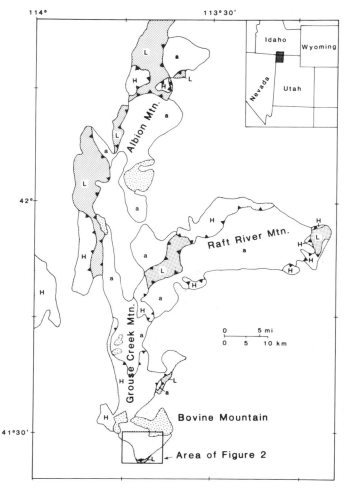

Figure 1. Generalized geologic map of the Raft River-Grouse Creek-Albion Mountains, located on inset at top right. Subdivisions include "a"—parautochthonous basement, "L"—lower allochthonous plate (lower and middle Paleozoic rocks in Raft River-Grouse Creek Mountains, Proterozoic Z or Paleozoic rocks in Albion Mountains), "H"—higher allochthons of Pennsylvanian and stratigraphically higher units; unlabeled tick pattern represents Tertiary intrusives. Bovine Mountain shown at south end. Compiled from Compton and others (1977), Miller (1980), and Armstrong (unpublished manuscript).

along bedding plane faults. An important characteristic of the metamorphic terrane is the common younger-on-older geometry across these low-angle faults. An early metamorphic fold set trends east-northeast in the Raft River Mountains, swinging to north and northeast trends in the Grouse Creek Mountains. These folds are overturned to the north and west. A middle allochthonous sheet was emplaced along a widespread low-angle detachment fault, mainly at a horizon in the Mississippian shales, before or early in the first metamorphism. This stage of deformation apparently pre-dates 38 m.y. ago. A second pulse of metamorphism and deformation in the Raft River-Grouse Creek Mountains produced west- to northwest-trending folds and younger northeast-trending folds. This metamorphic deformation was in progress during the Oligocene and Miocene. Widespread low-angle faulting as young as late Miocene has been documented (Compton and others, 1977; Compton, this volume; Todd, this volume).

Farther north in the metamorphic terrane, the first phase of metamorphic folding in the Albion Mountains (Fig. 1) created northeast-trending folds that are overturned to the northwest (Armstrong, 1968). Major low-angle younger-on-older faults formed before or during early phases of the first metamorphism (Miller, 1980). The first metamorphic deformation in the Albion Mountains occurred in the Mesozoic (Miller, 1980; Armstrong, unpublished manuscript). During the second metamorphic deformation in the Albion Mountains, northwest-trending folds with eastward overturning formed. Miller (1980) concluded that this metamorphic deformation pre-dated mid-Cenozoic intrusions.

The following descriptions and conclusions regarding the structural geology of Bovine Mountain are based on map-scale relations (1:24,000) and preliminary examination of small-scale structures. The stratigraphic section (described briefly below) occurs in three structural plates bounded by two major faults oriented at low angles to bedding. However, the geometry of those faults is greatly complicated by a major recumbent fold (the north-trending Bovine fold), which dominates the map area. Most exposures of all three structural plates are in the overturned limb of that fold. Because of confusions in terminology that result from identifying structural plates in an overturned sequence, the plates are referred to as A, B, and C in ascending stratigraphic sequence. Three fold sets have been recognized in preliminary study of small-scale structures, although only two sets are recognized from map relations. The sequence of folds and faults will be discussed, although future study of mesoscopic structures will undoubtedly modify and clarify relative timing.

A key to establishing relative and absolute timing relations of structures is their relationship to a stock, the Immigrant Pass pluton, on the north side of Bovine Mountain (Fig. 1). This pluton, forming the northeastern part of the

graphy of the upper Paleozoic Oquirrh Group at Bovine Mountain has been described by Jordan (1979) and Jordan and Douglass (1980). A U.S. Geological Survey map of the central and southern Grouse Creek Mountains, including Bovine Mountain, is being prepared by R. R. Compton, V. R. Todd, and T. E. Jordan.

The metamorphic core complex to the north, as reported by Compton and others (1977), Compton and Todd (1979), and Todd (1980), is characterized by widespread metamorphism and generally thinned stratigraphic units. Thinning has been accomplished by both ductile flow during metamorphism and by omission of section

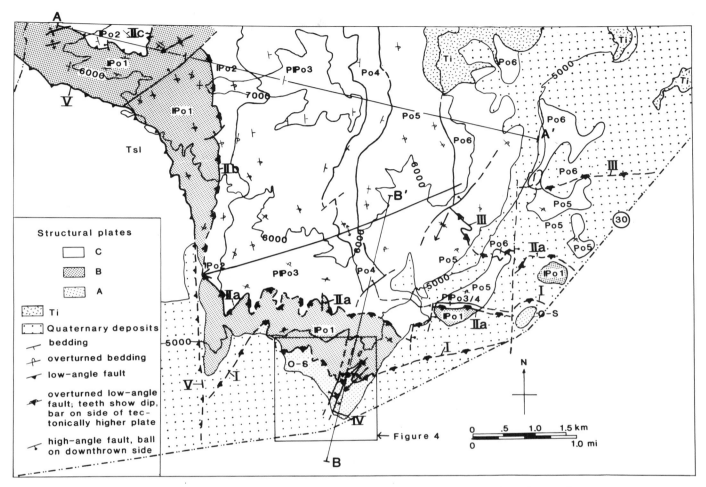

Figure 2. Structural map of the southern two-thirds of Bovine Mountain, with Highway 30 at southern boundary. O-S refers to Ordovician and Silurian strata, and Oquirrh is mapped as six members, IPo1 to Po6 (PIPo3/4 refers to an area where members 3 and 4 cannot be discriminated). Immigrant Pass pluton indicated as Ti. Elevation contours in feet. A-A' and B-B' are lines of cross sections for Figure 3. Area of Figure 4 outlined at south end.

map area, is of granodiorite to quartz monzonite composition, with characteristic 1 to 2 cm potassium feldspar phenocrysts and only biotite as a mafic phase. It has been dated as approximately 38 m.y. old (Compton and others, 1977). Although it provides the opportunity to constrain absolute ages of deformation, it also largely (but not completely) obscures direct structural connections between Bovine Mountain and the southern Grouse Creek Mountains (Compton, mapping in preparation).

Relationships described here suggest that low-angle faulting juxtaposed at least two and possibly all three structural plates before overturning by the Bovine fold. Inclusion of unmetamorphosed lower and middle Paleozoic rocks in one of the structural plates at Bovine Mountain contrasts with the metamorphic/structural sequence of the Raft River and Grouse Creek Mountains to the north. This difference suggests that Bovine Mountain was located at a distance from the main site of mid-Tertiary metamorphism.

STRATIGRAPHIC FRAMEWORK

Bovine Mountain exposes lower and middle Paleozoic strata and the thick late Paleozoic Oquirrh Group. It is not known whether the Bovine Mountain sequence includes in the subsurface Precambrian-Cambrian rocks typical of the rifted continental margin sequence observed to the south, east, and west, or whether the Bovine Mountain section lacks such rocks, as in the Raft River Mountains (see discussion in Compton and others, 1977; Crittenden, 1979; Compton and Todd, 1979; Miller, this volume). The lower and middle Paleozoic rocks were deposited in the Cordilleran miogeocline. During Mississippian time, sediments typical of the Antler synorogenic basin were deposited. Middle Pennsylvanian to Lower Permian strata accumulated in an apparently axial position in the fault-bounded Oquirrh basin (Jordan and Douglass, 1980). Upper Permian and Mesozoic rocks are not preserved in the map area, but nearby (and not structurally continuous) exposures suggest nearly continuous marine deposition through Early Triassic time. Tertiary nonmarine clastic and volcanic rocks, partly of Miocene age, are widespread (Compton and others, 1977; Compton, this volume; Todd, this volume).

Lower and Middle Paleozoic Units

Unmetamorphosed lower and middle Paleozoic strata crop out only in an overturned sequence at the south end of Bovine Mountain where they are partly obscured by Quaternary lake terrace deposits (Fig. 3). Incomplete sections of Ordovician Pogonip Group, Eureka Quartzite, and Fish Haven Dolomite, and Silurian Laketown Dolomite are present. Devonian units have apparently been eliminated by faulting. Unit assignments are based on lithologic similarity to those units throughout northwestern Utah.

The Ordovician Garden City Formation and Swan Peak Quartzite/Crystal Peak Dolomite (undifferentiated)

have been mapped in the Pogonip Group. The Garden City Formation is a thinly bedded, light- to medium-gray weathering limestone with silty partings. It is at least 125 m thick, but the base is not exposed and its upper contact is inferred to be a zone of bedding plane slip because the stratigraphically overlying unit is locally missing (Fig. 3). The Swan Peak Quartzite/Crystal Peak Dolomite (undifferentiated) consists of about 45 m of interbedded sandstone, quartzite, and dolomite. The very fine-grained, light- to medium-gray sandstone and quartzite weather tannish-brown and are thinly bedded. The thin- to medium-bedded dolomite, containing varying quantities of quartz sand and weathering medium-gray, is locally mottled.

The Eureka Quartzite is extremely pure vitreous quartzite, which is white where fresh and weathers brown. Intense fracturing obscures bedding, and mapping to date has not clarified whether its boundaries with the Crystal Peak and Fish Haven Dolomites are depositional or tectonic or both. The 105 m thickness at Bovine Mountain is somewhat thinner than Eureka Quartzite sections elsewhere in northwestern Utah (Doelling, 1980).

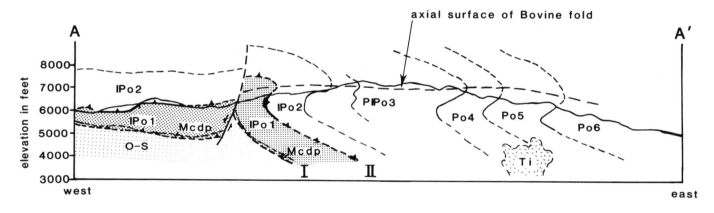

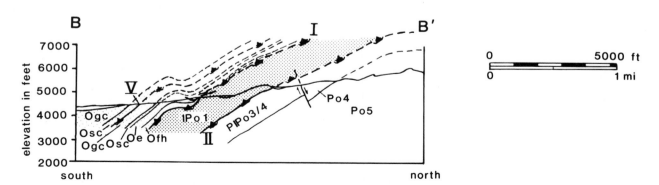

Figure 3. Cross sections through Bovine Mountain with no vertical exaggeration, as located on Figure 2. A-A' illustrates the overturned Bovine fold. In B-B', rocks in the overturned limb are tilted to the south by a large east-northeast-trending fold.

The Fish Haven Dolomite is about 50 m thick and its stratigraphically higher contact is probably a fault. It is mainly dark-gray, finely crystalline dolomite with local dark chert nodules, and it is commonly fossiliferous. Bedding is often difficult to discern in the Fish Haven Dolomite, but it is typically thinly bedded.

In the inverted sequence, structurally below the Fish Haven Dolomite is a 9-m-thick zone of light-brown weathering, finely crystalline dolomite that is characterized by tight folds. This distinctive tectonic zone is a useful mapping horizon and is here considered to mark the base of the Silurian Laketown Dolomite. The Laketown Dolomite is variably light- to dark-gray, massive to laminated, finely crystalline to sugary dolomite. It is characteristically unfossiliferous. A wide range of estimated thicknesses in exposures on the south flank of Bovine Mountain, from 0 to 340 m, suggests that its boundaries are faults.

Upper Paleozoic Units

At Bovine Mountain, the Mississippian to Pennsylvanian(?) Chainman Shale-Diamond Peak Formation (undifferentiated) is represented by only a thin (maximum 8-m-thick) tectonic sliver of dark-brown siltstone, impure sandstone, and chert- and argillite-pebble conglomerate.

The Middle(?) Pennsylvanian to Lower Permian (Wolfcampian) Oquirrh Group constitutes by far most of Bovine Mountain. The largely overturned sequence includes six informal mapping units, totalling over 4000 m in thickness (Jordan, 1979; Jordan and Douglass, 1980). From base to top these are 1) medium to thickly bedded limestone with common thin sandstone beds and chert stringers, local siltstone, and conglomerate beds of chert, quartzite, limestone, and fossil clasts, and very minor dolomite beds (estimated thickness 1000 m); 2) thinly bedded silty limestone interbedded with cleaner, more thickly bedded limestone, commonly very bioclastic and locally with chert- and carbonate-clast conglomerates (about 500 m thick); 3) medium to thickly bedded structureless brown-weathering sandstone, with less-resistant siltstone and sandy limestone interbeds locally; 4) interbedded medium to thickly bedded structureless sandstone and laminated siltstone; 5) laminated sandstone and less resistant interbedded conglomerate (3, 4 and 5 aggregating 2000 m); and 6) thinly bedded clean to sandy and silty limestone (800 m in a sequence with abundant mesocopic folds, top not exposed). The uppermost unit, with a single (late Wolfcampian) fusulinid date, may grade into the Pequop Formation, overlying the Oquirrh Group.

STRUCTURAL FRAMEWORK

Within the largely inverted stratigraphic section of Bovine Mountain (Fig. 3), two extensive low-angle faults separate three structural plates. A key question is whether these are early faults which have been overturned or whether the faults deformed a previously overturned sequence. Field relations across the faults will be described first, and then the folds will be discussed.

Low-Angle Faults

In ascending stratigraphic order, structural plate A consists of Ordovician to Silurian units; B, of lower Oquirrh limestone (member 1); and C, of younger Oquirrh units (members 2 to 6). Plates A and B are separated by low-angle fault I, plates B and C, by low-angle fault II (Fig. 2). Three other low-angle faults (III to V) that were mapped at Bovine Mountain will also be briefly described.

Fault I crops out only on the south side of the map area (Fig. 2) where it trends eastward between rocks of plate A and plate B. The mapped fault trace is nearly 6 km long. About 60 percent of it, however, is completely covered by Quaternary lake deposits (especially at its east end), and only along about 10 percent of the trace are units both immediately above and below the fault exposed; nowhere is the fault itself exposed. At its west end, the projected trace is overlapped by fault V. Where most completely exposed, a thin unit (up to 8 m thick), consisting variably of siliceous argillite, dark brownish-gray sandstone, and limestone- and chert-pebble conglomerate, occurs along the trace of the fault. These rocks are interpreted as structural slivers of Mississippian Chainman Shale-Diamond Peak Formation (undifferentiated), although they could instead be remnants of a terrigenous clastic lens in Oquirrh member 1. Where Silurian Laketown Dolomite and lowest Oquirrh limestone are well-exposed across a few-meter-wide poorly exposed fault zone, a 2- to 3-m-thick zone in the Laketown Dolomite has thin dolomite veins but is not brecciated, a bed in the dolomite about 50 m from the fault is chevron folded, and the Oquirrh limestone varies from having a 10-m-wide zone of extensive thick calcite veins to appearing no more strained than other typical parts of Oquirrh member 1. Chainman Shale(?) argillite is foliated, whereas sandstone in the fault zone has numerous anastomosing slickensided surfaces. The sinuous trace of fault I is approximately parallel to the trace of bedding in the moderately dipping (40 to 60°) flanking units. Fault I is considered to be oriented at a low angle to bedding (Fig. 4). Devonian strata, typically at least 500 m thick in northwestern Utah, have been structurally eliminated along this fault in the map area. Silurian Laketown Dolomite and Ordovician Fish Haven Dolomite, typically comprising about 500 m in thickness in northwestern Utah, are also everywhere thinned and locally missing along fault I at Bovine Mountain (Fig. 4).

Fault II separates Oquirrh member 1 limestone of

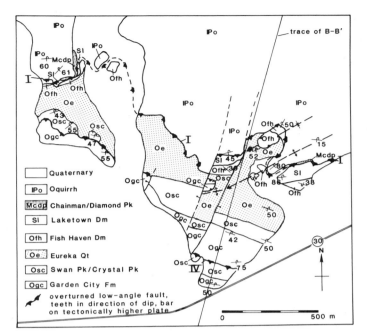

Figure 4. Detail of geology at south end of Bovine Mountain, as located on Figure 2. Contact between Silurian Laketown and Ordovician Fish Haven Dolomites is inferred to be faulted, as are portions of the contact between Garden City Limestone and Swan Peak Quartzite/Crystal Peak Dolomite (undifferentiated), but these faults are not shown due to space limitations.

plate B from Oquirrh member 2 limestone and member 3 sandstone (plate C). Fault II will be discussed as three sub-segments (a, b, c) because the continuity of these segments is open to interpretation. Fault IIa crops out on the south side of Bovine Mountain where its trace trends easterly for about 7.5 km. Whereas the eastern part of fault IIa is largely covered by Quaternary terrace deposits, the western part is better exposed, especially where it crosses ridges. Fault IIa cross-cuts southeast-trending beds in plate C but approximately parallels beds in plate B. Toward the west, it is more difficult to locate the fault because it places member 1 limestone against member 2 with similar limestone interbeds. Beds within a few meters above and below fault IIa have some slickensides but are not brecciated.

Along the 4-km-long, north-trending boundary labeled IIb, the nature of the boundary between Oquirrh members 1 and 2 is enigmatic. The contact parallels moderately to steeply dipping (40 to 90°) bedding in both units and marks an abrupt change in outcrop character. To the west, massive overturned beds of member 1 have variable textures but are locally mottled and banded marbles. In contrast, limestone, silty limestone, sandy limestone, and multi-lithologic conglomerate of member 2 have a strong, spaced cleavage. Locally, outcrops of member 1 within a few meters of the unexposed contact have spaced cleavage, small-scale folds, and thrusts with a few meters of offset.

A low-angle fault (IIc) has also been mapped for about

2 km between Oquirrh members 1 and 2 in the northwestern map area, where the stratigraphic section is upright. This fault was mapped on the basis of local minor bedding divergences across the contact, although toward the west the fault follows a bedding plane.

Because the stratigraphically lower plate of each segment is Oquirrh member 1 and the upper plate units juxtaposed along the segments are progressively lower in the section westward, I consider these faults to be segments of a single fault. According to this interpretation, segments IIa, IIb, and IIc form a single fault that has been deformed by folding about the Bovine fold and younger folds, producing the present exposure pattern. The fault would have originally been largely along a bedding plane between Oquirrh members 1 and 2, with a ramp across part of plate C now exposed at the south end of Bovine Mountain. Alternatively, fault IIa may cross-cut fault IIb at the south end of IIb. Exposure of that critical junction is poor, but there is no obvious change in plate B rocks across a possible western extension of fault IIa. The strongest evidence that IIb is also a fault is that both northwest and southeast of it, IIa and IIc *are* low-angle faults that occur at the same stratigraphic horizon as IIb.

Two other faults at low angles to bedding, one in the inverted plate C and one in the inverted plate A (Fig. 2, faults III and IV, respectively), are exposed locally. Low-angle fault V, which places Miocene rocks onto Paleozoic strata on the western side of the map area (Fig. 2), is discussed by Compton (this volume).

Low-angle fault III places overturned, thinly bedded sandstone of Oquirrh member 5 over overturned, thinly bedded limestone of Oquirrh member 6 and is locally well exposed in the southeast part of Bovine Mountain. As it is traced northward toward the pluton, the nature of this contact is obscured by poor exposure and contact metamorphism (Fig. 2).

Ordovician Garden City Formation and Swan Peak Quartzite/Crystal Peak Dolomite (undifferentiated) are repeated along a fault in the southernmost outcrop of structural plate A (Fig. 4). Because the trace of the fault parallels bedding and all known east-trending faults at Bovine Mountain are at very low angles to bedding, this is inferred to be a low-angle fault (IV). Faults at low angles to average bedding attitudes in plate A also occur between the Garden City Formation and Swan Peak Quartzite and between the Fish Haven Dolomite and Laketown Dolomite. These are thought to be minor bedding plane slip horizons, but exposure lengths are too short to evaluate their importance.

Two small klippen rest on nearly horizontal faults in moderately dipping rocks of structural plate A at the south end of Bovine Mountain (Fig. 4). These faults have several centimeters of gouge, and adjacent beds are brecciated and locally have deep fault grooves. These faults postdate northeast-trending folds (discussed below).

Folds

Two major fold sets, an earlier one of north trend and a later one of northeast trend, are well-expressed in map relations. Of three mesoscopic fold sets, the isoclinal first set trends east and the younger sets probably correspond to the map-scale fold sets. The fold sets are described from oldest to youngest.

Map-scale folds: A north-trending recumbent eastward-verging anticline, here referred to as the Bovine fold, is responsible for widespread overturning of the section on Bovine Mountain. This fold is difficult to represent on a map (Fig. 2) because its nearly horizontal axial surface is variably exposed by topographic irregularities. Only the east-trending main ridgeline and the northwestern part of the map area (down-dropped by a younger fault) are in the higher upright limb of that fold, as illustrated by cross section A-A' (Fig. 3).

Nowhere is an upright lower limb of the fold visible, although it has been inferred in cross section A-A' (Fig. 3). In the sides of deep valleys there are north-trending folds apparently parasitic on the Bovine fold, with horizontal axial surfaces and 100 to 300-m wavelengths, in Oquirrh member 3 sandstone, whereas at the same elevation the unit 2-3 contact is vertical and not involved in those intermediate-scale folds. Those intermediate-scale folds adjacent to vertical beds are interpreted to be near an axial surface between the exposed overturned limb and a lower upright limb. Therefore, the overturned Bovine anticline may be paired with a syncline in the subsurface, with about

1 km distance between their hinges. The interlimb angle of the Bovine fold is about 60°.

A set of northeast- to east-northeast-trending folds was also mapped (Figs. 2, 3, 4, 5). The major fold of this set, whose axis extends over 7 km across the center of the map area (Fig. 2) separates a northern area, with west-dipping bedding in the overturned limb of the Bovine fold, from a southern area, with south-dipping bedding. This map pattern is interpreted to result from folding the axial surface of the Bovine fold about the younger, broad, east-northeast-trending fold (Fig. 5). The younger fold has an amplitude exceeding 1 km and an upright or steeply-north-dipping axial plane. Similarly oriented folds with wavelengths on the order of 1 km and a few 100 m of amplitude are important locally. The fault I surface has been folded about a pair of these northeast-trending folds (Fig. 4).

Map-scale folds in Oquirrh member 1 limestone in the northwestern part of the map area are complex, probably due to superposition of several phases of deformation. Several of the better defined folds trend east-northeast, are of a comparable scale to the fold in fault I that was noted above, have interlimb angles of about 90°, and have axial surfaces that dip to the north at intermediate angles. These folds may be part of the same set of northeast-trending folds described above.

Small-scale folds: Three sets of small-scale folds were identified in a preliminary study of minor structures in a .03 km² area of plate B in Oquirrh member 1 limestone on the south side of Bovine Mountain (Fig. 2). No equally well-developed sequence of folds has yet been documented in the same unit on the northwestern side of the mountain nor in any other unit.

The oldest mesoscopic folds (F1) in member 1 limestone are outlined by folded chert layers (layers about 1 to 4 cm thick). F1 fold axes trend approximately to the east and plunge gently to moderately to the west (Fig. 6). A penetrative foliation (S1) is axial planar to F1 and subparallel to bedding; it has not been examined microscopically. F1 folds, with 3- to 6-cm amplitude and 10- to 20-cm wavelength, are isoclinal and recumbent. They are strongly asymmetric, with consistent vergences over large domains of plate B, suggestive of the possible presence of an east-trending map-scale recumbent antiform (not recognized from the stratigraphy). Except for Oquirrh member 1 limestone, small-scale folds with the same character as F1 have been seen only in the Garden City Formation. There an axial planar cleavage subparallel to bedding is weakly developed.

Sparse second generation (F2) folds in Oquirrh member 1 fold the penetrative S1 foliation (Fig. 7). F2 folds trend north to north-northwest and plunge moderately to the south (Fig. 6). F2 folds are more geometrically variable than F1 folds, with amplitudes ranging from 5 cm to .5 m. They are generally more symmetric than F1, with axial

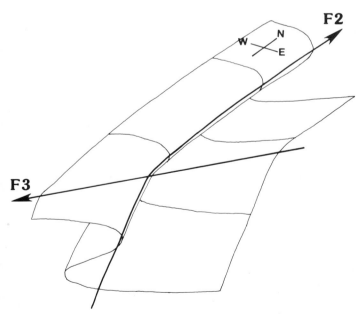

Figure 5. Perspective drawing of a surface deformed by both the overturned north-trending Bovine fold (F2) and subsequently refolded by a major northeast-trending fold (F3), producing the main outcrop pattern of Figure 2.

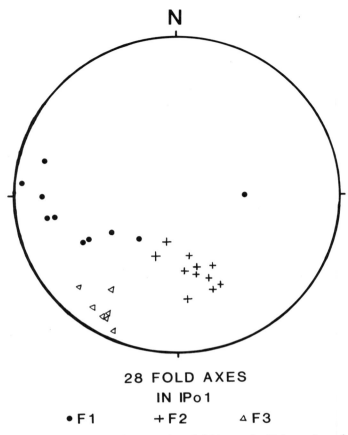

N

**28 FOLD AXES
IN IPo1**

● F1   + F2   △ F3

Figure 6. Lower hemisphere plot of fold axes in IPo1 member of Oquirrh Group on the southern flank of Bovine. Data are from 28 folds whose assignment to F1, F2, or F3 could be made with confidence based on folding of S surfaces and geometry.

planes at a high angle to bedding, and with approximately 20 to 50° interlimb angles. Locally an axial planar foliation (S2) is developed, but it is not penetrative on the scale of plate B.

F3 folds are distinguished from F1 and F2 folds in Oquirrh member 1 limestone by their more open style (interlimb angle 90° and greater), size (wavelengths greater than about 10 m, amplitudes about 3 m), and spaced axial planar cleavage. These folds clearly fold F1 and F2 structures (and S1 and S2) and trend northeast, plunging gently to the southwest (Fig. 6). Axial planes are upright or dip moderately to the northwest, suggesting southeast vergence.

F3-style folds are recognized widely on Bovine Mountain, especially in thinly bedded sandstone and limestone on the eastern side. Amplitudes of about 20 cm and wavelengths of .5 to 1 m are typical of these folds, which uniformly trend northeast and plunge about 30° to the southwest. A spaced axial planar cleavage is pronounced in limestones and weakly developed in sandstones. Within about 1 km of the pluton contacts, similarly oriented F3 folds are recumbent and isoclinal, with even smaller scale coaxial folds (few cm wavelengths) superimposed on

somewhat larger scale F3 limbs (wavelengths of a few meters). In such folds, axial planar foliation is defined by stylolites of very finely crystalline opaque material.

Relationship between map-scale and small-scale folds: Data directly relating the three observed sets of small-scale folds to the two sets of map-scale folds are incomplete. Relationships will be discussed from youngest to oldest, to take best advantage of the data. Small-scale F3 folds occur in the hinge regions of map-scale northeast-trending upright folds (folded fault I contact, Fig. 4) with parallel axial surfaces. This spatial coincidence and similarity in form and orientation imply that the younger map-scale folds formed coevally with F3 small-scale folds. In the remainder of the paper, both map-scale and small-scale (F3) northeast-trending, open, largely upright folds will be referred to as "younger northeast-trending folds."

It is thus reasonable to seek relationships between small-scale F1 and F2 and the map-scale Bovine fold. Intermediate-scale folds (described above) thought to be in the hinge region of the Bovine fold occur in thick-bedded sandstone of Oquirrh member 3, in which small-scale folds have not been observed. Rotation of the small-scale F2 fold axis data of Figure 6 about the major (map-scale) northeast-trending fold indicates that small-scale F2 folds in plate B were coaxial with the Bovine fold axis prior to northeast-trending folding. Thus, the small-scale F2 folds may have developed concurrently with the Bovine fold. F1 folds may thus pre-date the Bovine fold.

High-Angle Faults

Relatively few high-angle faults have been recognized in the Bovine Mountain area. The continuity of mapped contacts suggests that little significant high-angle faulting occurred, but monotonous and repetitious lithologies and snow cover on available air photos probably have caused oversight of many minor faults. Those mapped trend north and northeast. In general, faults trending northeast have greater throw (as much as 1 km, down to the north) than those that trend more to the north (a few tens of meters, with variable sense of throw) (Fig. 2). One major north-trending fault on the east side of Bovine Mountain offsets low-angle faults I, II, and III. Because offset contacts dip moderately, dip-slip offset of about 700 m with the east side down-dropped could have produced the present map relations.

STRUCTURAL SEQUENCE

Two major low-angle faults and two sets of map-scale folds have been described. A relative timing sequence can be partly drawn from these. Relations with the Immigrant Pass pluton set useful, but incomplete, limits on the absolute ages of the deformational events (Table 1).

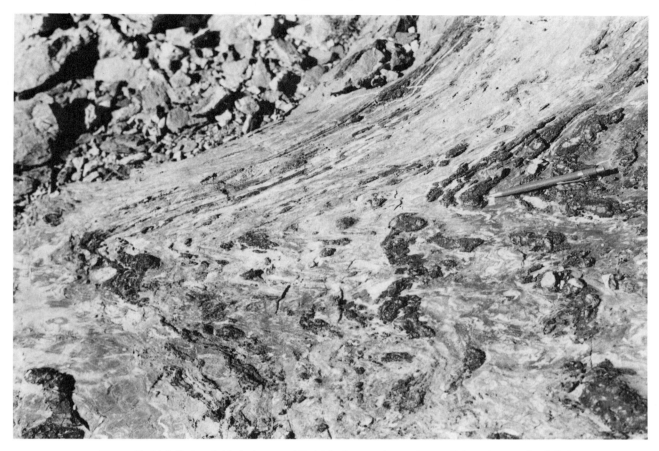

Figure 7. S1 foliation folded about an F2 fold whose axis trends parallel to the pencil, slightly into the plane of the photo. S2 is weakly developed in the core of the fold.

The first small-scale folds, F1, are not known to correlate to any map-scale folds. The north-trending Bovine fold may have been generated at the same time as small-scale F2 folds. The youngest mapped folds, trending northeast, formed coevally with common small-scale folds. Mapping to date indicates that F1 does not fold any of the known faults, although F1 is abundant in the proximity of faults I and II. Fault II is probably folded about the Bovine fold. Faults I and II are both folded by young, northeast-trending folds.

TABLE 1. BOVINE MOUNTAIN DEFORMATION SEQUENCE

Immigrant Pass pluton	Map-scale folds	Small-scale folds	Low-angle faults
			fault V*
23 m.y.- primary cooling or reheating			
	NE-trending	F3	
			Fault I? (reactivated)?
38 m.y. intrusion			
	Bovine fold	F2?	
		F1?	fault II? fault I?

*(Compton, this volume.)

There are several uncertainties in this structural sequence. First, although fault I is folded by the young, northeast-trending fold set, its relationship to the Bovine fold is unclear. Second, there are no unequivocal crosscutting relations between the young, northeast-trending fold set and fault II. The fault IIc trace is folded by northeast-trending folds, but those complex folds in the northwestern part of Figure 2 are not well enough understood to be sure that they are indeed part of the younger fold set. Although the largest northeast-trending fold of the area is thought to control the change in trends from fault IIa to IIb, that relationship cannot be proven in the absence of direct proof that IIa and IIb are continuous. Third, it cannot be certain that fault II is folded by the Bovine fold, without independent proof that contact IIb is a fault and that it is the same fault as IIa.

The age of development of small-scale metamorphic F1 and F2 folds in plate B is not clear. F1 and F2 folds are refolded by F3 folds; F3 is not penetrative, but F1 is. Thus, it seems likely that at least F1 significantly pre-dated F3.

The Bovine fold and the younger, northeast-trending fold set can be dated relative to the Immigrant Pass pluton in the northeast part of the study area and its extension to the west along the northern flank of Bovine Mountain (Figs. 1 and 2). First, the Bovine fold is clearly intruded by the Immigrant Pass pluton. Second, small-scale northeast-trending folds are more abundant and appear to be more ductile (isoclinal, recumbent) near the contact with the pluton, suggesting that F3 (and map-scale, northeast-trending folds) formed at the time of emplacement of the stock or before it cooled. Bent minerals in the granodiorite indicate strain after the pluton cooled (Compton and others, 1977), but field observations indicate that it is only very weakly foliated near its margins and map relations suggest no major deformation of the pluton.

Rb-Sr dating indicates that a lobe of the pluton to the west of Bovine Mountain has an age of 38 m.y. (Compton and others, 1977). K-Ar dating on another western lobe indicates an age of 23 m.y. (Armstrong, 1970). The thermal history of the pluton is not fully known, but the available isotopic data indicate either that the pluton remained heated from 38 to 23 m.y. or that it was reheated about 23 m.a. Therefore, the overturned Bovine fold formed before 38 m.y. (Late Eocene), and the young northeast-trending folds probably developed in mid-Tertiary time (latest Eocene to early Miocene).

No cross-cutting relations between low-angle faults I and II and the Immigrant Pass pluton are known, however. We must infer their ages relative to the major folds. The contact between Oquirrh members 5 and 6 is metamorphosed in the contact aureole of the Eocene stock, and along strike to the southeast that contact is low-angle fault III. But there is no direct evidence indicating whether fault III had only minor offset and died out along strike before reaching the vicinity of the pluton (and thus could be post-intrusive) or whether the metamorphosed contact is in fact an important metamorphosed bedding plane fault (and thus fault III was preintrusive).

Table 1 summarizes the structural sequence of Bovine Mountain based on existing data. We may conclude that before the Late Eocene (38 m.y.) the large north-trending eastward-overturned Bovine anticline developed. We may also be quite confident that faults I and II had detached and transported younger (higher) rocks over older (lower) rocks before mid-Tertiary (approximately 38 to 23 m.y. limits) folding. The relative time of motion along fault I is of great interest because its geometry, omitting Devonian units, may require that it had many kilometers of offset (this may also be true of other low-angle faults mapped at Bovine Mountain, but there is little direct evidence of the amount of offset). Unfortunately, there is little data bearing on its age (Table 1).

Faults I and II formed in the Mesozoic to early Cenozoic (pre-38 m.y.) or in the mid-Cenozoic (38 to 23 m.y.). Fault II has been interpreted to pre-date the Bovine fold, and thus to have been a bedding plane fault with younger-on-older translation and local omission of section

in the hanging wall (plate C). If fault I also pre-dated the Bovine fold, it would also have been a younger-on-older detachment that omitted significant section from its foot-wall (plate A). If fault I formed after the Bovine fold, it apparently moved (initially) structurally higher rocks over (initially) structurally lower rocks in the overturned limb of a pre-existing or coeval fold (the equivalent of a younger-on-older detachment, but in already inverted rocks).

METAMORPHISM

Only preliminary data exist on metamorphic conditions of units at Bovine Mountain, and existing samples are not well-distributed in the map area. In plate B a single sample from a sandstone (located in the northwest part of Figure 2, in the overturned limb of the Bovine fold) was examined in thin section. Very fine grained muscovite defines two foliation directions, and many quartz grains have been strained sufficiently to recrystallize to subgrains with stable boundaries. Fifteen samples in plate C (collected near the trace of section A-A', Fig. 2) all contain metamorphic muscovite; commonly it is very fine grained and often forms two directions of foliation. Quartz subgrains are common but not ubiquitous and locally calcite crystals are twinned. Within about 1 km distance of the pluton near the east end of A-A', five samples of strongly metamorphosed member 6 are strongly foliated with common muscovite and opaque mineral segregations (graphite?) forming a stylolitic cleavage. Although quartz strain is variable, most samples show extensive subgrain development. Four plate C samples from the southeast quadrant of Figure 2 have twinned calcite. In addition, there is minor muscovite in two samples, and minor development of quartz subgrains in just one sample. In quartzites of plate A, quartz grains have pronounced undulatory extinction and local deformation lamellae. No metamorphic minerals have been noted in those samples.

These preliminary data seem to indicate that metamorphic minerals and recrystallization are more common in the northern part of Bovine Mountain than in the southern part, perhaps because of the proximity to the Immigrant Pass pluton. The metamorphism of plate B rocks may be of similar grade as that of plate C, although it will be essential to compare similar composition samples from the south flank of the mountain. Quartz grains in plate A rocks are notably less recrystallized than in samples from plates B and C, but it is very difficult to attribute that contrast to any particular tectonic differences between the plates because several important variables are not constrained by present data. First, the plate A quartzites have different compositions than Oquirrh Group samples. Second, water contents may have varied with compositions. Third, plate B and C rocks were not adequately sampled on the southern side of the mountain, at about the same distance from the Immigrant Pass pluton as the plate A samples.

INTERPRETATION AND REGIONAL RELATIONS

The structural geometry of Bovine Mountain is complex, due to the superposition of low-angle faulting and large-scale overturning in a north-trending fold. As described above (Table 1), an early event in the structural sequence was probably the detachment faulting and younger-on-older translation along fault II of part of the Oquirrh Group (members 2 to 6) over member 1 limestone. This event was followed by large-scale folding with eastward-directed overturning, creating the Bovine fold. About 38 m.y. ago, the Immigrant Pass pluton intruded the folded sequence and, soon after that, a northeast-trending set of folds developed. Younger high-angle normal faults and extensive low-angle detachments (Compton, this volume) developed in the late Tertiary. Fault I, localized in the Mississippian strata and faulting out Devonian units, developed prior to northeast-trending folds. Its relationship to the total structural sequence may be better understood in the context of regional relations.

In the Grouse Creek and Raft River Mountains north of Bovine Mountain, there is a widespread low-angle detachment fault, largely located in the Mississippian strata, that separates an overlying allochthon of Oquirrh Group rocks from an underlying plate of metamorphosed lower and middle Paleozoic units. Devonian strata have been faulted out of the section throughout the Grouse Creek-Raft River Mountains, except in an area extending from just northwest of Bovine Mountain to 10 km farther north. In that area Devonian units are mapped below the detachment in Mississippian units (Compton and others, 1977). In the Raft River and Grouse Creek Mountains, important motion occurred along the fault before or early in the metamorphic sequence (Compton and others, 1977; Compton and Todd, 1979). Important offset also occurred along it after metamorphism, creating a contrast in metamorphic grade across the fault. Because of the similar stratigraphic position of fault I at Bovine Mountain, the very extensive distribution of the fault mapped by Compton and others (1977) to the north, and its possible extension to the east over a wide region (Allmendinger and Jordan, 1981), it is reasonable to correlate fault I with a regional detachment in the Mississippian shales. If that is valid, two lines of reasoning then suggest that fault I pre-dated the Bovine fold. First, it is not reasonable that a regional fault would have remained at the same stratigraphic horizon where strata were strongly folded. Second, the inferred style of motion on fault I (younger-on-older) is geometrically similar to that on fault II, suggesting that they were coeval. Because fault II was inferred above to be folded about the

Bovine fold, it is therefore also reasonable that fault I pre-dated the Bovine fold.

There is at least one significant difference between the regional detachment fault in Mississippian strata mapped by Compton and others (1977) and fault I at Bovine. In the Raft River and Grouse Creek Mountains, the units below that fault (their lower allochthonous sheet) are recrystallized and metamorphosed to greenschist and amphibolite grades. At Bovine Mountain, however, equivalent stratigraphic units are not recrystallized and are not known to contain metamorphic minerals. Therefore, plate A south of Bovine Mountain might be exotic to the metamorphic core complex, or Bovine Mountain might have occupied a position marginal to the metamorphic axis or belt at the time of metamorphism of the lower allochthonous sheet of Compton and others (1977). The low-grade metamorphism of the Oquirrh Group rocks of plate C has not been studied sufficiently to indicate when the metamorphic muscovite formed relative to folding or intrusion of the Immigrant Pass Pluton, so it does not yet constrain models of the timing or regional relations of fault I.

Regardless of the regional correlation of fault I, the omission of 500 to 1000 m of stratigraphic section along this fault nearly paralleling bedding requires either 1) that it faulted a sequence from which Devonian and some Silurian and Ordovician rocks were already structurally removed, or 2) that it had significant translation. If the first option is true of fault I, then the previous structure that it deformed may have been the regional detachment mapped north of Bovine Mountain.

Fault II is of interest relative to the development of the metamorphic core complex. The Oquirrh Group and other units are drastically thinned in the Raft River and Grouse Creek Mountains relative to the same units at Bovine Mountain and in less metamorphosed ranges to the south and east (Compton and others, 1977). Low-angle fault II has been inferred to have formed as a younger-on-older detachment, with strata locally faulted out from plate C. The lateral extent of this fault beyond Bovine Mountain (if it continues) is not known. Fewer subdivisions of the Oquirrh Group have been mapped in the Grouse Creek Mountains than at Bovine Mountain, although most of the same lithologic units have been recognized (Stanford University field camp, 1976; Jordan, unpublished data) so that fault II might have been overlooked. However, if fault II extends to the north, or if geometrically similar faults occur in the Oquirrh Group to the north, then they may omit section widely, possibly contributing to drastic thinning of the Oquirrh Group. Compton and others (1977) recognized that omission of strata along low-angle faults must contribute to thinning of the units in the metamorphic complex, in addition to the effect of flattening during solid state flow.

In the Raft River and Grouse Creek Mountains,

Oligocene to Miocene metamorphic deformation produced west to northwest-trending folds and younger northeast-trending folds (Compton and others, 1977). The northeast-trending folds are of similar orientation and are approximately coeval with the largely upright, northeast-trending folds at Bovine Mountain, suggesting correlation of these fold sets. Understanding the cooling history of the Immigrant Pass pluton and the exact relations of the northeast-trending folds to that pluton will possibly aid in establishing this correlation.

First metamorphic folds in the metamorphic core complex to the north, thought to be pre-38 m.y., may warrant comparison with structures at Bovine Mountain which pre-date the Immigrant Pass pluton, but correlations are not clear. In the southern Grouse Creek Mountains, Raft River Mountains, and Albion Mountains, most of the first metamorphic folds trend north-northeast to east and are overturned to the west and north (Compton and others, 1977; Miller, 1980). At Bovine Mountain, the Bovine fold trends north, similar to the first set in some parts of the metamorphic core complex, but the eastward-overturning at Bovine Mountain is opposite to that observed farther north. Approximately north-trending folds with eastward overturning are characteristic of second metamorphic (but pre-mid-Cenozoic) folds in the Albion Mountains (Miller, 1980).

In summary, certain observations from the Bovine Mountain area are of regional interest. First, the Bovine fold overturned a major stratigraphic section by west-to-east translation before intrusion of a Late Eocene pluton, but it is not clear how the pre-Late Eocene fold sets correlate to the early metamorphic deformation in areas to the north. Second, it is suggested that fault II is folded, and

therefore, that large-scale detachment faulting occurred before 38 m.y., the minimum age of Bovine fold development. Third, plate A is stratigraphically similar to but of different metamorphic grade than Compton and others' (1977) lower allochthonous sheet; that allochthon and plate A may form parts of the same major structural plate. Plate A was in place (relative to other rocks of Bovine Mountain) before F3 folding, which was apparently coeval with the mid-Tertiary second metamorphism of the region north of Bovine. Plate A apparently escaped metamorphism in the mid-Tertiary (due to distance from the metamorphic axis?). In a similar manner, plate A and the rest of the Bovine Mountain sequence may have been only slightly affected by many of the deformation events that distinguish the metamorphic core complex from the region around it.

ACKNOWLEDGMENTS

I thank Robert R. Compton, David M. Miller, and Richard W. Allmendinger for field cooperation and encouragement. With Victoria R. Todd and Michael Carr, they provided critical, patient, and extremely helpful reviews of an earlier draft of this manuscript. I thank the Stanford University 1976 field camp for collaboration during an early stage of this project, and Ed and Diane Mott and family for their hospitality during field work. My mapping of Bovine Mountain has been supported by National Science Foundation Grant EAR 80-18758, Cornell Program for Study of the Continents (COPSTOC), and an American Association of University Women Educational Foundation American Fellowship. Cornell Contribution number 714.

REFERENCES CITED

Allmendinger, R. W., and Jordan, T. E., 1981, Mesozoic evolution, hinterland of the Sevier orogenic belt: Geology, v. 9, p. 308–313.

Armstrong, R. L., 1968, Mantled gneiss domes in the Albion Range, southern Idaho: Geological Society of America Bulletin, v. 79, p. 1295–1314.

——, 1970, Geochronology of Tertiary igneous rocks, eastern Basin and Range province, western Utah, eastern Nevada, and vicinity, U.S.A.: Geochimica et Cosmochimica Acta, v. 34, p. 203–232.

——, unpublished manuscript, Geology of the Basin Quadrangle, Idaho: Albion Mountains and Middle Mountain metamorphic core complex and surrounding Cenozoic rocks.

Baker, W. H., 1959, Geologic setting and origin of the Grouse Creek pluton, Box Elder County, Utah [Ph.D. thesis]: Salt Lake City, University of Utah, 175 p.

Compton, R. R., 1982, Displaced Miocene rocks on the west flank of the Raft River-Grouse Creek core complex, Utah, *in* Miller, D. M., and others, eds., Tectonic and Stratigraphic Studies in the Eastern Great Basin: Geological Society of America Memoir 157 (this volume).

Compton, R. R., and Todd, V. R., 1979, Reply to Discussion *on* Oligocene and Miocene metamorphism, folding, and low-angle faulting in northwestern Utah: Geological Society of America Bulletin, v. 90, p. 305–309.

Compton, R. R., Todd, V. R., Zartman, R. E., and Naeser, C. W., 1977, Oligocene and Miocene metamorphism, folding, and low-angle faulting in northwestern Utah: Geological Society of America Bulletin, v. 88, p. 1237–1250.

Crittenden, M. D., Jr., 1979, Discussion *on* Oligocene and Miocene metamorphism, folding, and low-angle faulting in northwestern Utah: Geological Society of America Bulletin, v. 90, p. 305–309.

Doelling, H. H., 1980, Geology and Mineral resources of Box Elder County, Utah: Utah Geological and Mineral Survey, Bulletin 115, 251 p.

Jordan, T. E., 1979, Lithofacies of the Upper Pennsylvanian and Lower Permian western Oquirrh Group, northwest Utah: Utah Geology, v. 6, no. 2, p. 41–56.

Jordan, T. E., and Douglass, R. C., 1980, Paleogeography and structural development of the Late Pennsylvanian to early Permian Oquirrh Basin, northwestern Utah, *in* Fouch, T. C., and Magathan, E. R., eds., Paleozoic Paleogeography of the West-Central United States; West-Central United States Paleogeography Symposium 1: Rocky Mountain Section Society of Economic Paleontologists and Mineralogists, Denver, Colorado, p. 217–238.

Miller, D. M., 1980, Structural geology of the northern Albion Mountains, south-central Idaho *in* Crittenden, M. D., Jr., Coney, P. J., and Davis, G. H., eds., Cordilleran Metamorphic Core Complexes: Geological Society of America Memoir 153, p. 399–423.

——, 1982, Allochthonous quartzite sequence in the Albion Mountains, Idaho and proposed Proterozoic Z and Lower Cambrian correlatives in the Pilot Range, Utah and Nevada, *in* Miller, D. M., and others, eds., Tectonic and Stratigraphic Studies in the Eastern Great Basin: Geological Society of America Memoir 157 (this volume).

Todd, V. R., 1980, Structure and petrology of a Tertiary gneiss complex in northwestern Utah, *in* Crittenden, M. D., Jr., Coney, P. J., and Davis, G. H., eds., Cordilleran Metamorphic Core Complexes: Geological Society of America Memoir 153, p. 349–383.

Todd, V. R., 1982, Miocene displacement of Tertiary and pre-Tertiary rocks in the Matlin Mountains, northwestern Utah, *in* Miller, D. M., and others, eds., Tectonic and Stratigraphic Studies in the Eastern Great Basin: Geological Society of America Memoir 157 (this volume).

MANUSCRIPT ACCEPTED BY THE SOCIETY AUGUST 20, 1982

Geological Society of America
Memoir 157
1983

Structural evolution of the Raft River Basin, Idaho

H. R. Covington
U.S. Geological Survey
Box 25046, Mail Stop 954
Denver Federal Center
Denver, CO 80225

ABSTRACT

Recent geological mapping, geophysical studies, and deep drilling in the Raft River area, Idaho, have yielded information that is not consistent with fault-block development of the Raft River basin. Paleozoic and lower Mesozoic allochthonous rocks that occur in the surrounding Sublett, Black Pine, Albion, and Raft River Mountains do not occur beneath the Cenozoic basin fill deposits of the Raft River Valley, nor do Cenozoic volcanic rocks that form the adjacent Cotterel and Jim Sage Mountains. Range-front faults have not been identified along the margins of ranges flanking the Raft River Valley. Normal faults found in the Cotterel and Jim Sage Mountains are inferred to be concave-upward extensional structures that involve only Tertiary volcanic rocks and basin-filling sediments. Concave-upward faults within the Raft River basin have been identified in seismic reflection profiles. Fault displacement of the basement rocks beneath the Raft River Valley has not been documented. Structural development of the Raft River basin based on gravity-induced tectonic denudation of nearby metamorphic core complexes is suggested.

INTRODUCTION

The Raft River Valley lies in a north-trending Cenozoic basin near the northern limits of the Basin and Range province. The valley opens northward onto the Snake River Plain and is flanked on the east by the Sublett and Black Pine Mountains; on the west by the Cotterel, Jim Sage, and Albion Mountains; and on the south by the Raft River Mountains (Fig. 1).

Long known as a thermal area (Stearns and others, 1938), part of the southern Raft River Valley, near Bridge, was designated the Frazier Known Geothermal Resource Area (Godwin and others, 1971) by the U.S. Geological Survey in 1971. In 1973, the U.S. Geological Survey and the U.S. Department of Energy (formerly the U.S. Energy Research and Development Administration) began a cooperative multidisciplinary investigation of the Raft River geothermal system in order to provide a scientific framework for the evaluation of a geothermal resource. These investigations included surface geological and geophysical studies and a drilling program designed to aid subsurface studies as well as to develop the geothermal resource. The results of these studies have been summarized by Williams

and others (1976), Mabey and others (1978), Keys and Sullivan (1979), and Covington (1980).

The geothermal resource is contained in a fracture-dominated reservoir near the base of the Cenozoic basin fill. Recognition of an extensive fracture system within the basin fill beneath the Raft River Valley requires a different structural model for development of the basin than the fault-block model originally proposed by Williams and others (1976). Recent geological mapping, geophysical studies, and deep drilling in the Raft River area have yielded information that suggests a model based on gravity-induced tectonic denudation of nearby metamorphic core complexes for development of the Raft River basin.

This paper presents evidence for such a model and describes the sequence of tectonic events in late Cenozoic time.

GEOLOGIC FRAMEWORK

Albion and Raft River Mountains

The Precambrian basement complex that underlies the Raft River area is exposed in five domes within the Albion

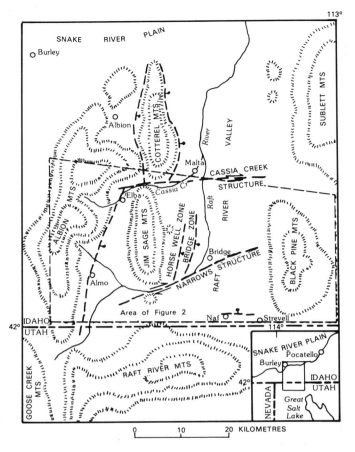

Figure 1. INDEX MAP, showing area of geologic map (Fig. 2) and major inferred structures. Bar and ball on downthrown side of faults, arrow indicates direction of slip.

and Raft River Mountains. The cores of these domes consist of 2.5-b.y.-old (Armstrong, 1976; Compton and others, 1977) Archean granite and granite gneiss unconformably overlain by Proterozoic Z and lower Paleozoic schists and quartzites of the regional autochthon. Above the autochthon, two allochthonous sheets are exposed in the Raft River Mountains (Compton, 1972, 1975; Compton and others, 1977) and possibly as many as four allochthonous sheets are exposed in the Albion Mountains (Armstrong, 1968; Miller, 1980). The lower allochthonous sheets generally consist of metamorphosed lower Paleozoic rocks, whereas the upper or highest sheet is slightly metamorphosed or nonmetamorphosed upper Paleozoic and lower Mesozoic rocks. Exposures of the allochthonous sheets are restricted almost wholly to the west flank of the Albion Mountains (Armstrong, 1968; Miller, 1980) and the west end of the Raft River Mountains (Compton, 1972, 1975). An exception is a group of small klippen of nonmetamorphosed upper Paleozoic rocks of the upper allochthonous sheet located along the north and east flanks of the Raft River Mountains (Compton, 1975; Compton and others, 1977). Rock exposures on the east flank of the Albion

Mountains and the north flank of the eastern Raft River Mountains are dominantly those of the autochthon (Armstrong, 1968; Compton, 1972, 1975; Miller, 1980). The unconformity between the crystalline basement and the overlying metasedimentary rocks culminates at 2,900 m above sea level in both the Albion (Armstrong and others, 1978) and Raft River Mountains (Compton, 1975). Deep drilling in the Raft River Valley has identified the same unconformity at about 300 m below sea level (Covington, 1977a-d, 1979a, b). These evaluations yield a minimum structural relief of about 3,200 m, and an average slope of about 6.5 degrees for the top of the Archean crystalline basement. Miller (1980) records slopes in excess of 30 degrees on the flanks of the Big Bertha dome, central Albion Mountains. Normal faults are common in the Albion and Raft River Mountains, but most have small displacements and are not of major structural importance (Armstrong, 1968; Compton and others, 1977; Miller, 1980).

Despite numerous radiometric dates throughout the region, timing of events in the Albion and Raft River Mountains is still open to question. It is generally agreed, however, that: (1) the crystalline basement is 2.5 b.y. old; (2) regional west to east thrusting occurred during the early Tertiary; (3) an early metamorphic event ended in late Cretaceous or early Tertiary time; and (4) a late metamorphic event was active until Miocene time.

Sublett and Black Pine Mountains

Weakly metamorphosed to unmetamorphosed allochthonous miogeoclinal rocks of late Paleozoic to early Mesozoic age make up the Sublett and Black Pine Mountains. These rocks are exposed in two or possibly three structural plates in the Sublett Mountains (R. L. Armstrong, unpublished data, 1977), two structural plates in the northern Black Pine Mountains, and three structural plates in the southern Black Pine Mountains (Smith, in press). A high-angle reverse fault, the West Dry Canyon fault (Fig. 2), transects the Black Pine Mountains near West Dry Canyon, separating the mountain range into two distinct structural and lithologic blocks (Smith, in press). Conodont collections from north and south of the West Dry Canyon fault (Fig. 2) indicate that the two structural blocks that form the Black Pine Mountains may have had different thermal histories (Smith, in press). This fault is in close alignment with the Narrows structure (Figs. 1 and 2). Ash-flow tuffs and tuffaceous sediments of Tertiary age exposed along the margins of the two mountain ranges clearly are unconformable on the upper Paleozoic rocks, and locally they contain random bedding dips in excess of 30 degrees (R. L. Armstrong, unpublished data, 1977; Smith, in press). Normal faults found within the two mountain ranges are sparse and appear to be of small displacement. Steep gravity contours along the west sides of both mountain ranges

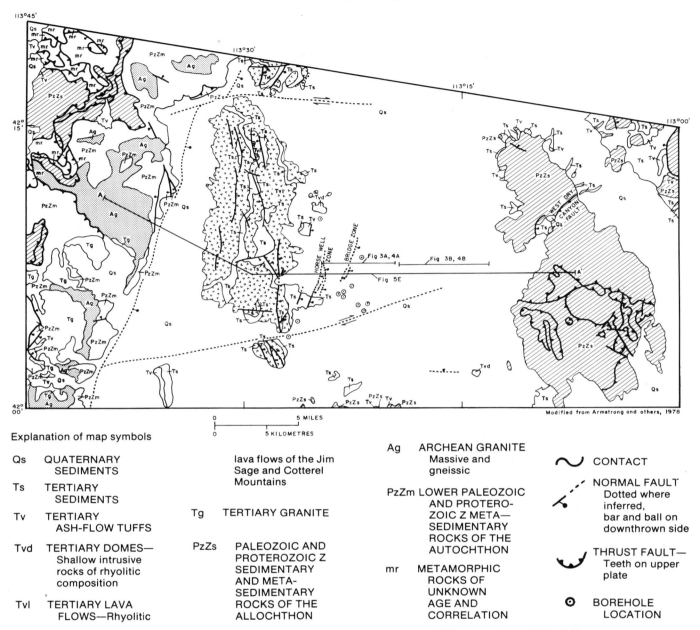

Figure 2. GENERALIZED GEOLOGIC MAP OF THE SOUTHERN RAFT RIVER VAL-
LEY, IDAHO, showing locations of boreholes, seismic reflection profiles (Figs. 3 and 4), and
line of interpretative structure section (Fig. 5E).

(Mabey and Wilson, 1973; Mabey and others, 1978) have
been interpreted as faults buried beneath the basin fill on
the west side of both ranges. A seismic reflection profile,
near the Black Pine Mountains in the Raft River Valley
(Figs. 2, 3B and 4B), shows that the basin floor slopes
westward away from the range. Rotation of the basin floor
appears to be associated with displacement on normal
faults that dip eastward toward the mountains.

Cotterel and Jim Sage Mountains

Two Tertiary rhyolite lava flows with a tuffaceous sedimentary unit between them in most places make up the
Cotterel and Jim Sage Mountains. Locally the middle sedimentary unit contains vitrophyre breccia and densely
welded ash-flow tuff (Williams and others, 1974). The two
mountain ranges define a north-trending anticline broken

by numerous normal faults. The east flank of the anticline dips 15 to 35 degrees toward the east, or Raft River basin, and the west flank dips 5 to 30 degrees toward the west. Most of the normal faults strike between N 30° E to N30°W, although there are also several faults that strike west-northwest to west (Williams and others, 1974; Pierce and others, in press). The sense of displacement on most of the north-trending faults is down to the east; however, along the crests of both ranges apical grabens occur in some places. Displacement on the faults is generally a few meters to a few tens of meters. The greatest offset is on the east side of the Cotterel Mountains, just north of Cassia Creek, where stratigraphic displacement is several hundred meters (Williams and others, in press). North-trending stratigraphic and structural relationships exposed in the Cotterel and Jim Sage Mountains are offset along the valley of Cassia Creek that separates the two ranges. The relationship suggests a right-lateral fault along Cassia Creek, here named the Cassia Creek structure (Figs. 1 and 2). Similarly, stratigraphic and structural relationships across the Raft River Narrows at the south end of the Jim Sage Mountains suggests right-lateral offset on the Narrows structure (Williams and others, 1976) (Figs. 1 and 2).

Rhyolite lava flows adjacent to the Raft River basin are restricted to the Cotterel and Jim Sage Mountains. The aerial extent of the lava flows is only slightly larger than that of the rock exposures: the flows do not extend far beneath the alluvial fan deposits. Radiometric ages from the upper rhyolite lava indicate that the flows are 9 to 11 m.y. old (Armstrong, 1976; Williams and others, 1976; Pierce and others, in press). Several small rhyolitic domes along the east flanks of the two mountain ranges and in the southeast corner of the Raft River Valley yield radiometric ages of 7 to 9 m.y. (Williams and others, 1976). An ash-flow tuff exposed on the east side of the Jim Sage Mountains also yields a radiometric age between 7 to 9 m.y. (Williams and others, 1976), as does an ash-flow tuff exposed in the upper Raft River Valley southwest of the Jim Sage Mountains (G. B. Dalyrmple, written communication, 1979).

Raft River Valley

The Raft River Valley is a north-trending basin filled with nearly 1,600 m of Cenozoic fluvial sediments that began accumulating during the early Miocene. Deep drill-hole data (Covington, 1977 a-d, 1978, 1979a, b; Oriel and others, 1978) indicate a general decrease in gravel content and an increase in open fractures and hydrothermal alteration downward. The drill-hole data also indicate that basin fill south of the Narrows structure is coarser and less well-indurated than is basin fill north of the structure. Correlation of depositional units within the basin fill is complicated by rapid lateral changes in both thickness and facies, by hydrothermal alteration, and by complex structures.

Seismic refraction studies show that velocities in the basin fill vary laterally and possibly vertically, corresponding to zones of hydrothermal alteration (Ackerman, 1979). Seismic reflection profiles indicate a complex depositional history for the basin (Figs. 3 and 4). Reflectors in the west and east parts of the basin dip basinward, whereas near the center of the basin reflectors are subhorizontal in the shallow part and gently dipping in the deeper part. Lateral discontinuities and terminations of reflectors indicate common faulting within the basin fill.

Deep drilling in the basin did not reveal lavas of the type exposed in the Cotterel and Jim Sage Mountains, nor were ash-flow tuffs or Paleozoic rocks of the types exposed in the other surrounding mountains found (Covington, 1977a-d, 1978, 1979a, b; Oriel and others, 1978). The base of the Cenozoic fill, marked by a breccia (Covington, 1979b), is in fault contact with schist and quartzite units of the autochthon that can be correlated with formations exposed in the Albion and Raft River Mountains (Covington, 1980). Seismic reflections (Figs. 3 and 4) known or inferred to represent basin-fill sediments dip into subjacent reflections interpreted to be representative of basement rocks.

Quaternary alluvial fans in the southern part of the valley are marked by linear features that were first interpreted to be fault scarps by Williams and others (1974). An east-west set of scarps in the southeast part of the valley, near Naf, probably indicates faulting with the north or basin side down. North-northeast trending scarps on the west side of the valley form two subparallel sets called the Horse Well and Bridge zones (Figs. 1 and 2). Mapping near the Raft River and seismic reflection profiles indicate that the Horse Well and Bridge zones are probably faults with the east or basin side down. Trenching across the zones shows no offset of Quaternary fan material and small displacements of the Tertiary beds (Williams and others, in press). Gravity contours indicate that the Horse Well and Bridge zones terminate near the Narrows structure (Mabey and others, 1978). A seismic refraction line across the Bridge zone shows no positive evidence of offset on the basement horizon (Ackerman, 1979). Seismic reflection profiles in the area of the Bridge zone show abundant east- and west-dipping faults in the Tertiary basin fill but no offset in the basement surface (Fig. 3). The faults seen in the reflection profiles are generally steep dipping near the top of the basin fill and flatten downward, becoming subparallel with the top of the basement complex (Fig. 4). Age of last movement on these faults is probably late Pliocene, based on age relations of Quaternary deposits and soils in the northern part of the valley and trenching (Pierce and others, in press; Williams and others, in press). The Cassia Creek structure and the Narrows structure transecting the Raft River Valley in east-west and east-northeast directions, respectively, have no surface expression in the Quaternary alluvium.

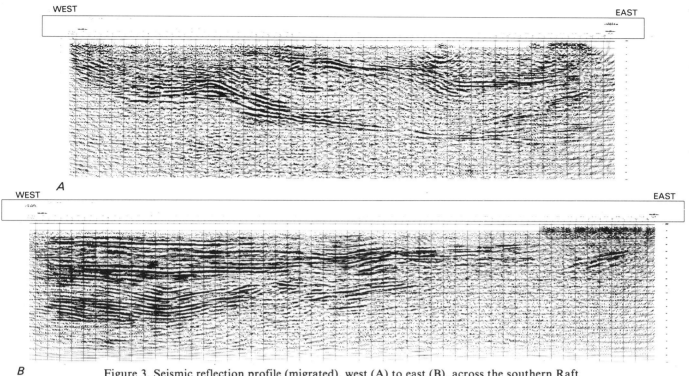

Figure 3. Seismic reflection profile (migrated), west (A) to east (B), across the southern Raft River Valley, Idaho.

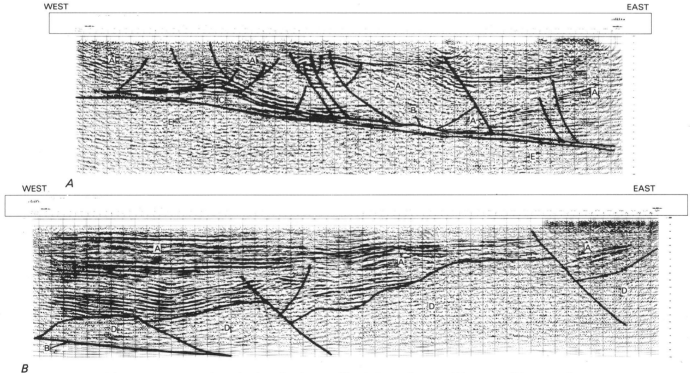

Figure 4. Interpretative seismic reflection profiles (migrated), west (A) to east (B) across the southern Raft River Valley, Idaho. A, Cenozoic basin fill; B, detachment surface; C, ductilly deformed lower Paleozoic and Proterozoic Z schists and quartzites; D, metamorphosed to nonmetamorphosed Paleozoic and Proterozoic Z rocks; and E, Archean granitic basement rocks.

Cassia Creek and Narrows Structures

Geologic relations and geophysical anomalies in the Raft River basin indicate the existence of steep east- and east-northeast trending transcurrent structures. The Cassia Creek structure offsets the alignment of stratigraphic and structural relationships between the Jim Sage and Cotterel Mountains along the valley of Cassia Creek (Figs. 1 and 2). Mapping shows a major component of right-lateral movement, and dip-slip with the north side down 200 m.

The Narrows structure passes through the lower Narrows of the Raft River at the south end of the Jim Sage Mountains and across the southern Raft River Valley toward the north end of the Black Pine Mountains (Figs. 1 and 2). On opposite sides of the Raft River Narrows, discordant relations in Tertiary sediments and rhyolite lavas indicate right-lateral offset on the Narrows structure. Within the southern Raft River Valley, there is no direct geological expression of the Narrows structure. It was noted above that none of the north-trending normal faults in the Jim Sage Mountains or Raft River Valley crosses the Narrows structure (Williams and others, 1976) and that the nature of the Tertiary basin fill is somewhat different on either side of the structure. Geophysical studies in the form of gravity, magnetic, seismic refraction, and d-c resistivity surveys indicate northeast trending anomalies in the Tertiary rocks east of the Raft River Narrows (Mabey and others, 1978). The data indicate the existence of major geologic changes across the zone, but they do not define a discrete structure. Seismic refraction studies (Ackerman, 1979) and seismic reflection profiles show extreme complexity of the Tertiary basin fill, but do not indicate the presence of a basement feature that might coincide with the Narrows structure. Deep drilling within the zone of the Narrows structure also shows chaotic basin fill with no apparent disturbance of the basement rocks.

DISCUSSION

Structurally, the gneiss domes that form the "cores" of the Albion and Raft River Mountains are prominent features in the area. Prior to gneiss dome development, regional west-to-east overthrusting placed thick sheets of Paleozoic and lower Mesozoic rock over the entire area. Presently these allochthonous sheets primarily occur on the west side of the Albion Mountains, the west end of the Raft River Mountains and in the Sublett and Black Pine Mountains. Only small patches or "klippen" are found on the east flank of the Albion Mountains and along the north and east flanks of the Raft River Mountains. No Paleozoic or lower Mesozoic rocks have been identified within the Raft River basin from deep drilling in the basin. Seismic reflection profiles indicate that Paleozoic rocks may exist beneath basin fill along the east margins of the valley near the

Sublett and Black Pine Mountains, where no drill-hole data are available (Figs. 3 and 4). In order to explain these relationships by block faulting, at least 5,000 m of Paleozoic and lower Mesozoic rock need to be removed from the Raft River basin area during an episode of prebasin block uplift and erosion. There is no known sedimentary record of this event, nor have the required faults been identified.

The rhyolitic lavas that form the Cotterel and Jim Sage Mountains are highly restricted in their east-west distribution and yet extend 55 km in a north-south direction. Drilling in the Raft River Valley has not identified these rocks within the sedimentary basin fill, nor have ash-flow tuffs found on the flanks of the surrounding mountains been identified within the Raft River basin fill. The absence of these volcanic rocks within the Raft River basin fill also is inconsistent with a block-fault model because the basin areas where these rocks are missing would have to have been a positive topographic and structural feature during volcanism. Tectonic denudation of the nearby gneiss domes presents a much simpler explanation for the distribution of Paleozoic, lower Mesozoic, and Tertiary volcanic rocks in the Raft River area.

High-angle normal faulting within the Albion, Raft River, Sublett, and Black Pine Mountains played a minor role and no large displacements have been observed. Range-front faults have not been mapped on the west side of the Sublett or Black Pine Mountains. A seismic reflection profile in the eastern part of the Raft River basin suggests an eastward-thinning wedge of basin-fill sediments. Lateral terminations and discontinuities in reflections are interpreted as faults that displace strata down toward the Black Pine Mountains (Figs. 3B and 4B). If the faults are low-angle extensional structures dipping eastward toward the mountains (Fig. 4B), a tectonic denudation model easily explains the observed features. Normal faults are abundant in the Jim Sage and Cotterel Mountains and were inferred along the west side of the southern Raft River Valley. Faults along the west margin of the valley were first thought by Williams and others (1974) to be buried range-front faults based on displacement toward the basin and a steep gravity profile. Seismic reflections (Fig. 3), seismic refraction, and drilling in the basin could not confirm basement displacement along the east side of the Jim Sage Mountains. The seismic reflection profiles indicate concave-upward extensional structures near the west side of the Raft River basin that involve only the Tertiary basin fill (Figs. 3A and 4A). Deep drilling within the basin has identified a breccia zone throughout the southern Raft River basin at the Tertiary-Precambrian contact. These structural features support the tectonic denudation model for basin development.

The Cassia Creek structure appears to be a near-vertical detachment surface analogous with a tear fault within the Cenozoic rocks that has allowed differential

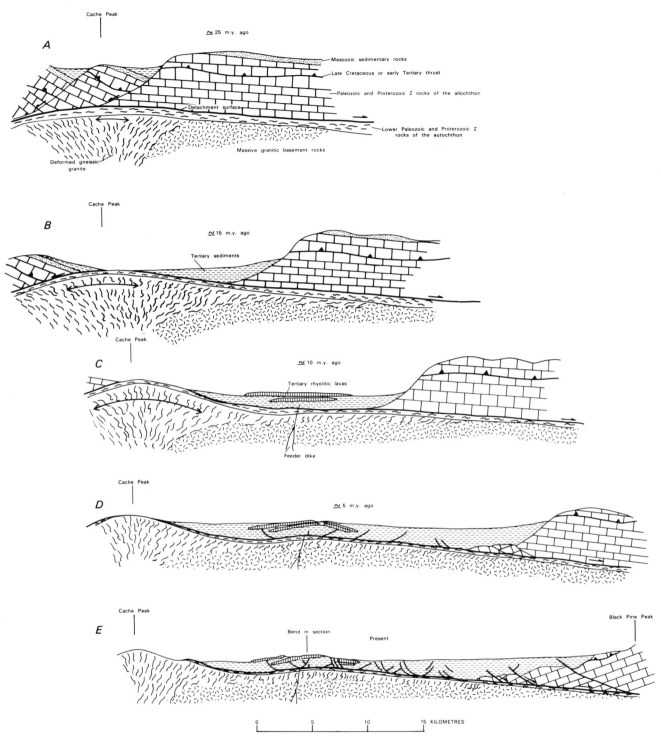

Figure 5. Time-sequential, schematic structure sections across the southern Raft River basin, Idaho, showing interpreted stages of basin evolution. A, Metamorphism and expansion of basement rocks, ductile thinning of schists and quartzites of the autochthon, formation of a detachment surface and beginning of gravity-gliding of cover rocks away from rising dome; B, continued thermal expansion of basement rocks and gravity gliding of cover rocks away from dome, deposition of clastic sediments on flank of domes; C, eruption of rhyolitic lavas in restricted basin between rising domes and eastward moving gravity-glide blocks; D, sagging of basement complex near domes, extension of basin-filling sediments and volcanic rocks and Paleozoic-Proterozoic Z glide blocks along concave-upward normal faults; and E, interpretative structure section (A-A′, Fig. 2) from Cache Peak in the Albion Mountains to Black Pine Peak in the Black Pine Mountains. Paleozoic and Proterozoic Z rocks are unit D in Figure 4B.

translation of the Cotterel and Jim Sage Mountains (Figs. 1 and 2). The Narrows structure is also a near-vertical detachment surface within the Cenozoic rocks that has allowed some differential translation at the southern end of the Jim Sage Mountains (Figs. 1 and 2). The West Dry Canyon fault (Fig. 2), separating the northern and southern Black Pine Mountain blocks, is apparently an extension of the Narrows structure. Geological and structural features found in the surrounding mountains and within the Raft River basin, described above, can be explained best by using a model based on gravity-induced tectonic denudation of the nearby gneiss dome complexes.

INTERPRETATION

The sequence of late Cenozoic tectonic events related to uplift of the metamorphic core complexes and subsequent tectonic denudation is shown in an interpretative mode in Figure 5A-E.

Extensive subhorizontal allochthonous sheets of Paleozoic and lower Mesozoic rocks totaling more than 10,000 m in thickness were in place over the entire region by middle Oligocene time as inferred from Armstrong (1968) and Compton and others (1977). In the late Oligocene, a regional increase in thermal activity produced plutons at depth and initiated regional metamorphism and the development of gneiss domes beneath the present Albion and Raft River Mountains. As the domes rose, metamorphic fluids produced by the increase in regional thermal activity increased the fluid pore pressure near the base of the Paleozoic and lower Mesozoic cover rocks. At some point early in dome development, fluid pore pressure exceeded lithostatic load and a detachment surface formed along a previous thrust surface at the top of the autochthon (Fig. 5A). Gravity-induced gliding of the entire overlying section of allochthonous Paleozoic and lower Mesozoic cover rocks began in a generally eastward direction away from the domes, in a manner similar to that described by Davis and Coney (1979). As a result of sustained high fluid pore pressure and a rapid rate of dome uplift, the allochthonous cover rocks moved away from the domes as large coherent blocks (Figs. 5A, B, C) rather than as thin slices extending along low-angle extensional faults. Clastic sediments were deposited on the flanks of the domes in the wake of these relatively high-standing gravity-glide blocks (Fig. 5B). By the late Miocene, the trailing edge of the gravity-glide blocks had moved about 25 km eastward from their original position above the gneiss domes. Coinciding with the northeastward passage of the thermal pulse now beneath Yellowstone National Park (Christiansen and Lipman, 1972), rhyolitic lavas and associated ash-flow tuffs and tuffaceous sediments of the Jim Sage Volcanic Member of the Salt Lake Formation (Williams and others, in press) were deposited in this narrow basin (Fig. 5C).

The basin-fill sediments and volcanic rocks moved eastward along the detachment surface with the Paleozoic and lower Mesozoic cover rock for 1 to 3 m.y. Between 9 and 7 m.y. ago, the volcanic rocks and locally associated sediments ceased to move eastward along the detachment surface; this cessation may have been the result of local sagging in the basement surface, irregularities in the detachment surface due to earlier regional sagging of the basement surface, a reduction in fluid pore pressure, or a combination of factors. The Narrows and the Cassia Creek structures developed as transverse faults to compensate for contrasting distance and rate of movement to the east.

During late Miocene time, ash-flow tuffs were deposited across the existing valleys and onto the flanks of the existing mountains. Meanwhile, voluminous sediments were being deposited into the expanding basin between the eastward-moving gravity-glide blocks and the rising gneiss domes. Sagging of parts of the basement surface near the domes began as thermal activity decreased. This sagging caused the broken anticlinal shape of the Jim Sage and Cotterel Mountains and created a deep sediment trap between the eastward-moving gravity-glide blocks and the domes (Fig. 5D). With continued rising of the basement rocks and eastward movement of the large coherent blocks of Paleozoic and lower Mesozoic rock, extension of the basin-filling sediments and volcanic rocks along concave-upward normal faults was initiated (Fig. 5D). The concave-upward faults have steep dips (60° to 90°) at the surface flattening downward until the faults merge with the detachment surface at the base of the Cenozoic basin fill. These extensional faults are interpreted as growth-type faults in which displacement increases in progressively older strata. Merging of the faults downward into a single zone results in large displacements along the basin fill-basement detachment surface. By the Pliocene, mechanics of the gravity-glide detachment surface had changed sufficiently to bring about extension of the eastward-moving Paleozoic and lower Mesozoic rocks along low-angle extensional faults (Fig. 5D). Extension of the Raft River basin cover terrain, including both the Paleozoic and lower Mesozoic rocks and the basin-filling sediments and volcanic rocks, continued through the Pliocene (Williams and others, in press) and still may be active (Fig. 5E).

ACKNOWLEDGEMENTS

I thank K. L. Pierce for thoughtful discussion during formulation of this paper, R. L. Armstrong for providing unpublished mapping of the Basin and Sublett quadrangles, J. F. Smith, Jr. for providing unpublished mapping of the Strevell quadrangle, and H. D. Ackermann for providing seismic reflection profiles of the Raft River basin. I also would like to thank R. E. Anderson, P. L. Williams, E. G. Frost, D. M. Miller, and K. A. Howard for very helpful comments and suggestions during review.

REFERENCES CITED

Ackermann, H. D., 1979, Seismic refraction study of the Raft River goethermal area, Idaho: Geophysics, v. 44, p. 216–255.

Armstrong, R. L., 1968, Mantled gneiss domes in the Albion Mountains, southern Idaho: Geological Society of America Bulletin, v. 79, p. 1295–1314.

——, 1976, The geochronometry of Idaho: Isochron/West, no. 15, p. 1–33.

Armstrong, R. L., Smith, J. F., Jr., Covington, H. R., and Williams, P. L., 1978, Preliminary geologic map of the west half of the Pocatello 1° x 2° quadrangle, Idaho: U.S. Geological Survey Open-File Report 78–533, scale 1:250,000, 1 sheet.

Christiansen, R. L., and Lipman, P. W., 1972, Cenozoic volcanism and plate tectonic evolution of the western United States II; Late Cenozoic: Royal Society of London Philosophical Transactions, ser. A, v. 271, p. 240–284.

Compton, R. R., 1972, Geologic map of the Yost quadrangle, Box Elder County, Utah and Cassia County, Idaho: U.S. Geological Survey Miscellaneous Geologic Investigations Map I–672, scale 1:31,680.

——, 1975, Geologic map of the Park Valley quadrangle, Box Elder County, Utah and Cassia County, Idaho: U.S. Geological Survey Miscellaneous Geologic Investigations Map I–873, scale 1:31,680.

Compton, R. R., Todd, V. R., Zartman, R. E., and Naeser, C. W., 1977, Oligocene and Miocene metamorphism, folding and low-angle faulting in northwestern Utah: Geological Society of America Bulletin, v. 88, p. 1237–1250.

Covington, H. R., 1977a, Deep drilling data, Raft River geothermal area, Idaho; Raft River geothermal exploration well no. 1: U.S. Geological Survey Open-File Report 77–226, 1 sheet.

——, 1977b, Deep drilling data, Raft River geothermal area, Idaho; Raft River geothermal exploration well no. 2: U.S. Geological Survey Open-File Report 77–243, 1 sheet.

——, 1977c, Deep drilling data, Raft River geothermal area, Idaho; Raft River geothermal exploration well no. 3: U.S. Geological Survey Open-File Report 77–616, 1 sheet.

——, 1977d, Deep drilling data, Raft River geothermal area, Idaho; Raft River geothermal exploration well no. 3, Sidetrack C: U.S. Geological Survey Open-File Report 77–883, 1 sheet.

——, 1978, Deep drilling data, Raft River geothermal area, Idaho; Raft River geothermal exploration well no. 4: U.S. Geological Survey Open-File Report 78–91, 1 sheet.

——, 1979a, Deep drilling data, Raft River geothermal area, Idaho; Raft River geothermal production well no. 5: U.S. Geological Survey Open-File Report 79–382, 1 sheet.

——, 1979b, Deep drilling data, Raft River geothermal area, Idaho; Raft River geothermal production well no. 4: U.S. Geological Survey Open-File Report 79–662, 1 sheet.

——, 1980, Subsurface geology of the Raft River geothermal area, Idaho: Geothermal Resources Council Transactions, v. 4, p. 113–115.

Davis, G. H., and Coney, P. J., 1979, Geologic development of the Cordilleran metamorphic core complexes: Geology, v. 7, p. 120–124.

Godwin, L. H., Haigler, L. B., Rioux, R. L., White, D. E., Muffler, L.J.P., and Wayland, R. G., 1971, Classification of public lands valuable for geothermal steam and associated geothermal resources: U.S. Geological Survey Circular 647, p. 17.

Keys, S. W., and Sullivan, J. K., 1979, Role of borehole geophysics in defining the physical characteristics of the Raft River geothermal reservoir, Idaho: Geophysics, v. 44, no. 6, p. 1116–1141.

Mabey, D. R., and Wilson, C. W., 1973, Regional gravity and magnetic surveys in the Albion Mountains area of southern Idaho: U.S. Geological Survey Open-File Report, 12 p.

Mabey, D. R., Hoover, D. B., O'Donnell, J. E., and Wilson, C. W., 1978, Reconnaissance geophysical studies of the geothermal system in southern Raft River valley, Idaho: Geophysics, v. 43, no. 7, p. 1470–1484.

Miller, D. M., 1980, Structural geology of the northern Albion Mountains, south-central Idaho, *in* Crittenden, M. D., Jr., and others, eds., Cordilleran metamorphic core complexes: Geological Society of America Memoir 153, p. 399–423.

Oriel, S. S., Williams, P. L., Covington, H. R., Keys, W. S., and Shaver, K. C., 1978, Deep drilling data, Raft River geothermal area, Idaho; Standard American Oil Company Malta, Naf and Strevell petroleum test boreholes: U.S. Geological Survey Open-File Report 78–361, 2 sheets.

Pierce, K. L., Covington, H. R., Williams, P. L., and McIntyre, D. H., 1983, Geologic map of the Cotterel Mountains and northern Raft River Valley, Idaho: U.S. Geological Survey Miscellaneous Geologic Investigations Map I–1450, scale 1:48,000 (in press).

Smith, J. F., Jr., 1982, Geologic map of the Strevell quadrangle, Cassia County, Idaho: U.S. Geological Survey Miscellaneous Investigations Map I–1403, scale 1:62,500 (in press).

Stearns, H. T., Crandall, Lynn, and Steward, W. G., 1938, Geology and groundwater resources of the Snake River Plain in southeastern Idaho: U.S. Geological Survey Water-Supply Paper 774, 268 p.

Williams, P. L., Pierce, K. L., McIntyre, D. H., and Schmidt, P. W., 1974, Preliminary geologic map of the southern Raft River Valley, Cassia County, Idaho: U.S. Geological Survey Open-File Report, scale 1:24,000.

Williams, P. L., Mabey, D. R., Zohdy, A.A.R., Ackermann, Hans, Hoover, D. B., Pierce, K. L., and Oriel, S. S., 1976, Geology and geophysics of the southern Raft River valley geothermal area, Idaho, U.S.A., *in* Proceedings, United Nations Symposium on the Development and Use of Geothermal Resources, 2d, San Francisco, California, May 20–29, 1975, v. 2, p. 1273–1282.

Williams, P. L., Covington, H. R., and Pierce, K. L., 1982, Cenozoic stratigraphy and tectonic evolution of the Raft River basin, Idaho, *in* Cenozoic geology of Idaho: Idaho Bureau of Mines and Geology (in press).

Manuscript Accepted by the Society August 20, 1982

Geological Society of America
Memoir 157
1983

Late Miocene displacement of Pre-Tertiary and Tertiary rocks in the Matlin Mountains, northwestern Utah

V. R. Todd
U.S. Geological Survey
345 Middlefield Road
Menlo Park, California 94025

ABSTRACT

The Matlin Mountains, an area of low hills on the northern edge of the Great Salt Lake Desert, expose a rootless sequence of upper Paleozoic, lower Mesozoic, and upper Miocene rocks that have been affected by folds and by older-on-younger low-angle faults. The area lies 5 to 10 km east of the Grouse Creek Mountains, the southernmost range of a metamorphic core complex which extends from south-central Idaho into northwestern Utah. In this core complex, metamorphism that was accompanied by intrusion, younger-on-older low-angle faulting, recumbent folding, and horizontal extension of rock units did not end until Miocene time.

The Matlin Mountains expose the coarse Tertiary sedimentary rocks and the displaced sheets that were shed from uplifts that followed metamorphism and intrusion in the core complex. These structurally interlayered sedimentary deposits and rock sheets overlie the zone of transition between metamorphically attenuated Paleozoic rocks in the core complex and unmetamorphosed, undeformed Paleozoic formations of normal thickness less than 15 km to the east. The Matlin Mountains consist of two structurally distinctive terranes: an eastern rooted terrane in which Miocene(?) sedimentary rocks unconformably overlie upper Paleozoic carbonate rocks, and a western displaced terrane that is composed of five brecciated displaced sheets of pre-Tertiary rocks, each resting upon Tertiary sedimentary rocks; the lowest sheet overlies the rooted terrane. All of the Tertiary sedimentary rocks contain abundant silicic volcanic ash probably derived from nearby volcanic centers. The displaced sheets strike north-northeast, are nearly flat-lying, and consist of steeply dipping, locally folded strata. Two major low-angle faults are inferred to divide the four lowest sheets into two superimposed composite plates, each of which has a unique lithologic, metamorphic, and structural character. The fifth and highest displaced sheet was emplaced after both of these plates were in their present position.

The Miocene(?) sedimentary rocks and displaced sheets of the structurally lower composite plate probably were derived from an ancestral range in the site of the Grouse Creek Mountains and moved from west to east in late Miocene time. The higher plate may once have been continuous with a complex upper allochthon of late Miocene age that crops out widely on the west side of the Grouse Creek and Raft River Mountains. The source of the upper Miocene sedimentary rocks and the displaced sheets of the higher composite plate lay somewhere west of the Grouse Creek Mountains and the plate apparently travelled eastward at some time after 11.1 to 10.5 m.y. B.P.

Ductile structures that formed during, and after, brecciation in the most completely exposed displaced sheet suggest that the sheets moved initially by slow creep under moderate load. In spite of locally intense brecciation, the orientation in the sheets of sedimentary and metamorphic structures that were inherited from the source areas remained remarkably consistent. In the Matlin Mountains, the sheets apparently moved over an eroded Tertiary surface and (or) into a depositional basin. Clastic dikes preserved in the brecciated base of one sheet suggest that fractures were propagated by relatively high fluid pressures derived from underlying sedimentary water, thus allowing the sheet to disintegrate in place (hydrofracturing). Locally, movement of the sheets was facilitated by thin layers of silicic volcanic ash and (or) tuffaceous lacustrine beds.

Little is known about the source areas of the displaced sheets, but compositions, textures, and fabrics of the associated upper Miocene sedimentary rocks and the lithologies and geometries of the sheets point to the existence of postmetamorphic, broad domal uplifts whose locations shifted temporally and spatially but generally lay to the west, near the Utah-Nevada state line. The postmetamorphic low-angle faults in the core complex cut down-section to the Precambrian basement as well as carrying displaced sheets laterally for tens of kilometers. The presence of rugged highlands to the west suggests a gravitational sliding mechanism and the close spatial and temporal association of volcanic activity suggests that uplifts were genetically related to subjacent intrusions. The older-on-younger geometry of the Matlin sheets and their local tight folding and reverse faulting indicate that this area represents the distal, eastern toe of a distending complex allochthon to the west. Basin and Range high-angle faulting did not begin in the area until Pliocene time, after postmetamorphic uplifts and low-angle faulting had ceased.

INTRODUCTION

The Matlin Mountains expose a late Cenozoic record of postmetamorphic older-on-younger faulting that may provide some constraints on models of tectonic denudation of metamorphic core complexes. Five thin, variably brecciated sheets composed of upper Paleozoic and lower Mesozoic rocks were displaced over, and interlayered structurally with, upper Miocene alluvial fan and lacustrine deposits. The displaced sheets, now superimposed and rootless, are exposed in a group of low hills on the northern edge of the Great Salt Lake Desert. The area, some 300 km^2 in size, lies about 15 km south of the Raft River Mountains and 5 to 10 km east of the Grouse Creek Mountains in northwestern Utah (Fig. 1). In the days of steam locomotives, there were water stops on the Union Pacific railroad at two of the hills, a wavecut terrace named Red Dome and a high ridge with two prominent peaks known as the Matlin Mountains. In this report, the entire area will be referred to as "the Matlin Mountains."

The hills owe their present elevation and elongation to Pliocene and younger Basin and Range faults that strike from northwest to northeast. Topographic relief ranges from 60 to 370 m and averages about 120 m; the southernmost peak of the Matlin Mountains ridge is the highest point in the area, 1820 m above sea level. The hills lie about 10 km inside the highest shoreline of Pleistocene Lake Bonneville, within which they once stood as islands. As a consequence, their complex bedrock geology is partially obscured by deposits of Lake Bonneville.

The study area is adjacent to the Raft River-Grouse Creek-Albion Mountains metamorphic core complex (Crittenden and others, 1978). The core complex extends from northwestern Utah into south-central Idaho and consists of Precambrian and Paleozoic rocks that have undergone high-grade metamorphism, folding, decollement faulting, and great metamorphic thinning since Mesozoic time (R. L. Armstrong, 1968; Compton and others, 1977; Compton, 1980; D. M. Miller, 1980; Todd, 1980) (Fig. 1). The displaced sheets of the Matlin Mountains consist of highly deformed, variably metamorphosed Paleozoic and Mesozoic rocks that are in part correlative with those of the nearby metamorphic core complex.

This study was undertaken to determine the late development of the Grouse Creek portion of the metamorphic core complex, specifically, to determine the nature of its contact with surrounding mountain ranges that have been described as consisting of unmetamorphosed rocks with relatively simple structural histories (Olsen, 1956; Schaeffer and Anderson, 1960; Stifel, 1964). The data indicate that east of the Grouse Creek Mountains, metamorphism and deformation die out over a distance of less than 15 km, but the contact between the metamorphosed and unmetamorphosed areas is covered by allochthonous sheets of late Miocene age. Repeatedly, sheets containing deformed Paleozoic and Mesozoic rocks that were derived from the metamorphic core complex were displaced eastward into the flanking Tertiary basins (Fig. 2A). The data further indicate that the metamorphic core complex extends westward from the Grouse Creek Mountains for an unknown distance under Cenozoic cover.

The corresponding Paleozoic and Mesozoic rock formations in the eastern Matlin Mountains are relatively thicker and unmetamorphosed, and they do not appear to have been displaced in late Miocene time; these rocks formed the bedrock outcrops over which the displaced sheets moved (Fig. 2A, B). They will be referred to here as "rooted" to distinguish them from the autochthonous rocks of the adjoining Grouse Creek and Raft River Mountains. Indeed, it seems almost certain that this "rooted" section is continuous beneath the intervening displaced terrane and Quaternary cover with the middle allochthon exposed in the mountain ranges to the north and west (Fig. 1) and is, therefore, not autochthonous. This report describes the displaced sheets and their enclosing rocks and summarizes evidence about their age, origin, and mechanism of emplacement.

RELATION TO RAFT RIVER-GROUSE CREEK-ALBION MOUNTAINS CORE COMPLEX

The Raft River-Grouse Creek-Albion Mountains metamorphic core complex, the remnants of which are exposed in the ranges north and west of the Matlin Mountains, was the source of most of the sedimentary deposits and displaced sheets that form the Matlin Mountains. This complex consists of autochthonous metamorphosed Precambrian adamellite, quartzite, and schist overlain by three allochthonous sheets composed of metasedimentary and sedimentary rocks and intruded by Tertiary granitic plutons (Fig. 1). The three allochthons are arranged in younger-on-older order and, prior to late Cenozoic folding and high-angle faulting, were nearly horizontal. The lower sheet consists of upper Precambrian(?), Cambrian(?), and Ordovician metasedimentary rocks; the middle sheet, of Silurian to Lower Permian metasedimentary and sedimentary rocks; and the upper, of unmetamorphosed Upper Permian to Lower Triassic sedimentary rocks and Miocene volcanic and sedimentary rocks.

The metamorphic rocks exhibit two, and locally three, sets of concordant folds and lineations, the second of which (northwest- to west-trending) was closely associated with the development of a gneiss dome in the central Grouse Creek Mountains. Because the intensity of first-stage metamorphic lineations varies systematically with vertical position in the gneiss dome and because the first metamorphism probably ended before 38 m.y. (Compton and others, 1977), the gneiss dome apparently began to form in early Tertiary time. A culmination stage was reached in late Oligocene time, coincident with intrusion of an adamellite pluton.

A number of factors led Compton and others (1977) to conclude that deformation in the autochthon and the metamorphosed allochthons was triggered by thermal upwelling and domal uplift. These included: 1) the isoclinal, recumbent style of folding and radial symmetry of fold

vergences; 2) extreme metamorphic thinning of the allochthons; and 3) association of areas of most intense metamorphism and deformation with Tertiary plutons. Compton (1980) determined that second-stage metamorphic folds formed during flattening (near-vertical maximum compressive stress) and horizontal extension in the Raft River Mountains. Fission-track and Rb-Sr mineral ages for rocks of the Grouse Creek Mountains indicate that the autochthon remained at metamorphic temperatures until late Tertiary time (8 to 12 m.y. B.P.) (Compton and others, 1977).

ROCK UNITS

In the Matlin Mountains, hills underlain by pre-Tertiary and Tertiary strata are surrounded by valleys filled with Lake Bonneville sedimentary deposits (Fig. 2A). In the western part of the area, the pre-Tertiary rocks occur as eroded remnants of five thin, variably brecciated displaced sheets that typically overlie Tertiary rocks and include more than one stratigraphic unit. These sheets are labeled I to V from lowest to highest. To the east, the lowermost displaced sheet (I) rests upon rooted pre-Tertiary rocks that are unconformably overlain by a distinctive eastern facies of Tertiary sedimentary rocks. Except for slivers of Cambrian(?) and Ordovician(?) units at the base of one displaced sheet, no rocks older than Mississippian occur in either the displaced sheets or the rooted section. Upper Mississippian through Middle to Upper Triassic(?) rocks occur in the displaced sheets, and Pennsylvanian through Upper Permian rocks in the eastern rooted section. As presently exposed, the rooted Permian sequence in the Matlin Mountains is on the order of 3100 m (10,000 ft) thick, as determined from vertical cross sections.

The Tertiary rocks consist of upper Miocene, and perhaps older, interbedded coarse fanglomerate, sandstone, and tuffaceous lacustrine beds that are in both fault and depositional contact with the Paleozoic and Mesozoic rocks. Several kilometers east of the area, remnants of a Pliocene basalt flow overlie Paleozoic rocks and their Tertiary cover. Pleistocene Lake Bonneville deposits include coarse-grained shoreline and near-shore sediments and finer-grained sediments laid down in deeper parts of the basin.

Paleozoic rocks

Metamorphosed Chainman or Diamond Peak Formation. Slivers of black siltstone and shale, and minor pebble conglomerate, all metamorphosed under greenschist-facies conditions, occur beneath displaced sheet IV where it overlies Tertiary rocks; these lithologies are here assigned to the Upper Mississippian metamorphosed Chainman or Diamond Peak Formation (Brew and Gordon, 1971). Tectonic

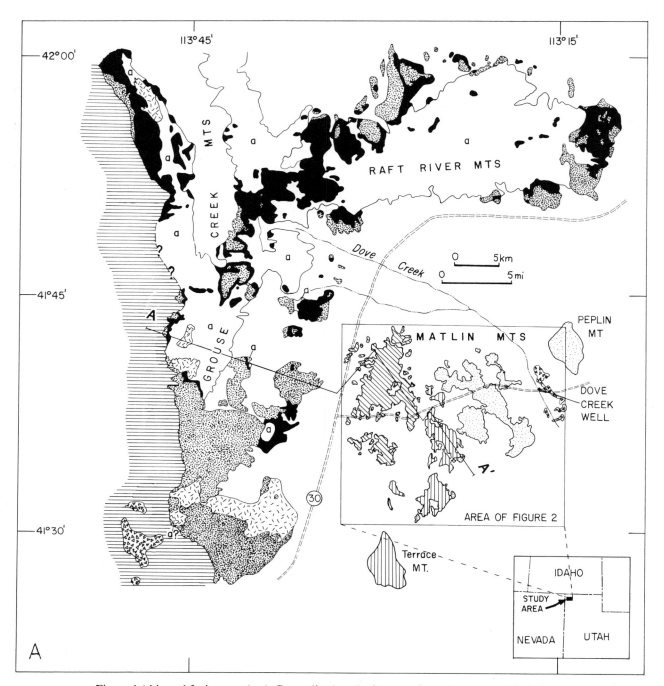

Figure 1 (this and facing page). *A*. Generalized geologic map of southern part of Raft River-Grouse Creek-Albion Mountains metamorphic core complex, northwestern Utah, including Matlin Mountains. Albion Mountains not shown. Geologic map is modified from Compton and others (1977); base of upper allochthon is approximate, from Compton (this volume). Solid lines bounding mountain ranges are depositional contacts with Cenozoic rocks and deposits. With the exception of intrusive and extrusive rocks, all other contacts are low-angle faults. A-A' is line of section of Figure 1B. High-angle faults shown on cross section only. *B*. Section A-A' across central Grouse Creek and Matlin Mountains, horizontal and vertical scales equal. Section includes subsidiary sheet of middle allochthon which moved at same time as uppermost allochthon (east flank of Grouse Creek Mountains, patterned as upper allochthon), projected into line of section from southern Grouse Creek Mountains (after Compton, this volume).

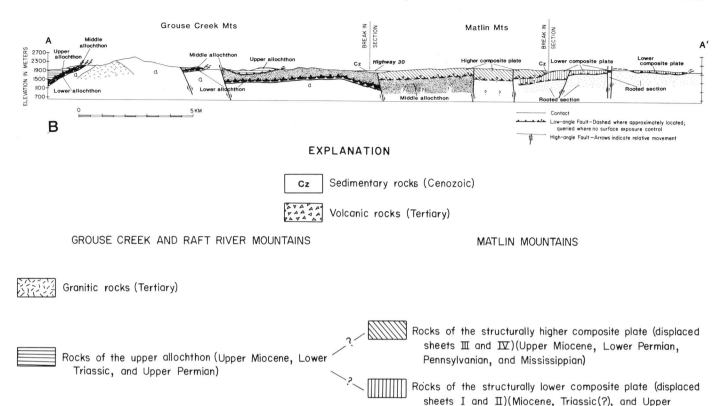

EXPLANATION

Cz Sedimentary rocks (Cenozoic)

Volcanic rocks (Tertiary)

GROUSE CREEK AND RAFT RIVER MOUNTAINS

MATLIN MOUNTAINS

Granitic rocks (Tertiary)

Rocks of the upper allochthon (Upper Miocene, Lower Triassic, and Upper Permian)

Rocks of the structurally higher composite plate (displaced sheets III and IV)(Upper Miocene, Lower Permian, Pennsylvanian, and Mississippian)

Rocks of the structurally lower composite plate (displaced sheets I and II)(Miocene, Triassic(?), and Upper Permian)

Rocks of the middle allochthon (Lower Permian through —?— Silurian)

Rocks of the rooted section (Upper Permian through Upper Pennsylvanian)

Rocks of the lower allochthon (Ordovician, Cambrian (?), and Upper Precambrian (?))

a Rocks of the autochthon (Precambrian)

slivers of black phyllite are also found within sheet IV along a low-angle fault (not shown in Figure 2) between metamorphosed limestone and sandstone subunits of the Oquirrh Formation. The structural position of the metamorphosed Chainman or Diamond Peak Formation in the Matlin Mountains is similar to the position of this unit in the metamorphic core complex, where phyllitic rocks commonly served as movement zones during low-angle faulting. In the Grouse Creek Mountains, the unit grades upward locally into metamorphosed limestone of the Oquirrh Formation (Todd, 1973).

Oquirrh Formation. The Pennsylvanian and Permian Oquirrh Formation originally named by Gilluly (1932, p. 34), consists of three informally designated subunits in the Raft River and Grouse Creek Mountains. In ascending order, these are a limestone subunit, a sandstone subunit, and a thinly-bedded limestone and sandstone subunit (Compton, 1972, 1975). In the central and southern Grouse

Creek Mountains, these subunits have undergone metamorphism under greenschist-facies conditions. Jordan (this volume) recognizes six subunits within the Oquirrh Formation in the Bovine Mountain section at the southern end of the Grouse Creek Mountains.

Metamorphosed equivalents of the limestone and sandstone subunits of the Oquirrh Formation occur in the southern part of displaced sheet IV (Fig. 2C). The sandstone subunit also occurs in a section at least 1500 m thick in the eastern rooted section. Metamorphosed thinly-bedded limestone and sandstone strata in the central part of sheet IV may belong to the uppermost Oquirrh subunit of the Grouse Creek Mountains. However, similar intervals of limestone and sandstone occur in the upper part of the limestone subunit, as well as in the thick Bovine Mountain section. Because depositional contacts are not preserved in sheet IV, it is not certain to which Oquirrh Formation subunit the thinly-bedded rocks of sheet IV belong.

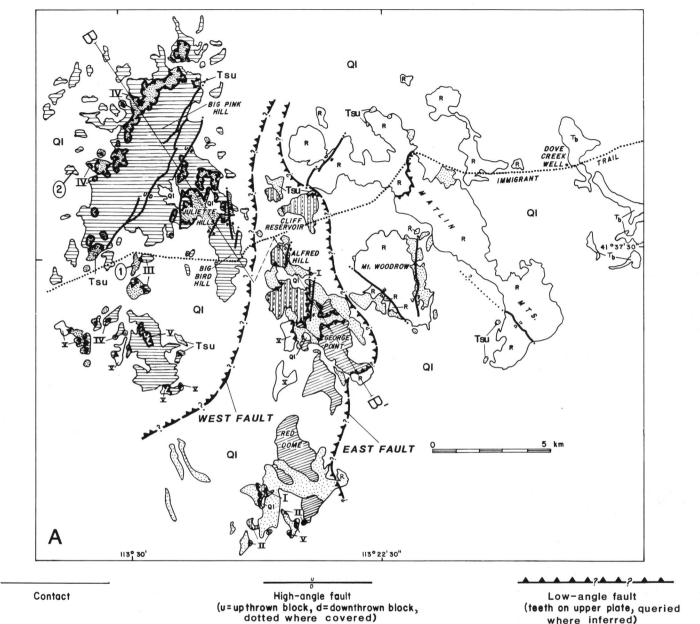

Figure 2 (this and following two pages). Geologic map and cross section of Matlin Mountains, northwestern Utah. *A.* Generalized structural map showing distribution of displaced sheets, rooted Paleozoic rocks, and Tertiary rocks (Tertiary rocks of sheet III not differentiated). Dashed, queried low-angle faults are inferred faults between displaced and rooted terranes (East Fault), and between two composite plates of the displaced terrane (West Fault). Eastern, lower composite plate consists of displaced sheets I and II and associated Tertiary rocks and western, higher plate of sheets III and IV. Point 1, clastic dikes; point 2, deformed Tertiary lake beds. *B.* Cross section B–B′, vertical scale exaggerated 2 x; some Lake Bonneville deposits omitted for clarity. *C.* Distribution of formations in sheets II and IV; also shown are Tertiary sedimentary rocks that underlie (Tfl) and overlie (Tfu) displaced sheet II; tectonic slivers of Paleozoic formations at base of sheet IV not shown; inset shows area of Figure 3. *D.* Distribution of formations in rooted section and in displaced sheet I (west of East Fault). *E.* Distribution of formations in displaced sheet III; Tertiary rocks comprise Big Pink sequence; inset shows area of Figure 4. Base and symbols used in *C–E* are same as those in *A.*

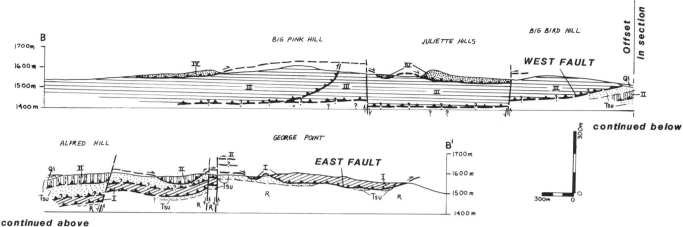

continued above

B

EXPLANATION FOR FIGURES 2A,B

 QI — Lake Bonneville deposits (Quaternary)

 Tb — Basalt (Tertiary)

 Tsu — Sedimentary rocks, undivided (Tertiary)

 V — Displaced sheet V, consists of Permian and Triassic (?) sedimentary rocks

 IV — Displaced sheet IV, consists of Mississippian, Pennsylvanian, and Permian metasedimentary rocks

 III — Displaced sheet III, consists of Permian metasedimentary rocks and Upper Miocene sedimentary rocks

 II — Displaced sheet II, consists of Permian sedimentary rocks

 I — Displaced sheet I, consists of Permian and Triassic (?) sedimentary rocks

R — Rooted section, consists of Pennsylvanian and Permian sedimentary rocks

In displaced sheet IV, the limestone subunit of the Oquirrh Formation consists of metamorphosed argillaceous, silty to very fine sandy limestone that is locally fossiliferous or fetid. Some beds are composed almost entirely of crinoid fragments. Carbonaceous material and pyrite are locally abundant, and recrystallized sandy laminations and sandstone interbeds are common. The sandstone subunit of the Oquirrh Formation in sheet IV is metamorphosed to quartzite, deformed, and very thin compared with that in the rooted section. Associated with this brown-weathering metaquartzite is dark phyllitic sandstone, which may correspond to the argillaceous sandstone that occurs as interbeds in the sandstone subunit of the Oquirrh Formation in the rooted section. The metaquartzite is in low-angle fault contact with marble (limestone subunit) of the Oquirrh Formation in displaced sheet IV; in the Grouse Creek Mountains, it lies depositionally above the marble.

In the eastern rooted section, the sandstone subunit of the Oquirrh Formation (Fig. 2D) is represented by calcareous quartzite, orthoquartzite, and calcareous sandstone. Limestone interbedded with the sandstone subunit in the rooted section contains late Pennsylvanian(?) and Early Permian (Wolfcampian) fusilinids (R. C. Douglass, U.S. Geological Survey, written communication, 1973). The dark brown-, pink- and buff-weathering quartzite appears light gray on fresh surfaces and consists of well-sorted fine- to medium-grained sand that is composed of quartz, K-feldspar, and plagioclase. The rock is massively bedded, but faint, fine dark laminae can be seen locally. Dark-gray sandy limestone concretions up to one-half meter thick are common in the subunit. Some are ovoid and discrete, but other concretions resemble rudely bedded, rough-surfaced sandy limestone beds. The quartzite is interbedded with thin bedded, sandy, siliceous limestone that is locally fossiliferous and also with argillaceous, platy, calcareous sandstone. Limestone composes approximately one-fourth to one-third of the subunit. The base of the subunit is covered, but the exposed thickness is at least 1500 m (measured in a cross section of the nose of a large, northeast-plunging anticline).

Interbedded limestone and sandstone unit. Flaggy, thinly interbedded (12- to 50-mm beds) sandy limestone and calcareous sandstone that weather to pink, red, yellow, and lilac form a distinctive unit informally named "the patterned unit." Quartz-filled hairline fractures divide the

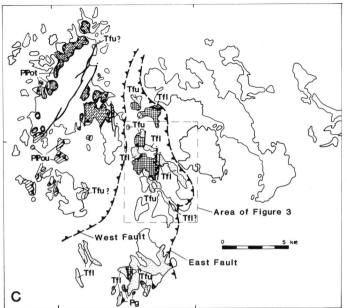

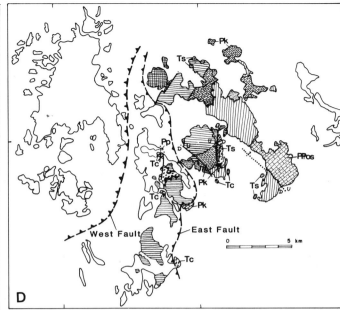

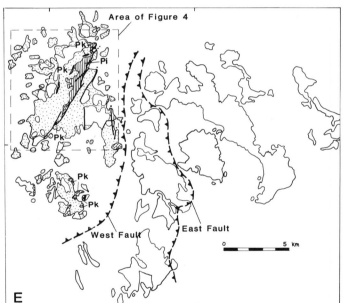

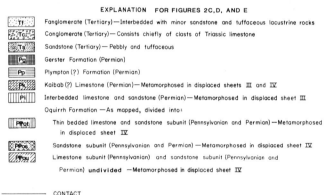

EXPLANATION FOR FIGURES 2C,D, AND E

Fanglomerate (Tertiary)—Interbedded with minor sandstone and tuffaceous lacustrine rocks

Conglomerate (Tertiary)—Consists chiefly of clasts of Triassic limestone

Sandstone (Tertiary)— Pebbly and tuffaceous

Gerster Formation (Permian)

Plympton(?) Formation (Permian)

Kaibab(?) Limestone (Permian)—Metamorphosed in displaced sheets III and IV

Interbedded limestone and sandstone (Permian)—Metamorphosed in displaced sheet III

Oquirrh Formation — As mapped, divided into:

Thin bedded limestone and sandstone subunit (Pennsylvanian and Permian)—Metamorphosed in displaced sheet IV

Sandstone subunit (Pennsylvanian and Permian)—Metamorphosed in displaced sheet IV

Limestone subunit (Pennsylvanian) and sandstone subunit (Pennsylvanian and Permian) **undivided** —Metamorphosed in displaced sheet IV

CONTACT

FAULT—Dashed where inferred, dotted where concealed, U=upthrown block, D=downthrown block

LOW ANGLE FAULT—Dashed where inferred, teeth on upper plate

blage that indicates an Early Permian (Leonardian) age (R. C. Douglass, U.S. Geological Survey, written communication, 1973).

The typical limestone of this unit is fine-grained silty and sandy limestone that is dark gray to black on fresh surfaces and is locally crossbedded; in thin section it seems to be a dolomitic limestone (micrite) with very fine quartz sand, carbonaceous material which defines bedding, and minor chert. The limestone is interbedded with, and gradational into, dolomitic silty, cherty fine-grained sandstone; the two lithologies are similar in thin section, except that the silty sandstone contains more chert and is more extensively dolomitized. Also included in the patterned unit are rhythmically and thinly-interbedded brown-weathering chert and silty, cherty, gray-weathering limestone; non-flaggy pink-weathering calcareous sandstone; and pale buff sandstone, the latter with ovoid bodies of sandy limestone. Sandstone interbeds are most numerous above the contact with the underlying sandstone subunit of the Oquirrh For-

thin beds or flags into angular multicolored blocks, giving the rock its distinctive pattern. The patterned unit occurs both in the eastern rooted section where it has a minimum thickness of 750 m (measured in a cross section of the nose of a northeast-plunging syncline) and in displaced sheet III, where it has undergone low-grade metamorphism (Fig. 2D, E). It may also occur in sheet IV where rocks similar to those of the patterned unit are interbedded locally with metamorphosed Kaibab(?) Limestone. In the eastern rooted section, the patterned unit grades down-section into Oquirrh sandstone of late Pennsylvanian(?) and Early Permian (Wolfcampian) ages. Local fossiliferous limestone beds in the patterned unit contain a foraminiferal assem-

mation. Similarly, all of the lithologies of the patterned unit occur close below the contact as interbeds in the Oquirrh sandstone subunit. The patterned unit is lithologically similar to the Arcturus Formation (Lawson, 1906) and to the Pequop Formation (Steele, 1960), both of Wolfcampian and Leonardian ages. Stevens (1965) considered these two formations to be lithosomes. Because work is being done on the lower Permian stratigraphy of northeastern Nevada and northwestern Utah and new nomenclature may be proposed, no formational assignment of the patterned unit will be made here.

Kaibab(?) Limestone. The patterned unit grades upward into, and is in part interbedded with, dark gray limestone and dolomitic limestone (micrite) which is argillaceous or silty to very finely sandy. The carbonate beds are thinly interbedded with sandstone and quartzite, all locally cherty. This limestone unit occurs in the rooted section and metamorphosed equivalents occur in the western part of the Matlin Mountains in displaced sheets III and IV (Fig. 2C-E). Coarser-grained fossiliferous horizons are also present. Irregular sand bodies, nodular and lenticular chert concretions, and platy limestone with argillaceous to silty partings are also common. The lower part of the unit contains interbedded oolitic and algal limestone, crossbedded sandy limestone, and limestone with distinctive subvertical chert pipes. The detrital sand content increases up-section in the unit until limestone and sandstone are approximately equal in abundance. This part of the section is marked by a distinctive fossiliferous limestone horizon containing *Timanodictya*(?) species, an unusual ramose bryozoa of Early Permian age, identified by O. L. Karklins (U.S. Geological Survey, written communication, 1973). Overlying the interbedded limestone and sandstone sequence is an interval of prominently outcropping reddish-brown-weathering sandstone and quartzite. Brachiopods from fossiliferous limestone horizons in the lower part of the unit indicate an Early Permian (Leonardian) age (M. Gordon, Jr. and B. Wardlaw, U.S. Geological Survey, written communication, 1975, 1976). The entire unit is about 800 m thick in the eastern rooted section. It is here assigned to the Kaibab(?) Limestone as its lithology is similar to the Kaibab Limestone as described by Hose and Repenning (1959) in the Confusion Range, west-central Utah.

Plympton(?) Formation. The Kaibab(?) Limestone grades upward with increasing dolomite into rocks that are here assigned to the Plympton(?) Formation. In the Matlin Mountains the Plympton(?) Formation consists of pale gray dolomite and dolomitic limestone, and pale gray and black chert, all of which are silty to sandy and gradational. The dolomite typically is finely and evenly laminated, and it contains siliceous blebs, stringers, and irregular to ovoid bodies of chert. Both the dolomite and dolomitic limestone are interbedded with silicified siltstone and sandstone. The sandstone locally has upright crossbeds. Locally, the do-

lomite and sandstone weather pink and red. The unit also contains minor interbeds of fossiliferous dark gray limestone. The Plympton(?) Formation crops out in the rooted section and in displaced sheet I, which overlies the rooted section (Fig. 2D). The minimum thickness inferred from a cross section in the rooted section is 800 m.

R. K. Hose (oral communication, 1973) suggested the lithologic similarity of these rocks to the Plympton Formation of west-central Utah described by Hose and Repenning (1959). The recognition of the formation in the Matlin Mountains is supported by an Upper Permian fossil assemblage consisting of molluscs and a large brachiopod fragment (J. T. Dutro, Jr., U.S. Geological Survey, written communication, 1974) and by the unit's stratigraphic position in the rooted section. The top of the Plympton(?) Formation is not exposed in the Matlin Mountains but on Terrace Mountain, 5 km to the south, dolomite similar to that of the Plympton(?) Formation grades into Gerster Formation with increasing limestone and chert.

The assignment of these rocks to the Upper Permian Plympton(?) Formation, however, presents a problem. Probable Middle to Upper Triassic (Anisian to Carnian) brachiopods (J. T. Dutro, Jr., U.S. Geological Survey, written communication, 1974; P. R. Hoover, written communication, 1976) were found in outcrops that appear to lie stratigraphically above, and are lithologically similar to, the outcrops containing the Late Permian fauna identified in displaced sheet I near informally named George Point (Figs. 2A, 3). Scattered occurrences of crossbeds in calcareous sandstone indicate that the entire section is upright. Thin cherty dolomite lying below the Triassic(?) fossils resembles Gerster Formation. This locality cannot represent a complete stratigraphic section from Upper Permian through Middle to Upper Triassic rocks because most of the Gerster Formation and the Dinwoody and Thaynes Formations are missing. As the Gerster, Dinwoody, and Thaynes Formations are recognized elsewhere in the region, these formations may once have been present here but removed by erosion or faulting. No evidence was seen for an unconformity, but the rocks of sheet I are locally brecciated which suggests the presence of bedding plane faults within the sheet.

Permian chert. Black chert occurs as tectonic lenses in two places in the Matlin Mountains. A lensoid pod of black chert occurs in a steep northeast-trending fault zone that separates the Plympton(?) and Gerster Formations in the northwestern part of the rooted section (Fig. 2D). This lens is on the order of 60 to 90 m thick; it is not shown on Figure 2D. A nearly horizontal lens of black chert occurs between Plympton(?) Formation of displaced sheet I and red-weathering Triassic limestone-bearing conglomerate of the rooted section in the area north of George Point (Fig. 3). In both places, the black chert is interbedded with lesser gray-weathering limestone. The age and stratigraphic correlation

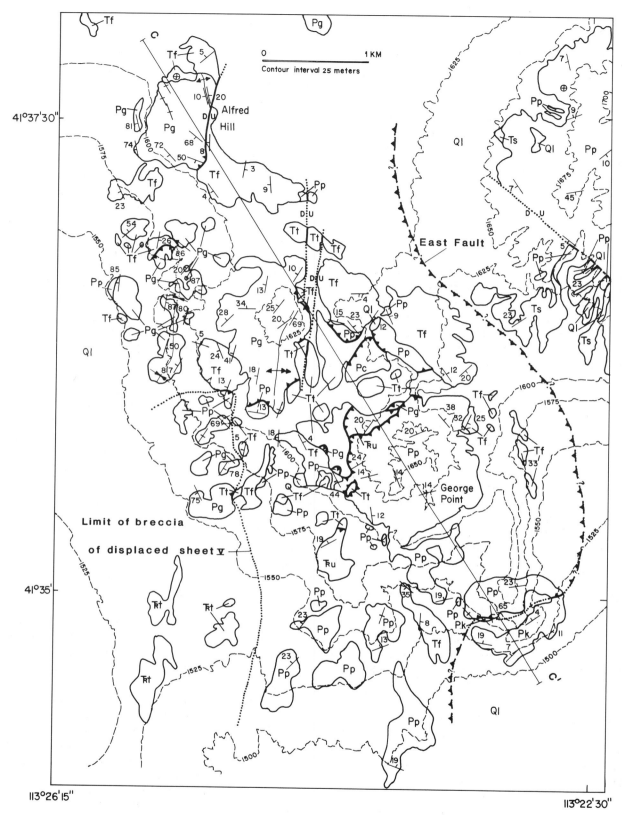

Figure 3 (this and facing page). Detailed geologic map and cross section of central part of Matlin Mountains, northwestern Utah (map area shown in Figure 2C). Queried, dashed low-angle fault is inferred fault contact between displaced sheets and rooted section (East Fault). Cross section C-C′ has slight vertical exaggeration; light dashed lines are traces of bedding; lake deposits omitted locally for clarity.

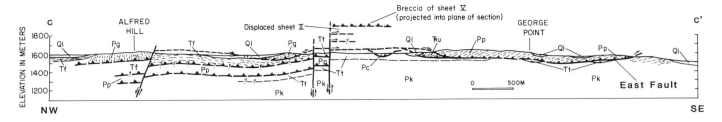

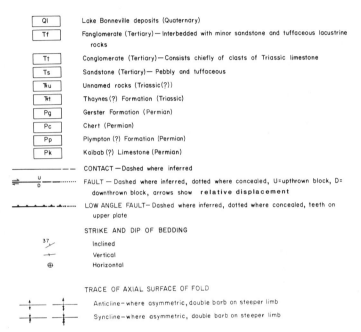

bioclastic and phosphatic, to sparsely sandy to argillaceous; most of it is silicified. Dolomite is clean or cherty, the latter concentrated in sandy areas.

The top and base of the Gerster Formation are not exposed in the Matlin Mountains. In the rooted section, the Gerster and Plympton(?) Formations are in fault contact. Slivers of exotic lithologies, such as black chert, along the fault suggest that other units or subunits may have been cut out of the section by faulting. At Terrace Mountain, where the entire section is vertical and may be allochthonous, the Gerster Formation grades westward (down-section?) into dolomite that resembles the Plympton Formation as described by Hose and Repenning in the Confusion Range (1959) and grades eastward (up-section?) into fine-grained gray limestone which may belong to the Lower Triassic Thaynes Formation (Boutwell, 1907). The original thickness of the Gerster Formation in the rooted section of the Matlin Mountains is inferred to be at least 700 m.

Mesozoic Rocks

Triassic rocks, or those presumed to be Triassic in age, are found in two localities in the Matlin Mountains. The rocks in one locality contain probable Middle and Upper Triassic (Anisian to Carnian) brachiopods (J. T. Dutro, Jr., U.S. Geological Survey, written communication, 1974; P. R. Hoover, written communication, 1976); the rocks of the other locality have not been dated but contain fresh- or brackish-water fossils that may be post-Paleozoic (E. L. Yochelson, U.S. Geological Survey, written communication, 1973).

Unnamed Triassic(?) rocks. The marine brachiopod locality is in the southeastern part of the Matlin Mountains in displaced sheet I, in which bedding dips about 20 degrees northwest (Fig. 3). The Triassic(?) rocks probably tectonically underlie sandy dolomite of the Plympton(?) Formation. The Triassic(?) rocks consist of interbedded, intergradational limestone, sandstone, chert, and dolomite. This sequence locally contains collapse breccia and questionable chert pebble conglomerate. Because these rocks are bounded by low-angle faults, the original thickness of this section and the nature of its contacts with the stratigraphically underlying and overlying formations are unknown.

Thaynes(?) Formation. The nonmarine Triassic(?) locality occurs in displaced sheet V in the southern part of the

of this unit is unknown. Since it occurs in association with the Plympton(?) and Gerster Formations (see tectonic slices of Gerster Formation in Figure 3), it is tentatively placed between these two units in this report. In the central Grouse Creek Mountains, similar black chert occurs in the upper allochthon where it is separated from Gerster Formation by a high-angle tear fault (Todd, 1973).

Gerster Formation. The Gerster Formation (Nolan, 1935) has been identified in the Matlin Mountains on the basis of lower Wordian (earliest Late Permian) fossils (B. Wardlaw and M. Gordon, Jr., U.S. Geological Survey, written communication, 1975, 1976). The formation crops out in a narrow north-trending zone in the central part of the Matlin Mountains in displaced sheet II and also in the rooted section (Fig. 2C, D). The unit consists of interbedded yellowish-gray-weathering cherty limestone and dolomite, and brown-weathering chert, the latter in beds up to 30 cm thick. Although virtually all the upper Paleozoic units in the Matlin Mountains are cherty to some degree, the Gerster Formation contains the thickest chert beds and chert-dominated sequences. The chert beds pinch and swell, and are locally nodular or interconnected. They have been extensively dolomitized. Limestone in the unit varies from

displaced terrane (Fig. 2A). These rocks consist of unmetamorphosed, silty to very finely sandy, dark-gray limestone, locally aphanitic or with algal debris. Sparse quartz sand is uniformly distributed and is yellow-, buff-, and reddish-brown-weathering on medium-gray-weathered limestone surfaces. Some of the limestone is platy and some contains fine, irregular sand laminae. The unit also includes lesser yellowish-gray-weathering, clean dolomite with faint light and dark gray lamination. Some limestone contains minor replacement dolomite and chert. As mentioned above, one sample of brecciated limestone contained possible fresh- or brackish-water worm tubes of post-Paleozoic age.

Tertiary Rocks

Tertiary rocks occur at four, possibly five, structural levels which are progressively lower to the east, and are separated by displaced sheets containing Paleozoic and Mesozoic rocks. The fifth, highest level is represented by a tiny patch of fanglomerate that lies on displaced sheet IV (Tsu, near point 1, Fig. 2A) and is not discussed below. The structurally lowest Tertiary rocks occur only in the eastern half of the area (Fig. 2D), where they include two facies. The first facies consists of pale-buff-weathering tuffaceous pebbly fine-grained sandstone, lacustrine limestone, and lesser conglomerate of local origin. The sandstone differs from the tuffaceous sandstone of the structurally higher, western Tertiary sequences in that the former is friable, less distinctly bedded, and forms sparse, irregular, and blocky outcrops. This eastern facies is overlain by the second facies, a red-weathering conglomerate dominated by clasts derived from Lower Triassic interbedded limestone and sandstone and representing both Dinwoody and Thaynes Formations (N. J. Silberling, U.S. Geological Survey, oral communication, 1973; J. T. Dutro, Jr., U.S. Geological Survey, written communication, 1974; P. R. Hoover, written communication, 1977) (Fig. 2D). This conglomerate also contains minor clasts of limestone with black chert layers, clasts of black chert, and clasts of the Gerster Formation. Blocks of Triassic limestone in this conglomerate range up to approximately 3 m in long dimension, which suggests a nearby upland source of Lower Triassic rocks. These two Tertiary facies depositionally overlie Pennsylvanian through Permian units of the rooted section. They are separated structurally from the western Tertiary rocks by displaced sheet I, which overlies the rooted section with a low-dipping fault contact.

Interbedded fanglomerate, sandstone, and tuffaceous lacustrine rocks which crop out sparsely in the southern and central parts of the Matlin Mountains lie between displaced sheets I and II, and similar rocks overlie sheet II (Fig. 2C). The fanglomerates consist of clasts of unmetamorphosed upper Paleozoic units, but interbedded sandstones contain sparse metamorphic quartz and mica clasts. The northeast-dipping beds which crop out around the southwestern rim of the Lake Bonneville terrace at Red Dome underlie displaced sheet II; they are predominantly white- and gray-weathering tuffaceous sandstone, siltstone, and claystone.

The structurally highest Tertiary rocks are widely exposed in the western part of the Matlin Mountains, where they form part of displaced sheet III and are in turn overlain by sheets IV and V (Fig. 2E). These rocks appear to have been folded broadly prior to emplacement of the overlying displaced sheets. A north-northeast-trending hill (informally called Big Pink for the pink-weathering conglomerate which underlies it), east of, and parallel to, Highway 30, between upper Dove Creek and the Immigrant Trail (Figs. 1 and 2A), exposes approximately 2000 m of Tertiary alluvial fan and lacustrine deposits. This section has been uplifted along coalescing north-northeast-trending reverse faults so that the basal unconformity with the underlying metamorphosed Lower Permian rocks is exposed (Fig. 4). The distinctly bedded basal fanglomerate consists of alternating beds of angular limestone and sandstone clasts, such that the rock has a crude lithologic layering that superficially resembles the underlying brecciated interbedded limestone and sandstone. The trace of the contact indicates that a surface of low to moderate relief had developed on the underlying Lower Permian rocks prior to deposition of the Big Pink sequence. The top of the Big Pink sequence is not exposed. As a result of folding and faulting, the strata dip northwestward; they are overlapped to the north and west by deposits of Lake Bonneville.

The upper part of the section consists dominantly of lacustrine sandstone, siltstone, claystone, tufa, limestone and terrace conglomerate, all of which contain admixtures of volcanic ash. These relatively fine-grained rocks are interbedded with well-cemented fanglomerate and sandstone throughout the section (Fig. 5), indicating alternating periods of torrential and quiet deposition. A sample of exceptionally pure vitric tuff from the upper part of the Big Pink sequence yielded concordant late Miocene ages for volcanic zircon (11.1 ± 1.7 m.y., fission track method, C. W. Naeser, U.S. Geological Survey, written communication, 1973) and volcanic plagioclase (10.5 ± 0.8 m.y., K-Ar method, E. H. McKee, U.S. Geological Survey, 1973).

In general, the clast content of the Tertiary rocks represents an inverted stratigraphy of the Raft River-Grouse Creek-Albion Mountains metamorphic core complex. The lower part of the Big Pink sequence contains clasts of unmetamorphosed carbonate rocks, sandstone, chert, and argillite that resemble the upper Paleozoic and Lower Triassic lithologies exposed in an unmetamorphosed uppermost allochthon in the Grouse Creek Mountains and the upper Paleozoic lithologies exposed in the eastern rooted section. Clasts of metamorphosed limestone and calcareous sandstone, and black phyllite (=middle allochthon of Grouse Creek Mountains) appear up-section, followed by metaquartzite, schist, strained quartz and mica (=lower allochthon of the core complex) and traces of mica-clouded plagioclase, the latter probably derived from metamorphosed Precambrian adamellite of the autochthon of the core complex. In the northern part of Big Pink Hill, conglomerate contains clasts of white and green metaquartzite and staurolite schist, lithologies which at present are found only in the northern Grouse Creek and Raft River Mountains. The upper part of the section also contains clasts of Tertiary sandstone and claystone which indicates that parts of the basin were undergoing uplift and erosion at the time these lacustrine beds were being deposited. Roughly a dozen current indicators in the lake beds suggest southwest to northeast and west-to-east transport, evidence which is supported by the clast lithologies; metamorphosed Paleozoic rocks crop out only in the metamorphic core complex to the west.

Basalt. The eroded remnants of a basalt flow or flows which once may have been continuous with basalt presently exposed southeast of the eastern Raft River Mountains in road cuts of Highway 30 (R. Compton, oral communication, 1973) occur near Dove Creek well, several kilometers east of the study area (Fig. 1). Lava flowed over upper Paleozoic carbonate rocks and overlying Tertiary sedimentary rocks on the west side of Peplin Mountain, but contacts are obscured by Lake Bonneville sedimentary deposits. A K-Ar whole rock age of 3.2 ± 0.2 m.y. was obtained for a sample of basalt collected near Dove Creek well (W. C. Hoggatt, U.S. Geological Survey, 1982).

Quaternary Deposits

Lake Bonneville deposits and landforms. A spectacular array of wave-cut sea cliffs and bedrock terraces are developed on south- and southeast-facing hillslopes in the Matlin Mountains; these hills faced what were the greatest reaches of open water in pluvial Lake Bonneville. Less prominent shoreline features are seen on eastern hillslopes. Outlines of quiet bays which once lay between these headlands are marked at present by successions of arcuate gravel beaches and curving gravel spit deposits composed of well-sorted, moderately rounded cobbles and pebbles. Broad wave-cut platforms such as Red Dome are partially veneered by gravel. Offshore shallow bedrock knobs and benches were the sites of tufa deposition; white silt-clay lake beds, composed largely of reworked Tertiary tuffaceous material, occur in the deeper parts of the basin. These fine-grained materials have been worked locally by wind into dunes.

Modern alluvium. Modern or present-day alluvial deposits consist of material transported by ephemeral streams and by torrential slopewash, material that is derived largely from Tertiary sedimentary rocks and unconsolidated Lake Bonneville sediments. Modern alluvium forms a thin veneer upon older deposits and floors the narrow channels incised in them. The volume of this material is insignificant compared with that of the Lake Bonneville deposits.

STRUCTURAL FRAMEWORK

The Matlin Mountains can be divided into two structurally distinct parts: an eastern area of rooted upper Paleozoic marine carbonate rocks overlain depositionally by Tertiary strata, and a western area composed of five nearly flat-lying superimposed displaced sheets which are structurally interlayered with Tertiary rocks. For brevity, the eastern and western parts of the area will be referred to as the "rooted section" and the "displaced terrane," respectively. The rocks of the rooted section, although locally folded, faulted, and brecciated, are unmetamorphosed and, in comparison with rocks of similar age and lithology in the displaced sheets, are relatively undeformed.

The displaced sheets are composed of highly deformed, locally metamorphosed rocks similar in age and lithology to those of the rooted section. Of the five sheets, all rest on Tertiary rocks, three are overlain by Tertiary rocks, and one incorporates Tertiary rocks. Displaced sheet I, the structurally lowest sheet, overlies the rooted section with a flat fault contact. This basal low-angle fault is exposed in two places and bounds the displaced terrane on the east (East Fault) (Fig. 2A). Its continuity beneath Lake Bonneville deposits is indicated by the juxtaposition of different facies of Tertiary sedimentary rocks across the fault and by the juxtaposition of displaced pre-Tertiary rocks west of the fault against rooted pre-Tertiary rocks on the east side of the fault. Although it is nowhere exposed, a second major low-angle fault can be inferred to lie 3 to 6 km west of, and subparallel to, the East Fault on the basis of the juxtaposition of displaced sheets of different lithology, metamorphism, and structural character across a narrow, north-trending zone. For convenience, this low-angle fault is called the West Fault (Fig. 2A, B). Thus, the displaced terrane appears to be composed of two low-angle fault plates, each consisting of two displaced sheets. Apparently, sheet V was emplaced over both composite plates.

The assignment of rocks that crop out discontinuously to a given displaced sheet was made on the basis of similarities in lithology, metamorphism, deformation, and structural position. The restored sheets are typically elongate in a north-northeast direction and are flat-lying. Progressively lower-lying sheets are exposed to the east, suggesting that the entire sequence has a gentle westward dip (Fig. 2B).

The discontinuity of displaced sheets and interlayered Tertiary strata in the eastern, structurally lower composite plate indicates a complex history with repeated episodes of

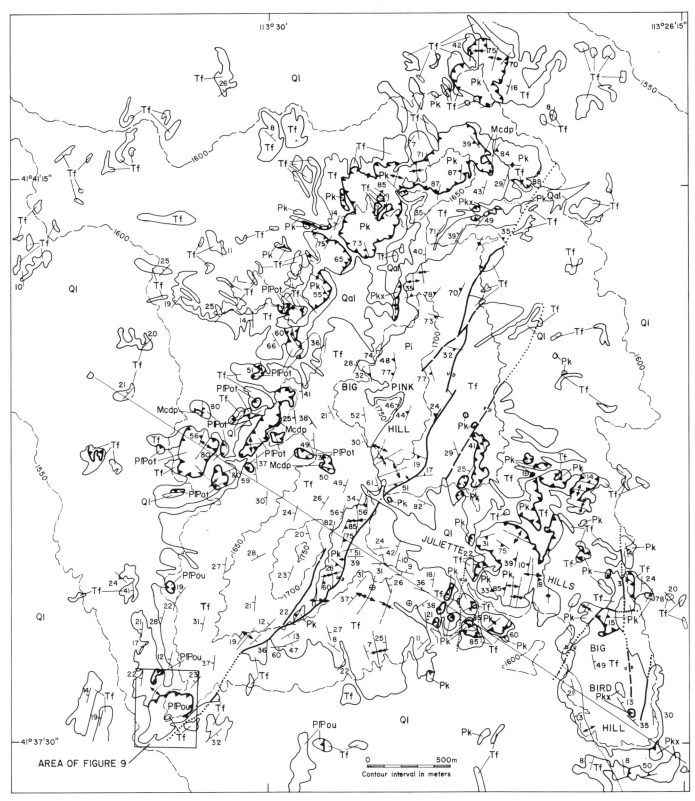

Figure 4 (this and facing pages). Detailed geologic map and cross section of northwestern part of Matlin Mountains, northwestern Utah (map area shown in Figure 2E). Cross section D-D' has equal horizontal and vertical scales; light dashed lines are traces of bedding; lake deposits omitted locally for clarity.

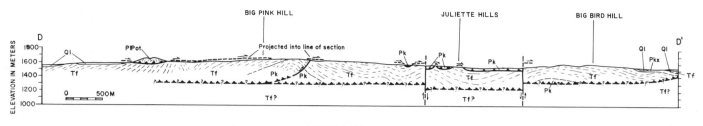

EXPLANATION

Qal	Alluvium (Quaternary)

Ql	Lake Bonneville deposits (Quaternary)

Tf	Fanglomerate (Tertiary)—Interbedded with sandstone and tuffaceous lacustrine rocks

Pkx	Kaibab(?) Limestone (Permian)—Sedimentary breccia, metamorphosed in displaced sheet III

Pk	Kaibab(?) Limestone (Permian)—Metamorphosed in displaced sheets III and IV

Pi	Interbedded limestone and sandstone (Permian)—Metamorphosed in displaced sheet III

Oquirrh Formation — As mapped, divided into:

PIPot	Thinly bedded limestone and sandstone subunit (Pennsylvanian and Permian) metamorphosed in displaced sheet IV

PIPou	Limestone and sandstone, undivided (Pennsylvanian and Permian) metamorphosed in displaced sheet IV

Mcdp	Metamorphosed Chainman or Diamond Peak Formation (Mississippian)

———————— CONTACT

—————⇌——ᵤ⁄d···· FAULT—Dashed where inferred, dotted where concealed, u=upthrown block, d=downthrown block, arrows show relative movement

▲▲▲▲▲▲▲▲▲▲ LOW-ANGLE FAULT—Dashed where inferred, teeth on upper plate

STRIKE AND DIP OF BEDDING

¹³↗ Inclined

⊕ Horizontal

STRIKE AND DIP OF FOLIATION

⁷⁵↗ Inclined

↗ Vertical

TRACE OF AXIAL SURFACE OF FOLD

Anticline—where asymmetric, double barb on steeper limb

Syncline—where asymmetric, double barb on steeper limb

low-angle faulting, deposition, and erosion(?) (Fig. 3). The Tertiary rocks either were carried piggyback on the underlying displaced sheets to their present position or themselves form displaced sheets, for they not only differ lithologically from the rooted Tertiary section immediately to the east but also structurally overlie the rooted section and appear to end abruptly at the East Fault (Fig. 2C). In the western, structurally higher composite plate, geometric relations indicate that Tertiary strata were displaced along with the pre-Tertiary rocks which they unconformably overlay by a low-angle fault which cuts both Tertiary and pre-Tertiary rocks (Fig. 4).

There is evidence that folding and (or) tilting took place before, probably during, and after low-angle faulting. Typically, bedding and (or) metamorphic foliation in the pre-Tertiary rocks of a sheet have variable strikes and gentle, moderate, or steep dips (up to 90 degrees) that suggest either the existence of large, upright and overturned folds that were truncated by the underlying faults, or rotation of strata during movement along listric normal faults, or some

A

B

Figure 5. Tertiary rocks of Matlin Mountains, northwestern Utah. *A.* Northwest-dipping, well-indurated fanglomerate (dark, planar beds) and interbedded, friable, tuffaceous lacustrine sandstone (light-colored layers), Big Pink sequence. Trees are full-grown junipers (10 to 12 ft). *B.* Weathered surface on debris flow fanglomerate of Big Pink sequence; hammer for scale. Light-colored clasts are chiefly sandy carbonate rocks; dark cobbles are metamorphosed sandstone of the Oquirrh Formation.

combination of the two (Figs. 3, 4). Locally, sequences of displaced sheets and the Tertiary strata interlayered with them were folded broadly on north-northeast and northwest axes. Steep Basin and Range faults cut all of these structures.

Rooted Section

Unmetamorphosed rocks ranging in age from late Pennsylvanian to Late Permian occur in a northwest-dipping (10 to 50 degrees) section in the eastern part of the Matlin Mountains (Fig. 2D). They are overlain unconformably by interbedded Tertiary buff-weathering tuffaceous sandstone and conglomerate of local derivation which has a normal sedimentary contact. In the area around informally named Mount Woodrow (Fig. 2A), the tuffaceous rocks are overlain by locally coarse red-weathering conglomerate containing abundant clasts of Lower Triassic limestone and minor Gerster Formation clasts. The rocks of the rooted section have not been deformed penetratively; however, locally they have been folded at outcrop scale on northeast to east-northeast and northwest, west-northwest, to east-west axes; parallel crenulation lineations are developed over short distances in the folded areas. The rooted terrane has undergone broad folding (wavelengths several km) on moderately plunging northeast axes; these folds verge to the southeast. Locally, they appear to have folded northwest-trending folds with wavelengths of 100 to 200 meters. The folded terrane subsequently was cut by one low-angle fault, which may have developed from an oversteepened fold, and by at least four high-angle Basin and Range faults.

Displaced Terrane

Displaced sheets I through V and the intervening Tertiary rocks are discussed in the following paragraphs in structural order from the lowest to the highest sheet.

Sheet I. Displaced sheet I consists of a narrow north-northeast-trending belt of outcrops about 11 km long and 3.5 km across at its widest point in the south-central part of the Matlin Mountains (Fig. 2D). Although it has been eroded, faulted, and partly covered by Tertiary and Quaternary deposits, the sheet probably had a minimum area of about 40 km^2. The sheet consists of an upright, northwest-dipping (15 to 40 degrees) sequence of Plympton(?) and Gerster Formations and unnamed Triassic(?) rocks and has a maximum thickness of 125 m. Sheet I overlies red-weathering, Triassic limestone-bearing conglomerate and Kaibab(?) Limestone of the rooted section (Fig. 3). The Kaibab(?) Limestone is considered part of the rooted section because to the east, the red-weathering conglomerate and underlying buff tuffaceous sandstone have a normal, depositional relation with the rooted Kaibab(?) Limestone and overlying upper Paleozoic rocks. Locally, tectonic lenses of black chert and Gerster Formation are present between displaced sheet I and red conglomerate (Fig. 3).

Displaced sheet I is overlain by fanglomerate and by Gerster Formation of sheet II. In one place, the fanglomerate overlies the low-angle fault contact between sheet I and the red-weathering conglomerate (Fig. 3). Dips of bedding in the overlying fanglomerate range from 0 to 50 degrees, with many between 20 and 40 degrees; the fanglomerate appears to have been folded on northerly axes. Locally, fanglomerate beds appear to dip into their underlying contact with displaced sheet I, indicating that this contact is a low-angle fault (faulted unconformity(?)).

Dolomite of the Plympton(?) Formation, which is the dominant formation of sheet I, is unmetamorphosed; although locally brecciated, it is, in general, intact and less deformed than the rocks of the other displaced sheets. The northwest dip of the strata might suggest tilting during southeast-directed listric normal faulting. However, locally the rocks are folded into upright anticlines and synclines with wavelengths between 0.5 and 1 km. The traces of the axial planes of these folds curve from northerly to northeast trends suggesting refolding of north-northeast (?) folds on northwest axes. Eastward dips of bedding locally are as steep as 78 degrees. In the area near the East Fault, bedding in the dolomite is vertical and brecciated in several places. The restricted presence of the Gerster Formation and Triassic(?) rocks in the northern part of the sheet suggests that the Permian and Triassic section has been greatly telescoped here, perhaps by undetected low-angle faults parallel to bedding.

The source area of displaced sheet I is unknown. The Plympton(?) and Gerster Formations which compose it occur widely in northwest Utah and northeast Nevada. Both formations crop out north of sheet I in the rooted section, 5 km to the south on Terrace Mountain, 25 km to the southeast in the Terrace and Hogup Mountains, and over 50 km to the northwest in the Goose Creek Mountains. The Permian rocks in the Goose Creek Mountains may be part of the complex upper Miocene allochthon mapped by Compton (this volume) west of the Grouse Creek and Raft River Mountains (Fig. 1). Middle and Upper Triassic rocks are not reported in northwestern Utah, south-central Idaho, or northeastern Nevada. The occurrence in the southern part of the rooted section of coarse conglomerate containing Upper Permian and Lower Triassic clasts suggests the existence of a highland composed of these rocks to the southwest, south, or southeast, prior to the emplacement of displaced sheet I. However, it is unlikely that the Terrace and Hogup Mountains were the source for sheet I because of their distance and because they were blocked out by Basin and Range faults and may not have existed in Miocene time. Terrace Mountain may itself be a part of the displaced terrane (Todd, unpublished data). Since displaced sheets have not been found east of the Matlin Mountains, it seems likely that the source area for sheet I (and the overlying allochthonous(?) Tertiary fanglomerate) lay to the west.

Sheet II. Displaced sheet II crops out in a north-northeast-trending zone in the central part of the Matlin Mountains (Fig. 2C). The sheet crops out almost continuously from beneath Tertiary and Quaternary deposits and a klippe of sheet V for 6 km from north to south and has a maximum width after high-angle faulting of 2 to 2.5 km. Scattered klippen in the southern part of the study area

indicate that the original length of the sheet was over 15 km and that its area was at least 50 km². Its maximum thickness is estimated as 120 m. Displaced sheet II consists mainly of the Gerster Formation; minor Plympton(?) Formation is present in the western part of the sheet where bedding dips 85 degrees to the west with tops to the east (Fig. 3). The sheet is deformed and, locally, limestone has recrystallized to coarse marble, but the metamorphic effects are very mild in comparison with those in the structurally higher displaced sheets. Although brecciation is common, especially in thick chert beds and near the fault contact with the underlying fanglomerate, most of the sheet consists of closely jointed, intact bedrock. Gentle to vertical bedding and opposed dips indicate the presence of asymmetrical and overturned folds with axial traces that curve from northerly to easterly trends. Wavelengths of these folds range from several hundred meters to less than 1 km, and the folds verge to east and west. The scale of these structures relative to the thickness of the sheet indicates that fold limbs are truncated by the underlying low-angle fault. The hill underlain by Gerster Formation between informally named Alfred Hill and George Point (Fig. 3) consists of a north-striking antiform with a gently dipping western limb and steeply dipping eastern limb (vergence to southeast). The curving pattern made by axial traces of these folds suggests refolding on northwest axes.

Displaced sheet II lies on Tertiary fanglomerate, the Plympton(?) Formation of sheet I and red-weathering Triassic-limestone-bearing Tertiary conglomerate. Either the underlying fanglomerate and sheet I were eroded locally prior to the emplacement of sheet II or they were removed tectonically (Fig. 3). Sheet II is overlain by Tertiary fanglomerate which dips 0 to 65 degrees away from the sheet. In the west-central part of Figure 3, this fanglomerate either was deposited on an eroded surface developed on sheets I and II and the rooted section, or is in low-angle fault contact with them. If the base of this fanglomerate is not a fault, then its deposition occurred somewhere to the west(?) and it was carried piggyback to its present location because it differs markedly in composition and clast size from the Tertiary rocks which overlie the rooted section 2 km to the east. In addition, this fanglomerate structurally overlies the red-weathering conglomerate which is the stratigraphically higher member of the rooted Tertiary section. Furthermore, both the upper and lower fanglomerates in the Alfred Hill-George Point area end abruptly at the east fault. The presence of tectonic slivers along the basal East Fault and the disruption of bedding in the red-weathering conglomerate below the fault suggest that the discontinuity of sheets in this area may be entirely tectonic.

As mentioned above, the Gerster Formation occurs in the upper allochthon of the Grouse Creek Mountains, in the Terrace and Hogup Mountains, and on Terrace Mountain. Terrace Mountain is underlain by a nearly vertical

A

Figure 6. Metamorphic folds in displaced sheets. *A.* Chevron folds in metamorphosed Lower Permian limestone and sandstone (dark-weathering beds) that lie unconformably below Big Pink sequence in displaced sheet III; hammer for scale. Fold axes plunge gently to north-northeast, axial planes verge to northwest. *B.* Isoclinal folds in metamorphosed limestone and sandstone of the thinly-bedded limestone and sandstone subunit of the Oquirrh Formation in displaced sheet IV; 6-inch rule for scale. Fold axes plunge gently to north; axial planes verge

east. Metasandstone beds are dark-weathering. Axial plane cleavage developed locally in sandy marble.

B

section composed of the Plympton(?), Gerster, and Thaynes(?) Formations which in reconnaissance appears to have an allochthonous relation with surrounding tuffaceous Tertiary rocks that locally dip steeply. The lithology, structure, and local mild metamorphism of the Permian rocks of Terrace Mountain resemble those in sheet II; in fact, Terrace Mountain may be part of sheet II. Displaced sheet II most likely came from the same source area as did sheet I. If the large southeast-verging antiform formed during movement of the sheet, then at one time, the direction of transport of sheet II may have been toward the southeast.

Sheet III. Displaced sheet III crops out almost continuously for 16 km from north to south and for 9 km from east to west at its widest point in the western part of the Matlin Mountains. Prior to emplacement of sheets IV and V and Lake Bonneville deposition, the sheet probably had a minimum area of 150 km². its estimated maximum thickness is between 200 and 300 m. Sheet III consists of Permian carbonate rocks and Tertiary sedimentary rocks which overlie them unconformably. The latter consist of interbedded fanglomerate and tuffaceous lacustrine sandstone—the Big Pink sequence—and the former consist of metamorphosed "patterned unit" and Kaibab(?) Limestone, and minor unmetamorphosed Gerster Formation (Fig. 2E). The Lower Permian rocks have undergone greenschist-facies metamorphism: the conodont color alteration index (7-½) indicates that postdeposition temperatures in the limestone exceeded 400 degrees centigrade (J.

Repetski, U.S. Geological Survey, written communication, 1982). Although pervasively fractured and locally brecciated, these rocks are intact. They have been tightly folded in chevron and box folds; thin siliceous sandstone interbeds have been systematically fractured and rotated within a matrix of more ductile marble (Fig. 6A). Generally northerly strikes of bedding and dips to both east and west as steep as 90 degrees suggest the presence of large-scale upright chevron folds, locally asymmetrical to the east and west. The inferred trends of the axial planes of these folds suggest refolding of north-northeast folds on northwest axes.

Although the contact between displaced sheets II and III is not exposed, it is inferred to be a westward-dipping low-angle fault (West Fault) (Fig. 2A, B). Poorly exposed Tertiary tuffaceous sandstone appears to underlie sheet III in the core of a north-northwest-trending anticline south of the Immigrant Trail and in the northern part of Big Pink Hill (Fig. 2A; Fig. 4). If these Tertiary beds were originally continuous with the Tertiary strata that overlie sheet II, then these occurrences are windows to sheet II.

There is also indirect evidence that the Permian and Tertiary rocks assigned to sheet III are allochthonous. In the northern part of sheet III, at least 2000 m of Tertiary strata that overlie metamorphosed Lower Permian rocks dip northwest from 50 to 70 degrees near the base of the section, to 5 to 10 degrees in the stratigraphically higher exposures located about 5 km west of Big Pink Hill (Fig. 4). About 5 km further west, in the eastern Grouse Creek

Mountains, the lower and middle allochthons are unconformably overlain by thin (less than 150 m), flat-lying fanglomerate composed of clasts of the underlying middle allochthon, whose youngest rocks are Pennsylvanian-Permian. A cross section across the valley between the Grouse Creek and Matlin Mountains indicates that there is a high-angle fault between these two Tertiary sections (Fig. 1B). The Big Pink sequence cannot simply be the eastern limb of a fold which continues across the valley under alluvium because there is insufficient room in the eastern Grouse Creek Tertiary section to accommodate the western limb of such a fold. The room problem is alleviated if the Big Pink sequence has been down-dropped against the middle allochthon along a high-angle fault. This interpretation requires that the Lower Permian Kaibab(?) Limestone and "patterned unit" which underlie the Big Pink sequence are part of the middle allochthon of the Grouse Creek and Raft River Mountains. However, these units do not occur in the middle allochthon.

In the area south of the Immigrant Trail, about 115 m of Tertiary strata consisting of fanglomerate with minor interbedded tuffaceous sandstone are continuous with the Big Pink sequence; the fine-grained upper part of the section appears to be missing. The southward thinning of the Big Pink sequence may be due to original stratigraphic thinning, erosion, and (or) removal by low-angle faulting. If the Big Pink sequence originally thinned drastically to the south and west, then it might have been continuous with the fanglomerate in the eastern Grouse Creek Mountains.

However, deriving the Big Pink sequence in situ from an ancestral Grouse Creek range in the position of the present range leads to a more serious problem. The lower part of the sequence consists largely of debris flow deposits which must have formed at or near a steep mountain front. The upper part of the sequence contains volcanic material dated as 11.1 to 10.5 m.y. old, material that is interbedded with clasts of the middle allochthon, lower allochthon, and autochthon of the adjacent metamorphic core complex. Rb-Sr mineral ages for biotite from autochthonous rocks of the Grouse Creek Mountains indicate that this part of the terrane was at temperatures of approximately 300 degrees centigrade 8 to 12 m.y. ago (Compton and others, 1977). Even if uplift and cooling of the autochthon took place very rapidly after 12 m.y. B.P., it would not be possible to derive the Big Pink sequence from an ancestral Grouse Creek range; judging from the present relations, the lower and middle allochthons of the Grouse Creek Mountains were still intact in Pliocene time when Basin and Range faulting began, that is, were not exposed to erosion until Pliocene time. Furthermore, the northern part of the Big Pink sequence contains clasts of green metaquartzite which crops out at present only in the autochthon of the Raft River Mountains. Yet, the Raft River Mountains did

not exist as such until Pliocene time. The source of the Big Pink sequence must have been a steep mountainous area that exposed rocks of the metamorphic core complex. If current indicators were not rotated during emplacement of sheet III, then this area lay to the west or southwest of the present Grouse Creek Mountains.

The metamorphism and tight folding of the pre-Tertiary rocks of displaced sheet III indicate that its source lay to the west in the metamorphic core complex. These folds are cut by the upper Miocene unconformity, and therefore, they probably bear no direct relation to the movement direction of sheet III. The Tertiary strata and the underlying Permian rocks were folded together on northerly axes (Fig. 4). If Basin and Range fault displacements are restored, it appears that one of these folds became oversteepened to the east and developed into an older-on-younger fault, the reverse fault which bounds the east side of Big Pink Hill (Fig. 4). This fault has a larger displacement than those of the short, straight Basin and Range faults that lie directly to the east. In fact, the displacement cannot be accounted for unless the reverse fault flattens at depth, in which case it must merge with the low-angle fault beneath sheet III, as shown in Figures 2B and 4. Although sheet III therefore probably consists of two subsidiary sheets separated by the reverse fault, displacement along this fault is minor compared with those of the East and West low-angle faults. The two subsidiary sheets have therefore been treated as a single sheet.

The northwest-dipping Tertiary section of Big Pink Hill might at first glance seem to be a good candidate for a package of strata that have been transported southeastward by listric normal faulting. However, as indicated in Figure 4, the strata in the eastern part of sheet III dip eastward. Furthermore, rotation of the reverse fault into the geometry of a listric normal fault results in an older-on-younger relation with the dip of strata opposite to that expected after listric normal faulting.

The reverse fault is clearly older than the adjacent Basin and Range faults; it does not offset sheet IV, but sheet IV is offset by the Basin and Range faults (Fig. 2B, 4). Thus sheet III was folded, and continued lateral compression had led to the formation of a reverse fault before sheet IV was emplaced. Basin and Range faults subsequently cut all of these structures; movement on the reverse fault may have been reactivated at this time. If, as seems likely, the north-trending folds and the reverse fault formed during movement of sheet III, then it moved from west to east.

Sheet IV. Sheet IV is the best exposed of the five displaced sheets. It is overlain by Quaternary lake deposits and, in one place, by Tertiary fanglomerate. Sheet IV occurs only in the western half of the displaced terrane, where it rests upon a surface of low relief developed on the interbedded fanglomerate, sandstone, and tuffaceous lake beds of late Miocene age which are part of displaced sheet III

Figure 7. View north of west-dipping contact (dashed line) between klippe composed of metamorphosed limestone and sandstone subunits of the Oquirrh Formation (displaced sheet IV) and Tertiary rocks (Big Pink sequence); Grouse Creek Mountains visible in distance; light-colored, tuffaceous lake beds in foreground (trees). (See Figure 9 map and cross section for structure of hill.)

(Fig. 2C). Low hills to the west and east of Big Pink Hill consist of klippen of dark-brown-weathering pre-Tertiary rocks in nearly flat contact over tilted, light-colored tuffaceous beds (Fig. 7). Displaced sheet IV has a maximum thickness of 75 m, and its erosional remnants occupy an area of approximately 100 km². Although sheet IV is pervasively fractured and the basal few meters of the sheet are brecciated, it has remained intact. Breccia near the base grades upward into rock that has broken across bedding surfaces into rather uniform subangular fragments through which bedding and concordant metamorphic foliation can be traced (Fig. 8). Most of the breccia fragments are less than 0.4 m in longest dimension and many are less than 0.15 m.

Outcrops of sheet IV lying just north of and to the south of, latitude 41°37′30″ N consist of phyllite of the metamorphosed Chainman or Diamond Peak Formation, and marble and quartzite of the metamorphosed Oquirrh Formation (Fig. 2C; Fig. 4). To the north of this latitude, on the west side of the reverse fault, the rocks of sheet IV vary from interbedded sandy marble and quartzite in the south (metamorphosed thinly bedded limestone and sandstone subunit (?) of the Oquirrh Formation) to marble in the north (metamorphosed Kaibab(?) Limestone). The hills east of the reverse fault (informally named Juliette Hills, Fig. 2A) also consist of the metamorphosed Kaibab(?) Limestone. Thus, in displaced sheet IV there appears to be, from south to north, a Pennsylvanian into Lower Permian sequence.

Metamorphic folds and lineations in the rocks of displaced sheet IV have undergone little rotation from the orientations observed in similar rocks in the Grouse Creek Mountains. Tight metamorphic folds of outcrop scale whose axes trend northerly to northeast and northwest, west-northwest, to east-west were seen commonly (Fig. 6B). Folds such as these must have formed at metamorphic

temperatures prior to emplacement of the sheet. Two superimposed metamorphic lineations which have wavelengths of less than 1 cm are invariably present in the rocks of sheet IV—northeast to north-northeast (L_1 of the Grouse Creek Mountains?) and west-northwest (L_2 of the Grouse Creek Mountains?) (Table 1). The rocks of sheet IV apparently were folded into large, upright folds after metamorphism and prior to, and (or) during, emplacement of the sheet. Foliation and concordant bedding have strikes that curve broadly from north to east and gentle-to-vertical dips that suggest the occurrence of large upright and asymmetrical folds with steep axial planes (Fig. 4). Upright folds with similar trends which did not involve recrystallization and probably formed under shallow conditions are exposed in the hills immediately north of, and to the south of, latitude 41°37′30″ N. Here, displaced sheet IV consists of isoclinally folded, low-angle faulted phyllite and metaconglomerate of the metamorphosed Chainman or Diamond Peak Formation and marble and quartzite of the metamorphosed Oquirrh Formation (Fig. 9). This structure has been thrown into a series of upright, northeast-trending folds with brecciated hinges, which apparently were refolded on northwest axes. The resulting map pattern resembles that observed in the outcrops of displaced sheet IV to the north, and it seems likely that refolding was responsible for the variable attitudes of bedding and foliation there. Figure 9 indicates that the northwest folding took place when the rocks were still sufficiently ductile to deform in a systematic manner: F_1 axes locally continued to bend through almost 180 degrees. Further evidence for pseudoductility of the rocks of sheet IV are the boudinaged and folded segregations of silica and calcite which originally filled fractures in the sheet (Fig. 8). These segregations range up to many meters in diameter and indicate that the brecciated sheet was flushed with solutions. These deformed segregations

TABLE 1. TECTONIC HISTORY OF RAFT RIVER - GROUSE CREEK - MATLIN MOUNTAINS, NORTHWESTERN UTAH

million years	RAFT RIVER MOUNTAINS	NORTHERN GROUSE CREEK MOUNTAINS	SOUTHERN GROUSE CREEK MOUNTAINS	CENTRAL GROUSE CREEK MOUNTAINS	MATLIN MOUNTAINS
0	Basin and Range faults	Basin and Range faults	Basin and Range faults	Basin and Range faults	Basin and Range faults
	Broad east-west Raft River anticline	Broad east-west folds		Northwest-to west-trending folds	Northwest folds
					Basalt . . . . 3.2 ± 0.2
5	Broad north-south folds; middle allochthon thrust over Tertiary	Broad north-south folds		Broad north-trending folds	Broad north-trending folds; Displacement of sheet V
		Sedimentation* in NW	Complex upper allochthon (subsidiary sheets of middle allochthon moved over Tertiary rocks)	Complex upper allochthon (subsidiary sheets of middle allochthon moved over Tertiary rocks)	Displacement of sheets III and IV; reverse fault
		Erosion			
8.5 - 9.0 . . . . Dacite		Welded dacite tuff			
10	Sedimentation	Sedimentation in NE (8000 ft +)		Autochthon cools below 300° C	Sedimentation (Big Pink . . . . 10.5 - 11.1 sequence)*
	Diabase and minette				
	Low-angle faults with up to 30 km eastward transport	Imbricate thrusts (vergence NE, E)	Rhyolitic volcanism . . . . 11.7 ± 0.4	Rapid uplift and erosion of middle allochthon	Displacement of sheets I and II; sedimentation
			Sedimentation*		
15	F₄ (vergence east; late-or postmetamorphic)	F₃ (vergence E; late-to postmetamorphic)	Basaltic volcanism . . . . 14.4 ± 0.4	F₃, imbricate thrusts (vergence E, SE; late-metamorphic)	Folding of rooted section?
20	Autochthon cooled below 400° C ≤ 20 m.y.			Doming and slow uplift	
	F₃ (vergence SE)				
25				Red Butte pluton (culmination of gneiss dome) . . . . . . 25 · · 24.9 ± 0.5	
30	F₂ (vergence N)	F₂ (vergence NE)		F₂ (vergence SW)	
34		Silicic volcanism and sedimentation			
35					
38.2 ± 2.0		Immigrant Pass Pluton			
40					
			F₂ (vergence N)		
45	?	?	?	?	
	F₁ (vergence N; local development in western part only)	F₁ (vergence NW, W)	F₁ (vergence W, NW)	F₁ (vergence NW)	
50					

Legend:

· · · · · · · · Dated event

| Range of tectonic event other than folding

|| Range of folding event, dashed where uncertain, synmetamorphic unless noted otherwise; low-angle faulting that preceded or accompanied folding not shown

F₁ First folding
F₂ Second folding
F₃ Third folding
F₄ Fourth folding

Remarks: Parts of the "Sedimentation" events in the Grouse Creek and Matlin Mountains took place elsewhere and deposits were emplaced into present sites as allochthonous sheets. Such sedimentation events are denoted by *.

SOURCES OF DATA:
Raft River Mountains - Compton, 1975
Northern Grouse Creek Mountains - Compton, 1972
Southern Grouse Creek Mountains - Compton (this volume)
Central Grouse Creek Mountains - Todd, 1980
Matlin Mountains - Todd (this volume)

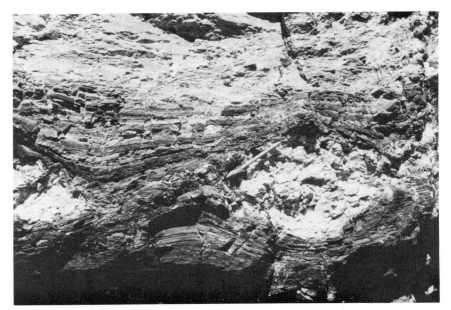

Figure 8 (this and facing page). Structure of brecciated displaced sheets. *A.* Boudinaged quartz-carbonate veins in brecciated basal part of displaced sheet IV; pencil for scale. Light gray rock at top of photo is marble; dark layers are interbedded quartzite and sandy marble. *B.* Clastic dike in chaotic breccia composed of metamorphosed, thinly-bedded limestone and sandstone of the limestone subunit of the Oquirrh Formation, basal part of sheet IV. Penknife lies above subrounded small boulder and cobbles of resistant metamorphosed Oquirrh sandstone; other, less resistant clasts in dike weather to lower relief than breccia. Dike is one of four, closely spaced, north-trending clastic dikes.

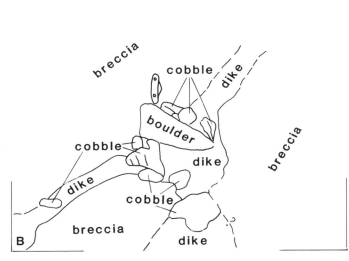

further suggest that this part of the sheet underwent at least two stages of deformation: early fracturing followed by deposition of silica and calcite, and subsequent pseudoductile flow of breccia during which the silica-calcite veins were deformed. Folds such as those in Figure 9 predate the final emplacement of the sheet, for if the limbs are projected downward they must be truncated by the underlying fault.

Many of the generalizations about the mode of emplacement of the displaced sheets are based upon what is known about sheet IV and its relations to the underlying Tertiary rocks. The contact between the brecciated basal part of displaced sheet IV and underlying Tertiary rocks

typically is sharp, and the Tertiary rocks are, in general, undisturbed. Locally in the basal brecciated zone, limestone of the sheet flowed in a ductile manner and siliceous sandstone beds and chert nodules broke apart and their fragments rotated chaotically in the limestone matrix. In some places, exotic blocks of limestone appear in the basal breccia zone. Thin slivers of brecciated white metaquartzite (metamorphosed Eureka(?) Quartzite) and tan-weathering metadolomite (metamorphosed Pogonip(?) Group or metamorphosed Fish Haven(?) Dolomite) are present along the contact. At one locality, L. Hintze (written communication, 1978) observed what he believed to be slivers of Cam-

C. Folded breccia composed of interbedded quartzite and sandy marble in displaced sheet IV; outcrop is about 3 m high. *D.* Steep (tear?) fault in brecciated Oquirrh Formation of displaced sheet IV (penknife for scale, center of photo). Note right-lateral bending of limestone and sandstone beds which are preserved despite brecciation.

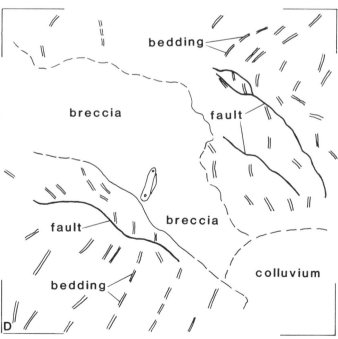

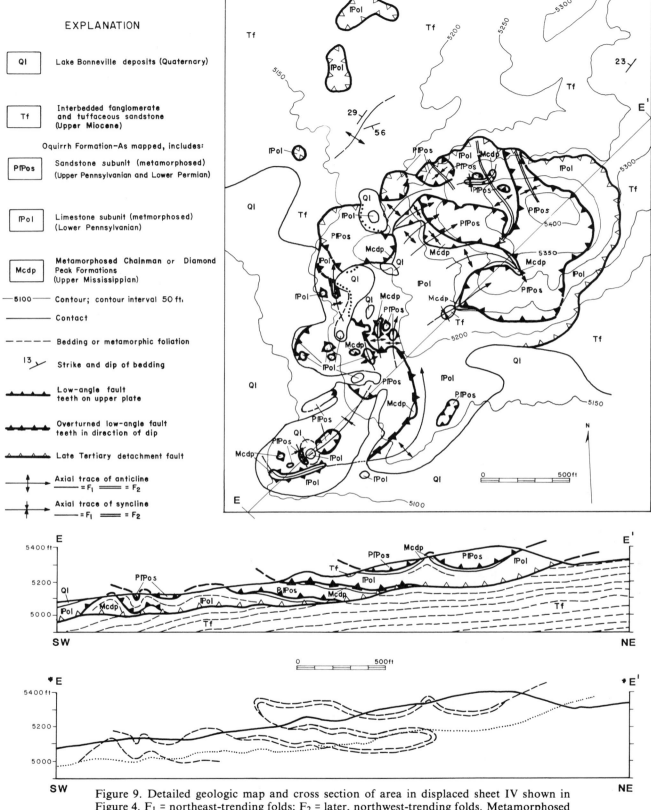

EXPLANATION

Ql — Lake Bonneville deposits (Quaternary)

Tf — Interbedded fanglomerate and tuffaceous sandstone (Upper Miocene)

Oquirrh Formation—As mapped, includes:

PIPos — Sandstone subunit (metamorphosed) (Upper Pennsylvanian and Lower Permian)

IPol — Limestone subunit (metamorphosed) (Lower Pennsylvanian)

Mcdp — Metamorphosed Chainman or Diamond Peak Formations (Upper Mississippian)

—5100— Contour; contour interval 50 ft.

———— Contact

——— Bedding or metamorphic foliation

13 ⌐ Strike and dip of bedding

▲▲▲ Low-angle fault teeth on upper plate

▲▲▲ Overturned low-angle fault teeth in direction of dip

△△△ Late Tertiary detachment fault

↕ Axial trace of anticline ——— = F₁ ═══ = F₂

↕ Axial trace of syncline ——— = F₁ ═══ = F₂

Figure 9. Detailed geologic map and cross section of area in displaced sheet IV shown in Figure 4. F₁ = northeast-trending folds; F₂ = later, northwest-trending folds. Metamorphosed Chainman or Diamond Peak Formation is fault-bounded but for clarity, some contacts not shown as faults. Cross section E-E′ has equal horizontal and vertical scales; *E-*E′ is interpretive cross section, dotted line represents late Tertiary detachment fault. (Compare *E-*E′ with view in Figure 7.)

Figure 10. View south from Matlin Mountains toward Great Salt Lake Desert. Newfoundland Mountains in far distance. Terrace Mountain in right-middle distance. In near distance are broad Red Dome terrace and dark hills capped by breccia of displaced sheets IV and V. Dark hills are underlain and surrounded by poorly exposed Tertiary beds.

brian(?) and Ordovician units, including Pogonip(?) Group and Eureka(?) Quartzite, in pseudostratigraphic order at the base of sheet IV overlying a clayey paleosol developed on volcanic ash. However, throughout most of the area, soil and (or) colluvial gravel are not developed on the Tertiary rocks that underlie sheet IV. Lenses of phyllite of the metamorphosed Chainman or Diamond Peak Formation are present in many places at the base of the sheet (Fig. 4). Elsewhere, thin discontinuous layers of greenish-white tuff lie at the base of sheet IV. The breccia zone above the fault commonly is well-indurated by deposition of silica, calcite, or caliche-like material and the underlying Tertiary rocks are silicified locally. The fault surface truncates northwest-dipping Tertiary strata that locally include massive, well-indurated fanglomerate beds up to 10 m thick; in some places, a thin layer of tuffaceous claystone separates sheet IV from the underlying massive fanglomerate. These lenses of fine-grained tuffaceous rock suggest that mechanically weak and (or) unconsolidated Tertiary beds were sheared off and dragged along the fault surface or possibly that airfall deposits of volcanic ash lubricated it. In one place, the Tertiary rocks beneath the advancing displaced sheet were wet (lake beds?) as indicated by clastic dikes in the basal part of the sheet (Fig. 2A, pt. 1 and Fig. 8). Locally, complex high- and low-angle faults occur in tuffaceous lake beds (Fig. 2A, pt. 2), and dips of lake beds are steepened (folded?) from 30 to 40 degrees to 75 to 80 degrees beneath displaced sheet IV.

Two thin lenses of breccia similar in lithology to the Lower Permian rocks of displaced sheet IV lie concor-

dantly within the northwest-dipping Tertiary Big Pink sequence. The northeast-dipping section of Tertiary rocks that underlies informally named Big Bird Hill and is probably continuous under Quaternary cover with the Big Pink sequence also contains a thin discontinuous lens of breccia that consists of rocks of Lower Permian age and is interlayered concordantly with the Tertiary beds. One of the two Big Pink breccia lenses may have been continuous between these two areas prior to erosion.

The source of the rocks of sheet IV is readily pinpointed as the Raft River-Grouse Creek-Albion Mountains metamorphic core complex. The attenuated, recumbently folded, low-angle-faulted rocks of the metamorphosed Oquirrh Formation which are shown in Figure 9 are identical to rocks and structures in the middle allochthon as it is presently exposed in the central Grouse Creek Mountains (Fig. 1). Greenstone dikes of unknown age similar to those that occur only in the middle allochthon of the Grouse Creek Mountains were also found in displaced sheet IV in the Matlin Mountains. The metamorphosed Lower Permian thinly bedded limestone and sandstone subunit(?) of the Oquirrh Formation which forms part of sheet IV occurs only in the middle allochthon of the central and southern Grouse Creek Mountains. As mentioned above, the sheet arrived with the metamorphic structures in its rocks rotated only slightly from the orientations of those in similar rocks in the metamorphic core complex. The slivers of Lower Paleozoic(?) metaquartzite and marble beneath sheet IV must have come from the core complex. The presence of metamorphosed Kaibab(?) Limestone in sheet IV indicates that this Lower Permian unit was present in the metamorphic core complex even though it is not found in the Grouse Creek or Raft River Mountains.

The postmetamorphic folds in displaced sheet IV (Fig. 9) may have formed during its emplacement. The trends and inferred ages of these folds imply an early period of northwest-southeast compression followed by compression in a northeast-southwest direction on fold axes that plunge southeast. If the source area lay to the west, then presumably sheet IV was transported toward the southeast at an early stage and underwent further movement to the northeast at a later stage. A single, tight, southeast-oversteepened fold in the underlying Tertiary rocks near the klippe shown in Figure 9 may have formed during the early period of movement of sheet IV.

Sheet V. The structurally highest displaced sheet is present only in the southern half of the Matlin Mountains, and after erosion, it is the thinnest (60 m) of the five sheets. Erosional remnants of sheet V encompass an area of about 18 km^2 in the southwest part of the Matlin Mountains, where almost totally brecciated, unmetamorphosed, thinbedded sandy limestone (Thaynes(?) Formation) rests with a flat fault contact on displaced sheet IV and, where sheet IV has been removed tectonically or by erosion, on rocks of the Tertiary Big Pink sequence (Fig. 2A).

Klippen that consist of totally brecciated Thaynes(?) Formation, locally with Plympton(?) and Gerster(?) Formations, occur in the eastern part of the displaced terrane in two places; west of George Point, and south of Red Dome, these klippen rest upon fanglomerate and tuffaceous lake beds that overlie displaced sheets I and II. If these scattered occurrences were originally part of a single displaced sheet, then its area was greater than 100 km^2. However, these two occurrences may represent two separate displaced masses. Sheet V is best exposed in the group of low hills about 2.5 km south of the Immigrant Trail, accessible by a jeep trail which turns south off the Immigrant Trail 2 km east of Highway 30. The sheet is overlapped by deposits of Lake Bonneville.

Displaced sheet V can be recognized by one or more of the following characteristics: (1) its distinctive limestone (Thaynes(?) Formation), which is virtually restricted to this sheet; 2) almost total brecciation and boxworks fractures filled by white quartz; and 3) its structural position above the other four displaced sheets. The basal contact of sheet V is not as well exposed as that of sheet IV chiefly because the southern part of the Matlin Mountains was under water during high stands of Lake Bonneville, but where the contact is exposed, the underlying Tertiary strata appear to be undisturbed. Hills capped by breccia were carved by waves into fantastic shapes and subsequently were desert-varnished to a dark brown color. As a result, remnants of sheet V commonly can be spotted by their dark, rubbly appearance (Fig. 10). Because of the extreme brecciation, relict

structures are rarely preserved; locally, however, breccia fragments are only slightly disoriented and relict limestone beds can be seen to dip steeply. Indistinctly layered breccia also dips steeply but discordantly to relict bedding, as though the breccia was folded or tilted during or after emplacement of the sheet.

The nearest exposures of Lower Triassic Thaynes Formation are in the complex upper allochthon of the Grouse Creek and Raft River Mountains. As mentioned above, a source of clastic debris derived from the Dinwoody and Thaynes Formations must have lain southwest, south, or southeast of the Matlin Mountains during Miocene time as indicated by the occurrence of coarse, Triassic limestone-bearing Tertiary conglomerate in the southeastern part of the area. Thaynes(?) Formation may also occur on the east side of Terrace Mountain (Fig. 1). The fact that displaced sheet V is present only in the southern half of the Matlin Mountains suggests a southerly source for the rocks of the sheet.

Major Low-Angle Faults

Two of the five low-angle faults that form the boundaries of the displaced sheets apparently divided the displaced terrane into two composite plates which have had different histories and moved separately. One of these faults, the East Fault, is the basal fault which separates the displaced terrane from the rooted section. The other, the West Fault, probably dips westward and joins the basal fault at depth (Fig. 1). These two faults juxtapose plates that differ significantly in lithology, metamorphism, and structural character; therefore, they may have had the largest displacements of the faults in the displaced terrane.

East Fault. Although the East Fault is covered for most of its length, its trace is marked by the 20-km-long, sinuous, northerly boundary between the displaced terrane and the rooted section (Fig. 2A). The fault dips 20 degrees to the north-northwest south of George Point; its overall dip must be westward. West of this fault, pre-Tertiary and Tertiary rocks have been displaced repeatedly with the result that they form nearly flat-lying, interlayered sequences. East of this fault, Tertiary sedimentary rocks overlie a relatively undeformed section of upper Paleozoic rocks unconformably. In addition to structural differences between the rooted and displaced terranes, the Tertiary rocks which lie east and west of the fault differ markedly in source of clastic debris and in depositional environment.

West Fault. The rocks of the displaced terrane differ significantly in lithology, metamorphism, and structural character across a narrow, north-northeast-trending zone which must represent a buried fault (Fig. 2A). The inferred length (20 km) of this fault, two to four times the length of the Basin and Range faults in the area, and its parallelism to the East Fault suggest that the western fault is also a low-angle fault.

If, on the other hand, the sequence of pre-Tertiary and Tertiary rocks from sheet I to sheet IV was once continuous across the displaced terrane and if the West Fault is a high-angle fault which broke and offset this sequence, then the most likely sense of displacement was down-dropping of the block west of the fault relative to the eastern block. The rooted section is exposed 3 to 6 km east of the fault, and its probable structural equivalent—the middle allochthon of the Grouse Creek Mountains—crops out 9 km west of the fault. Thus, the middle allochthon lies at a deeper structural level in the block between the West Fault and the normal fault that must underlie the valley between the Grouse Creek and Matlin Mountains (Fig. 1). If sheets III and IV once were present above sheets I and II on the eastern block and were removed by erosion on an upthrown fault block, then approximately 600 km of rock would have to have been removed, for no erosional remnants of sheets III and IV are found east of the fault. Yet, inferred offsets on the adjacent Basin and Range faults are only on the order of 100 to 200 m. Clearly, the block west of the fault has not been uplifted by Basin and Range faulting, and sheets I and II eroded from it; the window through sheets I and II, 3.5 km east of the fault, shows that the rooted section, not sheets III and IV, underlies sheets I and II. Sheet V is present on both sides of the West Fault and presumably was emplaced after the two composite plates had their present relative position. If sheets III and IV were once present east of the fault, the base of sheet V must have cut down-section structurally across them (Fig. 2A). However, it seems more likely that these sheets were never present east of the West Fault. Basin and Range faults are not observed to cut sheet V, but they do cut sheets I through IV and probably postdate low-angle faulting. The young age of Basin and Range faults further indicates that the West Fault is a low-angle fault which formed during the main episode of low-angle faulting.

The West Fault is considered to dip westward and to join the basal fault because, as noted above, rocks of sheets III and IV are not seen in the window through sheets I and II in the area northwest of George Point.

High-Angle Faults

High-angle faults in the Matlin Mountains are chiefly Basin and Range faults that have cut virtually all of the structures of the displaced terrane as well as the large folds in the rooted section. Basin and Range faults began to form in Pliocene time in the Raft River-Grouse Creek region, and faulting continues to the present time (Compton and others, 1977). Eight short (1 to 3 km), northerly high-angle faults have cut the displaced terrane forming a series of narrow (1 to 2 km) horsts and grabens (Fig. 2A, B). Their

straight traces indicate that these faults dip steeply. The branching reverse fault that forms the eastern margin of Big Pink Hill appears to be an older-on-younger fault that formed during low-angle faulting.

At least four Basin and Range faults cut the rooted section. Two of the four faults strike northwest, one strikes north, and one strikes northeast. These faults are longer (4 to 8(?) km) and the horsts and graben they have blocked out are wider (several km) than those of the displaced terrane. Since much of the Matlin Mountains is covered by deposits of Lake Bonneville, there undoubtedly may be undetected faults.

Mechanics of Low-Angle Faulting

The size and systematic structures of the displaced sheets indicate that they were emplaced as coherent sheets. The evidence concerning the way in which the sheets moved comes chiefly from sheet IV, which has the best exposure. Apparently, the sheet was deformed under a moderate load which may have consisted of pre-Tertiary formations that lay stratigraphically above the metamorphosed Kaibab(?) Limestone and (or) once superjacent Tertiary rocks. The flow of marble between and around broken fragments of more resistant quartzite beds and the boudinage and folding of fracture-filling silica and calcite segregations imply ductile deformation at some stage during emplacement. The ductile structures, which formed during and after brecciation, indicate that the sheet moved by slow creep rather than by catastrophic sliding. Locally, lower Paleozoic units were peeled off the lower allochthon of the adjacent metamorphic core complex, thinned to a fraction of a meter, brecciated, and carried beneath the slowly moving sheet in pseudostratigraphic order. Locally, thin lenses of phyllite, tuffaceous claystone, and vitric tuff were caught up and smeared out in the fault zone suggesting that mechanically weak and (or) wet materials served to lubricate movement.

Critical clues to the understanding of how the displaced sheets moved may be provided by the clastic dikes and the widespread calcite and silica veins in the basal breccia of sheet IV. The clastic dikes are found in only one locality, above the lacustrine beds of sheet III, but may have been more common; these delicate structures would be destroyed easily by post-dike movements and the poorly rounded clasts in the disrupted dikes would resemble the surrounding breccia fragments. The dikes and veins indicate that water derived from the underlying Tertiary sedimentary rocks became trapped under pressure at the base of the sheet. In spite of the general absence of paleosols and (or) colluvium beneath the fault surface, sheet IV probably moved over an erosional surface; although some scraping off of unconsolidated deposits probably occurred locally, it is unlikely that the advancing sheet sheared off the tops of the broad folds in the underlying Tertiary rocks. The presence of the clastic dikes in the basal breccia and the fabric of the breccia suggest that fracturing of the sheet was largely accomplished by hydrofracturing. Water and sediment from the underlying Tertiary rocks apparently entered pervasive, throughgoing, steep fractures in the base of sheet IV and fluid pressure caused the fractures to propagate by pushing apart their walls. The systematic, slight rotations of the breccia fragments which are typical of sheet IV indicate that the sheet disintegrated in place. Relatively high fluid pressure at the base would provide the buoyancy necessary for this thin, laterally extensive body of rock to move, by slow creep, as a coherent sheet for a considerable distance.

Locally, fluid pressures may have been high enough so that parts of sheet IV traveled on a watery cushion. The association of vitric tuff with the fault beneath sheet IV suggests that hot volcanic ash and gas may have become trapped beneath a newly detached mass in the source area. In either case, the displaced sheet would have experienced a period of catastrophic movement.

Tabular bodies of brecciated rock that overlie or are enclosed within basin deposits have been interpreted as catastrophic avalanches or landslides (Shreve, 1968; Krieger, 1977). These landslides have features similar to those of displaced sheets IV and V of the Matlin Mountains—great thinness relative to areal extent, presence of relatively unrotated breccia fragments ("jigsaw puzzle" effect or crackle brecciation), and relict stratigraphy. However, such an origin was rejected for the Matlin sheets because of similarities between the displaced sheets and postmetamorphic allochthons in the core complex and because paleogeographic reconstructions deny the presence of reasonably close source areas at the time the sheets were emplaced. In the Grouse Creek Mountains, subsidiary plates of the middle and upper allochthons are as thin as, and locally as brecciated as, displaced sheet IV. In each case, the plates rest upon Tertiary sedimentary rocks which Compton (this volume) considers to be tectonic slices. Furthermore, while the attenuation and brecciation of the lower Paleozoic formations that are present beneath sheet IV are extreme, they are not different from the local thinning and brecciation of these units where they overlie the Tertiary gneiss dome in the central Grouse Creek Mountains. The concordant lenses of monolithologic breccia in the Big Pink sequence are more likely candidates for a landslide origin. In an active tectonic region of high relief, debris flows, landslides, and gravity slides would probably all occur and in some situations, they might be gradational or indistinguishable. Much of sheet V consists of chaotic breccia, and calcite and silica have formed boxworks along fractures. Sheet V has some of the characteristics of a mass that flowed turbulently (Shreve, 1968) and it may have originated as a large landslide.

ORIGIN OF THE DISPLACED TERRANE

The late Cenozoic history of uplift, erosion, and deposition in the Grouse Creek and Raft River Mountains is summarized in Table 1. This history is important because it provides the framework within which low-angle faulting in the Matlin Mountains must be considered. At the present time, the highest peaks of the Grouse Creek and Raft River ranges consist of Precambrian rocks of the autochthon (Fig. 1). Basin and Range faulting in the Grouse Creek Mountains and east-west arching of the Raft River Mountains began in Pliocene time (Compton and others, 1977); the unconsolidated gravels which lie on the flanks of the ranges were formed by Pliocene and younger erosion. These gravels consist chiefly of clasts of the autochthon and lower allochthon which presumably overlie older gravels derived from the middle and upper allochthons.

Prior to onset of Basin and Range faulting in the Grouse Creek Mountains, the autochthon was covered by the lower and middle allochthons, which can be reconstructed from klippen that lie around the flanks of the range, and probably also by the complex upper allochthon that is presently exposed only on the west side of the range (Fig. 1). The upper allochthon, which consists of unmetamorphosed upper Miocene, Upper Permian, and Lower Triassic rocks, moved into its present position sometime after 11.7 m.y. B.P. (Compton, this volume). Upper Paleozoic and (or) lower Mesozoic rocks that overlay or formed part of the middle allochthon before the arrival of the upper allochthon must already have been eroded or removed by low-angle faulting because they were not present beneath the upper allochthon at the time when Basin and Range faulting exposed the present structural sequence. Rb-Sr isotopic studies indicate that in late Miocene time, the rocks of the Grouse Creek autochthon were buried to depths of about 10 km or less than 10 km if the thermal gradient was higher than average in this region, which was probably the case given the evidence of late Miocene volcanic activity in the Grouse Creek Mountains (Table 1). The maximum total thickness of the lower and middle allochthons is roughly 1 km; therefore, up to 9 km of rock had to have been removed from the region above the middle allochthon of the metamorphic core complex in the period between 12 m.y. and about 5 m.y. (Table 1).

Compton and others (1977) cited evidence for the existence of a broad domal uplift centered in the vicinity of the present central Grouse Creek Mountains from at least late Oligocene to early Miocene time. In late Oligocene time, the central Grouse Creek Mountains were the site of intrusion, metamorphism, and development of a gneiss dome (Todd, 1980). The radial distribution of overturned folds and the implied thermal upwelling both suggest that a topographic high existed in this area between late Oligocene and late Miocene times, at which time the thermal system cooled in the autochthon and the locus of thermal activity shifted southwestward (Fig. 1 and Table 1). The existence of such an uplift would explain the removal of the thick section of rocks that must have overlain and (or) been part of the middle allochthon in late Miocene time.

The radiometric age of vitric tuff in the Big Pink sequence provides a constraint on the sequence of tectonic events in the Matlin Mountains: low-angle faulting of displaced sheets I and II, deposition of the associated fanglomerates, and deposition of the Tertiary rocks of the rooted section occurred before 11.1 to 10.5 m.y. B.P.; deposition of the Big Pink sequence took place about 11.1 to 10.5 m.y. B.P., at a time when the Grouse Creek autochthon and lower allochthon were still buried to depths of at least several km; and low-angle faulting of the Big Pink sequence and the underlying metamorphosed Lower Permian units occurred after 11.1 to 10.5 m.y. B.P. (Table 1). The emplacement of sheets III and IV must have been close in time since the rocks of the Big Pink sequence were still wet when sheet IV was emplaced. It seems likely that the structurally lower composite plate (sheets I and II and the associated fanglomerates) formed during the late Miocene episode of uplift and erosion of the ancestral Grouse Creek range. The Grouse Creek Mountains probably were low by the time the structurally higher composite plate (sheets III and IV) arrived, at some time after 11.1 to 10.5 m.y. B.P. The higher composite plate may have moved at a slightly later time than the complex upper allochthon of the Grouse Creek and Raft River Mountains. A subsidiary low-angle fault plate in the complex upper allochthon moved between 14.4 and 11.7 m.y. (Compton, this volume) suggesting that the episode of postmetamorphic detachment faulting in the region may have lasted for several million years. It seems likely that the composite plates of the Matlin Mountains were emplaced during this regional episode of detachment faulting. Since there is at present no near highland composed of Lower Triassic rocks, the movement of sheet V after both composite plates were in place probably occurred during a late stage of this episode. The tuffaceous sandstone and fanglomerate which locally overlap the Grouse Creek Mountains in a normal depositional relation are then younger than the allochthonous and rooted Tertiary sedimentary rocks of the Grouse Creek and Matlin Mountains, latest Miocene or early Pliocene in age.

Source of the Displaced Sheets

In summary, the displaced terrane is separated from the middle allochthon of the Grouse Creek Mountains by two west-dipping, low-angle faults, the East and West Faults (Fig. 1). These faults divide the terrane into two composite plates, a structurally lower plate that consists of unmetamorphosed Upper Permian and Triassic(?) rocks, and Tertiary sedimentary rocks that contain clasts of these

rocks (displaced sheets I and II); and a higher plate composed chiefly of metamorphosed Lower Permian, Pennsylvanian, and Mississippian rocks, and Tertiary sedimentary rocks whose clasts include unmetamorphosed Permian and Triassic rocks, metamorphosed upper and lower Paleozoic rocks, and metamorphosed Precambrian rocks (sheets III and IV). Displaced sheet V overlies both composite plates. As shown in Figure 1, the East and West Faults probably join at depth, and, to the west, the basal fault has been down-dropped against the middle allochthon of the Grouse Creek Mountains by a buried Basin and Range normal fault.

The previous section summarized the evidence for uplift in the ancestral Grouse Creek Mountains that was particularly rapid about 12 m.y. B.P. and was accompanied by the erosion and low-angle faulting of the rocks that must have been present above the Oquirrh Formation in the middle allochthon of the metamorphic core complex. Displaced sheets I and II and their associated Tertiary sedimentary rocks probably were derived from this uplift. It is important to note that sheets I and II were joined when they moved to their present position. This joining is shown by the fanglomerate which overlies sheets I and II and which could not have been deposited where it presently lies. The sheets were in their present position relative to one another when the fanglomerate was deposited and subsequently, the entire sequence was transported. The direction of movement of the composite plate was not necessarily the same as that of the individual sheets; however, since the rocks of the plate probably originated in the ancestral Grouse Creek Mountains, the net movement probably was from west to east. In this event, the displacement of the composite plate was on the order of 25 km.

The history of the composite plate which includes sheets III and IV was different in detail from that of the underlying plate. This history began with the uplift and erosion of metamorphosed Lower Permian rocks and was followed by the deposition, about 11.1 to 10.5 m.y. ago, of the Big Pink sequence. These sedimentary strata were folded and, locally, were overturned and faulted over one another before and (or) during the emplacement of sheet III, which was followed closely by the emplacement of sheet IV. The clast stratigraphy of the Big Pink sequence reflects the progressive unroofing of a core complex terrane in which the middle allochthon included, or was covered by, unmetamorphosed upper Paleozoic and lower Mesozoic rocks. For reasons outlined earlier, this terrane was not the ancestral Grouse Creek Mountains. If the overturning and faulting of folds in sheet III was related to its movement, then the sheet moved toward the east. If sheet III is restored to a position somewhere west of the Grouse Creek Mountains, then it is possible to describe the probable source of the Big Pink sequence. Hundreds of meters of coarse debris were shed to the east and northeast from a steep mountainous area that existed in this vicinity about 11.1 to 10.5 m.y. ago. Lenses of metamorphosed Lower Permian breccia interlayered with this debris probably came from the same source. Folds in sheet IV indicate two periods of compression but do not uniquely identify the direction of tectonic transport. However, the composition of sheet IV is identical in part to the pre-Tertiary rocks of sheet III, which suggests that both sheets had the same source. The presence in sheet IV of rocks older than Lower Permian suggests progressive unroofing at the source.

As was noted in the case of the structurally lower composite plate, sheets III and IV probably traveled together to their present site because klippen of sheet IV end abruptly at the West Fault. Therefore, the displacement and movement directions of sheets III and IV did not necessarily coincide with that of the composite plate. Although sheet III may have moved eastward at an early time in its history, the direction in which it moved while part of the composite plate is unknown. It seems likely that the plate also traveled toward the east because, as far as is known, no denudation faults of late Miocene age occur in northwestern Utah east of the Matlin Mountains. Furthermore, the direction of tectonic transport of postmetamorphic low-angle faults and folds in the core complex as a whole was toward the east (Compton and others, 1977).

In general, these speculations about the source areas and directions of transport of the upper Miocene gravels and displaced sheets of the Matlin Mountains agree with the data for the complex upper allochthon of the Grouse Creek and Raft River Mountains (Compton, this volume). According to Compton, the coarse gravels of the upper allochthon were shed south and southwest from an unusually steep highland beginning as early as 14 m.y. B.P. Sheets of unmetamorphosed Upper Permian, Lower Triassic, and upper Miocene rocks moved to the north-northeast at an early stage of detachment faulting; these movements were followed by transport of the entire complex allochthon to the east-northeast at some time after 11.7 m.y. B.P. Inferred transport directions for the higher composite plate of the Matlin Mountains indicate that a steep mountainous area located somewhere west of the Grouse Creek Mountains in late Miocene time shed detritus to the east and northeast (the Big Pink sequence). This uplift may have been more deeply eroded than the older one which was the source for the gravels of the upper allochthon of the Grouse Creek Mountains because the Big Pink gravels contain clasts of the lower allochthon and autochthon whereas the oldest known clasts in the Tertiary rocks of the upper allochthon are composed of the Oquirrh Formation (middle allochthon). Compton's data indicate that, at some time after deposition of the upper Miocene sedimentary rocks, allochthonous sheets of pre-Tertiary and Tertiary rocks moved toward the north-northeast from the area that had been the basin of deposition. In the case of the higher com-

posite plate of the Matlin Mountains, the basin in which the Big Pink sedimentary sequence was deposited apparently became emergent and the sedimentary rocks together with the underlying Lower Permian rocks were transported to the east as displaced sheet III. Thus, both studies suggest that the locus of uplift shifted in late Miocene time.

The minimum displacement of the upper allochthon in the Grouse Creek and Raft River Mountains is 5 km (Compton, this volume). If the structurally higher composite plate of the Matlin Mountains was once continuous with this upper allochthon, then the entire complex is allochthonous over an area that measures thousands of km^2, and the lateral displacement of the composite plate containing the Big Pink sequence was a minimum of 30 to 35 km.

CONCLUSIONS

Concepts and models regarding the origin of the regionally extensive Tertiary decollement, or detachment, faults and their relations, if any, to metamorphism and ductile deformation in the core complexes of the Great Basin have continued to evolve since the 1977 Penrose conference on metamorphic core complexes and the subsequent publication of a Geologic Society of America Memoir on this subject (1980). Articles in the memoir volume brought into sharp focus both the similarities and the differences in geologic setting, geometry, and timing of tectonic events among the core complexes and detachment terranes of the western Cordillera. Wernicke (1981) briefly summarized pertinent studies of the past two decades and reviewed the unsolved problems concerning the relations among the three major aspects of the core complexes: deformation (including mylonitization) of autochthonous rocks, decollement or detachment faulting of cover rocks, and uplift (doming or arching) of individual ranges.

Data resulting from ongoing study of the Grouse Creek-Raft River-Matlin Mountains terrane seem to set some constraints on the postmetamorphic low-angle faulting in this part of the Great Basin. The recent work of Compton (this volume) and the data of this paper point to the existence in Late Miocene time of a long-standing rugged highland in the vicinity of the present Utah-Nevada border; a domal geometry is suggested for this uplift by the regular changes in the composition of the gravels shed from the uplift. The interlayering of allochthonous sheets of pre-Tertiary rocks with these gravels strongly suggests that this and other domal uplifts were the sources of allochthons which moved downslope into the Tertiary basins. The last movements apparently involved emergence of the late Miocene basins to form source areas for a complex allochthon that may have traveled eastward for distances greater than 35 km. There is no evidence as yet concerning the geometry of the breakaway faults in the source area, but the local emplacement of the youngest rocks in the region (upper

Miocene) upon the oldest rocks (Precambrian) indicates that the faults cut deeply down-section as well as displacing strata for tens of kilometers laterally. This geometry points to listric normal faulting or some combination of high- and low-angle faults in the source region. Compton's study (this volume) and the local presence of silicic ash layers along low-angle faults in the Matlin Mountains suggest a genetic relation between volcanic activity, subjacent intrusion(?), and low-angle faulting, i.e., a model that involves shallow crustal extension sited over Tertiary intrusions.

The postmetamorphic low-angle faults in the Grouse Creek Mountains, with local exceptions, emplace younger strata on older, but the geometry of faults of similar age in the Matlin Mountains is, without exception, older-on-younger. In both areas, the rocks above the basal low-angle fault are folded, and in the Matlin Mountains, folds are broken by a reverse fault. In the Matlins, the low-angle faults of the displaced terrane appear to mark places where the toes of the upper Miocene allochthons overrode the surface. The total thickness of the displaced sheets, about 0.5 km, and the deposition of gravels across eroded or tectonically thinned sheets indicate that faulting took place at shallow depths. The complex allochthon that is exposed west of the Grouse Creek Mountains was probably closer to the source area and thus has an overall geometry more characteristic of distension, whereas the part exposed in the Matlin Mountains probably represents the distal, leading edge of the allochthon and is characterized by compression.

In their paper on the gravity of northwestern Utah, Cook and others (1964) showed a relatively steep southeastward gravity gradient (relief of 15 to 20 mgal) that extended southwestward from the Wildcat Hills, located a few kilometers southeast of the eastern Raft River Mountains, through the eastern Matlin Mountains, Red Dome, and Terrace Mountain, to the southeastern tip of the Grouse Creek Mountains. This feature coincides almost exactly with the inferred basal East Fault and must follow the leading edge of the displaced terrane; it represents a sharp eastward increase in Bouguer anomalies across the boundary between the anomalous western terrane with its intersheeted carbonate and Tertiary sedimentary rocks and the normal eastern carbonate section with its overlying thin Tertiary sedimentary cover.

Although movement of the upper allochthon and of subsidiary sheets in the middle allochthon took place after metamorphism had ceased in the present Grouse Creek and Matlin Mountains, the chronology of Table 1 indicates that ductile and brittle tectonic events followed one another closely. The evidence for intrusion and doming in late Oligocene and early Miocene time cited by Compton and others (1977) and Compton (1980) seems well-established; the "dome" apparently consisted of a broad, gently convex ductile shear zone located at metamorphic depths, a zone below which Precambrian basement remained relatively undeformed and in, and above which, flow of rocks away from the domal apex took place by ductile low-angle faulting and recumbent folding. Included in this zone were rocks of the upper part of the autochthon and that part of the middle allochthon which was in place by early Oligocene time. As mentioned earlier, the autochthon of the Grouse Creek Mountains did not cool below metamorphic temperatures until late Miocene time. By some time after 10.5 m.y., but before 5 m.y., gravels and postmetamorphic allochthonous sheets were emplaced upon the metamorphosed rocks of the lower and middle allochthons and, locally, upon those of the autochthon. The implication is that a topographic high did not develop in the area of the present Grouse Creek Mountains until about 12 m.y. ago, after which time erosion and denudation faulting proceeded rapidly. The late Miocene uplift in the ancestral Grouse Creek Mountains seems to have been the delayed result of earlier heating and intrusion. Thus, denudation faulting there can be viewed as an indirect but inevitable consequence of earlier, deeper ductile events.

The late Tertiary tectonic history of the Raft River-Grouse Creek Mountains core complex and the peripheral Matlin Mountains can be best explained by the "gravity sliding with associated intrusion" model. Unlike some other core complexes, there is an eastern area of shortening that is correlative in age and direction of transport (west to east) with a presumed extensional terrane to the west. There are still many unanswered questions, in particular, the cause of intrusion and uplift and the reason for the apparent shifts in locus of uplift. Furthermore, this scenario for northwestern Utah does not explain the data of other core complexes.

ACKNOWLEDGMENTS

The research reported in this paper was done as part of a postdoctoral research associateship at the U.S. Geological Survey under the guidance of Max D. Crittenden, Jr. His interest in and enthusiasm for the project were major factors in the completion of this report. In the early days of the study, Max and his wife, Mabel, tramped with me over the funny little hills with the dark brown rocks at the top and white rocks at the bottom, and later Max patiently edited the many drafts of the evolving manuscript. His tactful criticisms and suggestions were always helpful, and many field trips and discussions with Max not only broadened my knowledge of tectonics but were valued experiences.

Many other colleagues have aided in the completion of this report, in particular R. R. Compton, D. M. Miller, and A. S. Snoke. Diane and Ed Mott and their family provided a home away from home, and Robin Wightman, Mimi Mowshowitz, Jon Eanet, and Wendy Hillhouse helped me in the field.

REFERENCES CITED

Armstrong, R. L., 1968, Mantled gneiss domes in the Albion Range, southern Idaho: Geological Society of America Bulletin, v. 79, p. 1295–1314.

Boutwell, J. M., 1907, "Stratigraphy and structure of the Park City Mining District, Utah": Journal of Geology, v. 15, p. 434–458.

Brew, D. A., and Gordon, M., Jr., 1971, Mississippian stratigraphy of the Diamond Peak area, Eureka County, Nevada: U.S. Geological Survey Professional Paper 661, 84 p.

Compton, R. R., 1972, Geologic map of the Yost quadrangle, Box Elder County, Utah, and Cassia County, Idaho: U.S. Geological Survey Miscellaneous Geologic Investigations Map I–672.

——, 1975, Geologic map of the Park Valley quadrangle, Box Elder County, Utah, and Cassia County, Idaho: U.S. Geological Survey Miscellaneous Geologic Investigations Map I–873.

——, 1980, Fabrics and strains in quartzites of a metamorphic core complex, Raft River Mountains, Utah, *in* Cordilleran Metamorphic Core Complexes: Geological Society of America Memoir 153, p. 385–398.

——, 1983, Displaced Miocene rocks on the west flank of the Raft River-Grouse Creek core complex, Utah, *in* Tectonic and Stratigraphic Studies in the eastern Great Basin: Geological Society of America Memoir 157 (this volume).

Compton, R. R., Todd, V. R., Zartman, R. E., and Naeser, C. W., 1977, Oligocene and Miocene metamorphism, folding, and low-angle faulting in northwestern Utah: Geological Society of America Bulletin, v. 88, p. 1237–1250.

Cook, K. L., Halverson, M. D., Stepp, J. C., and Berg, J. W., Jr., 1964, Regional gravity survey of the northern Great Salt Lake Desert and adjacent areas in Utah, Nevada and Idaho: Geological Society of America Bulletin, v. 75, p. 715–740.

Crittenden, M. D., Jr., Coney, P. J., and Davis, G. H., 1978, Penrose conference report: Tectonic significance of metamorphic core complexes in the North American Cordillera: Geology, v. 6, p. 79–80.

Gilluly, J., 1932, Geology and ore deposits of the Stockton and Fairfield quadrangles, Utah: U.S. Geological Survey Professional Paper 173, 171 p.

Hose, R. K., and Repenning, C. A., 1959, Stratigraphy of Pennsylvanian, Permian, and Lower Triassic rocks of the Confusion Range, west-central Utah: American Association of Petroleum Geologists Bulletin, v. 43, p 2167–2196.

Jordan, T. E., 1983, Structural geometry and sequence, Bovine Mountain, northwest Utah, *in* Tectonic and Stratigraphic Studies in the eastern Great Basin: Geological Society of America Memoir 157 (this volume).

Krieger, M. H., 1977, Large landslides, composed of megabreccia, interbedded in Miocene basin deposits, southeastern Arizona: U.S. Geological Survey Professional Paper 1008, 25 p.

Lawson, A. C., 1906, The copper deposits of the Robinson Mining district, Nevada: California University Department of Geology Bulletin 4, p. 287–357.

Miller, D. M., 1980, Structural geology of the northern Albion Mountains, south-central Idaho: *in* Cordilleran Metamorphic Core Complexes: Geological Society of America Memoir 153, p. 399–423.

Nolan, T. B., 1935, The Gold Hill mining district, Utah: U.S. Geological Survey Professional Paper 177, 172 p.

Olsen, R. H., 1956, Geology of the Promontory Range, *in* Guidebook to the Geology of Utah, no. 11: Geology of parts of northwestern Utah: Utah Geological and Mineralogical Survey, Salt Lake City, p. 41–75.

Schaeffer, F. E., and Anderson, W. L., 1960, Geology of the Silver Island Mountains, Box Elder and Tooele Counties, Utah, and Elko County, Nevada: Guidebook to the Geology of Utah, no. 15: Utah Geological and Mineralogical Survey, Salt Lake City, 185 p.

Shreve, R. L., 1968, The Blackhawk landslide: Geological Society of America Special Paper 108, 47 p.

Steele, G., 1960, Pennsylvanian-Permian stratigraphy of east-central Nevada and adjacent Utah, *in* Geology of east-central Nevada: Intermountain Association of Petroleum Geologists, 11th Annual Field Conference, 1980 Guidebook, p. 91–113.

Stevens, C. H., 1965, Pre-Kaibab Permian stratigraphy and history of Butte Basin, Nevada and Utah: American Association of Petroleum Geologists Bulletin, v. 49, no. 2, p. 139–156.

Stifel, P. B., 1964, Geology of the Terrace and Hogup Mountains, Box Elder County, Utah [Ph.D. thesis]: Salt Lake City, Utah, University of Utah, 173 p.

Todd, V. R., 1973, Structure and petrology of metamorphosed rocks in central Grouse Creek Mountains, Box Elder County, Utah [Ph.D. thesis]: Stanford, California, Stanford University, 316 p.

——, 1980, Structure and petrology of a Tertiary gneiss complex in northwestern Utah: *in* Cordilleran Metamorphic Core Complexes: Geological Society of America Memoir 153, p. 349–383.

Wernicke, B., 1981, Low-angle normal faults in the Basin and Range Province: nappe tectonics in an extending orogen: Nature, v. 291, p. 645–648.

Manuscript Accepted by the Society August 20, 1982

Geological Society of America
Memoir 157
1983

Displaced Miocene rocks on the west flank of the Raft River-Grouse Creek core complex, Utah

Robert R. Compton
Department of Geology
Stanford University
Stanford, California 94305

ABSTRACT

Mapping on the west side of the Raft River-Grouse Creek core complex has shown that Upper Permian, Triassic, and Middle to Late Miocene rocks compose an extensive allochthonous sheet that was displaced laterally as well as down-section onto the complex. The sheet is underlain locally by two thin allochthonous sheets of Permian rocks that were displaced at about the same time. That all these sheets moved either northeastward or southwestward is suggested by the trends of fold hinge-lines and by the strike of strike-slip faults confined to the sheets. The Miocene sequence is at least 2000 m thick and consists of poorly sorted fluvial conglomerate intercalated with numerous thick beds of silicic vitric tuff and smaller amounts of monolithologic breccia, lacustrine limestone, basalt, and intrusive as well as extrusive rhyolite. K-Ar dates on a basalt flow and a rhyolite plug show that the Triassic and part of the Permian rocks were infaulted with the Miocene rocks between 14.4 and 11.7 m.y. ago, and that the composite sheet was displaced later than 11.7 m.y. ago.

INTRODUCTION

The Grouse Creek Mountains and adjoining Vipont Mountains expose the western part of a broad metamorphic core complex. The part of the complex exposed in Utah (Fig. 1) consists of an autochthon of metamorphosed Precambrian and Cambrian(?) rocks overlain by (1) a lower allochthonous sheet of metamorphosed Precambrian to Ordovician rocks, (2) a middle allochthonous sheet of locally metamorphosed Devonian to Lower Permian rocks, and (3) an upper allochthonous sheet of Lower and Upper Permian and Triassic rocks (Compton and others, 1977). Parts of the core complex shown in Figure 2 were intruded by a granodiorite pluton 38.2 ± 2.0 m.y. ago and, 4 km north of the area of Figure 2, by an adamellite pluton 24.9 ± 0.6 m.y. ago. These intrusive events date some structural relations within the complex (Compton and others, 1977). Mapping completed recently along the west side of Grouse Creek Mountains and in adjoining parts of Grouse Creek Valley (area 1 in Fig. 1) indicates that the Miocene rocks of the valley lie on the older rocks of the range along a fault that dips 20° or so to the west. This relation is particularly interesting because the displaced body of Miocene rocks

includes an infaulted plate of Triassic and Upper Permian sedimentary rocks that is bounded by two low-angle faults (Fig. 2). Mapping to the north by Todd (1973) and local mapping by the writer suggest that the fault relation between the core complex and the composite sheet of Miocene, Triassic, and Upper Permian rocks continues at least to Vipont Mountains, a total distance of 60 km. The Miocene rocks thus prove to be a major component of the upper allochthonous sheet of the core complex. The purposes of this paper are to describe the Miocene rocks, their structural relations to older rocks, and their possible genetic connection to the core complex.

PRE-MIOCENE ROCKS

Rocks older than Miocene make up almost all of the Grouse Creek Mountains as well as the thin allochthonous plate interleaved with the Miocene rocks. Rocks older than Permian are similar to units described for the Raft River Mountains (Compton, 1972, 1975) and central Grouse Creek Mountains (Todd, 1973, 1980) and will be summar-

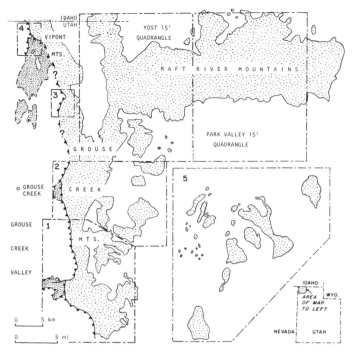

Figure 1. Map of the Raft River-Grouse Creek core complex, showing Cenozoic rocks and deposits unpatterned, Lower Permian and Triassic rocks of the western margin closely stippled, and other rocks lightly stippled. 1 = area described in this paper; 2 = area mapped by Todd (1973); 3 and 4 = areas mapped or reconnoitered by the writer, as referred to in this paper; and 5 = area mapped by Todd (this volume) in which rocks were displaced during Miocene time. Small map on right shows location of map on left.

ized here only as to their specific occurrences in the area of Figure 2.

The autochthon of the core complex is exposed in the northeastern part of the area (Fig. 2) and forms large inclusions (not shown in Fig. 2) in the southwestern part of the granodiorite pluton. Except for one small outcrop that may be Precambrian granitic gneiss, the autochthonous rocks

are, in order of stratigraphic age and structural sequence: 1) Elba Quartzite (Precambrian), here more thinly bedded and including more schist than usual; 2) schist of Stevens Spring (Precambrian), a muscovitic and locally graphitic metashale; and 3) quartzite of Clarks Basin (Cambrian?), a flaggy metaquartzite with thin interlayers of coarse muscovite schist.

The lower allochthonous sheet of the core complex also includes the quartzite of Clarks Basin, and above that: 1) schist of Mahogany Peaks (Cambrian?), a dark garnet-biotite-muscovite schist, locally with staurolite; 2) metamorphosed rocks of the Pogonip Group (Ordovician), chiefly calcite and dolomite marbles with micas and other metamorphic silicate minerals; 3) metamorphosed Eureka Quartzite (Ordovician); and 4) laminated gray metadolomite thought to be Fish Haven Dolomite (Ordovician).

A major detachment zone separates the lower and middle allochthonous sheets. In the northeastern part of Figure 2, this zone is characterized by slices of metamorphosed phyllitic Chainman Shale (Mississippian) and Guilmette Limestone (Devonian) lying on greatly thinned Fish Haven Dolomite and Eureka Quartzite of the lower allochthonous sheet, or on slices repeating parts of these Ordovician units. In the central and southern parts of Figure 2, the middle sheet includes units older than those making up this sheet in the Raft River Mountains and central Grouse Creek Mountains—specifically, metamorphosed Simonson Dolomite (Devonian) and metamorphosed Guilmette Limestone. The Devonian formations are overlain by units seen widely in the core complex: metamorphosed Chainman Shale and Diamond Peak Conglomerate, and metamorphosed to unmetamorphosed members of the Oquirrh Formation (Pennsylvanian and Lower Permian). In addition to the lower Oquirrh member, which is limestone, and the middle member, which is chiefly sandstone, the formation in this part of Grouse Creek Mountains includes an upper member of thinly bedded limestone, limestone and siltstone which are locally metamorphosed.

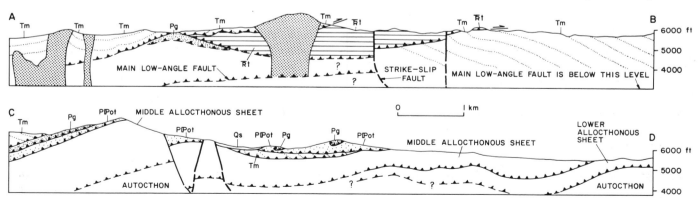

Figure 2. *Facing page,* geologic map of the southern Grouse Creek Mountains and part of Grouse Creek Valley, generalized from Todd and others (in preparation). *Above,* vertical geologic sections along the two lines shown on the map, facing page.

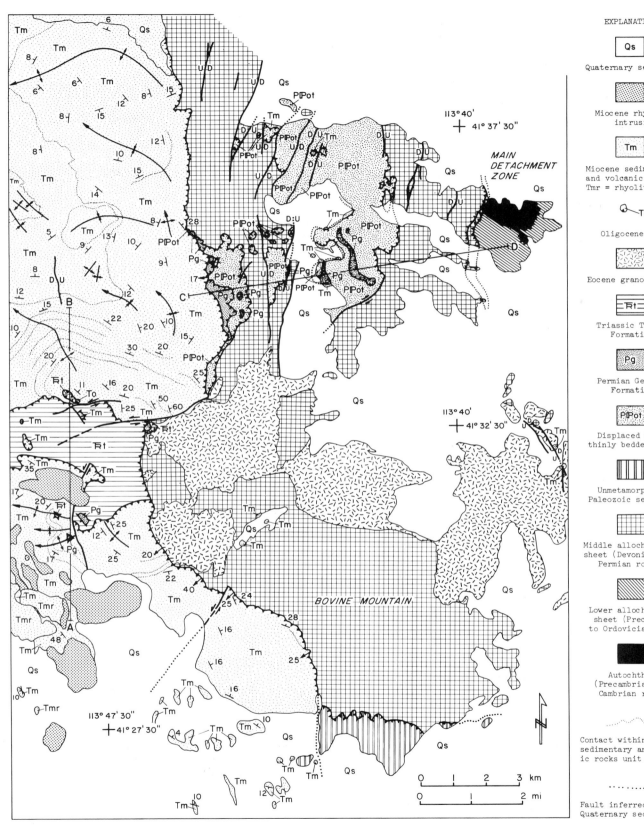

EXPLANATION

Qs

Quaternary sediments

Miocene rhyolite
intrusion

Tm

Miocene sedimentary
and volcanic rocks;
Tmr = rhyolite flow

To

Oligocene tuff

Eocene granodiorite

Triassic Thaynes
Formation

Pg

Permian Gerster
Formation

PlPot

Displaced Oquirrh
thinly bedded member

Unmetamorphosed
Paleozoic sequence

Middle allochthonous
sheet (Devonian to
Permian rocks)

Lower allochthonous
sheet (Precambrian
to Ordovician rocks)

Autochthon
(Precambrian and
Cambrian rocks)

Contact within Miocene
sedimentary and volcan-
ic rocks unit

Fault inferred beneath
Quaternary sediments

This unit is of interest here because part of it was displaced as a thin allochthonous sheet during the Miocene. A continuous sequence of Oquirrh rocks is well exposed on and around Bovine Mountain in the southern part of the area, and it has been subdivided and described in detail by Jordan (1979a and b). An anomaly to all parts of the core complex studied to date is an allochthonous body of unmetamorphosed Paleozoic rocks lying against the southern flank of Bovine Mountain (Fig. 2). This body includes formations ranging in age from Early Ordovician to Pennsylvanian. The sequence is structurally inverted, with Oquirrh limestone at the base and Pogonip Limestone at the top (Jordan, this volume).

The three remaining pre-Miocene units are everywhere separated from other units by low-angle faults. The oldest of these, the Upper Permian Gerster Formation, consists of gray cherty limestone. It forms thin allochthonous sheets that lie on the thinly bedded upper Oquirrh member and are exposed in windows through Triassic and Miocene allochthonous rocks in the west-central part of the area (Fig. 2). The Triassic rocks are mainly thinly bedded silty to fine sandy limestone of the Thaynes Formation and include maroon to gray limestone and mudstone near their base.

A small body of crystal-vitric tuff of Oligocene age (Table 1) lies at the eastern tip of the small klippe of Triassic rocks just north of the main Triassic outcrop (Fig. 2). Judged from its position and limited occurrence, the tuff body is probably a tectonic fragment carried beneath the sheet of Triassic rocks. The tuff differs texturally from all the tuffs considered to be Miocene in that it contains abundant 0.2 to 2 mm phenocrysts of quartz, sanidine, plagioclase, and biotite. The vitric matrix is altered to clays. The only other rocks in the vicinity with similar age are conglomerates containing land snails of early Tertiary age,

located 28 km to the north (A. R. Young, as reported by Heylmun, 1965).

MIOCENE ROCKS

Sedimentary and volcanic rocks correlative broadly with the Salt Lake Formation (Heylmun, 1965) occur extensively around the Raft River Mountains and Grouse Creek Mountains. Tuff on the east side of Grouse Creek Mountains is of approximately the same age as the rocks described here (Todd, this volume). The Miocene rocks mapped in Grouse Creek Valley are separated into four blocks by faults, and the rocks vary so much along strike that they cannot be correlated in detail from one block to another. The strata in all four blocks contact older rocks along low-angle faults and are overlain unconformably by Quaternary alluvium or Lake Bonneville deposits. Approximate thicknesses measured from geologic maps of 1:24,000 scale are: 1800 m in the large northern block; 1100 m in the west-central block; roughly 1000 m in the east-central block; and 1000 m in the southern block.

The two principal rocks in all four blocks are poorly sorted conglomerate and silicic vitric tuff. The conglomerate crops out only locally and was mapped largely on the basis of abundant pebbly, cobbly float. Exposed beds are typically 0.5 to 5 m thick. Clast imbrications measured at seven widely distributed localities indicate flow to the south and southwest. Maximum clast sizes are typically in the cobble range but in some cases are small boulders or pebbles. Most of the conglomerate is clast-supported and locally interbedded with moderately to well-sorted lenses of sand and fine pebble conglomerate. These rocks are commonly calcite-cemented, and the cement in some cases suggests Miocene caliche. Thick beds of matrix-supported

TABLE 1.   K-Ar AGES OF CENOZOIC ROCKS ON THE SOUTHWESTERN FLANK OF GROUSE CREEK MOUNTAINS, UTAH*

Sample #	Latitude	Longitude	Name	Mat'l dated	Mean K_2O (wt. %)	$*Ar^{40}$ moles/gm $x10^{-10}$	$*Ar^{40}$ %	Apparent age (m.y.)	± (m.y.)
9W-35-4-3	41° 31' 23"	113° 48' 12"	rhyolite	san	8.65	1.46	85.51	11.7	0.4
RC-73-1	41° 30' 26"	113° 48' 22"	basalt	bas	1.308	0.272	52.18	14.4	0.4
RC-73-2A	41° 32' 59"	113° 48' 13"	crystal-vitric tuff	bio "	7.28 "	3.86 3.48	57.57 47.50	36.4 32.9	1.1 1.0
RC-73-2B	"	"	"	san "	12.22 "	5.95 5.91	99.74 93.24	33.5 33.3	1.0 1.0

*Potassium analyses by L. B. Schlocker, by flame photometer using lithium metaborate fusion. Argon analyzed by standard isotope dilution and mass spectrometry techniques (Dalrymple and Lanphere, 1969); analysts, V. R. Todd and W. C. Hoggatt. Uncertainties in reported ages (±3%) represent analytical uncertainty at one standard deviation.

Constants used in calculations: $\lambda_\epsilon + \lambda_{\epsilon'}$ = .581 x 10^{-10} year^{-1}; λ_β = 4.962 x 10^{-10} year^{-1}; K^{40}/K_{total} = 1.167 x 10^{-4} mol/mol.

conglomerate are held together by clays; these beds tend to disintegrate easily and may thus be more abundant than suggested by natural exposures. All the conglomerates are gray, tan, or some mixture of these colors.

The clast compositions of the conglomerates have not been studied systematically, but field observations suggest major differences among the four blocks of Miocene rocks, as well as stratigraphic variations within any one block. The lowest 900 m of the northern block consist of clasts of the younger of the pre-Miocene units, largely Gerster limestone and chert, perhaps with limestone from the upper, thinly bedded member of the Oquirrh Formation. Approximately 900 m above the base of the northern block, Oquirrh sandstone appears as clasts and is a constituent of most of the conglomerate from that level up. Limestone clasts of the lowest Oquirrh member are difficult to distinguish at the outcrop from some younger limestones; however, the uppermost 500 m of conglomerate in the northern block appears to contain Oquirrh limestones, some of which are pale gray to white marble. The conglomerate of the west-central block is relatively fine (pebbly) and is composed dominantly of Gerster clasts, whereas the conglomerate of the east-central block is cobbly and contains Triassic and probably Oquirrh clasts. Finally, the conglomerate of the southern block contains abundant clasts of what appear to be all of the Oquirrh member rocks and, uniquely, scarce clasts of gabbro, diorite, quartz-bearing diorite, and granodiorite(?) that are unlike the rocks in the Eocene pluton nearby or the Oligocene granitic stocks in the central Grouse Greek Mountains (Compton and others, 1977).

The second major Miocene rock, silicic vitric tuff, is of two varieties: relatively pure white to gray layers 1 to 20 m thick that tend to crop out prominently, and brown sand and argillaceous layers that do not. Both kinds of tuff consist predominantly of fresh glass shards. White and gray tuffs contain only a few percent of quartz and feldspar crystal fragments and scattered lithic pebbles, whereas brown tuff includes moderate to large amounts of nonvolcanic sand and silt, and commonly grades into conglomerate. The purer tuffs are size-laminated, generally with shards 1 to 3 mm long making up the coarser laminae. White and gray tuffs typically have sharp lower contacts over brown tuff or conglomerate and tend to grade upward into brown tuff. Pumice lapilli occur at several localities, typically as size-sorted beds in finer tuff and locally showing distinct reverse grading. The only coarser pyroclastic rocks exposed are near the two largest plugs shown in Figure 2, and they contain pumice and vitrophyre clasts up to 30 cm in diameter.

Fossil molluscs were collected by the Stanford summer field classes from three localities in tuffs approximately 700 m above the oldest rocks exposed in the northern block. Professor James R. Firby of the University of Nevada,

Reno, kindly studied them and identified the molluscs *Viviparus turneri* Hannibal, *Pisidium* sp. (compare *P. lesliae* Firby), *Carinifex* sp. (cf. *C. brevispira*), *Planorbis utahensis* (Meek), and *Amnicola* sp., as well as ostacods and a water rush, an asemblage that indicates Late Miocene (late Barstovian or Clarendonian) as the most likely age (Firby, written communication, 1974). McClelland (1976) reported a mammal fauna of Barstovian age (16 to 12 m.y. B.P.), collected from similar rocks 10 km north of the northwest corner of the area of Figure 2.

Besides conglomerate and tuff, the following Miocene rocks occur stratigraphically as noted:

1. *Monolithologic breccia* forms fourteen lenses in the lower 900 m of the northern sequence and one lens in the southern sequence. The deposits are typically 4 to 10 m thick and 0.4 to 1 km in exposed length and are composed of nonporous breccia in which bedding is locally continuous through slabs several tens of meters across. Most of the breccias are composed of limestone and chert like those of the Gerster Formation; one near the lowest exposure of the nothern sequence is of gabbro, and one 900 m above the base of that same sequence is of quartzite of the middle Oquirrh member. Most of the breccia lenses are intercalated in conglomerate, and some lie in tuff.

2. *Limestone* 50 to 80 m thick lies 150 m above the lowest part of the west-central sequence. The rock is tan, locally cherty, contains abundatn silt and clay, and forms beds 0.1 to 0.3 m thick with clayey partings. Chert and limestone containing ostracods compose beds 1 to 3 cm thick in tuff approximately 300 m above the lowest beds in the northern block.

3. *Basalt* forms two separate flows, 35 and 5 m thick, intercalated in conglomerate in the lower 150 m of the west-central sequence. The only phenocrysts are scarce, resorbed crystals of plagioclase, and the black groundmass is rich in plagioclase, suggesting that the rock is a high-alumina basalt. The lower, thicker flow has yielded a K-Ar age of approximately 14.4 m.y. (Table 1).

4. *Rhyolite* flows form prominent outcrops in the upper part of the west-central sequence, composing two thick lenticular bodies separated by a thin layer of tuff (Fig. 2). The rhyolite is typically flow-layered or with a planar fabric composed of abundant small vesicles. It carries 4 to 8 percent of quartz, sanidine, plagioclase, and biotite phenocrysts in a pale-toned devitrified groundmass.

5. *Intrusive rhyolite* occurs as a crescent-shaped group of plugs and thick dikes in the southwestern part of the area (Fig. 2). Contacts and outermost flow layering are typically vertical or dip steeply toward the center of each intrusion, whereas flow layering dips less steeply in the central and higher parts of each intrusion. This outward fanning of layers suggests that the lower parts of expanding surficial domes once lay a short distance above the present levels of exposure. Additional indications of near-surface exposure

in one plug are columnar joints that are approximately horizontal at the outer vertical walls and curve upward within the intrusion. Fresh vitrophyre forms the marginal rock at a few places, locally carrying pebbles from the conglomerate, but more typically the marginal rock is brecciated and vapor-altered pink rhyolite. Fresh rock from the inner parts of the intrusions is nonvesicular, pale gray, and aphanitic except for 5 to 8 percent of quartz, sanidine, plagioclase, and biotite phenocrysts which are generally less than 3 mm in diameter. The rhyolite in one intrusion is continuous with a flow to the west and southwest of it, and coarse breccias near plugs, already mentioned, suggest that the plugs fed parts of the pyroclastic sequences. A K-Ar age determination on a fresh sample from the northernmost plug gave an age of approximately 11.7 m.y. (Table 1).

Interpetation

The Miocene rocks were apparently deposited on a fan or alluvial slope that was fed largely from the north and northeast, and graded southwestward into a basin with local shallow lakes. Thick debris-flow deposits and monolithologic breccias indicate that the areas feeding the fans had unusually steep slopes, especially when the lower part of the northern sequence was being deposited. Explosive volcanic activity produced abundant silicic vitric ash, most of which was reworked by streams and in lakes. Vents in the southwest part of the area produced at least part of the pyroclastic materials and at least two rhyolite flows. The plugs and dikes now exposed there suggests that the feeding systems later intruded their way to higher levels.

STRUCTURES

The most extensive structure is the low-angle detachment fault along which Miocene and Triassic rocks lie on older rocks in the western part of the mapped area. This fault cuts across folds and faults in the Miocene rocks as well as folds and faults in the older, underlying units. These older structures will be described first, in order of age.

The oldest structures are associated with the autochthon and the lower and middle allochthonous sheets of the core complex, and they are similar to structures mapped over most of the Raft River and Grouse Creek Mountains (Compton and others, 1977). The two allochthonous sheets and several low-angle faults within them are cut by the Late Eocene pluton in the southern half of the mapped area, as are two generations of folds formed during metamorphism. The folds of the older of these two generations trend north to northeast and verge west to northwest. Folds with the same trend in the southern part of the area verge east to southeast and thus may or may not be of the same age (Jordan, this volume). Second generation folds typically trend west to N 60° W and verge

north. Folds of this generation are cut by the southwestern lobe of the Late Eocene pluton, but folds with the same trend and vergence are impressed on a 24.9-m.y.-old pluton only 4 km north of the area of Figure 2. It thus appears that some folds with this trend and vergence in the area of Figure 2 are older than 38 m.y. and that others are probably 25 m.y. old and younger, perhaps as young as 20 m.y. (Compton and others, 1977). The Eocene pluton is not megascopically lineated but has microscopically kinked biotite and feldspar, suggesting that it was emplaced and crystallized between two episodes of folding.

Allochthonous Sheet of Upper Oquirrh and Gerster Rocks

The upper Oquirrh member is a contiguous part of the middle allochthonous sheet in the southeast part of the area, where it is cut and metamorphosed by the Late Eocene pluton. In the north-central and central parts of the area, however, it forms a separate sheet and lies almost against the same pluton yet is not metamorphosed by it. In fact, the sheet appears to lie locally on Miocene tuffs and must thus have been emplaced to its present position after Middle Miocene time. The crucial outcrops of Miocene rocks are each labeled with a symbol (Tm) in Figure 2. They are shown more completely on a 1:31,680 scale map with topographic base (Todd and others, in preparation). Most outcrops are subdued, and some are no more than tuff fragments brought up in animal burrows; all are based on actual occurrence of tuff, however, rather than on appearance of soil or topography. The best exposed are the three patches near the north-central edge of the map, each of which is in contact with thinly bedded Oquirrh rocks that lie topographically above the tuff. Bedding in the most northerly patch dips 35° north and bedding in the more westerly of the closely spaced pair of patches dips 43° north-northwest—attitudes distinctly discordant with the gently inclined contact between the tuff and the Oquirrh sheet. The cluster of tuff outcrops 3 to 5 km further south indicates a roughly horizontal contact with overlying Oquirrh rocks. The sinuous dotted line between the outcrops is an inferred trace of the fault hidden by Quaternary sediments; it is based partly on topography, opalized rocks, and soil. In contrast to these occurrences, the tuff patches near the east-central margin of the area and in the central part of the area, 4 km northwest of Bovine Mountain, lie on older rocks and are locally displaced against them on high-angle faults.

An important high-angle fault along the southeast margin of the sheet of thinly bedded Oquirrh rocks in the central part of the area is evidently a strike-slip fault. It dips steeply under the sheet and turns abruptly into the low-angle fault along the northeastern margin of the sheet.

Rocks of the Gerster Formation constitute a thin sheet, or fragments of a sheet, that lies on upper Oquirrh member rocks along a low-angle fault. The elongate outcrop of the Gerster Formation sheet in the central part of the area appears to be infolded with Oquirrh rocks. Other, smaller, northwest-trending folds involving the two units occur nearby as well as 3.5 km to the southwest. Small-scale folds and lineations within the two units trend either north to northeast or west to northwest. Some, at least, belong to the two generations of metamorphic folds already mentioned, for they are coaxial with metamorphic lineations in the lower part of the Oquirrh sheet.

Displaced Plate of Triassic and Gerster Rocks

The plate infaulted with Miocene rocks in the western part of the area consists mainly of Triassic rocks but includes a faulted slice of Gerster rocks near its southern margin (Fig. 2, Section AB). The faulted nature of this southern margin is proven by a well-exposed, irregular fault trace that cuts across bedding and folds in the overlying Miocene strata and appears in two fensters, one exposing Gerster and the other Triassic rocks just south of the most northerly rhyolite plug (Fig. 2). A high-angle strike-slip fault displaces the contact just north of the northernmost rhyolite plug, and because the fault is confined to the upper (Miocene) plate, its north-northeast orientation suggests that the Miocene plate moved either north-northeast or south-southwest with respect to the underlying Triassic-Gerster plate.

The low-angle fault at the northern margin of the Triassic-Gerster plate is well exposed in several places, including the small klippe of Triassic rocks just north of the main plate outcrop (Fig. 2). This fault cuts across bedding in both the Triassic and the underlying Miocene rocks. The north-northeast dips of strata in both units steepen from around 15-20° to 33-60° within 200 m of the fault trace. These dips suggest drag by north-northeast displacement of the Triassic plate relative to the underlying Miocene rocks. The fault trace is offset by two high-angle faults; the latter, however, continue into the underlying Miocene rocks, and their orientations suggest they were formed during the later displacement of the composite Triassic-Miocene plate, as will be noted shortly. The Triassic rocks are locally folded strongly on axes that are oriented either N 10° E or N 70-80° W, directions similar to the two principal sets of folds formed during metamorphism of the core complex units.

Folds in the Miocene Rocks

Hinge-traces of the principal folds in the Miocene are shown in Figure 2. A few more folds are doubtless present in the areas underlain by conglomerate, which crops out poorly. Although the hinge-lines have a variety of orientations and many curve considerably, most of them plunge

between 6 and 10 degrees toward the sector N 50-80° W, and the average plunge is 8 degrees in the direction of N 65° W. The folds are open and not strongly asymmetrical, but the extensively exposed anticline in the west-central part of the map has a more steeply inclined north-northeastern limb. In addition, northerly and northeasterly dips of Miocene beds are generally steeper near the main basal low-angle fault than they are further to the west (Fig. 2).

The Main Low-angle Fault Beneath the Miocene Rocks

The trace of the youngest low-angle fault was examined carefully throughout its mapped length in Figure 2. At the five localities marked by dip arrows in Figure 2, the fault is either exposed or so nearly exposed that its dip could be determined to within a few degrees. At the most northerly locality, where the dip is 17°, gouge about 7 cm thick separates firm rock in both walls. At the two intermediate localities, where dips are 20° and 40°, a zone of gouge and sliced rocks is several meters thick and includes nearly black material suggestive of Mississippian Chainman Shale. At the more northerly of the two localities where the dip is 25°, gouge and broken rock are 1 to 3 m thick and include brecciated rhyolite identical to that in the plugs 3 to 7 km to the west and southwest. Outside of these exposures, the trace of the fault indicates a dip of between 20° and 25°.

The low-angle fault trace is offset in two places by high-angle strike-slip faults, already mentioned, that displace the hanging wall rocks but not the footwall rocks. The fault located in the south-central part of the area strikes N 30° E and has caused a 200-m separation in the trace of the main low-angle fault. The one located in the west-central part of the area strikes N 65° E and has caused a 2-km separation in the low-angle fault trace.

Outside of these two offsets, the low-angle fault has a simple linear trace, cutting across Miocene and Triassic units at similar angles and across Miocene beds and subunits at a large range of angles. A notable relation is the westward bulge of the trace at the latitude of the granodiorite pluton; here the fault follows the intrusive perimeter closely, lying largely in Paleozoic carbonate rocks just outside the intrusion.

Mapping in three areas north of Figure 2 suggests that the fault continues with the same sorts of relations all the way to the Idaho state line, a distance of 60 km from the southern tip of Grouse Creek Mountains. Todd (1973) mapped it from the northern boundary of Figure 2 for 10 km to the north (area 2, Fig. 1). Although the trace is largely hidden by Quaternary alluvium in this sector, bedding and folds in the Miocene rocks are clearly cut by the fault and large faulted slices of Triassic and Gerster rocks lie just under it. The next mapped segment of the fault is 9 km north of Todd's mapping (area 3, Fig. 1). Here the Tertiary rocks consist of red to tan conglomerate from which A. R. Young is reported by Heylmun (1965) to have collected land snails indicative of an early Tertiary age. The writer found these beds to be folded and to dip into underlying Cambrian(?) and Ordovician metamorphic rocks of the core complex, meeting them along a fault that dips as little as 15 degrees west. The fault zone, locally exposed in road cuts, consists of pale gray to nearly black gouge and broken rock slices, appearing much like the thicker parts of the fault zone to the south. The writer has not mapped the relations between the Permian strata and Miocene farther to the north (between areas 3 and 4, Fig. 1), but reconnaissance combined with the geologic map of Box Elder County, Utah (Doelling, 1980), suggests that the fault continues north-northwestward, lying between Permian rocks and the metamorphosed rocks of the core complex. It thus appears that the Permian and Tertiary rocks of this sector have been faulted as a single unit over the core complex, just as the Triassic and Miocene are to the south. Finally, the writer mapped 5 km of the contact between these same Permian rocks and Miocene rocks that lie west of them and found it to be another low-angle fault dipping to the west (area 4, Fig. 1). This fault appears to merge with the main low-angle fault at a point 3 km south of the state line.

Basin and Range Structures

High-angle normal faults, younger than the low-angle faulting, form a subparallel north-trending group in the north-central and central parts of the area and occur locally elsewhere. The relative offsets along the more westerly faults of the main group tend to be down on the east, whereas those for the more easterly faults are down on the west, thus forming a structural graben that widens northward in the central Grouse Creek Mountains (Todd, 1973, 1980). The most southeasterly of these faults appears to bring middle plate rocks against alluvium; in the central and northern Grouse Creek Mountains, some faults of the same group cut late Quaternary alluvium and are marked in one case by a scarplet (Todd, 1973; Compton, 1972). The one high-angle fault in the west-central part of Figure 2, and the two near the east-central boundary, definitely cut Quaternary alluvium. High-angle extensional faulting in this area has thus taken place in the late Quaternary, and relations elsewhere in the core complex suggest it did not begin until Pliocene time (Compton and others, 1977).

A consequence of the formation of Grouse Creek Mountains in the Pliocene is the westward tilting of older structures along the west flank. Prior to this tilting, the main low-angle fault described in this paper, and the subparallel allochthonous sheets beneath it, were more nearly horizontal and could have dipped eastward.

INTERPRETATION

Mapped relations show that a sequence of Upper Miocene rocks, more than 2000 m thick, was infaulted with a plate of Triassic and Upper Permian rocks, and that this composite body was then displaced along a low-angle fault over part of the Raft River-Grouse Creek core complex. Both periods of low-angle faulting were younger than 14 m.y. as shown by a dated basalt within the Miocene sequence. The infaulting of Miocene rocks with the Triassic and Upper Permian rocks took place before 11.7 m.y. ago, as shown by a dated rhyolite plug that cuts this set of low-angle faults. Rhyolite flows and coarse pyroclastic deposits near two of the rhyolite plugs indicate that part of the Miocene sequence was fed by these vents during an early stage of their activity. This relation and the presence of rhyolite fragments in the main fault zone suggest that the detachment and displacement of the composite sheet took place later than 11.7 m.y. ago.

Steepening (apparent drag) of both the Triassic and Miocene beds near their low-angle fault boundary suggests northward or north-northeastward displacement of the Triassic-Gerster plate relative to the underlying Miocene rocks. A high-angle strike-slip fault that displaces Miocene but not Triassic rocks suggests a similar direction of displacement of a separate plate of Miocene rocks over the southern part of the Triassic-Gerster plate.

The trends of fold hinge-lines in the Miocene rocks and the strike of the two high-angle strike-slip faults that displace the main low-angle fault indicate that the composite sheet moved either southwestward or northeastward relative to the core complex. Although far from conclusive, local steepening (apparent drag) of beds in the Miocene rocks near the fault suggests movement toward the northeast. One asymmetric anticline in the Miocene suggests the same direction, as does the displaced brecciated rhyolite 7 km northeast of the southernmost rhyolite plug. This direction is toward, rather than away from, the topographic high implied by southward and southwestward flow of the streams that deposited the Miocene conglomerates.

The displacement of the separate sheets of Gerster Formation and upper Oquirrh member rocks was probably more or less concurrent with the displacement of the overlying composite sheet, as suggested by the Miocene rocks that appear to lie under the sheet and by the northeast strike of a high-angle strike-slip fault that forms the margin of one of the sheets. The latter fault indicates that the sheet moved either northeastward or southwestward relative to the deeper parts of the core complex.

The northerly dips of the tuff under the sheet of Oquirrh rocks, which are as much as 43° to the north, show that the tuff was not a thin surficial deposit overrun by the sheets. It is either a thick, folded body deposited where it now lies, or a separate allochthonous sheet, perhaps one emplaced northward at the same time that the Miocene rocks became infaulted with the Triassic and Upper Permian rocks to the west.

The minimum displacement of the main composite sheet relative to the core complex is 5 km, as suggested by folds and subunits that are cut off along the fault. Rock units similar in age and composition to those making up the sheet occur widely to the southwest, west, and northwest of the area described here; however, the writer knows too little of their details to suggest any original attachment of the allochthonous rocks. It is possible that similar rocks originally lay east and northeast of the area.

CONCLUSIONS

In spite of the young age of its displacement, the sheet of Miocene, Triassic, and Lower Permian rocks is not a denudation slide like those envisioned by Armstrong (1972) or described in detail by Pierce (1957, 1973). The original fracture broke through a variety of Paleozoic rocks over an area measured in at least hundreds of square kilometers, and the detached mass moved coherently, segmented by a few high-angle strike-slip faults. Its movement appears unrelated to the present basins and ranges. It seems particularly impressive that the strength of the granodiorite pluton controlled the position and shape of the fault in that part of the area. Separate sheets of Gerster Formation and upper Oquirrh member rocks were probably in motion at the same time, and they are broadly concordant with the complex. Like the older allochthonous sheets of the core complex, all of these younger sheets were displaced downsection as well as laterally, so that the original stratigraphic sequence has become attenuated. Together, the sheets make up what was originally called the upper allochthonous sheet of the core complex (Compton and others, 1977).

As discussed by Todd in this volume, it is possible that the displaced sheets are parts of young displaced sheets that she has mapped east of the Grouse Creek Mountains (area 5, Fig. 1). The two occurrences leave little doubt that the cover of the core complex has been repeatedly detached and displaced since sometime in the Mesozoic Era. On the west side of the Grouse Creek Mountains, the latest two displacements can be dated as Middle to Late Miocene in age.

ACKNOWLEDGEMENTS

The mapping and companionship of the Stanford field geology classes of 1973 and 1976 are greatly appreciated, as is Teresa Jordan's mapping of Bovine Mountain and surroundings. Crucial assistance in my work came from Claudia Owen, who helped map the Miocene rocks and their faulted boundary in 1979; from Victoria Todd and Wendy Hoggatt, who contributed the age determinations; and from Max Crittenden, who obtained field support from the U.S. Geological Survey and has always had a lively and exceedingly helpful interest in the geology. Richard Al-

lmendinger, David Miller, and Terry Shackleford read the manuscript and helped greatly in improving it.

REFERENCES CITED

Armstrong, R. L., 1972, Low-angle (denudation) faults, hinterland of the Sevier orogenic belt, eastern Nevada and western Utah: Geological Society of America Bulletin, v. 83, p. 1729–1754.
Compton, R. R., 1972, Geologic map of the Yost Quadrangle, Box Elder County, Utah, and Cassia County, Idaho: U.S. Geological Survey Miscellaneous Geologic Investigations Map I–672.
——1975, Geologic map of the Park Valley quadrangle, Box Elder County, Utah, and Cassia County, Idaho: U.S. Geological Survey Miscellaneous Investigations Map I–873.
Compton, R. R., Todd, V. R., Zartman, R. E., and Naeser, C. W., 1977, Oligocene and Miocene metamorphism, folding, and low-angle faulting in northwestern Utah: Geological Society of America Bulletin, v. 88, p. 1237–1250.
Doelling, H. H., 1980, Geology and mineral resources of Box Elder County, Utah: Utah Geological and Mineral Survey Bulletin 115, 251 p.
Heylmun, E. B., 1965, Reconnaissance of the Tertiary sedimentary rocks in western Utah: Utah Geological and Mineral Survey Bulletin 75, 38 p.
Jordan, T. E., 1979, Evolution of the Late Pennsylvanian-Early Permian western Oquirrh basin, Utah [Ph.D. dissertation] Stanford, California, Stanford University, 253 p.
——1979b, Lithofacies of the Upper Pennsylvanian-Lower Permian western Oquirrh Group, northwestern Utah: Utah Geology, v. 6, no. 2, p. 41–56.
McClelland, P. H., 1976, New evidence for age of Cenozoic Salt Lake beds in northeastern Great Basin: American Association of Petroleum Geologists Bulletin, v. 60 (12), p. 2185.
Pierce, W. G., 1957, Heart Mountain and South Fork detachment thrusts of Wyoming: American Association of Petroleum Geologists Bulletin, v. 41, no. 4, p. 591–626.
——1973, Principal features of the Heart Mountain fault and the mechanism problem, in Scholten, R., and DeJong, D., eds., Gravity and tectonics: New York, John Wiley & Sons, Inc., 502 pp.
Todd, V. R., 1973, Structure and petrology of metamorphosed rocks in central Grouse Creek Mountains, Box Elder County, Utah: [Ph.D. dissertation] Stanford, California, Stanford University, 316 p.
——1980, Structure and petrology of a Tertiary gneiss complex in northwestern Utah: Geological Society of America Memoir 153, p. 349–383.
Todd, V. R., Compton, R. R., and Jordan, T. E., in preparation, Geologic map of central and southern Grouse Creek Mountains, Box Elder County, Utah: U.S. Geological Survey Miscellaneous Investigations Map.

MANUSCRIPT ACCEPTED BY THE SOCIETY AUGUST 20, 1982

Printed in U.S.A.

Geological Society of America
Memoir 157
1983

Stratigraphic relations of Permian units, Cassia Mountains, Idaho

James W. Mytton
U.S. Geological Survey
Box 25046, Denver Federal Center
Denver, Colorado 80225

William A. Morgan
Conoco, Inc.
Exploration Research
Ponca City, Oklahoma 74603

Bruce R. Wardlaw
U.S. Geological Survey
Box 25046, Denver Federal Center
Denver, Colorado 80225

ABSTRACT

Four new formations, the Wahlstrom Hollow, Third Fork, Badger Gulch, and Trapper Creek, are proposed for Lower Permian rocks that underlie the Park City-Phosphoria sequence in the Cassia Mountains, Idaho. The Permian units represent two regressive-transgressive phases deposited in three depocenters—the Oquirrh (Wolfcampian), the Cassia (lower Leonardian), and the Sublett (upper Leonardian-lower Guadalupian) basins—and record the history of the development of each separate basin. Six regional biostratigraphic conodont zones based on species of *Neostreptognathodus* aid in recognition of depositional patterns and timing of deposition of Permian rocks in the region.

INTRODUCTION

The stratigraphic relations of Permian rocks exposed in the Cassia Mountains of southern Idaho, and their depositional environments and tectonic settings are presented. Four new stratigraphic units that underlie the Park City and Phosphoria Formations are introduced and their lithologies described in detail. They are, in ascending order: the Wahlstrom Hollow, Third Fork, Badger Gulch, and Trapper Creek Formations. Comparative biostratigraphy of the Permian units with equivalents in nearby ranges, specifically the Pequop and Leach Mountains in Nevada, is discussed. The Permian section in the Cassia Mountains based on conodont zonation, is late Wolfcampian (Sterlitamakian Stage of Furnish, 1973) through early Guadalupian (Wordian). Conodont systematics are reviewed by Wardlaw in a separate section of the paper.

Previous investigations of Permian rocks in the Cassia Mountains include those of Youngquist and Haegle (1956), and Laudon and others (1956) who recognized the Phosphoria Formation, notably the Rex Chert Member. Mapping by Mytton in 1976 led to the subdivision of the pre-Park City-Phosphoria sequence. A detailed stratigraphic study by Morgan (1977, 1980a, b) of the Wahlstrom Hollow, Third Fork, and Badger Gulch Formations utilizing fusulinids and conodonts resulted in assignment of an Early Permian age for these units. Morgan also described the depositional settings for each of these units. Regional stratigraphic and paleontological studies of Wardlaw and others (1979, 1980) confirmed the presence of the Grandeur Tongue of the Park City Formation and a modified Meade Peak Phosphatic Shale Member of the

Phosphoria Formation, and established their relation with Park City-Phosphoria equivalents in Idaho, Nevada, and Utah.

GEOLOGIC SETTING

The Cassia Mountains are located in the Basin and Range province. They are bounded by the Snake River Plain on the north and broad valleys on the east and west.

Permian units make up the core of the Cassia Mountains and are surrounded and partially covered by late Miocene air-fall and welded tuffs. The volcanic units dip gently away from the highest elevations of the Cassias and give the area a dome-like appearance. The Cassias have been affected by pre- and post-Miocene extensional faulting. Low-angle faulting, common in Paleozoic rocks in other ranges in the region, has not been observed in exposed Permian rocks in the Cassias.

An affinity to Basin and Range tectonics is reflected in the northerly trend and linearity of ranges that project through the volcanic cover in the southern part of the Cassia Mountains area and their extensions into Nevada. The Triassic Dinwoody Formation, the youngest pre-Tertiary unit in the area, has been preserved in a south-plunging synclinal fold in the uplifted block located in the south central part of the area.

West of the Cassia Mountains, no pre-Tertiary rocks have been reported and current mapping in southwestern Idaho indicates that none project through the volcanic cover (H. Covington, oral communication, 1980). Ordovician rocks are exposed in a small area at the northwestern edge of the Cassias (Youngquist and Heagle, 1956; Coats, 1980). Mississippian rocks have been mapped in the same general area and occur above a thrust plate under which lie the same Ordovician units (Coats, 1980; P. L. Williams, H. R. Covington, J. W. Mytton, and W. A. Morgan, unpublished mapping, 1976-1982).

PHYSICAL STRATIGRAPHY

Wahlstrom Hollow Formation

The Wahlstrom Hollow Formation is the name herein given to a succession of silty carbonates located south of the confluence of Third Fork and Fourth Fork of Rock Creek (Fig. 1). It supersedes the Wahlstrom Hollow sequence of Morgan (1980b, p. 308), and the description below is mostly taken from that work. The formation is exposed in deeper valleys and at higher elevations in upthrown fault blocks in the central and northeastern parts of the Cassia Mountains.

The Wahlstrom Hollow Formation consists of light- to dark-gray or brown to reddish-colored, fine-grained, medium- to thick-bedded, fossiliferous silty limestone and calcareous siltstone, which weather to dark-gray or buff—or less commonly, various shades of pink. Spicules are a common and often abundant constituent of these rocks but are not visible in hand specimen. Pyrite is also common. Bedding is continuous where exposures permit lateral tracing, and contacts between beds are generally sharp and planar. Submarine channels are locally present.

Grayish-orange- to grayish-brown-weathering subspherical to elliptical siliceous nodules, which are aligned parallel to the bedding and stand out prominently in relief, are especially abundant within certain intervals. Grayish-orange-weathering siliceous laminae, parallel to the bedding, are a common feature of the more calcareous rocks.

Zones of recumbently folded, laterally discontinuous beds, interpreted as subaqueous soft-sediment slumps, are present throughout the Wahlstrom Hollow Formation. The most striking example in the type section is a 3-m (10-ft) interval 84 m (276 ft) above the base in which is exhibited laterally discontinuous folds whose axial planes lie subparallel to contiguous beds. The folds are 0.6 to 1.2 m (2 to 4 ft) thick on the outcrop face, and are bounded by flat-lying strata of the same lithology with no observable discontinuity. Their axial orientations are dominantly NE-SW. Contacts of the slumped strata with underlying beds are often irregular. Stringers and protuberances of a slumped bed may extend into the bed directly underlying it and produce of "pseudonodular" appearance. These "pseudonodules" often weather to a light tan and stand in marked relief against the dark-gray strata in which they are enclosed.

Sedimentary structures such as graded lamination, microripples and cross lamination, convolute lamination, scoured surfaces, and load structures are common to the Wahlstrom Hollow Formation. Normal and inverse graded thin beds and laminations are present, although inverse grading is extremely rare. Normal grading is of two types: 1) the abundance of very fine sand, coarse silt, and skeletal grains diminishes vertically over the space of an individual laminae or thin bed; and 2) very fine sand or silt-rich laminae alternate with micrite-rich laminae as coarser-grained laminae thin upward and subtly pass into unlaminated micrite. The latter type of grading, Piper (1972) referred to as "graded laminated."

Morgan (1980a, b) has recognized several recurrent microfacies within the type section of the Wahlstrom Hollow Formation with the aid of polished slabs, etched strips, and stained thin sections. The microfacies, based in part on the classification of Dunham (1962), include lime-mudstone, spiculitic wackestone-packstone, calcareous siltstone, spiculitic siltstone, and skeletal packstone-grainstone. There is an overall restriction on the lowest and highest stratigraphic occurrence of each lithofacies and a gradual coarsening upward progression. For example, the lime-mudstone lithofacies is present only within the basal

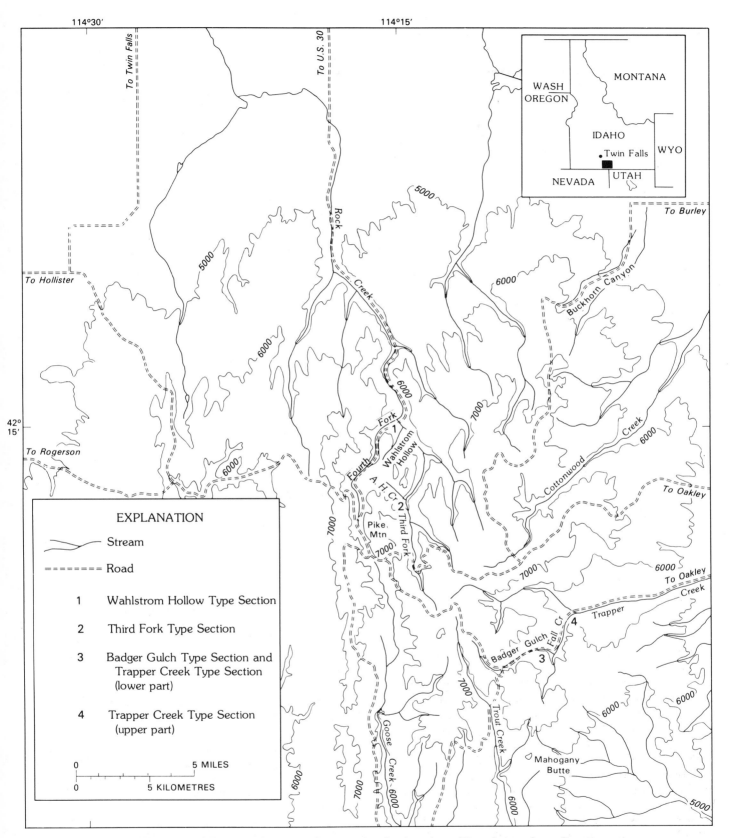

Figure 1. Map of the Cassia Mountains area showing location of Permian sections. Contour interval 1,000 ft.

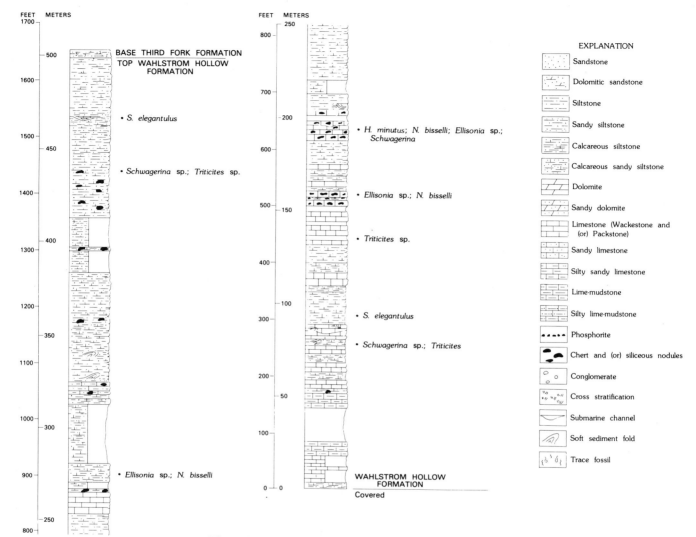

Figure 2. Type section of Wahlstrom Hollow Formation.

portion of the section and has a total thickness of 34 m (112 ft), whereas the upper 167 m (548 ft) of section is almost solely spiculitic siltstone. The skeletal packstone-grainstone lithofacies is unique in that it occurs as channel fill.

The type section of the Wahlstrom Hollow Formation is on the Grand View Peak, Rams Horn Ridge, Trapper Peak, and Pike Mountain 7.5-minute quadrangles. The base of the type section (Fig. 2) is in the NE 1/4, SW 1/4, NE 1/4, section 32, T. 13 S., R. 19 E., Cassia County, Idaho, south of the confluence of Third Fork and Fourth Fork of Rock Creek, and south of Rock Creek Road (Fig. 1). The base of the measured section begins on a NE-SW-trending ridge approximately 5.2 m (17 ft) above Third Fork. The section proceeds up the ridge (upsection) with repeated offsets along strike to the southeast, toward Wahlstrom Hollow, and ends in a saddle in the SW 1/4, NE 1/4, SW 1/4, section 32, T. 13 S., R. 19 E. below the Third Fork

Formation. Some minor structural folds are present at the type section, but true thicknesses have been calculated wherever possible. The lower part of the formation forms irregular, scraggly cliffs; the upper part, steep slopes with broken ledges.

WAHLSTROM HOLLOW FORMATION TYPE SECTION

Third Fork Formation (Permian): Float, calcareous sandy siltstone; buff, weathering into small buff-, pink-, and orange-colored chips; scattered meandering worm-like trace fossils on bedding planes; mostly silt with minor very fine-grained quartz.

Gradational strata with Wahlstrom Hollow Formation: Float, calcareous spiculitic, sandy siltstone; gray, weathering mostly buff; local wispy laminae; local meandering worm-like trace fossils on bedding planes; spicules less abundant than below. Thickness: 8.9 m (29.2 ft).

Wahlstrom Hollow Formation (Permian):

21. Spiculitic quartz siltstone, calcareous: upper 33.7 m (110.5 ft): gray, weathering tan; thick to very thick bedding with wispy, silty laminae; intermittent cover; minor tectonic folds near top. Lower 109.2 m (358.2 ft): gray to pink weathering tan to pink; thin to thick bedding appears to be in depositional packages consisting of a thick bed overlain by thin beds; some channeling, microcross-laminations, and micrograded laminations; some pelmatozoan columnals; locally, siliceous nodules parallel bedding and weather in relief to dark brown. At 95.1 m (311.9 ft) *Streptognathodus elegantulus* was recovered from a channel fill (Sample CAS 52-101). At 80.8 m (252.2 ft) fusulinids *Schwagerina* sp. and *Triticites* sp. (Sample CAS 52-97). Float and cover between 26.9 m (88.2 ft) and 39.1 m (128.2 ft), and between 40.5 m (132.7 ft) and 56.6 m (185.7 ft). Thickness: 142.9 m (468.7 ft).

20. Calcareous quartz siltstone; gray, weathering tan; medium to massive bedding (to 3.4 m thick); wavy laminations common. Thickness: 15.3 m (50.2 ft).

19. Spiculitic quartz siltstone, calcareous; gray to pink, weathering gray to pink; thin to massive bedding in packages similar to unit 21; soft-sediment folds (slumps) common. Thickness: 15.9 m (52.1 ft).

18. Spiculitic limestone (wackestone-packstone); gray, weathering tan; thin to thick bedded; thinly laminated; consists mostly of spicules in a silty (quartz), lime-mud matrix; local siliceous nodules. Thickness: 7.4 m (24.4 ft).

17. Calcareous quartz siltstone; dark gray, weathering gray or locally buff; thick to massive bedding, mostly featureless, some horizontal, cross, and convolute laminations; siliceous nodules prominent in lower 1.7 m (5.5 ft); some pelmatozoan columnals, conodont elements *Ellisonia* sp. and *Neogondolella bisselli* from 9.2 m (30.3 ft, Sample CAS 52-71); mostly float between 1.7 m (5.5 ft) and 4.4 m (14.5 ft), and between 15.8 m (52.0 ft) and 46 m (150.8 ft). Thickness: 53.7 m (176.0 ft).

16. Spiculitic limestone (wackestone-packstone); gray, weathering to pink and purple; thick to massive bedding; thin laminations, micrograded, crossbedded in part; spicules abundant; pelmatozoan fragments, brachiopod fragments, bellerophontid gastropods, and *Plagioglypta* sp. less abundant. Thickness: 12.0 m (39.3 ft).

15. Spiculitic quartz siltstone, calcareous; dark gray, weathering gray, pink, and purple; upper 32.9 m (107.9 ft) thin to massive beds in packages similar to unit 21; lower 9.1 m (30.0 ft) thick bedded; beds mostly featureless but some crossbeds and ripples; reddish-brown-weathering siliceous nodules in upper 32.9 m; scattered fossil hash; float between 9.2 m (30.0 ft) and 16.8 m (55.0 ft). Thickness: 49.7 m (162.9 ft).

14. Spiculitic limestone (wackestone-packstone) with minor spiculitic siltstone; gray, weathering light gray; medium to massive bedding; some channeling; abundant siliceous nodules, dark gray, weathering reddish brown in relief; intermittent cover; conodont elements *Hindeodus minutus, Neogondolella bisselli, Ellisonia* sp. and fusulinid *Schwagerina* recovered at 6.3 m (20.6 ft, Sample CAS 52-48). Thickness: 16.3 m (53.4 ft).

13. Spiculitic quartz siltstone, calcareous: gray, weathering light gray; thick bedding; silty quartz laminae weathering tan in relief; intermittent outcrops, float between 1.5 m (5.0 ft) and 7.0 m (23.0 ft). Thickness: 12.9 m (42.4 ft).

12. Mixed lithologies, interbedded spiculitic limestone (wackestone-packstone), spiculitic quartz siltstone, and calcareous siltstone; mostly dark gray, weathering gray; thin to thick bedding, thinly laminated in part; dark gray chert and slightly calcareous siliceous nodules, abundant in lower 5.2 m (17.0 ft). At 3.4 m (11.0 ft) conodont elements *Ellisonia* sp. and *Neogondolella bisselli* recovered (Sample CAS 52-40). Thickness: 20.1 m (65.8 ft).

11. Spiculitic limestone (wackestone-packstone); dark gray, weathering gray; medium to thick bedding; graded laminae rare, parallel and convolute laminae common; mostly sponge spicules in lime-mud matrix with lesser amounts of quartz silt; pelmatozoan columnals, brachiopod fragments, gastropods rare; dark-gray chert and slightly calcareous siliceous nodules common in upper 2.7 m (9 ft); intermittent cover. At 4.0 m (13.1 ft) one specimen of *Triticites*? sp. recovered (Sample CAS 52-31). Thickness: 23.4 m (76.6 ft).

10. Spiculitic quartz siltstone, calcareous; dark gray, weathering gray; thin to thick bedding; micrograded laminae, cross laminae, convolute laminae common. Thickness: 9.8 m (32.0 ft).

9. Calcareous quartz siltstone, same as unit 20. Thickness: 5.2 m (17.0 ft).

8. Spiculitic limestone (wackestone-packstone), similar to unit 11, but no siliceous nodules. Thickness: 4.5 m (14.8 ft).

7. Spiculitic quartz siltstone, calcareous; gray, weathering tan; bedding similar to unit 21. Thickness: 3.7 m (12.1 ft).

6. Calcareous quartz siltstone; dark gray, weathering purple, gray, and tan; thick to massive bedding; locally, calcite filled fractures; thinly laminated in part; intermittent cover near base. Thickness: 17.2 m (56.4 ft).

5. Spiculitic limestone (wackestone-packstone); dark gray, weathering gray; medium bedding, silty laminae stand in relief; micrograding, cross laminae, small-scale slumps common; some channeling; spicules abundant, brachiopod fragments common. Thickness: 8.1 m (26.7 ft).

4. Interbedded spiculitic limestone (wackestone-packstone) and calcareous siltstone; dark gray, weathering gray; medium to thick bedding; soft-sediment slumps recumbently folded, and "ball and pillow" structures near top;

normal and inverse graded laminae and microripples common; skeletal debris in fine-grained hash; near base rare dark-gray nodular chert. Thickness: 28.7 m (94.0 ft).

3. Lime-mudstone, silty (quartz); purplish gray, weathering gray; medium bedding; no internal structures seen; rare spicules and an unidentified spherical microfossil; cover and float between 7.7 m (25.3 ft) and 26.2 m (85.8 ft). Thickness: 34.2 m (112.0 ft).

2. Float, spiculitic limestone (wackestone-packstone); light gray, weathering tan; outcrops scarce, medium bedding?; consists of equal quantities sponge spicules, quartz silt, and lime-mud. Thickness: 14.7 m (48.3 ft).

1. Calcareous siltstone and chert, gray weathering gray; medium bedding; thinly laminated in part; section ends in cover above Third Fork Creek. Thickness: 3.7 m (12.0 ft).

Total thickness of the Wahlstrom Hollow Formation–499.4 m (1637.1 ft).

Section measured by William A. Morgan, 1976.

Third Fork Formation

Conformably overlying the Wahlstrom Hollow Formation is a succession of calcareous sandy siltstones characterized by ubiquitous trace fossils. The Third Fork Formation is the name herein given to this succession. It supersedes the Third Fork sequence of Morgan (1980b, p. 313). The Third Fork is present chiefly in the central part of the Cassia Mountains where it is exposed in upthrown fault blocks. Pike Mountain, one of the more prominent landmarks and among the highest in elevation, is made up entirely of rocks of the Third Fork. The formation is characterized not only by its trace fossils but also by the distinctive colors of its sandy siltstones, which on a fresh surface are gray to buff and pink, and weather gray, buff, pink, and orange. Very fine-grained angular quartz is the major component of the sandy siltstones; calcite is the major cement. Bed thicknesses appear to range from thin to thick but are difficult to determine because the formation is generally poorly exposed and thoroughly bioturbated. Because of extensive bioturbation, rocks often appear mottled and laminations are rare. Prominent trace fossils include various forms of *Scalarituba* sp., *Zoophycos* spp., and *Chondrites* sp. (Morgan, 1980b).

In addition to the siltstones, three distinctive beds of argillaceous spiculitic lime-mudstones and calcareous siltstones, separated from one another by siltstone, are present at the type section. The lime-mudstones are black and weather to gray or grayish yellow. The beds are less than 0.4 m thick and may serve as key beds in an otherwise monotonous sequence of siltstones.

The Third Fork Formation is generally poorly exposed and commonly forms slopes. The type section is located in the S½, sec. 17, T. 14 S., R. 19 E., Pike Mountain

and Trapper Creek 7.5-minute quadrangles, Cassia County, Idaho. The type section (Fig. 3) is on a ridge southwest of the confluence of A. H. Creek and Third Fork (Fig. 1). The measured section described below begins in the Badger Gulch Formation, at the top of a prominent outcrop near the base of that formation. The section proceeds down the ridge (downsection) and includes beds gradational between the Badger Gulch and Third Fork Formation (which we include in the Badger Gulch Formation), and the Third Fork Formation. The measured section ends at the contact between the Third Fork and the underlying Wahlstrom Hollow Formation, at an altitude of 6,350 ft.

THIRD FORK FORMATION TYPE SECTION AND LOWER PART, BADGER GULCH FORMATION TYPE SECTION

Badger Gulch Formation (Permian), entire formation not measured: Lime-mudstone, dark-gray to black, weathering black; locally siliceous; soft-sediment deformation and cut and fill structures; thin silty laminae, locally rhythmically banded, fine-grained (mudstone); rare small pelmatozoan columnals; sulfurous odor from fresh break; fractures common and filled with white calcite; upper 4 m well exposed, lower 41.2 m (135.0 ft) float. Thickness: 45.2 m (148.0 ft).

Gradational strata with Third Fork Formation: Float, calcareous sandy siltstone, dark-gray weathering pink and buff, commonly with weathering rind; grains angular, mostly quartz and an orange-colored mineral (identity unknown); sandstone is very fine-grained; near base of Badger Gulch Third Fork transition float weathers gray. Thickness: 35.4 m (116.2 ft).

Third Fork Formation (Permian):

12. Float, calcareous sandy siltstone, buff and pink, weathering buff, orange, and pink; scattered meandering worm-like trace fossils approximately 4 mm in diameter; mostly angular quartz grains and an unidentified orange mineral; minor very fine-grained sandstone. Thickness: 20.1 m (65.8 ft).

11. Silty lime-mudstone, black, weathering light-gray; distinctive yellow downward-branching trace fossils with fecal back-fills; fetid odor from fresh break. Thickness: 0.1 m (0.4 ft).

10. Float, calcareous sandy siltstone, gray, weathering gray and pink; locally indistinct laminae; locally meandering traces; mostly angular quartz and unidentified orange mineral. Thickness: 16.6 m (54.5 ft).

9. Float, calcareous sandy siltstone, gray, buff, and pink, weathering gray, pink, buff, and orange, few scattered *Scalarituba* in upper 21.3 m (70.0 ft), considerable increase in trace fossils in lower 50.6 m (165.9 ft) including *Scalarit-*

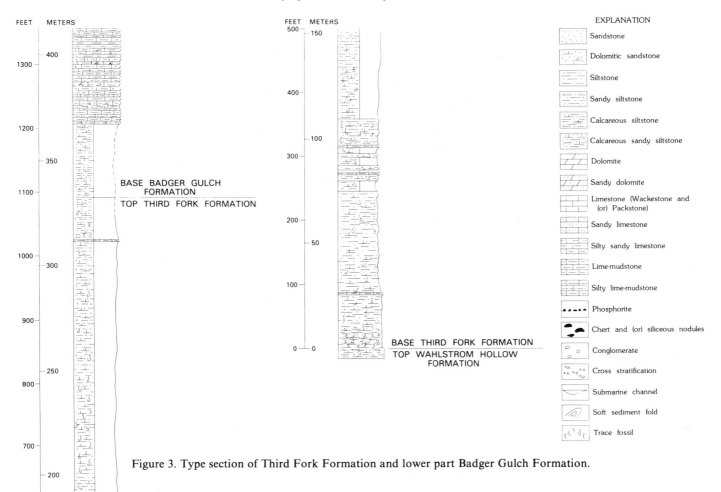

EXPLANATION

	Sandstone
	Dolomitic sandstone
	Siltstone
	Sandy siltstone
	Calcareous siltstone
	Calcareous sandy siltstone
	Dolomite
	Sandy dolomite
	Limestone (Wackestone and (or) Packstone)
	Sandy limestone
	Silty sandy limestone
	Lime-mudstone
	Silty lime-mudstone
	Phosphorite
	Chert and (or) siliceous nodules
	Conglomerate
	Cross stratification
	Submarine channel
	Soft sediment fold
	Trace fossil

BASE BADGER GULCH FORMATION
TOP THIRD FORK FORMATION

BASE THIRD FORK FORMATION
TOP WAHLSTROM HOLLOW FORMATION

Figure 3. Type section of Third Fork Formation and lower part Badger Gulch Formation.

uba, *Zoophycos,* and meandering worm-like form; mineralogically same as unit 10. Thickness: 93.3 m (305.9 ft).

8. Float, sandy siltstone, same as unit 9 except noncalcareous. Thickness: 68.8 m (222.9 ft).

7. Calcareous sandy siltstone, same as unit 9 with addition of *Chondrites;* upper 23.9 m (78.3 ft) float; lower 12.8 m (42.1 ft) massively bedded, appears to be homogenized from bioturbation. Thickness: 36.7 m (120.4 ft).

6. Silty lime-mudstone, same as unit 11. Thickness: 0.3 m (0.9 ft).

5. Calcareous sandy siltstone; same as unit 9; outcrops, low relief and intermittent. Thickness: 12.8 m (42.1 ft).

4. Lime-mudstone; similar to unit 11. Thickness: 0.1 m (0.4 ft).

3. Calcareous sandy siltstone; same as unit 9; mostly

float between 2.7 m (8.8 ft) and 8.0 m (26.3 ft) from top of unit. Thickness: 56.4 m (184.9 ft).

2. Silty limestone, dark gray, weathering gray to buff; very similar to calcareous sandy siltstone (unit 9) but more lime-mud. Thickness: 0.6 m (1.8 ft).

1. Calcareous sandy siltstone, gray, weathering gray; abundant traces: *Scalarituba, Zoophycos, Chondrites,* meandering worm-like forms; some thinly laminated calcareous siltstones with molds of small pelmatozoan columnals; parallel-laminated-to-burrowed beds near base of unit; mostly angular quartz grains and an unidentified orange mineral; base of formation is lowest calcareous siltstone bed. Thickness: 26.3 m (86.3 ft)

Total thickness of the Third Fork Formation—331.2 m (1086.3 ft).

Wahlstrom Hollow Formation (Permian): Contact with Third Fork Formation conformable and gradational. Silty limestone, dark gray, weathering gray; trace fossils less diverse and less abundant than Third Fork Formation and consist of a few meandering worm-like forms; silty laminae weather in relief; thin beds.

Section measured by William A. Morgan, James W. Mytton, and Jean N. Weaver, 1980.

Badger Gulch Formation

The Badger Gulch Formation is the name herein given to a succession of dark lime-mudstones and siliceous lime-mudstones located on the south side of Badger Gulch 1.7 km west of its confluence with Fall Creek. It supersedes the Badger Gulch sequence of Morgan (1980b, p. 318). Only the upper part of the formation is exposed at Badger Gulch. Its contact with the Third Fork is described with the

Third Fork type section. Exposures of the Badger Gulch occur in many areas of the Cassia Mountains, but are most striking in Badger Gulch where they form castellated cliffs with jagged spires.

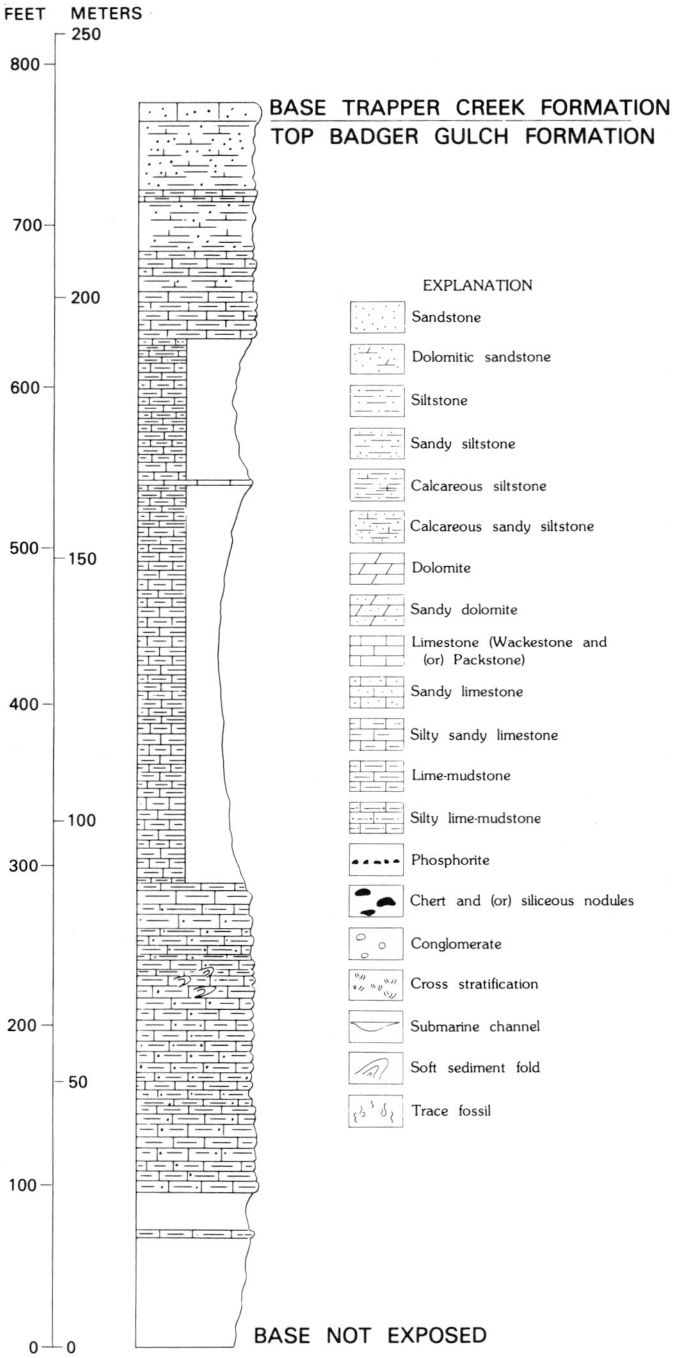

The Badger Gulch Formation is a predominantly brownish-gray to black very fine-grained, thin-bedded lime-mudstone. It contains variable amounts of silt-sized skeletal fragments replaced by, or recrystallized to, calcite; and quartz silt; and argillaceous material. The rock commonly exhibits a blotchy texture, the darker areas being more calcareous, the lighter areas more siliceous. Beds are extremely regular, with sharp, even contacts and are characterized by evenly spaced laminae, about 1 mm thick. Cleavage at high angles to the laminae is common. Beds break into angular plates which easily shatter. In many localities beds are highly contorted with fold axes generally plunging in a southernly direction. Silty and sandy interbeds are sparse in the upper part of the formation but more common in the lower part where lime-mudstones are variegated and more silty. Soft sediment deformation is not uncommon.

The type section of the upper part of the Badger Gulch Formation is in the NW¼ sec. 20, T. 15 S., R. 20 E., Mahogany Butte 7.5-minute quadrangle, Cassia County, Idaho. The proposed type section (Fig. 4) is located on a northwest-facing ridge on the south side of Badger Gulch (Fig. 1). The measured section described below begins in the Trapper Creek Formation at the base of a prominent cliff. The section proceeds down the ridge (downsection) and includes a gradational boundary between the Trapper Creek and the upper part of the Badger Gulch Formation. Although the gradational strata have stronger sedimentologic affinities to the Trapper Creek Formation, we include the transitional unit in the Badger Gulch Formation because, in mapping, its topographic expression is like that of the Badger Gulch. The lowermost part of the Badger Gulch is present at the type section of the Third Fork Formation in the Pike Mountain 7.5-minute quadrangle. Faulting and poor exposures make measurement of a complete stratigraphic section at one locality infeasible. The thickness of the composite section is 313.7 m (1029.6 ft).

UPPER PART, BADGER GULCH FORMATION TYPE SECTION AND LOWER PART, TRAPPER CREEK FORMATION TYPE SECTION

Trapper Creek Formation (Permian): Fossiliferous limestone, medium-gray, fine-grained, medium- to thick-bedded (average 0.3 m), silty to sandy with tan-weathering siliceous elliptical concretions. Crinoid ossicles, brachiopods, and gastropods (bellerophontids) common. Higher-up section, scaphopods, conodonts, and algae common. Talus is blocky. Thickness: 74.2 m (243.4 ft).

Figure 4. Type section of upper part, Badger Gulch Formation.

EXPLANATION

Sandstone

Dolomitic sandstone

Siltstone

Sandy siltstone

Calcareous siltstone

Calcareous sandy siltstone

Dolomite

Sandy dolomite

Limestone (Wackestone and (or) Packstone)

Sandy limestone

Silty sandy limestone

Lime-mudstone

Silty lime-mudstone

Phosphorite

Chert and (or) siliceous nodules

Conglomerate

Cross stratification

Submarine channel

Soft sediment fold

Trace fossil

BASE TRAPPER CREEK FORMATION
TOP BADGER GULCH FORMATION

BASE NOT EXPOSED

Badger Gulch Formation (Permian), partial section:
Gradational contact with Trapper Creek Formation:

10. Calcareous siltstone to very fine-grained sandstone, gray; weathers light gray; grades into medium- to dark-gray lime-mudstone, slightly silty; talus blocky in upper part to platy in lower part. Thickness: 34.0 m (111.7 ft).

9. Lime-mudstone, dark gray to brownish-gray to black, in part siliceous, silty, very thin even-bedded; weathers to thin tan and gray plates a few centimeters thick. Thickness: 7.0 m (23.1 ft).

8. Float, platy lime-mudstone. Thickness: 26.5 m (87.0 ft).

7. Lime-mudstone, dark gray, thin-bedded, slope- to ledge-forming. Thickness: 0.9 m (3.0 ft).

6. Float, platy lime-mudstone; same as unit 8.) Thickness 76.2 m (250.0 ft).

5. Lime-mudstone, dark gray, thin-bedded, slope- to ledge-forming. Thickness: 4.0 m (13.2 ft).

4. Lime-mudstone, dark gray, silty, somewhat siliceous and hard, aphanitic, thin- to medium-, even-bedded, ledge- to slope-forming. Thickness: 12.1 m (39.8 ft).

3. Lime-mudstone, dark gray; similar to unit 4 with soft-sediment folds in beds at top, ledge-forming. Thickness: 26.3 m (86.2 ft).

2. Lime-mudstone, dark gray; same as unit 4, cliff-forming. Thickness: 16.0 m (52.5 ft).

Covered interval. Thickness: 7.4 m (24.4 ft).

1. Lime-mudstone, same as unit 2. Thickness: 1.2 m (4.0 ft).

Covered interval to base of ridge. Thickness: 21.5 m (70.5 ft).

Total thickness of partial section—233.1 m (765.4 ft).

Section measured by William A. Morgan and James W. Mytton, 1980.

Trapper Creek Formation

The Trapper Creek Formation is the name herein given to the succession of thick-bedded limestones and cherty calcareous sandstones located on the east side of Fall Creek, just south of its confluence with Trapper Creek, and at Badger Gulch.

Exposures of the Trapper Creek Formation occur in many areas of the Cassia Mountains, although they are most prominent, as well as most accessible, in the narrow valley of Fall Creek south of its junction with Trapper Creek. Here the unit dips to the south and forms imposing cliffs rising above steep talus slopes. The Trapper Creek Formation consists of medium to dark brownish-gray, very fine-grained, medium- to thick-bedded, fossiliferous limestone (packstone to lime-mudstone), alternating with buff, calcareous siltstone and very fine-grained sandstone. The limestone generally is silty to sandy and characterized by

algal-like elliptical to concentric spheres. Lenticular bands of very fine-grained sandstone are common and more resistant to weathering than the limestone. Where the sandstone is silicified, it resembles chert. Calcareous zones between sandstone bands resemble concretionary nodules. Limestone beds contain light- to dark-gray nodular to lenticular chert. Some beds have abundant brachiopod fragments, scaphopods, bellerophontid gastropods, and bryozoans, although many are obliterated by silicification. Crinoid ossicles, notably pentacrinoids, are abundant in the lower part of the Trapper Creek Formation. Thin- to medium-bedded, laminated sandstone and siltstone grade laterally into orthoquartzite. Both sandstone and sandy limestone exhibit small-scale cross-stratification. The Trapper Creek is a prominent ledge- and cliff-former in contrast to the overlying Grandeur Tongue of the Park City Formation with which it is in contact; Morgan (1980b) referred the unit overlying the Badger Gulch Formation (the Trapper Creek Formation of this paper) to the Park City Group. The total thickness of the Trapper Creek Formation, based on partial sections above Fall Creek and Badger Gulch (Fig. 1), is 271.0 m (880.0 ft).

The type section of the Trapper Creek Formation is a composite section consisting of upper and lower parts at two separate locations (Fig. 1) within T. 15 S., R. 20 S., Severe Spring 7.5-minute quadrangle, Cassia County, Idaho. The lower part is exposed in Badger Gulch above the type section of the Badger Gulch Formation in sec. 20 and is 135.7 m (445.2 ft) thick. The upper part is exposed in prominent cliffs above Fall Creek, south of its confluence with Trapper Creek (center S½ SE¼ sec. 9). Though exposures are poor at the base of the section above Fall Creek, it appears from lithostratigraphic and biostratigraphic correlation between the partial sections that approximately 8 m (26.2 ft) of the formation are missing at the Fall Creek locality.

The partial section and samples collected for conodonts at Badger Gulch up to the first sandstone unit (unit 7, augmented by descriptions from Fall Creek) and partial section and conodont samples at Fall Creek above the first sandstone (unit 7, augmented by descriptions from Badger Gulch) constitute the composite section (Fig. 5).

TRAPPER CREEK FORMATION COMPOSITE TYPE SECTION

Grandeur Tongue of the Park City Formation (Permian): Dolomite, sandy, light-gray, and sandstone, medium-grained, light-gray, poorly exposed at base.

Trapper Creek Formation (Permian): Sharp contact with the overlying Grandeur generally marked by the abrupt change in slope from well-exposed, cliff-forming Trapper Creek to poorly-exposed, slope-forming Grandeur.

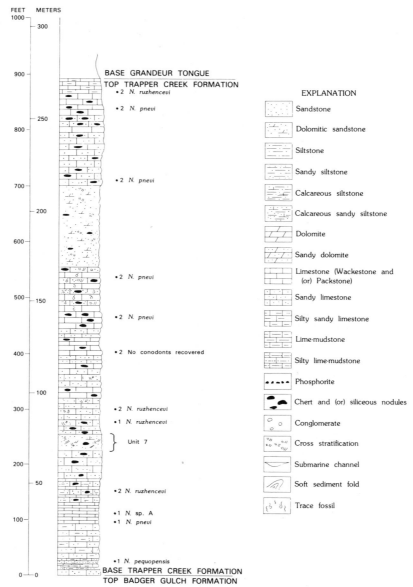

Figure 5. Type section (composite) of Trapper Creek Formation showing conodont samples. 1, samples from Badger Gulch section; 2, samples from Fall Creek section.

15. Limestone, light medium-gray, weathering yellow-gray; thin-bedded, argillaceous wackestone near top; medium-bedded crinoidal and algal cherty packstone near base. At 264.3 m (867.1 ft) (accumulated thickness) *Neostreptognathodus ruzhencevi* recovered (Sample 1CAS842) and at 255.8 m (839.2 ft) (accumulated thickness) *N. pnevi* recovered (Sample 1CAS814). Thickness: 16.2 m (53.0 ft).

14. Cherty sandy limestone, medium- to dark-gray, weathering brownish-gray; chert in small interbeds showing cross-lamination; sand concentrated in layers, and a medium-gray limestone consisting of skeletal fragments and algal plates (packstone) at 30.4 m (99.7 ft); bellero-

phontid gastropods at 24.0 m (78.7 ft). At 216.5 m (710.3 ft) (accumulated thickness) *Neostreptognathodus pnevi* recovered (Sample 1CAS685). Thickness: 40.5 m (133.0 ft).

13. Dolomitic sandstone, light-medium-gray, weathering brownish-gray; fine-grained; beds 0.3-1.2 m thick, with small-scale crossbedding, silicified layers accentuate crossbedding. Thickness: 47.9 m (157.0 ft).

12. Cherty sandy limestone, dark-medium-gray, weathering light brownish-gray; with chertified vertical burrows; sandy layers with large- to small-scale cross-lamination and horizontal burrows; some sandy layers silicified into brownish-gray chert; bedded chert rare. At

163.1 m (535.1 ft) (accumulated thickness) *Neostreptog-nathodus pnevi* recovered (Sample 1CAS510). Thickness: 21.3 m (70.0 ft).

11. Cherty limestone, dark medium-gray; chert brown-gray in 0.3 m irregular bands, hackly, becoming more common near top. Thickness: 9.1 m (30.0 ft).

10. Interbedded sandy limestone, dark medium-gray, weathering brown, with small-scale cross-lamination in sandier layers, fossiliferous with skeletal fragments and some clams; and, cherty limestone (packstone), dark medium-gray; dark- to medium-gray chert in layers suggesting cross-lamination; some beds only partially exposed forming small ledges and slopes. Thickness: 32.6 m (107.0 ft).

9. Limestone, dark-gray, interbedded with irregular layers of laminated brown-gray chert. Thickness: 5.5 m (18.0 ft).

8. Cherty sandy limestone (packstone), dark- to medium-gray, weathering brownish-gray; with sand concentrated in crossbedded layers, with scattered interbeds of calcareous sandstone; brownish-gray chert; fossiliferous in section at Badger Gulch with a bed of abundant bellero-phontid gastropods at 7.2 m (23.5 ft) and a bed of abundant algal plates at 14.8 m (48.6 ft). At 91.8 m (301.2 ft) (accumulated thickness) *Neostreptognathodus ruzhencevi* recovered (Sample 1CAS276) and also at 84.8 m (278.2 ft) (Sample W81558). Thickness: 19.8 m (65.0 ft).

7. Cherty calcareous sandstone, light brownish-gray, weathering light yellowish-gray; with small-scale crossbedding; with scattered thin interbeds (0.15 m) of medium- to dark-gray sandy limestone; forming slight slope above cliff. Thickness: 16.7 m (55.0 ft).

6. Skeletal sandy limestone (packstone), medium-gray, weathering light brownish-gray; medium- to thick-bedded; with lenticular chert; wavy bedded; with scattered interbeds of light-brownish-gray calcareous sandstone; cross-laminated; unit forming cliff. Thickness: 7.6 m (25.0 ft).

5. Interbedded skeletal limestone (wackestone), medium-gray, weathering light-gray; thin- to medium-bedded; with scattered ellipsoidal chert nodules; and silty sandy limestone, brownish-gray, weathering light-brownish-gray; thin- to thick-bedded; slightly silicified with chert in lenses along bedding planes and in siltier beds, wavy bedded; unit forming base of cliff. *Penniculauris* sp. (productid brachiopod) at 3.1 m (10.2 ft). At 49.0 m (160.8 ft) (accumulated thickness) *Neostreptognathodus ruzhencevi* recovered (Sample 1CAS136). Thickness: 13.7 m (45.0 ft).

4. Sandy skeletal limestone (wackestone), medium-gray, weathering light-gray; with a very fossiliferous bed at 33.8-34.1 m (110.9-111.9 ft) with abundant chonetid brachiopods. This unit in the section at Fall Creek contains ellipsoidal yellow-chert nodules with concentric laminations which are rare at the section at Badger Gulch. At 34.0 m (111.5 ft) (accumulated thickness) *Neostreptonathodus*

sp. A recovered (Sample W81557) and at 29.6 m (97.1 ft) *Neostreptognathodus pnevi* recovered (Sample W81556). Thickness: 11.3 m (37.0 ft).

3. Silty sandy limestone (wackestone), medium-gray, weathering light-gray; medium- to thick-bedded, with skeletal horizons of mostly crinoid debris. Thickness: 18.3 m (60.0 ft).

2. Limestone, thin-bedded and lenticular of 4 types in decreasing abundance: 1) brownish-gray silty sandy limestone (wackestone); 2) light-gray very sandy limestone (packstone); 3) medium-gray skeletal limestone (wackestone), very lenticular; and, 4) very thin-bedded, very silty lime-mudstone interbeds around limestone lenses. At 8.2 m (26.9 ft) (accumulated thickness) *Neostreptognathodus pequopensis* recovered (Sample W81555). Thickness: 7.0 m (23.0 ft).

1. Very sandy limestone (packstone), medium brownish-gray, weathering yellowish-brown; medium- to thick-bedded, forming low ledge. Thickness: 3.0 m (10.0 ft).

Total thickness of Trapper Creek Formation—270.7 m (888.0 ft).

Section at Fall Creek measured by Bruce R. Wardlaw and James W. Collinson, 1976; section at Badger Gulch measured by Bruce R. Wardlaw, Susan T. Miller, and Jeanne E. Wardlaw, 1981.

Grandeur Tongue of Park City Formation

The Grandeur Tongue is restricted to the southeastern part of the Cassia Mountains, where it is present above the Trapper Creek Formation, and to the southwest on a prominent ridge east of Trout Creek (Fig. 1). The Grandeur Tongue (Fig. 6) consists of white, fine- to medium-grained, finely laminated, thin- to medium-bedded sandstone that is silicified and dolomitic. The sandstone grades into cross-stratified orthoquartzite and sandy and silty laminated dolomite both vertically and laterally. White to light-gray, sandy, very fine-grained dolopackstone and dolowackestone with abundant fossil debris, notably brachiopod fragments and crinoid ossicles, alternate with sandstone in the lower part of the unit. Chert and cleaner dolomite increase upward in the section. The uppermost dark cherts of the unit are overlain by phosphate-bearing rocks of the Meade Park Phosphatic Shale Member of the Phosphoria Formation. The Grandeur forms steep to gentle slopes in contrast to the ledges and massive cliffs of the underlying Trapper Creek Formation. The thickness of the Grandeur Tongue is 152.0 to 183.0 m (500.0 to 600.0 ft).

Phosphoria Formation

The Phosphoria Formation in the Cassia Mountains includes both the Meade Peak Phosphatic Shale and Rex Chert Members. They are restricted to the same localities of

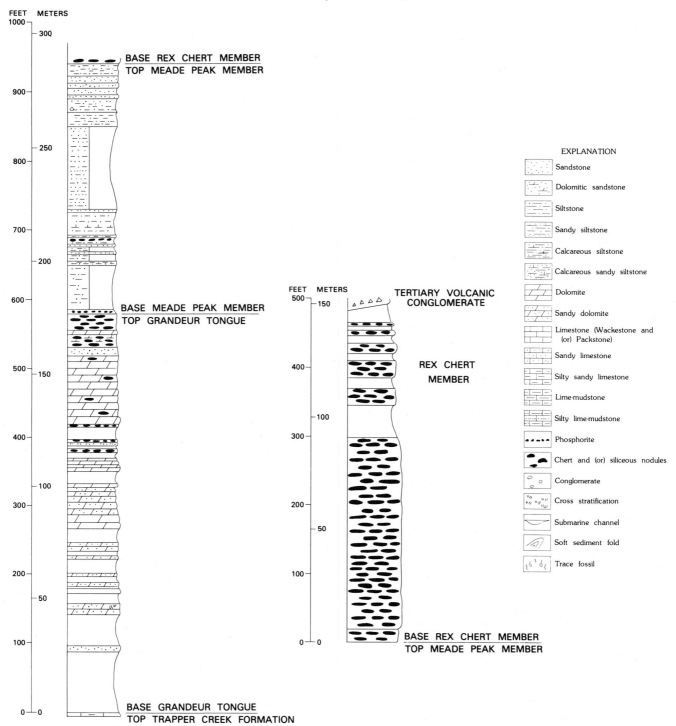

Figure 6. Stratigraphic section of Park City-Phosphoria Formations, east side of Fall Creek, south of Trapper Creek type section (upper part). Section measured by B. R. Wardlaw, J. W. Collinson, and E. K. Maughan, 1976, (Wardlaw and others, 1979a).

the Grandeur Tongue. These outcrops represent the westernmost exposure of the Phosphoria in southern Idaho.

Meade Peak Phosphatic Shale Member. The Meade Peak (Fig. 6) is distinguished by dark-brown thin-bedded siltstone with minute white-weathering phosphatic oolites or pellets. Interbeds of shale, chert, silty limestone and fine-grained sandstone are locally present. Dark-gray, partly silicified phosphorite with abundant fossil debris is prominent on the ridge above Trout Creek (Fig. 1). The phosphorite marks the first bed of the Meade Peak, is represented in most outcrops, and lies directly above a thick bed of chert included in the Grandeur Tongue.

The upper part of the Meade Peak is characterized by thin-bedded orthoquartzite, brown chert, and white to variegated platy silicified sandstone and siltstone. "Liesagang" rings are common on the surface of slabs of platy orthoquartzite, sandstone, and siltstone. Olive- to brownish-gray, partially silicified limestone (wackestone) to dolomitic limestone is locally present. This upper part of the Meade Peak was referred to as Rex Chert Member by Wardlaw and others (1979) and as the cherty shale member by Wardlaw and others (1980) because of the total lack of phosphorite and abundance of siltstone and sandstone. Though the stratigraphic correlation of this unit with the cherty shale member seems correct, the unit cannot be consistently separated from the underlying Meade Peak and, therefore, has been mapped as Meade Peak. The Meade Peak forms steep to gentle slopes in contrast to the massive, cliff-forming Rex Chert which lies above. The thickness of the Meade Peak is 107.0 to 122.0 m (350.0 to 400.0 ft).

Rex Chert Member. The Rex Chert Member is the uppermost unit of the Permian sequence in the Cassia Mountains. An outcrop of thin-bedded limestone has been observed in the southern part of the Cassias, just north of Mahogany Butte (Fig.1), but was not mapped. It may represent a unit overlying the Rex, possibly equivalent to the Gerster Limestone. The distribution of the Rex follows that of the Meade Peak and Grandeur, but it also is widespread in southeasternmost exposures of the Permian in the Cassia Mountains area. It is made up of medium-dark-gray generally medium- to thick-bedded chert (Fig. 6) with 10 to 15 cm bands. The chert weathers olive drab to brown. Some beds show lamination and cross-stratification. Some beds are spicular and others have sparse fossil debris. The chert is fractured, with chalcedony occupying the fractures. The unit generally is thin to medium bedded in its lowermost part, thick bedded to massive in its middle part, and medium to thick bedded in its upper part. The Rex Chert Member is a prominent cliff-former. Its thickness is at least 137.0 m (450.0 ft) but could be as much as 274.0 m (900.0 ft) or more.

DEPOSITIONAL ENVIRONMENTS

Details of Wahlstrom Hollow, Third Fork, and Badger Gulch sedimentation have been discussed by Morgan (1980b), and the following descriptions of these formations are largely a summary of this work.

Wahlstrom Hollow Formation

Most of this unit appears to have been deposited by turbidity currents, as evidenced by the association of graded beds and laminations, convolute laminations, mi-

croload structures, and microripples. The fine grain size may reflect both a fine-grained source and deposition from the distal parts of turbidity currents. Contour currents could have played a role in sedimentation of this formation, but the paucity of current indicators makes their importance difficult to assess.

Hemipelagic sedimentation may have been influential in the deposition of the lime-mudstone lithofacies. This facies is locally structureless, and contains randomly scattered quartz silt and spherical bodies of microspar (0.2 mm in diameter). The silt may be windblown and the spherical bodies are probably pelagic microfossils.

The only record of coarse-grained detritus being introduced into the area during Wahlstrom Hollow time is found within submarine channels. Channel-fill sediments generally consist of a wide variety of skeletal debris and lithoclasts.

Throughout deposition of the formation, bed conditions appear to have been unstable. Slumps and other soft-sediment deformation features are found throughout the type section.

The relatively dark color of the Wahlstrom Hollow and its lack of an indigenous, abundant, and diverse calcareous benthic fauna suggest that sedimentation occurred under dysaerobic conditions. In modern euxinic basins dysaerobic conditions exist within the pycnocline, a zone approximately 100 m (328 ft) thick in which water density increases with depth. Byers (1975, 1977) suggested a depth of about 150 m (492 ft) as the base of the pycnocline. If his estimate is correct, water depths may have ranged from 50 to 150 m (164 to 492 ft) during Wahlstrom Hollow time. Most sedimentation probably occurred in the deeper end of this depth range, with a gradual shallowing taking place toward the end of Wahlstrom Hollow sedimentation as the influx of quartz silt increased and Third Fork sedimentation began. The fact that the only conodont element not found exclusively in submarine channels is a neogondolellid also supports a deep-water interpretation for the Wahlstrom Hollow. Clark (1974), in a study of Lower Permian rocks in Nevada, found that flat platform elements, such as neogondolellids, were associated with deeper water trace fossils than thin high-blade elements. He concluded that species with flat platform elements were pelagic, deeper water forms that lived farther from shore than those with thin high-blade elements. Wardlaw and Collinson (1979b) also recognized that *Neogondolella* species from middle and Upper Permian rocks tend to be distributed in offshore deposits.

Third Fork Formation

Poor exposure of this formation makes interpretation of the processes responsible for its deposition difficult. The base of this unit is gradational with the underlying Wahlstrom Hollow Formation, and the gradual increase in silt deposition through Wahlstrom Hollow time culminated with sedimentation of the Third Fork Formation. The influx of silt was probably accompanied by decreasing water depths, which appear to have been shallowest near the onset of Third Fork sedimentation. Near the base of the Third Fork, parallel-laminated-to-burrowed beds suggest that the sea floor was within reach of storm base (Howard, 1971).

The trace fossils present, which are mainly infaunal deposit-feeders, and the fine grain size imply that sedimentation took place in a relatively low energy environment (Rhoads, 1975). The thoroughly bioturbated texture of this unit above the basal storm-generated beds implies relatively slow sedimentation (Howard, 1975).

In the upper part of the Third Fork, a few thin, dark-gray to black spiculitic mudstone beds occur. They contain single-chambered microfossils that are commonly spinose and are probably pelagic forms. It seems likely that this facies accumulated through hemipelagic sedimentation, and it appears to have been a portent of Badger Gulch deposition.

The high diversity of infaunal deposit-feeders and the light color of the siltstone beds indicate that most of Third Fork sedimentation took place under aerobic conditions. The dark-colored mudstone beds appear to represent interludes of oxygen-deficient conditions. Water depths are difficult to estimate, but the presence of storm-generated beds near the base of the section and the indication of well-oxygenated waters imply that the shallowest depths were probably reached near the beginning of Third Fork deposition. Water depths at this time may have been less than 50 m (164 ft), using the estimate of Byers (1975, 1977) for the depth of the pycnocline. As sedimentation progressed, water depths apparently increased. The trace fossils present represent the *Zoophycos* and *Nereites* associations of Seilacher (1978). Although the *Zoophycos* association overlaps with other associations, Seilacher considered it representative of intermediate depths, whereas the *Nereites* association signifies deeper water. Near the end of Third Fork time, dysaerobic conditions occasionally prevailed, and dark-colored mudstones were deposited.

Badger Gulch Formation.

The black, thinly laminated, fine-grained rocks of the Badger Gulch Formation were deposited under basinal conditions. The sharp, even bedding, and the presence of scattered, probably pelagic microfossils (radiolarians?) and quartz silt suggest slow and uniform sedimentation. Rare crinoid columnals found on bedding planes may be the remains of epiplanktic forms (Heckel, 1972). The origin of angular discontinuities present within this unit is uncertain, but similar features have been reported from other deep-

water deposits and interpreted as channel fills or slumps (see Wilson, 1975; Cook and Enos, 1977).

A stratified water column with anaerobic bottom waters existed during Badger Gulch time and excluded a benthic invertebrate fauna from the Cassia Mountains area. However, overlying waters were sufficiently oxygenated to sustain a normal marine fauna as indicated by the sparse allochthonous crinoid debris, sponge spicules, and pelagic microfossils found within the unit. Sedimentation of the Badger Gulch took place below the pycnocline, and water depths were probably in excess of 150 m (492 ft).

A return to aerobic conditions commenced with sedimentation transitional with the Trapper Creek Formation, as indicated by an abundant calcareous fauna and bioturbation of the sediment.

Trapper Creek Formation

The succession of strata of the Trapper Creek Formation represents a fairly rapid shallowing of marine waters following the deposition of the Badger Gulch sequence. The arenaceous carbonates and diverse fauna of the Trapper Creek imply well-oxygenated waters in an aerobic environment in which water depths would be less than 50 m (Morgan, 1980b). The faunal assemblage is typically normal marine, and specific organisms, notably algal structures, strongly imply shallow water, especially in the upper part of the unit.

The lower part of the Trapper Creek Formation contains more wackestone and lime-mud and is more crinoidal than the upper part. Therefore, the lower part indicates a transition from the deep water lime-mudstones of the Badger Gulch to the shallower water deposits of the upper Trapper Creek in which hummocky cross-stratification, indicative of storm-generated waves, and packstones consisting of benthic organisms are more common.

Grandeur Tongue

The sandy dolomites and cherts of the Grandeur Tongue characterize continued shallow-water sedimentation. The increase of chert and the decrease in terrigenous clastics near the top of the unit suggest deposition farther from shore. The unit is fossiliferous, commonly with only broken fragments, indicating a high-energy environment of deposition.

Meade Peak Phosphatic Shale Member.

Phosphorite marks the base of the Meade Peak and is indicative of high biologic activity caused by nutrient-rich upwelling currents sweeping across a shallow shelf (Sheldon, 1980; Wardlaw, 1980). Phosphorite is apparently absent throughout the majority of the Meade Peak, which is made up mostly of siltstone and fine-grained sandstone. Most of Meade Peak sedimentation, therefore, was characterized by a relatively large clastic influx that prevented further phosphorite formation. This implies that the Cassia Mountains area may have been near the edge of phosphorite deposition and close to a source of clastics during Meade Peak time.

Rex Chert Member

The Rex is characterized by well-banded medium- to dark-gray cherts with some spicules preserved. These rocks probably represent a continuation of shelf sedimentation from Meade Peak time (Wardlaw, 1980), but with a complete curtailment of the supply of clastics. The environment was probably aerobic to dysaerobic as evidenced by the sponge spicules and sparse skeletal debris in the cherts.

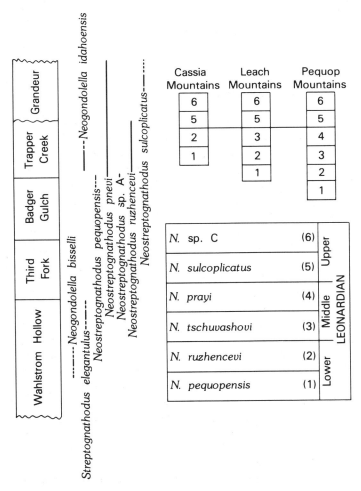

Figure 7. Range of conodonts in Cassia Mountains, Idaho; and, zonation of *Neostreptognathodus* species in Cassia Mountains, and Leach and Pequop Mountains, Nevada; localities shown on Figure 8.

BIOSTRATIGRAPHY

Through the course of this study, it became apparent that a useful conodont zonation based on *Neostreptognathodus* species existed for the Leonardian (Fig. 7). The upper three zones are well established (Wardlaw and Collinson, 1978, 1979a) and occur throughout the Great Basin. Zone 1 (*N. pequopensis*) has also been established (Behnken, 1975). *N. ruzhencevi* and *N. tschuvashovi* were described by Kozur and Mostler (1976) and are indicative of Kozur's lower subzone (Kozur, 1978) of the *N. pequopensis* Zone. We find no overlap of the ranges of *N. ruzhencevi* and *N. tschuvashovi* in our sections. *N. tschuvashovi* appears to be the predecessor to *N. prayi* and fits well into the sequential zonation of species.

Much of the problem of zonation schemes involving species of *Neostreptognathodus* is the confusion by several authors of many of the species. Wardlaw and Collinson (1979a) attempted to clarify the distribution of one species (*N. sulcoplicatus*). Wardlaw (in this paper) has illustrated and briefly described the conodont species used for this report to clarify their designations.

The stratigraphic intervals representing the regional zones of *N ruzhencevi* and *N. tschuvashovi* have not been studied in great detail and are commonly represented by marginal-marine sequences that do not yield significant conodont faunas. *N. ruzhencevi* and *N. tschuvashovi* have been found only in the same sections in the Pequop Mountains and Leach Mountains. For the purpose of this study, however, valuable information is gained by using these regional zones to demonstrate a hiatus in the section. Additional studies are needed to demonstrate the full extent and worth of these zones.

Age of the Permian Units

The lowest unit in the Cassia Mountains, the Wahlstrom Hollow Formation, is Wolfcampian. *Neogondolella bisselli*, diagnostic of the late Wolfcampian, occurs near the middle of the formation (Fig. 7; Morgan, 1980b). *Streptognathodus*, which is not known from rocks younger than Wolfcampian, occurs near the top of the formation, indicating that most of the Wahlstrom Hollow Formation is late Wolfcampian in age and probably not younger. Fusulinid faunas (Morgan, 1980b) also indicate a Wolfcampian age for the formation.

The Third Fork Formation and Badger Gulch Formation are poorly fossiliferous and difficult to date directly. The base of the overlying Trapper Creek Formation contains *N. pequopensis,* which occurs in the lowermost Leonardian and is diagnostic of the lowest zone of the Leonardian (Fig. 7). The Wolfcampian-Leonardian boundary, therefore, lies within the Third Fork-Badger Gulch sequence. Strictly for convenience, we have drawn the boundary at the Third Fork-Badger Gulch contact.

The Trapper Creek Formation is Leonardian and contains conodont faunas that represent the lower two zones (lower Leonardian, Zones 1 and 2) of the Leonardian (Fig. 7). Brachiopods, notably *Penniculauris* faunas, are common and also are Leonardian in age.

The Grandeur Tongue of the Park City Formation contains conodont faunas diagnostic of the *N. sulcoplicatus* Zone (Zone 5) of the lower upper Leonardian.

Faunas were not recovered from the Meade Peak or the Rex Chert Members of the Phosphoria Formation. Wardlaw and others (1979, 1980) have correlated these units with the Phosphoria throughout southeastern Idaho, northeastern Nevada, and northern Utah, and show the Meade Peake to be Leonardian (Zones 5 and 6) and the Rex to be Guadalupian (Wordian).

Regional Biostratigraphic Relations

Many of the units described for the Cassia Mountains have been correlated with units in the Leach Mountains and surrounding areas in Nevada (S. T. Miller, oral communication, 1981). These strata and those of the age-equivalent Pequop Formation of Steel (1960) in the central Pequop Mountains, Nevada, have been sampled for conodonts. In the Pequop Mountains, the Pequop Formation (of Steele, 1960), the Kaibab Limestone, and the basal part of the Plymptom Formation represent Leonardian strata; in the Leach Mountains, the Trapper Creek Formation, the Grandeur(?) Formation, and the Meade Peak Tongue represent the Leonardian strata (S. T. Miller, oral communication, 1981).

Central Pequop Mountains, Nevada. All regional zones based on *Neostreptognathodus* species (Fig. 7) appear to be present in the Leonardian sequence which includes the Pequop Formation (of Steele, 1960), the Kaibab Limestone, and the basal part of the Plympton Formation. Therefore, no major hiatus seems to exist in the Leonardian strata.

Leach Mountains, Nevada. Strata below the Trapper Creek Formation are structurally complex and rarely exposed, but are considered representative of the *N. pequopensis* Zone, and in part, equivalent to the rocks in the Cassia Mountains (S. T. Miller, oral communication, 1981). *N. ruzhencevi* is recovered from the upper two-thirds of the Trapper Creek Formation, much like it is in the Cassia Mountains. It is also found in the lower part of the Grandeur(?) Formation (of S. T. Miller, W. T. Fedewa, and S. Martindale, unpublished mapping, 1981). *N. tschuvashovi* occurs in the upper part of the Grandeur(?). *N. sulcoplicatus* occurs in the upper 15 m of that formation (Kaibab Limestone of Wardlaw and others, 1979, 1980) and in the overlying Meade Peak (mapped as a member of

the Murdock Mountain Formation by S. T. Miller, W. T. Fedewa, and S. Martindale, unpublished mapping, 1981). *N.* sp. C occurs throughout most of the Meade Peak. *N. prayi* faunas, common to the lower Kaibab Limestone in the Pequop Mountains, are completely missing.

Cassia Mountains. *N. pequopensis* occurs in the base of the Trapper Creek Formation. *N. ruzhencevi* occurs throughout the upper three-fourths of the formation. *N. sulcoplicatus* first occurs near the middle of the Grandeur. Meade Peak deposition appears to represent at least part of Zones 5 and 6 (Fig. 7; Wardlaw and others, 1980). *N. tschuvashovi* and *N. prayi* are missing, but a small barren interval at the base of the Grandeur may represent deposition during part of the *N. prayi* Zone.

From the Pequop Mountains northward to the Cassia Mountains, it appears that progressively more time is represented by the contact between the Kaibab-Grandeur and the strata below, listed in descending order, Grandeur(?) of S. T. Miller and others (unpublished mapping, 1981), Trapper Creek, and Pequop. It is apparent that the Trapper Creek-Grandeur contact marks the major gap in sedimentation within the Permian sequence in the Cassia Mountains.

REGIONAL SYNTHESIS

The Permian rocks of the Cassia Mountains reflect development of three depocenters through Early Permian time—the Oquirrh basin in late Wolfcampian, the Cassia basin in early Leonardian, and the Sublett basin in late Leonardian-early Guadalupian. Development and abandonment of the three "basins" show an active and complex Permian tectonic history. Post-Permian structural displacement (i.e., thrusting) of the Permian rocks of southern Idaho and adjacent Nevada and Utah does not significantly alter the facies relationships of these rocks and, therefore, is not dealt with in this paper.

Jordan and Douglass (1980) and Stevens (1977, 1979) included the Lower Permian rocks of the Cassia Mountains as part of the depositional sequence in the northern part of the Oquirrh basin (Fig. 8). The siltstones and limestones of the Wahlstrom Hollow and Third Fork Formations are similar to the upper Wolfcampian-Leonardian part of the Oquirrh Group in northwestern Utah (Jordan and Douglass, 1980). However, the ubiquitous sandstones common to the Oquirrh Group are missing in the Cassia Mountains. It is interesting, therefore, to compare the depositional history of the Oquirrh basin in Utah and Idaho to that of the Cassia Mountains area.

Oquirrh Basin. In early to middle Wolfcampian time a poorly oxygenated, deep-marine basin developed in northwest Utah in the Oquirrh basin (Jordan and Douglass, 1980). In late Wolfcampian-early Leonardian time, the basin shallowed and experienced thin-bedded siltstone dep-

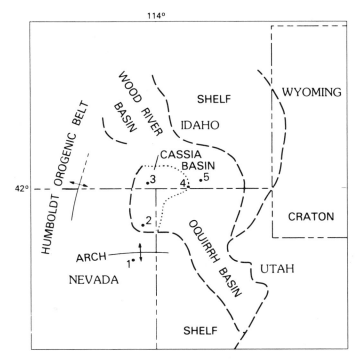

Figure 8. Paleogeographic map of northeastern part of eastern Great Basin showing early Wolfcampian-early Leonardian tectonic elements. Cassia basin extension shown by dotted lines. Numbered localities are: 1, Pequop Mountains; 2, Leach Mountains; 3, Cassia Mountains; 4, Black Pine Mountains; 5, Sublett Range.

osition. Commonly, the youngest beds of the Oquirrh Group are earliest Leonardian. These beds are overlain by the Grandeur Formation of late Leonardian age, and a probable hiatus is represented in the contact. The Pennsylvanian-Permian sequence of Yancey and others (1980) in the Sublett Range, Idaho (Fig. 8), is similar to the Oquirrh Group in Utah, especially in that the Oquirrh is overlain by Grandeur (Yancey, oral communication, 1980; post-Oquirrh of Yancey and others, 1980). The closest outcrops of Permian rocks in southeastern Idaho to the Cassia Mountains are in the Black Pine Mountains. The "southern block" of the Black Pine Mountains contains Oquirrh and no younger rocks (Smith, 1982). The "northern block" contains Oquirrh and younger equivalents that are of uncertain affinities (Smith, 1982), but may be related to the sequence in the Cassia Mountains.

Cassia Mountains. In late Wolfcampian time, thin-bedded silstones were deposited (upper part of the Wahlstrom Hollow Formation, Third Fork Formation) that were part of a shallowing sequence (Fig. 9; Morgan, 1980b). The sequence is the same as that found in the Oquirrh basin and strongly suggests that the Wolfcampian rocks in the Cassia Mountains were deposited as part of the same basin. Jordan and Douglass (1980) have described the Oquirrh basin as a northwest-trending structural basin that

is in part fault bounded and forming a graben. It appears that late Wolfcampian deposition in the Cassia Mountains represents deposition under similar conditions to that of the Oquirrh and possibly that it is the northern continuation of this graben. Soft-sediment slumporientations suggest that the strike of the paleoslope was northeast-southwest (Morgan, 1980b).

Latest Wolfcampian-earliest Leonardian deposition, however, differs markedly from the depositional sequence of the Oquirrh basin. The Cassia basin (Morgan, 1980b) began to develop during Third Fork time and sedimentation occurred in deeper water as evidenced by the episodic deposition of dark spiculitic calcareous mudstones. The basin was fully developed and euxinic during Badger Gulch time (earliest Leonardian). It then shallowed, and sandy limestone and calcareous sandstone were deposited. Oquirrh sedimentation ended with deposition of siltstone in northwestern Utah (Jordan and Douglass, 1980); nondeposition is probably represented by the Oquirrh-Grandeur contact. No lithologic equivalents to the Badger Gulch and Trapper Creek Formations are found in the Leonardian deposits of the Oquirrh basin in northern Utah. This strongly suggests that at earliest Leonardian time the Cassia basin (Fig. 8) became a separate entity from the Oquirrh basin. As the Oquirrh basin continued to shallow and fill, the Cassia basin experienced continued marine sedimentation dramatically marked by the development of basinal deposits of the Badger Gulch Formation. The Trapper Creek probably represents deposition younger than the youngest Oquirrh beds. The hiatus (Fig. 7) between the Trapper Creek Formation and the Grandeur Tongue marks

AGE	UNIT	SEDIMENTATION	
		TRANS-GRESSIVE	REGRESSIVE
GUADA-LUPIAN	REX		
LEONAR-DIAN	MEADE PEAK		
	GRANDEUR		
	TRAPPER CREEK		
	BADGER GULCH		
WOLFCAMP-IAN	THIRD FORK		
	WAHLSTROM HOLLOW		

Figure 9. Transgressive-regressive relations of Permian units in Cassia Mountains, Idaho.

nondeposition in the middle Leonardian in the Cassia Mountains area.

The presence of both the Badger Gulch and Trapper Creek Formations in the Leach Mountains (S. T. Miller, oral communication, 1981) implies that this area experienced a sedimentary history similar to, if perhaps shallower, than the Cassia Mountains area. It also appears that the Trapper Creek Formation is present approximately 24 km (15 mi) southeast of the Leach Mountains in the northern Pilot Range of Utah and Nevada (D. Miller, oral communication, 1981).

The late Leonardian-early Guadalupian Park City-Phosphoria sequence represents shelf deposition in the Sublett Basin (Fig. 10) of Maughan (1979; 1980). Phosphorites of the Meade Peak are poorly developed in the Cassia Mountains, and the unit is dominated by clastic rocks coarser than the common phosphatic mudstones found elsewhere in southeastern Idaho. This coarser texture implies closer proximity to a terrigenous source in contrast to the dominantly fine-grained rocks east of the Cassias. The configuration of the Sublett basin suggests that this source was west and southwest of the Cassia Mountains and possibly originated by rejuvenation of the Humboldt orogenic belt. Deposition of the Rex Chert Member marks a termination of this large clastic influx. A slight deepening of the shelf could have cut off the clastic supply.

A major hiatus, represented by the contact between the Rex Chert and the Triassic Dinwoody Formation, includes most of Late Permian time. This contact is rarely exposed in the Cassia Mountains, but appears to be very regular. Dinwoody deposition appears to conform to the Sublett basin (Collinson and Hasenmueller, 1977), which indicates the basin's continued existence in the Triassic.

Thermal alteration of Permian rocks. The thermal maturity of Permian rocks as assessed by conodont color alteration indices (Fig. 11; Epstein and others, 1977) shows no definite pattern. The regional (burial) alteration index appears to be 1.5 to 2.0, which is similiar to values found in northeastern Nevada for Permian rocks (Wardlaw and others, 1980, Fig. 14). Locally, the values are 3 and 4, and were probably elevated by the presence of thick Tertiary volcanics. Maughan (1979) reported thermal maturation values, determined by pyrolitic analysis, equivalent to CAI values of 4, for two localities in this area.

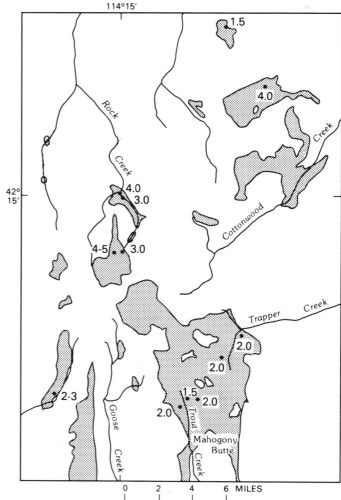

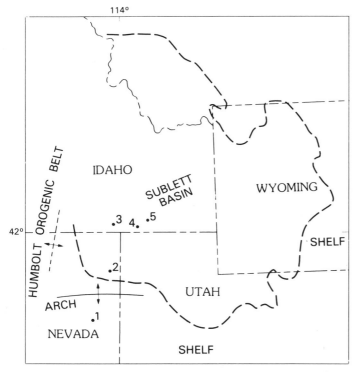

Figure 10. Paleogeographic map of northeastern part of eastern Great Basin showing late Leonardian-early Guadalupian tectonic elements. Numbered localities are the same as Figure 8.

Figure 11. Generalized outcrop pattern of Permian rocks of the Cassia Mountains showing sample localities and CAI values.

CONODONT PALEONTOLOGY

Bruce R. Wardlaw

The conodont faunas recovered from the Lower Permian rocks of the Cassia Mountains and the Leach Mountains have not been well documented in North America. However, many elements of the faunas are known from Eurasia (Kozur, 1978). To better understand these faunas, we briefly describe and illustrate them here.

The majority of the conodonts are platformed elements of *Neostreptognathodus*. These elements can be divided into two types: 1) symmetrical, those with the posterior termination of the free blade occurring at the median, not uniting with either carina (*N. pequopensis, N. pnevi, N. sulcoplicatus*); or 2) lateral, those with the posterior termination of the free blade occurring to the side, uniting with one of the carinae (*N. ruzhencevi, N. tschuvaschovi, N.* sp. A).

Neogondolella bisselli (Clark and Behnken)
Figure 12: 26
Gondolella bisselli Clark and Behnken, 1971, p. 421, Pl. 1, figs. 12-14; Kozur, 1978, Pl. 2, figs. 16a-c, Pl. 3, figs. 6, 7, 10; Kozur and Movschovitsch, 1979, Pl. 1, fig. 11, Pl. 2, figs. 1a-c.
Neogondolella bisselli (Clark and Behnken), Behnken, 1975, p. 306, Pl. 1, figs. 27, 31; Clark and others, 1979, Pl. 1, fig. 21.
Remarks: This species is characterized by a rounded cusp forming the posterior-most portion of the element, shallow, narrow, and poorly developed lateral furrows, rounded posterior "shoulders" of the platform margin and a platform that is widest on the posterior half and gradually narrows anteriorly extending for the entire length of the element; the margins are not greatly upturned.

Neogondolella idahoensis
(Youngquist, Hawley, and Miller)
Figure 12: 6-9
Gondolella idahoensis Youngquist, Hawley, and Miller, 1951, p. 361, Pl. 54, figs. 1-3, 14, 15; Clark and Ethington, 1962, p. 108, Pl. 2, figs. 15, 16; Clark and Mosher, 1966, p. 388, Pl. 47, figs. 9-12; Clark and Behnken, 1971, p. 431, Pl. 1, fig. 9.
Gondolella phosphoriensis Youngquist, Hawley, and Miller, 1951 (part), p. 362, Pl. 54, figs. 27-28.
Neogondolella idahoensis (Youngquist, Hawley, and Miller) Behnken, 1975, p. 306-307, Pl. 1, figs. 28-30.
Remarks: This species is characterized by its prominent cusp forming the posterior-most portion of the element, deep and narrow lateral furrows, and rounded posterior "shoulders" of the platform margins. The illustrated specimens are fairly typical of older examples of the species and

probably could be separated from younger examples of *N. idahoensis* from the Meade Peak Phosphatic Shale Member of the Phosphoria Formation largely on platform configuration. More material is presently deemed necessary to demonstrate this inference.

Neostreptognathodus pequopensis Behnken
Figure 12: 1
Neostreptognathodus pequopensis Behnken, 1975, p. 310, Pl. 1, figs. 19-22, 25; Sweet 1977, p. 239-240, Pl. 1, fig. 2; Kozur, 1978, Pl. 4, figs. 6, 9-12; Kozur and Movschovitsch, 1979, Pl. 2, figs. 7a-b; Clark and others, 1979, Pl. 1, fig. 18.
Remarks: This species is characterized by two carinae that bear nodes for their entire length; the carinae are separated by a very narrow furrow. *N. pequopensis* is "symmetrical" in that the free blade terminates at or near the median and does not unite with a carina.

Neostreptognathodus pnevi Kozur and Movschovitsch
Figure 12: 2-3
Neostreptognathodus pnevi Kozur and Movschovitsch, 1979, p. 115-116, Pl. 3, figs. 1-4.
Neostreptognathodus sulcoplicatus (Youngquist, Hawley, and Miller), Behnken, 1975, p. 311-312, Pl. 1, figs. 1, 2, 4 (not fig. 3).
Neostreptognathodus sp. D of Wardlaw and Collinson (1978, 1979); Clark and others, 1979, Pl. 1, fig. 13.
Remarks: This species is very variable and long-ranging (refer to Wardlaw and Collinson, 1978). Generally, the anterior-most portion of the carinae is smooth or ornamented by very reduced nodes or ridges. The posterior ends in a single large node. *N. pnevi* is "symmetrical" in that the free blade terminates at or near the median and does not unite with a carina. *N. sulcoplicatus* differs by having only transverse ridges that are along the entire length of the carinae, irregularly spaced posterior-most ridges, and a median furrow that widens near the posterior end to form a sulcus. *N. pnevi* differs from *N. pequopensis* by having the anterior-most portion of the carinae smooth or with reduced nodes.

Neostreptognathodus ruzhencevi Kozur
Figure 12: 15-19
Neostreptognathodus ruzhencevi Kozur, 1976, p. 16-17, Pl. 2, figs. 5-7; Kozur, 1978; Pl. 4, figs. 3, 4, 7; Kozur and Movschovitsch, 1979, Pl. 2, figs. 5, 6.
Remarks: This species is characterized by two carinae that bear short transverse ridges that commonly are more node-like posteriorly, especially in smaller specimens; the carinae generally are smooth or bear reduced ridges on the anterior-most portion; generally, the carina that unites with the free blade has a longer portion that is smooth. The platform bears a well-developed median furrow. The posterior ends in a single transverse node, and the next 1 to

Figure 12. All specimens X60 unless noted otherwise.

1, *Neostreptognathodus pequopensis* Behnken, upper view, USNM 356781 (W81555, Badger Gulch section, Trapper Creek Formation).

2, 3, *Neostreptognathodus pnevi* Kozur and Movschovitsch, 2, upper view, USNM 356782, 3, upper view, USNM 356783 (Tb2Ba, Deadline Ridge, Trapper Creek Formation).

4, *Streptognathodus elegantulus* Stauffer and Plummer, oblique upper view, USNM 356784 (CAS52-101, Wahlstrom Hollow section, Wahlstrom Hollow Formation).

5, *Neostreptognathodus sulcoplicatus* (Youngquist, Hawley, and Miller), upper view, USNM (1CAS1315, Fall Creek section, Grandeur Tongue).

6–9, *Neogondolella idahoensis* (Youngquist, Hawley, and Miller), 6, lateral view, USNM 356786; 7, oblique upper view, USNM 356787; 8, upper view, USNM 356788; 9, oblique upper view, USNM 356789 (TB2Ba, Deadline Ridge, Trapper Creek Formation).

10–14, *Neostreptognathodus* sp. A, 10, upper view, USNM 356790; 11, upper view, USNM 356791; 12, upper view, USNM 356792; 13–14, upper view and enlarged view of anterior of platform X400, showing free blade uniting with dextral carina, USNM 356793 (W81557, Badger Gulch section, Trapper Creek Formation).

15–19, *Neostreptognathodus ruzhencevi* Kozur, 15, upper view, USNM 356794 (1CAS842, Fall Creek section, Trapper Creek Formation); 16, upper view, USNM 356795; 17, upper view, USNM 356796; 18, upper view, USNM 356797; 19, upper view, USNM 356798 (1CAS136, Fall Creek section, Trapper Creek Formation).

20–25, *Neostreptognathodus tschuvashovi* Kozur, 20, upper view, small specimen, USNM 356799; 21, upper view, USNM 356800; 22, upper view, USNM 356801; 23, upper view, USNM 356802; 24, upper view, USNM 356803; 25, upper view, USNM 356804 (1LM422, Murdock Mountain section, Grandeur? of S. T. Miller and others, oral communication, 1981.

26, *Neogondolella bisselli* (Clark and Behnken), oblique upper view, UW 1664/2 (CAS52-48, Wahlstrom Hollow section, Wahlstrom Hollow Formation, Morgan, 1980b).

All localities from measured sections of this report except: TB2Ba, Deadline Ridge, center NW 1/4 sec. 1, T.16S., R.18E., Timber Butte 7.5-minute quadrangle, Twin Falls County, Idaho; and 1LM422, east face of Murdock Mountain from section of Wardlaw and Collinson, 86 m below Meade Peak Phosphatic Shale, N 1/2 NW 1/4 NW 1/4 sec. 31, T.39N., R.68E., Loray 7.5-minute quadrangle, Elko County, Nevada.

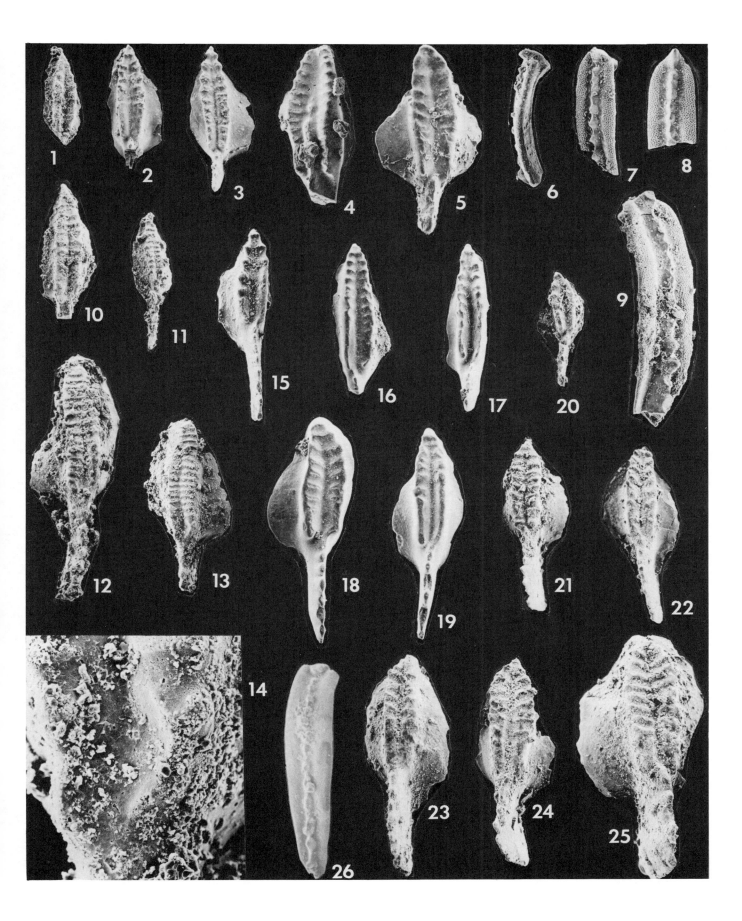

2 posterior ridges merge and cross the platform, closing the furrow, in all growth stages. *N. ruzhencevi* is "lateral" in that the free blade unites with one of the carinae, either sinistral or dextral. The carina opposite to the one uniting with the free blade ends sharply anteriorly, and the furrow curves and separates it from the free blade. *N. pnevi* can be easily separated in that it has a "symmetrical" posterior free blade termination.

Neostreptognathodus sulcoplicatus (Youngquist, Hawley, and Miller)
Figure 12: 5

Streptognathodus sulcoplicatus Youngquist, Hawley, and Miller, 1951, p. 363, Pl. 54, figs. 7-9, 16, 17, 22-24; Clark and Ethington, 1962, p. 111, Pl. 1, figs. 8, 9, 18, Pl. 2, figs. 3, 4, 7; Clark and Behnken, 1971, p. 430, Pl. 1, figs. 23, 24.
Neostreptognathodus sulcoplicatus (Youngquist, Hawley, and Miller), Clark, 1972, p. 155.
Remarks: This species is characterized by two widely separated carinae that bear transverse ridges for their entire length, and a wide furrow that widens farther posteriorly forming a sulcus. The transverse ridges are farther spaced from each other posteriorly. The posterior ends in a single transverse ridge or node. *N. sulcoplicatus* is "symmetrical" in that the free blade terminates at or near the median and does not unite with a carina.

Neostreptognathodus tschuvaschovi Kozur
Figure 12: 20-25

Neostreptognathodus tschuvaschovi Kozur, 1976, p. 17-18, Pl. 2, figs. 1-3; Kozur, 1978, Pl. 4, figs. 2, 8; Kozur and Movschovitsch, 1979, Pl. 2, figs. 2, 4.
Remarks: This species is characterized by two carinae that bear large transverse ridges for their entire length; the carinae are separated by a narrow furrow, and the ridges generally are laterally directed and become posteriorly directly nearing the posterior end. *N. tschuvaschovi* is "lateral" in that the free blade unites with one of the carinae, either sinistral or dextral.

Neostreptognathodus sp. A
Figure 12: 10-14

Remarks: This undescribed species is at present represented only by poorly preserved material from two sections of the lower part of the Trapper Creek Formation in the Cassia Mountains. It is characterized by two rather flattened carinae that bear thin but long transverse ridges along their entire length; the ridges merge posteriorly. The furrow is very narrow and is generally absent on the posterior one-fourth of the platform. *N.* sp. A is "lateral" in that the free blade unites with one of the carinae, either sinistral or dextral. The species appears to be a homeomorph for the genus *Idiognathoides*.

Streptognathodus elegantulus Stauffer and Plummer
Figure 12: 4

Streptognathodus elegantulus Stauffer and Plummer, 1932, p. 47-48, Pl. 4, figs. 6, 7, 22, 27; Ellison, 1941, p. 127, Pl. 22, figs. 1-6, 10; Branson, 1944, p. 327, Pl. 46, figs. 1-6, 10; Rhodes, 1952, p. 894, Pl. 127, figs. 3, 4, 8; Stone, 1959, p. 158, fig. 14; Jennings, 1959, p. 995, Pl. 124, fig. 6; Koike, 1967, p. 311, Pl. 3, figs. 13-15; Higgins and Boukaert, 1968, p. 46, Pl. 5, fig. 8, 10; Von Bitter, 1972, p. 52-53, Pl. 1, figs. 1a-e; Sweet, 1975, p. 367-368, Pl. 1, fig. 9.
Polygnathus pawhuskaensis Harris and Hollingsworth, 1933, p. 199, Pl. 1, figs. 12a, b.
Streptognathodus sulcatus Gunnell, 1933, p. 280, Pl. 32, fig. 10; Ellison, 1941, p. 130, Pl. 22, fig. 8; Branson, 1944, p. 327, Pl. 46, fig. 8.
Idiognathodus elegantulus (Stauffer and Plummer), Perlmutter, 1975, p. 99-100, Pl. 1, figs. 3-12.
Remarks: This species is characterized by its deep median trough and lack of accessory lobes. It ranges in North America from Desmoinesian (Pennsylvanian) to late Wolfcampian (Permian).

ACKNOWLEDGEMENTS

Edwin K. Maughan and William J. Mapel reviewed the manuscript and offered useful suggestions.

REFERENCES CITED

Behnken, F. H., 1975, Leonardian and Guadalupian (Permian) conodont biostratigraphy in western and southwestern United States: Journal of Paleontology, v. 49, p. 284–315.

Byers, C. W., 1975, Biofacies patterns in euxinic basins—general model: American Association of Petroleum Geologists/Society of Economic Paleontologists and Mineralogists Annual Meeting Abstracts, v. 2, p. 8.

——— 1977, Biofacies patterns in euxinic basins: a general model, *in* Cook, H. E., and Enos, P., eds., Deep-water carbonate environments: Society of Economic Paleontologists and Mineralogists Special Publication 25, p. 5–17.

Clark, D. L., 1974, Factors of Early Permian conodont paleoecology in Nevada: Journal of Paleontology, v. 48, p. 710–720.

Coats, R. R., 1980, The Roberts Mountains Thrust in central Twin Falls County, Idaho: Geological Society of America Abstracts with Programs, v. 12, no. 6, p. 270.

Collinson, J. W., and Hasenmueller, W. A., 1978, Early Triassic paleogeography and biostratigraphy of the Cordilleran miogeosyncline, *in* Howell, D. G., and McDougall, K. A., eds., Mesozoic paleogeography of the western United States: Pacific Section, Society of Economic Paleontologists and Mineralogists, Pacific Coast Paleogeography Symposium 2, p. 175–187.

Cook H. E., and Enos, P., eds., 1977, Deep-water carbonate environments: Society of Economic Paleontologists and Mineralogists Special Publication 25, 336 p.

Dunham, R. J., 1962, Classification of carbonate rocks according to depositional environments: American Association of Petroleum Geologists Memoir 1, p. 108–121.

Epstein, A. G., Epstein, J. B., and Harris, L. D., 1977, Conodont color alteration—an index to organic metamorphism: U.S. Geological Survey Professional Paper 995, 27 p.

Furnish, W. M., 1973, Permian stage names, *in* Logan, A., and Hills, L. V., eds., The Permian and Triassic Systems and their mutual boundary: Canadian Society Petroleum Geologists Memoir 2, p. 522–548.

Heckel, P. H., 1972, Recognition of ancient shallow marine environments, *in* Rigby, J. K., Hamblin, W. K., eds., Recognition of ancient sedimentary environments: Society of Economic Paleontologists and Mineralogists Special Publication 16, p. 226–287.

Howard, J. D., 1971, Comparison of the beach-to-offshore sequence in modern and ancient sediments, *in* Howrd, J. D., Valentine, J. W., and Warme, J. E., eds., Recent advances in paleoecology: American Geological Institute shortcourse lecture notes, p. 148–184.

——1975, The sedimentological significance of trace fossils, *in* Frey, R. W., ed., The study of trace fossils: New York, Springer-Verlag, p. 131–147.

Jordan, T. E., and Douglass, R. C., 1980, Paleogeography and structural development of the Late Pennsylvanian to Early Permian Oquirrh Basin, northwestern Utah, *in* Fouch, T. D., and Magathan, E. R., Paleozoic paleogeography of west-central United States: Rocky Mountain section Society of Economic Paleontologists and Mineralogists, West-central United States Paleogeography Symposium 1, p. 217–238.

Kozur, H., 1978, Beitrage zur Stratigraphie des Perms Teil II: Die Conodontenchronologie des Perms: Frieberger Forschungsheft, v. 334, p. 85–161.

Kozur, H., and Mostler, H., 1976, Neue Conodonten aus dem Jungpalaozoikum und der Trias: Geologisch-Palaontologische Mitteilungen Innsbruck, v. 6, p. 1–33.

Laudon, L. R., Thomasson, R., and Bancroft, G.A, 1956, The Cassia Mountains area, Idaho: Geologic reconnaissance northeastern Nevada and south-central Idaho, Phillips Petroleum Research Report, pt. 1, p. 125–135.

Maughan, E. K., 1979, Petroleum source rock evaluation of the Permian Park City Group in the northeastern Great Basin, Utah, Nevada and Idaho: Rocky Mountain Association of Geologists-Utah Geological Association, 1979 Basin and Range Symposium, p. 523–530.

——1980, Relation of phosphorite, organic carbon, and hydrocarbons in the Permian Phosphoria Formation, western United States of America: Bureau de Recherches Geologiques et Minieres Document no. 24, p. 63–91.

Morgan, W. A., 1977, Stratigraphy and paleoenvironments of three Permian units in the Cassia Mountains, central southern Idaho: [unpublished M.S. thesis]: Madison, University of Wisconsin, 135 p.

——1980a, Stratigraphy and paleoenvironments of three Permian units in the Cassia Mountains, central southern Idaho: Geological Society of America Abstracts with Programs, v. 12, no. 3, p. 143.

——1980b, Euxinic Early Permian sedimentation in the Cassia Basin of southern Idaho, *in* Fouch, T. D., and Magathan, E. R., eds., Paleozoic paleogeography of west-central United States: Rocky Mountain Section Society of Economic Paleontologists and Mineralogists, West-central United States Paleogeography Symposium 1, p. 305–326.

Piper, D.J.W., 1972, Turbidite origin of some laminated mudstones: Geological Magazine, v. 109, p. 115–126.

Rhoads, D. C., 1975, The paleoecological and environmental significance of trace fossils, *in* Frey, R. W., ed., The study of trace fossils, p. 147–160.

Seilacher, A. L., 1978, Use of trace fossils for recognizing depositional environments, *in* Basan, P. B., ed., Trace fossil concepts: Society of Economic Paleontologists and Mineralogists Short Course no. 5, p. 167–181.

Sheldon, R. P., 1980, Episodicity of phosphate deposition and deep ocean circulation—a hypothesis: Society of Economic Paleontologists and Mineralogists Special Publication 29, 239–247.

Smith, J. F., 1982, Geologic map of the Strevell 15-minute quadrangle, Cassia County, Idaho: U.S. Geological Survey Miscellaneous Investigations Map I 1403.

Steele, G., 1960, Pennsylvanian-Permian stratigraphy of east-central Nevada and adjacent Utah, *in* Geology of east-central Nevada: Intermountain Association of Petroleum Geologists, 11th Annual Field Conference Guidebook, p. 91–113.

Stevens, C. H., 1977, Permian depositional provinces and tectonics, western United States, *in* Stewart, J. H., Stevens, C. H., and Fritsche, A. E., eds., Paleozoic paleogeography of the western United States: Pacific Coast Section Society of Economic Paleontologists and Mineralogists, Pacific Coast Symposium I, p. 113–135.

——1979, Lower Permian of the central Cordilleran Miogeosyncline: Geological Society of America Bulletin, pt. II, v. 90, p. 381–455.

Wardlaw, B. R., 1980, Middle-Late Permian paleogeography of Idaho, Montana, Nevada, Utah, and Wyoming, *in* Fouch, T. D., and Magathan, E. R., eds., Paleozoic paleogeography of west central United States: Rocky Mountain Section Society of Economic Paleontologists and Mineralogists, West-central United States Paleogeography Symposium I, p. 353–361.

Wardlaw, B. R., and Collinson, J. W., 1978, Stratigraphic relations of Park City Group (Permian) in eastern Nevada and western Utah: American Association of Petroleum Geologists Bulletin, v. 62, p. 1171–1184.

——1979a, Biostratigraphic zonation of the Park City Group, *in* Wardlaw, B. R., ed., Studies of the Permian Phosphoria Formation and related rocks, Great Basin-Rocky Mountain region: U.S. Geological Survey Professional Paper 1163D, p. 17–22.

——1979b, Youngest Permian conodont faunas from the Great Basin and Rocky Mountain regions, *in* Sandberg, C. A., and Clark, D. L., eds., Conodont biostratigraphy of the Great Basin and Rocky Mountains: Brigham Young University Geology Studies, v. 26, part 3, p. 152–164.

Wardlaw, B. R., Collinson, J. W., and Maughan, E. K., 1979, Stratigraphy of Park City Group equivalents (Permian) in southern Idaho, northeastern Nevada, and northwestern Utah, *in* Wardlaw, B. R., ed., Studies of the Permian Phosphoria Formation and related rocks, Great Basin-Rocky Mountain region: U.S. Geological Survey Professional Paper 1163C, p. 9–16.

Wardlaw, B. R., Collinson, J. W., and Ketner, K. B., 1980, Regional relations of middle Permian rocks in Idaho, Nevada, and Utah: Rocky Mountain Association of Geologists-Utah Geological Association, 1979 Basin and Range Symposium, p. 277–283.

Wilson, J. L., 1975, Carbonate facies in geologic history: New York, Springer-Verlag, 471 p.

Yancy, T. E., Ishibashi, G. D., and Bingman, P. T., 1980, Carboniferous and Permian stratigraphy of the Sublett Range, south-central Idaho, *in* Fouch, T. D., and Magathan, E. R., eds., Paleozoic paleogeography of west-central United States: Rocky Mountain Section Society of Economic Paleontologists and Mineralogists, West-central United States: Paleogeography Symposium I, p. 259–269.

Youngquist, W., and Haegle, J. R., 1956, Geological reconnaissance of the Cassia Mountain region, Twin Falls and Cassia Counties, Idaho: Idaho Bureau of Mines and Geology Pamphlet 110.

MANUSCRIPT ACCEPTED BY THE SOCIETY AUGUST 20, 1982

Geological Society of America
Memoir 157
1983

Overlapping overthrust belts of
late Paleozoic and Mesozoic ages,
Northern Elko County, Nevada

Robert R. Coats
U.S. Geological Survey
345 Middlefield Road
Menlo Park, California 94025

John F. Riva
Department of Geology
Laval University,
Quebec, Québec GIK 7P4
Canada

ABSTRACT

For several decades it has been recognized that the extent to the east of lower Paleozoic eugeosynclinal rocks (Fig. 1) is greater in northern Elko County than elsewhere in Nevada. It was early recognized that these rocks are allochthonous; in some places it was demonstrable that they form the upper plate of the Roberts Mountains thrust and rest on eastern-facies (carbonate-clastic) lower Paleozoic rocks. Recent detailed mapping has shown that the Roberts Mountains thrust is not deflected to the east in central Elko County, as had been earlier supposed, but continues with a generally northward trend through the Mountain City quadrangle. Western-facies and overlap-assemblage rocks present to the east are the result of younger thrusting southward and eastward. Ages of younger thrusting include late Paleozoic, early Mesozoic, and post-Early Jurassic. These observations render superfluous explanations that require a deformation of the Roberts Mountains thrust to bring it to a northeasterly trend through north-central Elko County. We propose to account for the presence of the Roberts Mountains thrust in southern Twin Falls County, Idaho, by the presence of an east-trending right-lateral rift. This rift is now buried beneath Tertiary lava, with about 120 km of post-Devonian displacement, in southern Owyhee County, Idaho. Subsequent to the right-lateral movement, we propose that the Idaho block was translated still farther east and rotated counterclockwise about 30°. The compression in northern Elko County resulting from this rotation may be responsible for the east-west trends of the formations just south of the Idaho line, which condition the east-west trend of Mesozoic plutons in northern Elko County. Since these plutons are as old as Jurassic, the tectonism responsible may be Late Triassic and Early Jurassic. It is also responsible for the presence of western-facies and overlap-assemblage Paleozoic rocks thrust over eastern-facies and overlap-assemblage rocks in parts of northern Elko County in post-Antler time.

305

INTRODUCTION

In the cordillera of western North America, the time span from the end of the Devonian to the Late Cretaceous has been one of great, but episodic tectonic activity. The northeastern part of Elko County, the northeasternmost county of Nevada (Figs. 1, 2), is one place where the records of successive orogenies are sufficiently spread out to permit some segregation of the effects of different orogenies.

The purpose of this paper is to describe certain localities critical for an interpretation of the pre-Cenozoic tectonic history of northeastern Elko County and vicinity, and to offer explanatory hypotheses relating these exposures to each other and to the regional structure.

This is a region underlain by pre-Jurassic stratified rocks as old as Late Proterozoic and representing every system of the Paleozoic. The major sedimentary history ended in the Early Triassic. The Jurassic and Cretaceous were periods of plutonism; the distribution of the Mesozoic plutons is also significant in understanding the tectonic history. The diverse depositional environments and provenances of the pre-Jurassic rocks produced distinct facies differences between stratigraphic units. Some units accumulated close to one another, and others have been juxta-

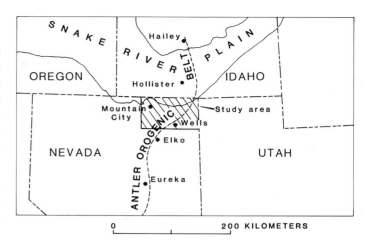

Figure 1. Map after Roberts and others (1965) showing approximate eastern limit of allochthonous western-facies lower Paleozoic rocks which Roberts and others interpreted as the eastern limit of the Antler orogenic belt in Late Devonian and Early Mississippian time. Newer data challenge their concept.

posed by tectonic disturbances, some of great magnitude, occurring at various times from the late Paleozoic to the late Mesozoic. The directions of movement of thrust plates range from northeast to east to southwest.

Much of the pre-Tertiary stratigraphic record, and

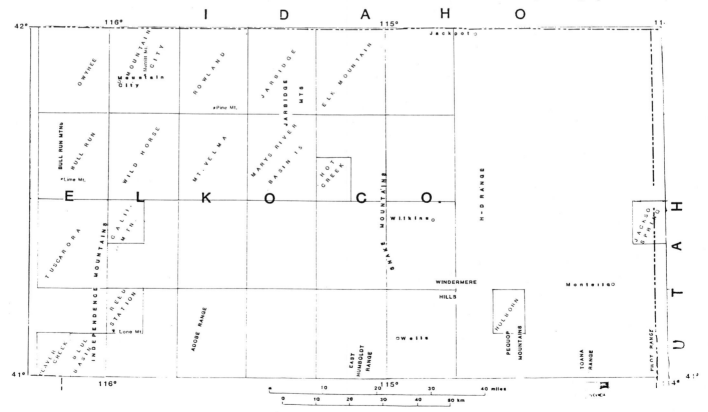

Figure 2. Index map of northern Elko County, Nevada, showing locations of some geographic features discussed, and some U.S. Geological Survey quadrangles.

thus the tectonic history that can be deduced from the relations of the pre-Tertiary rocks, is concealed by a mantle of Cenozoic volcanic and sedimentary rocks. For some units, few paleontological data are available. Extrapolations from reconnaissance of the scanty and widely scattered exposures have, in the past, resulted in the formulation of several hypotheses to explain the tectonic history of this area. These hypotheses are not all compatible with each other or with our observations in certain areas where critical outcrops are present.

In this paper we offer evidence and a new hypothesis to explain why western facies (eugeosynclinal) lower Paleozoic rocks occur farther east in this region than in the rest of Nevada (Fig. 1). To set the stage, we first briefly review the major pre-Tertiary orogenic episodes of the region and then previous explanations for the distribution of western-facies rocks. Next we describe localities critical to the interpretation of movement direction and ages of major thrusts: the east-directed Roberts Mountains thrust of presumed middle Paleozoic age, younger eastward-directed thrusts, southward-directed thrusts, and a NW-SE strike-slip fault. We finally present a regional tectonic model consistent with our observations and with geophysical and geological data from Idaho, according to which the Roberts Mountains thrust was translated eastward about 100 km in early Mesozoic time by a right-lateral rift, now beneath the volcanic rocks of the Snake River Plain. Continued eastward movement, combined with counterclockwise rotation of the Idaho block is thought to have provided the impetus for east-trending thrusts, metamorphism, and plutons in the northern 50 km of Elko County.

ANTLER, SONOMA, AND LATE MESOZOIC OROGENIES

The three major pre-Tertiary orogenies that are the most significant events in the long history of orogenic unrest of this area (Nolan, 1974) are the Antler orogeny of middle Paleozoic age, the Sonoma orogeny of Permian age, and the Mesozoic orogeny. A review of these is useful to put in context our observations that follow.

The Antler orogeny was named by Roberts (1951) for events that explain relations in the Antler Peak quadrangle of central Nevada. The regional aspects of the Antler were summarized in 1979 by P. B. King (*in* Nilsen and Stewart, 1980):

The Antler orogenic belt (sensu stricto) is a belt of rocks orogenically deformed in mid-Paleozoic time that extends north-north-eastward across central Nevada . . . It is composed of eugeosynclinal (ocean floor?) Devonian and older Paleozoic rocks that were deformed and thrust eastward over the adjacent outer shelf rocks, mainly (possibly not wholly) along the Roberts Mountains Thrust.

By far the greater part of the pre-Jurassic rocks in the region are allochthonous. The most widely recognized of the sequences of allochthonous rocks is that transported long distances eastward by the Roberts Mountains thrust, a sequence generally called the western or siliceous-volcanic facies. It includes the Valmy and the Vinini Formations and related rocks ranging in age from Early Ordovician to Early Devonian. Many workers have recognized in this area an eastern facies sometimes known as the carbonate facies (mostly autochthonous), and locally, a transitional facies, which tends to be richer in fine-grained clastics than the eastern-facies rocks, but poorer in chert and quartzite than the western-facies rocks. In some areas, gradations between eastern and transitional facies, locally oscillatory, have been recognized within a single formation in a single quadrangle (Michael Taylor, oral communication, 1976). The terms overlap facies, overlap assemblage, overlap succession, and postorogenic sequence have been used for the rocks that are as young (in northern Elko County) as Early Triassic and that generally rest unconformably on rocks of the eastern, transitional, and western facies.

The Sonoma orogeny of north-central Nevada was named by Silberling and Roberts (1962). Silberling (1973) refined the dating of the Sonoma orogeny and summarized his results as follows (in part):

(1) Upper Paleozoic oceanic deposits of the Havallah sequence have been carried eastward on the Golconda thrust for at least several tens of miles over partly correlative rocks of the continental shelf along an arcuate belt, a 200 mi length of which bounds northwestern Nevada from the rest of the state.

(2) This thrusting took place during latest Permian and Early Triassic time, its age being younger toward the south end of the orogenic belt in Nevada.

(3) The 'Havallah sequence' was accumulated in a deep seafloor environment, perhaps on the continental rise and slope, and at least some of these rocks may have been deposited directly on oceanic crust.

A number of east-moving thrusts in northern Elko County have been attributed to the Sonoma orogeny (Miller and others, 1981; Coats and Gordon, 1972), but no particular thrust can be identified with the Golconda thrust because of the differences in the geology of the lower plate from place to place.

The Mesozoic orogeny in northern Elko County is limited, so far as is known, to activity in Early Triassic to Early Cretaceous time. It probably includes folding and thrusting that might be correlated with the Nevadan orogeny as recognized in the Sierra Nevada and (or) with the earlier part of the Sevier orogeny of Utah (Armstrong 1972; Roberts and Crittenden, 1973). In northern Elko County, a broad zone of imbricate southward thrusting coincides in trend with and may have controlled the emplacement of Jurassic and Cretaceous plutons (Coats and others, 1965; Coats and McKee, 1972). Much of the dating of this tectonic episode depends on the dating of the plutons. The trend, though long recognized (Nolan, 1943; Bushnell, 1965; Coats and others, 1977; Coats, 1971; Coats and others, 1977), has not been adequately explained; one purpose of this paper is to offer an explanatory hypothesis.

DISTRIBUTION OF WESTERN-FACIES ROCKS IN NORTHERN ELKO COUNTY

Western-facies lower Paleozoic rocks occur farther east in northern Elko County than in the rest of Nevada (Fig. 1). This observation has prompted several explanations, all based on the common assumption that the distribution of these allochthonous rocks was an effect of the Antler orogeny in Devonian and (or) Mississippian time. The earliest explanation, offered by Roberts and others (1965), simply assumed an initial irregularity in the trend of the thrust (Fig. 1). A second hypothesis was that of an oroflexural feature (Stewart and Poole, 1974) that deformed an initially straight trend of the Antler orogenic belt. A third hypothesis was that a west-northwest-trending right-lateral strike-slip fault, the "Wells fault" (Thorman, 1968, 1970; Thorman and Ketner, 1979), has displaced the western-facies rocks and the Roberts Mountains thrust by 65 km in northern Elko County.

Our new data that demonstrate a generally northerly trend of the Roberts Mountains thrust in northern Elko County, coupled with the observations that the western-facies rocks in eastern Elko County are generally in the upper plates of southward-moving thrusts, or of thrusts that bring the western facies above post-Mississippian rocks, or both, make untenable the hypothesis of a northeast trend for the Antler orogenic belt across north-central Elko County. Those occurrences of western-facies rocks are instead due to younger thrusts, of late Paleozoic and Mesozoic age.

In Figure 3, largely generalized from Hope and Coats (1976), tectonic ages are assigned to some of the tectonic units in this study area. Some of the most significant localities on Figure 3 are indicated by circled numbers; these numbers correspond to parenthetical numbers in the following description. Geographic features are shown in Figure 2, and stratigraphic units in the area are summarized in Figures 7 and 8. Directions of motions on the thrusts are summarized in Figure 4, together with generalized outer limits of areas affected by thrusting of different orientations.

As indicated on Figure 3, the lower Paleozoic western-facies rocks of northern Elko County are exposed in a number of isolated areas of widely varying extent, separated by later thrusts and by overlapping younger rocks. The largest continuous areas are those in the Independence Range, which extends from south to north, nearly along the 116th meridian, in the H-D Range and Snake Mountains

that lie north of Wells, and in an east-west-trending belt of mountains that extends eastward from the 116th meridian in the latitude of Jarbidge.

Determination of the direction of movement of the thrusts is important but not always straightforward. In a few places the actual thrust surfaces may be examined, and minor details of the upper or lower plate, such as cleavage or drag folds, have been utilized for some of the directional information included in Figure 4. As indicated by King (in Nilsen and Stewart, 1980), transport in the Antler orogeny (on the Roberts Mountains thrust) was generally eastward; according to Ralph Roberts (oral communication, 1982), the same is true of the Sonoma orogeny. We infer that it is probably also generally true for much of the Mesozoic orogeny. As will be shown later, southward transport is a local but significant feature over a belt, the latitudinal extent of which may be measured in many tens of kilometers in northeastern Elko County. The presence of southward transport has significant implications for the regional tectonic framework.

MAJOR THRUSTS

The Roberts Mountains thrust

The Roberts Mountains thrust, as we here consider it, is represented by the westernmost exposures of the eastern-facies lower Paleozoic rocks, lying beneath a thrust plate of western-facies lower Paleozoic rocks. The western-facies rocks have been moved generally eastward as indicated by features of the thrust surface or by minor structures in the upper or lower plates. We are restricting the term "Roberts Mountains thrust" to thrust segments that show no evidence of post-Antler movement. By considering only the westernmost position of eastern-facies rocks, the problem of irregular klippen and fensters is avoided; we have found that such a definition of the thrust generally places it where the dip of the thrust is relatively steep. In none of our localities can a Devonian or Mississippian age of the thrusting be certified; however, we believe an Antler age is likely, based on regional considerations (e.g., Roberts and others, 1965). Note, however, that Ketner and Smith (1982) have recently offered evidence that the Roberts Mountains thrust, as it has been identified in central Elko County, south of the area here considered, may be younger than generally supposed. Post-Antler thrusts of eastward displacement have been recognized previously (Smith and Ketner, 1977).

The trend of the Roberts Mountains thrust, so defined, is northerly, not northeasterly, all across the western part of the area of Figure 3, to the point where the trace of the thrust disappears beneath the Tertiary volcanic rocks of the Mountain City quadrangle, which conceal it as far north as the edge of the Snake River Plain. The supposed north-

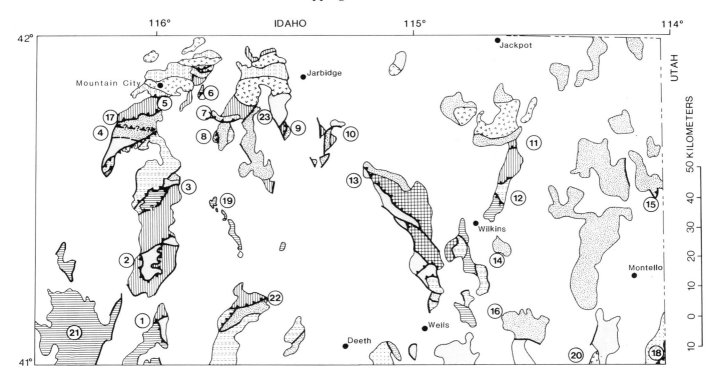

PALEOZOIC AND MESOZOIC TECTONIC UNITS IN N. E. NEVADA

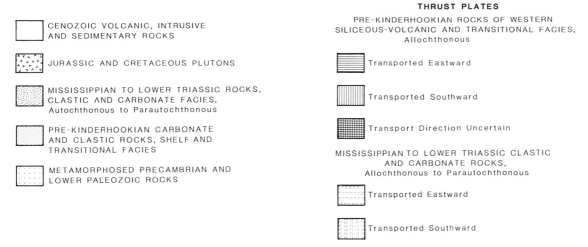

Figure 3. Map showing distribution of Paleozoic and Mesozoic tectonic units in northeastern Nevada (generalized from Hope and Coats, 1976). Circled numbers indicate localities discussed in text.

easterly trend of the Roberts Mountains thrust and the Antler orogenic belt across central Elko County was first suggested by Roberts and others (1965, Fig. 4, here reproduced as Fig. 1). This map was based on the widespread occurrence of western-facies lower Paleozoic rocks in this area, and the assumption that, wherever seen, these rocks are allochthonous and rest tectonically on eastern-facies lower Paleozoic rocks. The general northeasterly trend of the belt of western-facies lower Paleozoic rocks has been

accepted with little modification by later authors (Stewart and Poole, 1974; Thorman, 1970; Thorman and Ketner, 1979; Stewart, 1980).

Re-examination has shown that the western-facies lower Paleozoic rocks, instead of resting exclusively on eastern-facies lower Paleozoic rocks, commonly rest tectonically on younger Paleozoic or Triassic rocks, having been once moved eastward, probably in southern Owyhee County, Idaho (possibly on the Owyhee Rift), and subse-

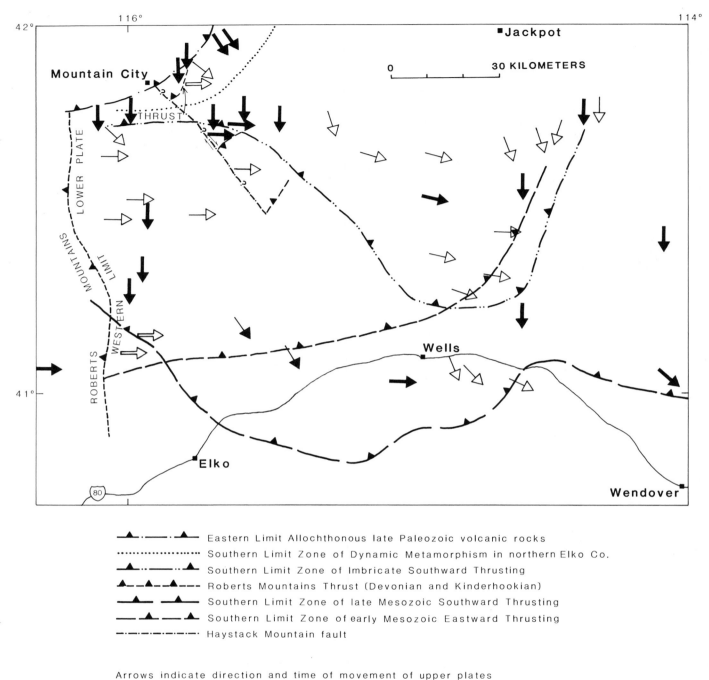

Figure 4. The directions and time of movement and generalized maximum extent of the larger tectonic units of the pre-Tertiary rocks of northern Elko County.

quently thrust southward into Elko County. The absence of the Roberts Mountains thrust from north-central Elko County removes the necessity for elaborate tectonic mechanisms to modify the course of the thrust in north-central Elko County, as it has never been present in that part of the county but trends northward from Lone Mountain to the southeastern part of the Mountain City quadrangle.

Localities where the Roberts Mountains thrust, so defined, has been seen are described below in order from south to north.

In the Lone Mountain area (locality (1), Fig. 3)[1] (Lovejoy, 1959) is the lowermost significant thrust on which western-facies lower Paleozoic rocks have moved into the area. Two younger thrusts lie above it; on each of them the western-facies rocks have moved in a different direction. In the southwestern portion of the area (the Blue Basin quadrangle), Lovejoy's (1959) and Ketner's (1974) maps show continuous strips of western-facies rocks, each representing a separately mappable unit of Ordovician, Silurian, or Devonian age, some repeated by several thrust faults that have a generally northerly trend in the northeastern part of the Blue Basin quadrangle. Rocks of the western facies occupy much of the southwestern part of the Reed Station quadrangle. As mapped by Lovejoy (1959), the southwestern part of the exposure of the western-facies rocks resembles in structure and rock distribution the area occupied by rocks of this facies in the Blue Basin quadrangle (Ketner, 1974). The northern part of the area of western-facies rocks may be divided into two terranes, separated by a thrust fault. The lower plate is exposed in a window in the northwest part of sec. 26, T. 38 N., R. 53 E. in the Reed Station quadrangle. In it outcrops of tuffaceous greenstone of obscure structure and fossiliferous Ordovician limestone resembling reef deposits trend nearly north-south. The western-facies rocks of the upper plate, mostly chert and shale, as indicated by Lovejoy, have generally easterly trends and very complicated structures. The western-facies rocks in the Reed Station quadrangle are probably parts of three different thrust sequences that were emplaced at diverse times. The limestone and tuff in the window are regarded as transitional to western-facies and as part of the upper plate of the Roberts Mountains thrust. The plate of western-facies rocks immediately above it was brought into this area by movement on the Roberts Mountains thrust and remobilized in later Paleozoic time or early Mesozoic time by southward thrusting on the Seetoya thrust of Kerr (1962); the repeated movement in differing directions accounts for the complex structure. The western-facies rocks of the western part of the Blue Basin quadrangle probably represent a third period of thrusting, most probably Mesozoic, from the west.

The Roberts Mountains thrust is not believed to be

exposed in the surface in the Independence Range to the north of the Reed Station quadrangle. The western-facies rocks in the Independence Range, while they may have been moved on the Roberts Mountains thrust, are considered to have been remobilized and moved southward by later thrusts. The position of the Roberts Mountains thrust is inferred from the mapping of Decker (1962) to lie in the subsurface beneath a cover of Tertiary rocks just west of Lime Mountain (4). Lime Mountain is the westernmost outcrop in northern Elko County of eastern-facies lower Paleozoic rocks having a predominantly northward strike. This attitude is consistent with these blocks being the outcrop of the lower plate of the Roberts Mountains thrust. The same lower Paleozoic rocks are present in the Bull Run Mountains, a short distance farther north, with predominantly eastward trends believed to be due to post-Pennsylvanian, perhaps Mesozoic, southward-directed thrusting. No outcrops of the Roberts Mountains thrust are found farther north in the Bull Run or Owyhee (Coats, 1971) quadrangles.

The northernmost locality in Elko County where the western-facies rocks can be shown to have been thrust eastward over eastern-facies lower Paleozoic rocks is in the Mountain City quadrangle, in sec. 6, T. 45 N., R. 55 E. (6), where the Ordovician Valmy Formation (Fig. 5) has been thrust eastward over the Ordovician Goodwin Limestone of the Pogonip Group (Fig. 5). Here the Valmy Formation, with its characteristic chert and gray quartzite, has been thrown into a set of isoclinal folds having vertical limbs, trending northeast, parallel to the Roberts Mountains thrust and parallel to the trend of the nearly isoclinal folds in the Goodwin Limestone in the footwall of the thrust (cf. Evans and Theodore, 1978). The folds in the Goodwin verge southeastward and do not appear to die out for at

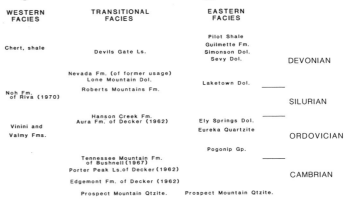

Figure 5. Correlation of some lower Paleozoic rocks in northern Elko County, Nev. The mention of a stratigraphic name (or the age shown for that unit) in this figure does not necessarily imply its current adoption (or acceptance) by the U.S. Geological Survey.

---

[1]Parenthetical numbers in text refer to circled numbers on Figure 3.

least 2 km to the southeast. Traced northeastward, the thrust, with its footwall formation, disappears under overlapping Tertiary volcanic rocks; the strike of the Valmy in the hanging wall of the thrust swings to north, and the dips remain steep until the Valmy disappears from view beneath Quaternary alluvium.

Near the southwest limit of its exposure in the Mountain City quadrangle, the Roberts Mountains thrust has been shifted right laterally a few tens of meters on a northwest-trending fault; the prolongation of the shifted part to the southwest is concealed beneath the overlapping Tertiary volcanic rocks. This minor right-lateral shift is believed to be the effect of a small divergent splinter of the northwest-trending Haystack Mountain fault of Coash (1967, Fig. 4 of this chapter). The Haystack Mountain fault in the Mount Velma quadrangle corresponds in trend with the bounding fault recognized here in the Mountain City quadrangle.

Post-Antler thrusts of eastward displacement

The outstanding thrust of eastward displacement is the Roberts Mountains thrust, presumed here to be of Antler age. Younger thrusts with similar directions of displacement can be documented in a few places. Eastward movement of Cretaceous age was previously documented by Smith and Ketner (1977) south of our study area. Thrusts to be described here are in the Beaver Peak (21), Divide Peak (19), Bull Run (3), and Wildhorse (8) areas, and in the Snake Mountains (13).

About 2 km west of Beaver Peak (21) in the Beaver Peak quadrangle flat-lying arenite that resembles the Edna Mountain Formation of Late Permian age rests in probable fault contact on the Valmy Formation. The summit of Beaver Peak is composed of cherty limestone of the Antler Peak Limestone, Pennsylvanian and Permian in age, that rests unconformably on the Valmy Formation with a nearly horizontal contact. Traced upward, the Antler Peak changes continuously from the flat-lying attitude to a 30-degree eastward dip at the summit. A possible explanation for this odd structure is that the uppermost part of a flat-lying plate composed of the Antler Peak has been given the eastward dip by drag by an unobserved Mesozoic thrust plate that formerly moved eastward above the summit of Beaver Peak and is present in the Valmy Formation beneath the plate composed of the Edna Mountain Formation that lies west of Beaver Peak. The trend of the thrust is inferred from the direction of dip of the Antler Peak Formation. The thrust may be related to the east-northeast-moving thrust in the Divide Peak area (19).

In the Divide Peak area (19) (Coats and Gordon, 1972), limy sedimentary rocks (probably part of the Park City Group rather than the Phosphoria Formation as suggested by Coats and Gordon, 1972) rest depositionally on

cherty and shaly rocks of the Valmy Formation. The Permian rocks deposited on the allochthonous Valmy indicate a pre-Late Permian, possibly Antler-age emplacement for the allochthonous Valmy. It is possible that the emplacement of the Valmy Formation was part of the Humboldt orogeny of Ketner (1977), but younger thrusting is indicated where the Edna Mountain Formation is locally thrust east-northeast over the Valmy, as indicated by mullions.

In the Bull Run quadrangle, in the Independence Range (3), Miller and others (1981) have mapped an extensive thrust plate of the Devonian(?), Mississippian, and Pennsylvanian Schoonover Formation of Fagan (1962), a sequence of deep-water sedimentary rocks with some greenstone, which has been thrust southeastward over chert and quartzite of the Valmy Formation, probably during the Sonoma orogeny.

In the extreme northeastern part of the Wildhorse quadrangle (8), chert and shale of Permian (probably the Phosphoria Formation) age are thrust southeastward over Misssssippian limy and shaly rocks. This west-dipping thrust, traced northward into the Mountain City quadrangle, is seen to be overridden by an east-trending thrust on which transitional-facies limestone and shale of the Tennessee Mountain Formation of Bushnell (1967) of Cambrian or Ordovician age and the unconformably overlying Antler Peak Limestone have been thrust southward.

Exposures of allochthonous western-facies rocks are scarce in the upper valley of Marys River, west of the Snake Mountains; little is known about the direction of movement of plates composed of western-facies rocks. It is tempting to speculate that the subdued relief in the valley of Marys River results from a lack of imbricate tectonic accumulations of resistant units such as occur to the west and east.

The Snake Mountains, H-D Range, and Windermere Hills to the east contain numerous thrust plates, some of eastward displacement but more of southward displacement, discussed for convenience in the following section.

Post-Antler thrusts of southward displacement

The evidence of southward movement of thrust plates comes from several areas. These will be described in a geographic sequence, beginning in the Independence Mountains in the latitude of Lone Mountain (Fig. 2), and tracing the sequence northward in the Independence Mountains to the Owyhee quadrangle thence eastward through the Mountain City quadrangle, Mount Velma quadrangle, and the Marys River Basin quadrangle, to the Snake Mountains and H-D Range, in which several successive southward- to southeastward-moved thrust plates have been recognized by Riva (1970 and this chapter). To the south, some of the thrust plates mapped by Riva in the H-D Range have also been recognized by Oversby (1972) in the

Windermere Hills. Sporadic thrust plates have been found almost as far east as the Utah border, in the Jackson Springs quadrangle and in the Pilot Range.

Two areas in the Independence Range require special mention. The southernmost of these is that near Lone Mountain (1), where two different facies are present: the lower plate is made up of one of these, and is discussed under the Roberts Mountains thrust; above it is a younger thrust, the upper plate of which consists of shale and chert, that possibly includes some western-facies Devonian rocks. This plate is apparently the same as the upper plate of the Seetoya thrust in the Tuscarora quadrangle.

The earliest recognition of a post-Antler southward-moving thrust in this area was that by Kerr (1962) who mapped in the Independence Range (Tuscarora quadrangle) (2) the Seetoya thrust. The Seetoya thrust brought western-facies rocks about S. 18° E. across overlap-facies rocks of the Waterpipe Canyon Formation of Kerr (1962), probably a correlative of the Chainman Formation, which rests unconformably on eastern-facies lower Paleozoic rocks. The upper plate of the Seetoya thrust of Kerr may be traced northward into the California Mountain quadrangle (3); the Valmy Formation of this plate is closely folded on east-west axes. It is overridden by another plate that came in from the west-northwest in Triassic time and brought in chert, sandstone, and mafic volcanic rocks of Mississippian and Pennsylvanian age (Miller and others, 1981). Still farther north, nearly on the border between the Bull Run and Owyhee quadrangles, the Valmy is overthrust on a thick series of limy rocks (the Storff and Chellis Formations of Decker (1962), and the Van Duzer Limestone) (4). One fossil location has been found in the Van Duzer, which yielded Pennsylvanian brachiopods and conodonts. The Storff and Chellis Formations were considered Silurian by Decker, but no fossils have been found; they are regarded as allochthonous and are omitted from Figure 8.

From the western edge of the Owyhee quadrangle eastward to the eastern side of the Jarbidge Wilderness in the Marys River Basin Northeast quadrangle, upper Paleozoic rocks (the "overlap assemblage" of Roberts and others, 1965) are tectonically involved with autochthonous or parautochthonous upper Paleozoic and Triassic rocks. Imbricate thrusting is characteristic of this terrane, and the movement direction is generally southeastward to southwestward.

Several examples of east-west-trending and north-dipping overthrusts that involve the overlap assemblage have been mapped in the Owyhee quadrangle (5). Here the Grossman Formation (Mississippian?) rests unconformably on the Valmy Formation, and the Banner and Nelson Formations of probable middle Mississippian age (Meramecian and Osagean) rest unconformably on the Grossman. The Mountain City Formation (Coats, 1969), correlated with the Chainman Shale of Mississippian age, is conformable or paraconformable on the Nelson Formation. These formations are truncated along the ridge north of Mill Creek in the Owyhee quadrangle by the Mountain City thrust, above which the Pennsylvanian(?) and Permian(?) Reservation Hill Formation is thrust southward.

Northeast of Mountain City, in the Mountain City quadrangle, the Valmy, Grossman, Banner, and Nelson Formations are present as a sequence of imbricate east-west-trending and north-dipping thrust slices, locally divided into separate thrust blocks by north-south-trending tear faults. Thrust sequences on one side of a tear fault differ from those on the other side. The same complex of thrust slices of east-west-trending, north-dipping stratified rocks continues northward, except where interrupted by Mesozoic plutons, to the overlapping contact of Tertiary volcanic rocks.

Along the border between the Rowland and Mount Velma quadrangles (7), the unconformable contact between the transitional-facies Cambrian and Ordovician Tennessee Mountain Formation of Bushnell (1965) and the overlap-assemblage Pennsylvanian and Permian Strathearn Formation (Sunflower Formation of Bushnell, 1965; Coash, 1967) is vertical or overturned to the south and trends nearly east-west. The Strathearn and Tennessee Mountain Formations must form the upper plate of a thrust, although the actual thrust surface is not observable in the Mount Velma quadrangle. It is, however, visible in the northeastern corner of the Wildhorse quadrangle, near the center of sec. 3, T. 44 N., R. 55 E. where the Strathearn lies in thrust contact on Permian shale and chert that carry a fauna similar to that of the Phosphoria Formation (8). This north-dipping thrust is post-Phosphoria in age and probably Mesozoic. The Permian (Phosphoria?) shale and chert is part of a thrust plate that lies on authochthonous Mississippian to Permian limy and shaly rocks. This thrust has a generally north-south trend and dips westward. The vergence of the minor structures in the Phosphoria-equivalent rocks is toward the east.

What is apparently the major thrust beneath the Tennessee Mountain and Strathearn Formations was mapped by Bushnell (1965) still farther east in the Rowland quadrangle (23) about 1.6 km east of the summit of Pine Mountain, where the Strathearn rests in thrust contact on the Prospect Mountain Quartzite (Bushnell, 1965). In the Mount Velma and Marys River Basin northwest quadrangles (9), about 1.6 km south of their common north corner, the Strathearn is thrust over probable Valmy equivalent, which itself is thrust over Mississippian to Permian limy, shaly, and cherty rocks. The vergence of these Mississippian to Permian sedimentary rocks is also toward the south, suggesting that the plate including the Valmy Formation has moved southward. In the Rowland quadrangle, westerly exposures of the transitional Tennessee Mountain Formation on the western edge of the quadrangle grade eastward into eastern-facies carbonate rocks and occupy a geographic position intermediate between the Cambrian Prospect Mountain Quartzite in the southwestern part of the Jarbidge quadrangle (Copper Mountain Quartzite of Bushnell, 1967) and the Ordovician Pogonip Group that crops out in the southeastern part of the Mountain City quadrangle. Both the Prospect Mountain Quartzite and the Pogonip Group are regarded as essentially autochthonous.

From the Mount Velma quadrangle eastward to the southern part of the Jarbidge Wilderness, hundreds of meters of generally dark-gray fine-grained clastic limestone, shale, sandstone, and siltstone form a nearly continuous belt several kilometers wide, except where interrupted by faulting or younger rocks. Ages of these rocks range from Late Mississippian to Early Permian. The rocks have been generally considered autochthonous to parautochthonous, though no depositional base has been found in this stretch. The upper boundary in many places is a thrust, with a generally east-west trend and with accessory structures indicating southward movement. In the Mount Velma quadrangle, the structurally higher plate contains the equivalent of the Pennsylvanian and Permian Strathearn Formation. In the northeast part of the Mount Velma quadrangle, the Valmy Formation is thrust southward over the autochthonous carbonate rocks and shale. Similar relations are present just south of the Jarbidge Wilderness, in the Marys River Basin Northwest quadrangle.

The Snake Mountains (13) (Fig. 3) are a broad, 65-km-long northwest-southeast-trending mountainous complex west of the H-D Range and Windermere Hills. Gardner (1968) and Bezzerides (1967) mapped the north-central part of the Snake Mountains, and Peterson (1968) mapped the Antelope Peak area in the south-central part. In general, the works of Gardner, Bezzerides, and Peterson were broadly coordinated, and their results differ only in details.

The main conclusions reached by Gardner, Peterson, and perhaps, Bezzerides are as follows: (1) A major thrust sequence of eastern-facies rocks, the Currant Creek thrust sequence, lies tectonically on poorly exposed Devonian carbonate rocks of the transitional facies on the western edge of the Snake Mountains. This thrust sequence consists of Ordovician, Silurian, and Devonian carbonate rocks. (2) The Currant Creek thrust sequence, in turn, is overthrust by a major plate formed by the Stormy Mountain thrust sequence composed, in Gardner's (1968) definition, only of undivided western-facies rocks of Ordovician and Silurian age. (3) A thick postorogenic clastic sequence composed of the Strathearn Formation, Buckskin Mountain Formation of Fails (1960), and Phosphoria Formation all of late Paleozoic age, and the Dinwoody and Thaynes Formations of Triassic age was interpreted by Gardner (1968) as overlapping both the Currant Creek and Stormy Mountain thrust sequences.

The interpretation that upper Paleozoic and lower Mesozoic rocks of the overlap assemblage in the Snake Mountains lie indiscriminately on two separate thrust plates also would mean that the thrust sequences in question have been moved into their present position during the Antler orogeny, and by consequence, the structural history of the Snake Mountains is much older and unrelated to that of the H-D Range and the Windermere Hills immediately to the east described by Riva (1970). To arrive at a better understanding of this problem, Riva, in 1970, visited the Hot Creek quadrangle of the northwestern end of the Snake Mountains (13) where Gardner (1968) had mapped the only occurrence of postorogenic clastic rocks overlapping both the Stormy Mountain and Currant Creek thrust sequences. Riva (1970) concluded that the overlap assemblage covered only the western-facies rocks of the Stormy Mountain sequence and was separated by a thrust from the tectonically underlying Currant Creek sequence. This relationship means that the overlap rocks are strictly allochthonous and were moved into their present position, together with the unconformably underlying western-facies rocks, not by the Antler orogeny but by one of the Mesozoic orogenic movements that emplaced the thrust plates in the Windermere Hills and the H-D Range. In its overall composition, the stratigraphy of the Stormy Mountain thrust sequence is remarkably similar to that of the Thousand Springs sequence of the Windermere Hills (Oversby, 1972), and the two sequences may well be considered or interpreted as large segments of a once-continuous thrust plate.

Gardner (1968) and Peterson (1968) assigned the western-facies rocks of the Stormy Mountain thrust sequence to the Valmy Formation and estimated a thickness in excess of 3,000 m. No attempts were made to separate Ordovician from Silurian rocks, although their lithologies are markedly distinct and the fossils (chiefly graptolites) abundant. The thickness reported for the Valmy is clearly excessive and is probably due to superposition of various thrust segments. Riva readily recognized at least three such segments in an exploratory tour through the western-facies rocks mapped by Peterson (1968) on the east flank of Antelope Peak, as well as considerable amounts of the typical quartzite of the Diamond Peak Formation and Chainman Shale, both apparently belonging to other distinct thrust sequences. Detailed mapping is obviously needed here to arrive at a reasonable understanding of the stratigraphy and structure of the western-assemblage rocks in the Snake Mountains.

On the eastern side of the southern Snake Mountains, a brief survey made by Riva in 1970 readily showed the following: a lower deformed slice of Ordovician shale, chert, and associated volcanic rocks together with siltstone of the Silurian Noh Formation of Riva (1970), all bearing abundant graptolites; this lower slice is overlain by a slice composed of quartzite of the Diamond Peak Formation and by another slice consisting of highly deformed chert of the Upper Permian Carlin Canyon Formation of Fails (1960). These slices are thrust on a basement of the Chainman Shale, repeating a relationship also observed in the south-central H-D Range (Riva, 1970) and the Windermere Hills (Oversby, 1972). Chert typical of the Carlin Canyon Formation was also extensively mapped by Oversby in one of the thrust slices forming the Thousand Springs sequence on the eastern flank of the Windermere Hills (Black Mountain); Riva has noted that it forms a highly deformed thrust slice at the northeast end of the Snake Mountains near the head of Jakes Creek, where the thrust overrides overlap-assemblage rocks of the Stormy Mountain sequence.

On the western side of the southern Snake Mountains is another massive thrust sequence composed of Devonian limestone and the Mississippian Diamond Peak Forma-

tion, which is quite similar to Oversby's (1972) Black Mountain sequence of the eastern Windermere Hills. R. A. Hope (oral communication, 1973) observed that this sequence is thrust upon a basement of undated siliceous shale that could well be the Chainman Shale. The Black Mountain sequence lies tectonically on the Chainman Shale in the Windermere Hills. A study of Gardner's (1968) and Peterson's (1968) geologic maps of the north-central Snake Mountains suggests that this sequence is the southward continuation of the Currant Creek thrust sequence, but contains younger strata than are present in the northern parts of the mountains. The Diamond Peak Formation clastic rocks at the top of the sequence certainly indicate that its emplacement postdates the Antler orogeny. At the northern end of the Snake Mountains, near O'Neill Pass, Bezzerides (1967) mapped a dike cutting the Triassic Din-

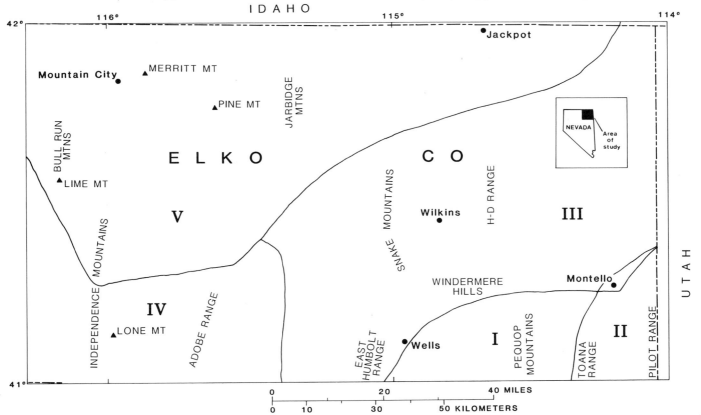

Figure 6 (this and following pages). Graphic glossary of some stratigraphic terms used in the mapping of Pennsylvanian, Permian, and Triassic rocks of Elko County. Index map shows by Roman numerals I-V, approximate areas in which the units in part B of the corresponding columns of the table have been used in mapping. Part A of each column indicates the commonly recognized units that correspond most closely to the units used in the mapping. Part B of each column indicates the units generally used in the mapping of areas indicated by Roman numerals on the index map herewith. Space devoted to each unit in the column is not necessarily a reflection of the duration of the time required for deposition or of the stratigraphic thickness. Apposition of units between columns is not intended to indicate precise time equivalency. Double underlining of names indicates units allochthonous, at least in part, in the area indicated by the column heading. Vertical lining indicates gaps due to erosion, nondeposition, or faulting. The mention of a stratigraphic name or the age shown for that unit does not necessarily imply its current adoption or acceptance by the U.S. Geological Survey.

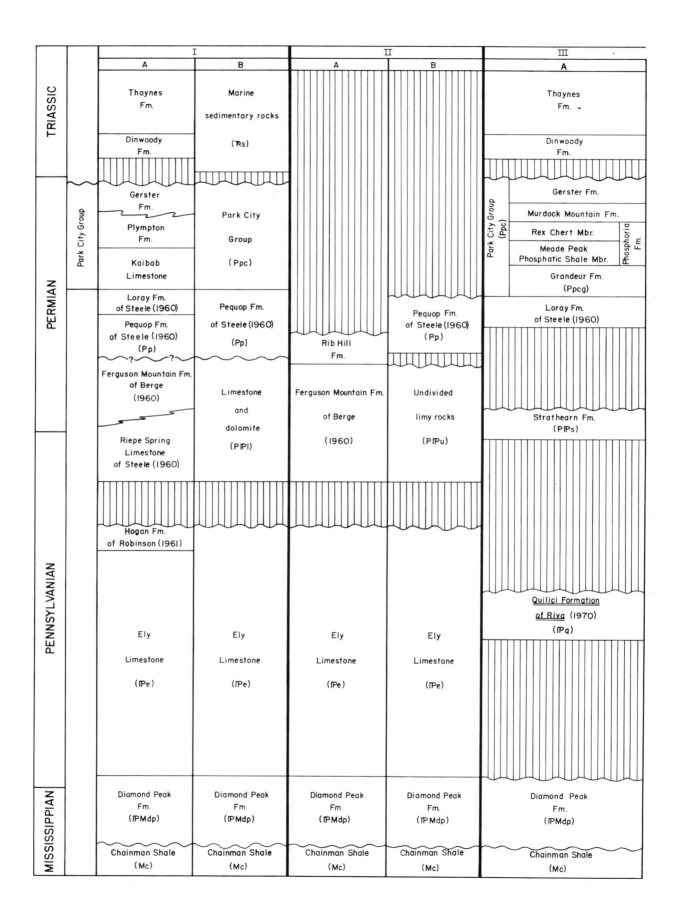

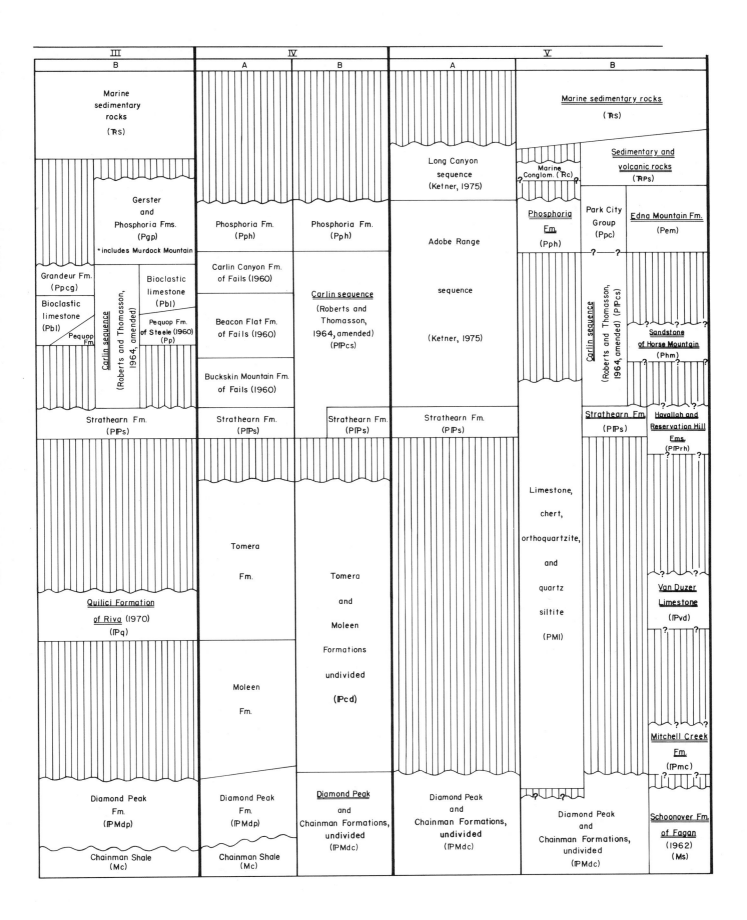

woody Formation and surmised that it was genetically related to the emplacement of the Contact pluton of Late Jurassic age (Coats and others, 1965). If this hypothesis is confirmed by radiometric dating, both major thrust sequences in the Snake Mountains must have been emplaced in Late Triassic or Early Jurassic time; the same reasoning will, by analogy, have to be applied to the complex thrust sequences in the Windermere Hills, as they are the forward extension of the major thrust plates in the Snake Mountains.

Gardner (1968) concluded from studies of the orientation of drag folds and major folds in the carbonate and western-facies rocks that thrusting in the Snake Mountains was from the northwest. Peterson (1968) reached the same conclusions on the basis of the orientation of major folds and a review of the regional geology. This direction of thrusting is in agreement with conclusions reached by Oversby (1972) for the Windermere Hills and by Riva (1970) for the upper plates in the H-D Range.

Thrusting in the H-D Range and the Windermere Hills can be attributed to one of three orogenic episodes (Riva, 1970, p. 2702–2703; Oversby, 1972, p. 2681–2683). The Antler orogeny was responsible for the displacement of the eugeosynclinal or western-facies rocks eastward over the belt extending from central Nevada to south-central Idaho. In the temporary resting places where the Antler orogeny had left them, the lower Paleozoic western facies rocks were overlapped by the upper Paleozoic overlap-assemblage rocks and by some of the Carlin sequence of Roberts and Thomasson (Fig. 6). The overlap assemblage rocks and the western-facies rocks beneath were thrust southeastward into the north-central H-D Range. In the Windermere Hills the overlap assemblage consists of the Strathearn, Buckskin Mountain, and Carlin Canyon Formations of Pennsylvanian and Permian age and the Lower Triassic Dinwoody and Thaynes Formations (Oversby, 1972, p. 2682–2683). In Jurassic or younger time the Ordovician eugeosynclinal rocks and the upper Paleozoic overlap-assemblage rocks were thrust eastward, as is suggested by the north-south trending folds overturned to the east (Riva, 1970, p. 2703). The source from which the Permian rocks of the Carlin sequence were thrust has not been recognized north of the North Fork of the Humboldt River; if they were once present in this area they have been buried or removed by later thrusting or erosion. In the H-D Range and the Windermere Hills, the thrusts have directions of motions between south and east.

In the H-D Range (11), Riva (1962a, 1962b, 1970, and this chapter) has recognized no less than six major and, perhaps, three minor thrust plates resting on a basement composed of the Mississippian Chainman Shale or an upper Paleozoic clastic sequence. The Chainman Shale itself is known to be part of a thrust plate in the northern Pequop Range, 80 km southeast of the H-D Range (Thorman, 1970). In tectonic style and stratigraphic makeup, the northern H-D Range differs distinctly from the southern part.

In the northern H-D range the lowest thrust plate, Plate A, has postorogenic clastic rocks of the Diamond Peak Formation lying unconformably on Ordovician and Silurian rocks of the western facies. This plate appears in several windows through the length of the range and is thrust on the Chainman Shale (Fig. 3). Plate A was thrust in an east-southeastward direction as evidenced by folds overturned to the east-southeast. The two plates next above, Plates B and C, show large east-west-trending folds overturned to the south and locally to the southeast; in the northern part of the range, drag folds indicate movement to the southeast. Plate C is an extensive but thin plate which underwent considerable deformation during its translation, and some segments apparently rotated and moved independently. The larger folds of Plate C are superposed on older folds, generally north-south trending, probably developed during the Antler orogeny at the site from which Plate C moved. Plate D, the fourth from the bottom, consists entirely of siliceous and volcanic western-facies lower Paleozoic rocks and differs markedly in composition from Plate B, which is made up of the transitional carbonate-siliceous-volcanic facies lower Paleozoic rocks, and from Plate C, which is similar to B but lacks the siliceous (chert) component. Plate D is entirely restricted to the northern end of the H-D Range where it forms a succession of imbricate slices with drag folds, mullions, and slickensides on the lower plate indicating a south-southeastward movement. Plate E, the highest in the north-central H-D Range, contains both western-facies rocks (shale, volcanic rocks, chert, and siltstone) and a thick upper Paleozoic overlap sequence which resembles in part rocks of the Elko-Carlin Canyon area (Dott, 1955; Fails, 1960) to the southwest. Large overturned folds at the base of this plate indicate movement to the southeast. Another plate, the "Unnamed Plate" of Riva (1970), here called Plate H, lies over Plates A, B, and C in the central part of the range and continues southward where it lies across the highest thrust plates in the southern H-D Range. This plate was apparently emplaced last and came from the west-northwest, rather than from the north-northwest as in the case of plates B, C, and D.

In the southern H-D Range (12) an extensive thrust plate, Plate F, forms the backbone of the southern part of the range and is thrust over Plate A. Plate F occurs as a succession of relatively undeformed east-dipping monoclinal segments of a stratigraphic sequence practically identical to that described by Fails (1960) from the Elko-Carlin Canyon area of western Elko County. It consists of a lower Paleozoic western facies similar to the Valmy Formation of north-central Nevada (Roberts and others, 1958) and of a thick overlap sequence, formed by the Strathearn Formation, the Buckskin Mountain Formation of Fails (1960) (with a 150-m-thick conglomeratic lens), and the Beacon Flat Formation of Fails (1960). Plate F is overthrust by two smaller plates, Plates G and H. Plate G consists largely of western-facies rocks with a thin overlap sequence; it is intensely deformed with drag folds locally indicating movement to the east-southeast. (This plate carries also a small segment of another plate consisting of highly deformed western-facies rocks hitherto seen only in Plates C and D in the northern H-D Range.) Plate H is essentially restricted to the northern portion of the southern H-D Range, but extends also to the central part (the "Unnamed Plate" of Riva, 1970) where it lies over rocks of Plates A, B, and C. It consists of a nearly undeformed and thick (at least 1,700 m) postorogenic sequence that can be largely, but not entirely, related to the sequence of Permian rocks described by Fails (1960) in the Elko-Carlin Canyon area, and of deformed segments of western-facies rocks, largely of the Valmy Formation. Drag folds in the overlap sequence suggest movement in an east-southeasterly direction.

In the north-central H-D Range, swarms of dikes radiating from the 150-160-m.y.-old Contact pluton (Coats and others, 1965) cut Plates A to E, but none was observed to cut Plates F to H in the south-central part of the range. This relationship is taken to indicate that thrust plates in the north-central H-D Range were emplaced prior to those in the southern part of the range, possibly in the Triassic or Early Jurassic. This hypothesis is supported by the stratigraphy of the western and overlap facies in the north-central part of the range, which differs significantly from the stratigraphy of plates in the southern H-D Range. The hypothesis also is supported by differences in the direction of thrusting, which was mostly to the south-southeast (except for plate A, which moved east-southeast) in the north-central area, rather than to the east-southeast as in the south. The thrust plates in the southern H-D Range are believed to have been emplaced during a later tectonic phase, possibly Cretaceous, and moved in from the west-northwest. The stratigraphic makeup of both the western-facies rocks and the overlap assemblage of the H-D Range shows, with some exceptions, a remarkable similarity to correlative units of north-central Nevada, and especially to the rocks of the Elko-Carlin Canyon area.

The suggestion of a post-Antler rejuvenation of the Roberts Mountains thrust attributed by Stewart (1980) to Riva (1970) is evidently based on a misunderstanding by Stewart of Riva's (1970) description, outlined above, of the tectonics in the northern H-D Range. The western-facies rocks may indeed have once been moved on the Roberts Mountains thrust, which is not now visible in this area. We believe, however, that this thrust left them in a position far northwest of their present one and that they owe their present position to later movement in a different direction, on different thrust surfaces, and probably also on the Owyhee Rift.

In the Windermere Hills northeast of Wells (14) (Figs. 1, 4), Oversby (1969, 1972, 1973) mapped a tectonic succession of two major, thrust-bounded stratigraphic sequences and three minor ones similar to those mapped in the southern H-D Range to the northeast, the Snake Mountains to the west-northwest, and the Pequop Range to the southeast. The lowest sequence, or Black Mountain thrust sequence, lies tectonically on the Chainman Shale, thus repeating a relationship already noted in the H-D Range and also seen in the southern Snake Mountains. This sequence consists of middle Paleozoic carbonate rocks of the Guilmette Formation (Devonian) and the Tripon Pass Limestone (Mississippian) of Oversby (1973) capped by a thick section of clastic rocks of the Diamond Peak Formation. Thrust over the Black Mountain sequence is the Thousand Springs tectonic sequence, a major tectonic unit formed by rethrusted segments of western-facies eugeosynclinal rocks and limestone of the transitional facies, all topped by an overlap assemblage consisting of the Strathearn Formation, and the Buckskin Mountain and Carlin Canyon Formations of Fails (1960), all of late Paleozoic age, and the Dinwoody and Thaynes Formations, of early Mesozoic (Triassic) age. Oversby (1972, p. 2680–81) identified among the western-facies rocks Ordovician and Silurian units distinguished by Riva (1970) in the north-central H-D Range, and Riva (this chapter) has since recognized similar units in much of the Snake Mountains to the west-northwest. The carbonate rocks of the transitional facies are similar to Devonian limestone mapped by Evans and Ketner (1971) on Swales Mountain 96 km to the west, north of Carlin Canyon. The overlap assemblage is similar to that seen in the Carlin Canyon area (Fails, 1960), except that the Beacon Flat Formation is absent between the Buckskin Mountain and Carlin Canyon Formations; the succession is also nearly identical to that described by Gardner (1968) and Bezzerides (1967) from the northern Snake Mountains and is similar to that of Plates F and H of the southern H-D Range, except that the Beacon Flat Formation is widely exposed there while the chert of the Carlin Canyon Formation and the Triassic limestone units are absent.

The three minor thrust sequences in the Windermere Hills top or are interspersed among segments of the major thrust sequences. They consist of segments of eugeosynclinal rocks or postorogenic sedimentary rocks which could not be referred definitely to the Thousand Springs sequence. In general, there are few outcrops of western-facies eugeosynclinal rocks east of U.S. Highway 93 in the Windermere Hills, and the western Windermere Hills and the H-D Range must be regarded as the easternmost large exposure of eugeosynclinal rocks of early Paleozoic age in Nevada.

Low relief in parts of the Windermere Hills has somewhat limited a detailed structural study of the area. Oversby (1969; 1972, p. 2684) interpreted the occurrence of small folds overturned to the southeast near basal thrusts as the result of thrusting from the northwest in post-Triassic time.

The tectonic sequences in the H-D Range, Snake Mountains, and Windermere Hills are similar. Southeast of the Windermere Hills, in the Wood Hills and the Pequop Mountains (Thorman, 1970), folds overturned to the southeast suggest that post-Early Triassic thrusting there was from the northwest.

At the north end of the Pequop Range in the Holborn quadrangle (16), Ralph Roberts and R. R. Coats discovered a small klippe of white to dark-gray quartzite, resembling closely the quartzite of the Valmy Formation. The quartzite is intensely brecciated, as is the underlying formation, probably the Pequop Formation of Steele (1959). In the valley to the east, outcrops of black chert resembling that of the Valmy Formation suggest disjunct fragments of the klippe.

To the east, just north of the Jackson Mine, in the Jackson Springs quadrangle and almost on the Utah-Nevada border northeast of Montello, Ralph Roberts (1979, oral communication) recognized black chert, apparently part of the Valmy Formation, resting in thrust contact on the Devonian Guilmette or Devils Gate Formations. The black chert is overlain by conglomerate of the Diamond Peak Formation, but it is uncertain whether the conglomerate is in thrust contact with the chert and the Devonian rocks, or overlaps the chert and is thus thrust with it over the Devonian rocks.

Relations are clearer a short distance farther south in the Jackson Springs quadrangle (15) where conglomerate of the Diamond Peak Formation is present as a thrust plate on the Guilmette and Devils Gate Formations, undivided. The thrust, which dips northward at an angle of about 4°, is characterized by thrust breccias in both upper and lower plates, and mullions, having a relief of a few meters, that trend nearly north.

Just south and southeast of the Gamble Ranch, a few kilometers north of Montello and a few kilometers southwest of the Jackson Mine, light-gray and greenish chert of uncertain age is in thrust contact on the Pequop Formation, in narrow outcrops in the valley bottom, and eastward at a higher level. Mullions observed suggest that the thrust movement direction was southward; internal structure in the chert, including minor fold axes, suggests a prior deformation accomplished during movement from the west.

In the Pilot Range (18), Miller and Lush (1981) mapped a decollement that separates metamorphic rocks of Proterozoic Z and Cambrian age below from Cambrian sedimentary rocks above. The decollement was determined by Miller and Lush as Jurassic or Early Cretaceous in age. Internal structures suggest a southeastward movement for the upper plate.

As summarized in Figures 3 and 4, evidence for southward and southeastward moving thrusts is found in northern Elko County across the entire width of the map area. Most of this thrusting is demonstrably younger than Mississippian, some can be dated as Late Triassic or Early Jurassic, and some is younger than 150 to 160 m.y.

NORTHWEST-TRENDING STRIKE-SLIP FAULT OF MESOZOIC AGE

The southwest boundary of the upper plate of the Roberts Mountains thrust in the Mountain City quadrangle has been called the Haystack Mountain fault (Coash, 1967). The Roberts Mountains thrust does not reappear southwest of the Haystack Mountain fault as mapped in the Mountain City quadrangle. Measurement of the total offset on the Haystack Mountain fault is difficult for lack of good control. The youngest pre-Triassic formation found on both sides of the fault, with trends that could reasonably be extrapolated to the fault, is the Strathearn Formation (Sunflower Formation of Bushnell, 1965; Coash, 1967) of Late Pennsylvanian and Early Permian age. On the northeast side of the fault, this formation terminates abruptly to the southwest, presumably against the now buried Haystack Mountain fault, on Jenkins Peaks in the southeast corner of the Mountain City quadrangle. The sole exposure of the Strathearn Formation (mapped by Coats, 1971, as the Antler Peak Limestone) on the southwest side of the projected course of the fault is near the headwaters of Fawn Creek, in the Owyhee quadrangle. If these two exposures were once continuous, and the difference in position is due solely to offset on the Haystack Mountain fault, then the post-Early Permian right-lateral offset was on the order of 20 km, probably in pre-Cretaceous time. As both exposures may be parts of thrust blocks that may have been displaced southward unknown distances in later Mesozoic time, this possible offset is, at best, tentative. Some dip-slip movement has also occurred in Tertiary time; in the Mountain City quadrangle, the southwest side of the fault is downthrown. The amount of throw ranges widely from place to place along the strike of the fault.

Near the southeast end of the mapped extent of the Haystack Mountain fault in the Mount Velma quadrangle, a considerable area of quartzite of the Valmy Formation is exposed and underlies most of the high peak of Mount Velma (Haystack Mountain) (Fig. 4). This mass of the Valmy Formation is interpreted as a klippe of the upper plate of the Roberts Mountains thrust. The lower plate is not visible anywhere here; the oldest rocks to the northeast of the Valmy Formation are younger Triassic; similar rocks are found to the southwest of the Valmy outcrops. The southwest boundary of the klippe is not a simple fault here, but apparently a splay of the Haystack Mountain fault; to the southeast the Haystack Mountain fault cannot be traced beyond the Bruneau River.

THE "WELLS FAULT"

Thorman (1970, p. 2441) postulated a major right-lateral tear fault, the "Wells fault" of Jurassic and Early Cretaceous age, which he described as transecting the Pequop Range as follows:

The main fault crosses the Pequop Mountains, just north of Pequop Pass, beneath a thick cover of Chainman shales and sandstones. It reaches the surface south of the pass and alternates between a low-angle thrust (Rocky Canyon thrust) and high-angle reverse faults (Meyers Canyon, No-Name Canyon, and Highway faults), so that it resembles stairs leading up to the south. Right-slip movement along the Wells fault is indicated by the juxtaposition of markedly different Paleozoic facies in the Pequop Mountains. Subplate IIH contains a western miogeosynclinal facies, while the underlying and adjacent subplates (IID, IIE, IIF) contain typical eastern miogeosynclinal facies.

The Wells fault is shown in Thorman's (1970) Figure 15 as just north of his Highway fault but is not shown in his more detailed map of the Pequop Mountains (Thorman, 1970, Fig. 2). Further sedimentologic evidence for the Wells fault is offered by Thorman and Ketner (1979).

The mapping of Pilger (1972) is difficult to reconcile with the southeastward extension of the Wells fault, as Thorman mapped it, through Silver Zone Pass in the northern Toano Range. The Wells fault, as nearly as can be determined from Thorman's (1970) Figure 15, would have to pass either through the area occupied by the Silver Zone Pass pluton (20), of probable Jurassic or Cretaceous age, or through the undivided Cambrian rocks postdating the Pioche Shale that lie north of the pluton. The same Middle and Upper Cambrian formations mapped by Pilger north of the pluton are to be found south of it, and the same distribution is true also of the younger Paleozoic rocks. Dr. Rex Pilger reported (oral communication, 1979) that there are no substantial facies differences in these rocks from one side of the pluton to the other. However, Keith Ketner reported (written communication, 1980) that he was able to discern facies differences within the formations exposed on the two sides of the Wells fault. The general trends of the strikes are similar both north and south of the pluton, except that local folding, with axes parallel to the plutonic contact, is present near the contact and is probably due to forceful intrusion of the pluton. In short, there is no room, in our opinion, for any major transcurrent northwest-trending tear fault in the northern Toana Range (20). The east-west fault in the northern Pequop Range is regarded here as the boundary of a south-moving thrust.

Sheehan (1979) pointed out that a right-slip fault is not needed to explain the regional distribution of Silurian rocks if the Roberts Mountains Formation was deposited along an east-west-trending shelf margin in northern Nevada. A similar depositional trend was independently suggested by Mullens (1980, p. 40), who preferred it to oroclinal folding as an explanation for the trend of the reef line in the Silurian limestone.

Stevens (1979, p. 431) noted that the Wells fault of Thorman (1970) "apparently fails to offset the zone of Leonardian massive rugose corals, at least not [by] 65 km, suggesting that offset on this fault is not of the magnitude envisaged by Thorman." Stevens (1981) also noted that there are large differences in the amount of offset of Paleozoic sedimentary facies of different ages in the region of the supposed Wells fault and that this pattern is incompatible with postdepositional tectonism as an explanation for the facies changes.

Facies trends in the upper Paleozoic rocks of the Confusion Range in western Utah (Hose and Repenning, 1959) south of the extrapolated trend of Thorman's (1970) Wells fault, do not exhibit any substantial offset from those of the Silver Island Mountains northeast of Wendover (Schaeffer and Anderson, 1960) an observation that suggests no substantial right-lateral translation has taken place between these two areas during Mesozoic time.

The Wells fault has been projected westward for considerable distances by subsequent writers (for example, Johnson and Sandberg, 1977), usually to account for some abrupt facies differences that, in our view, are better explained, even though well substantiated sedimentologically, as boundaries between relatively minor thrust plates of Mesozoic age. No physical evidence for any such right-lateral northwest-trending throughgoing wrench fault has been seen in the Independence Range on the projected extension of the Wells fault.

The eastward offset of the Roberts Mountains thrust, mentioned by Thorman (1968) as part of the evidence for the major right-lateral shift attributed to the Wells fault, is negated by the fact that the Roberts Mountains thrust crops out in the Mountain City quadrangle, nearly due north of its position in the Carlin-Robinson Mountain area (Smith and Ketner, 1977), and in the Lone Mountain area of Elko County (Lovejoy, 1959). This offset is perhaps concealed by Tertiary rocks west of Lime Mountain in the Bull Run quadrangle (Decker, 1962). We conclude that there is no room for any deep-seated west-northwest-trending right-lateral transverse fault between the Mount Velma quadrangle and the southwestern part of Elko County.

A consideration of the map patterns in the Pequop Mountains on the north and south sides of the supposed Wells fault suggests that the nearly orthogonal attitudes of the sets of strikes on the two sides, and the facies contrasts so carefully reported by Thorman (1970), are better explained by the hypothesis that the Wells fault is a thrust fault zone, separating a region of eastward compression on the south from one of later southward compression on the north. This hypothesis thus regards the north end of the Pequop Range as the southern margin of a southward-moving thrust plate, similar to those in the H-D Range described by Riva (1970).

REGIONAL TECTONIC MODEL

Figure 4 summarizes and attempts to synthesize information on the directions and ages of thrusting, and the inferred maximum extent of the larger pre-Tertiary tectonic units in northeastern Elko County. From this figure it can be seen that the western limit of the lower plate of the Roberts Mountains thrust trends north-south except where offset by the Haystack Mountain fault. Younger post-Antler thrusting to the south and east has affected much of the area.

As indicated earlier, the northwesternmost exposure of the Roberts Mountains thrust in Nevada is just south of Merritt Mountain, in the Mountain City quadrangle, a short distance east of Mountain City. What is apparently the southwesternmost exposure of the same thrust in Idaho is found on the west flank of the Cassia Mountains just east of Hollister (Fig. 1), in Twin Falls County, Idaho. At this place, gray quartzite, closely resembling quartzite of the Valmy Formation in texture and color but anomalously crossbedded, and chert, locally with silicified mafic volcanic fragments, are thrust eastward over Ordovician dolomitic limestone (Coats, 1980). The thrust here also has a northward trend. The locality was discovered by Youngquist and Haegele (1956), who regarded the quartzite as the Eureka Quartzite. The nearly 100-km physical separation of similar rocks having similar structural relations from the Mountain City area to the Cassia Mountains could be explained by an east-trending structural offset, probably concealed beneath overlapping southward-directed schuppen of upper Paleozoic rocks or beneath Tertiary volcanic rocks in southern Owyhee County, Idaho. The most reasonable explanation for this shift is the presence of a right-lateral strike-slip fault, here called the Owyhee rift, having a minimum displacement of about 100 km (Fig. 7).

The assumed rift fault must be post-Antler in age, that is, post-Kinderhookian. The Mississippian through Lower Triassic clastic and volcanic rocks that accumulated west of the longitude of Mountain City may have moved along this fault to positions north of the central part of Elko County, possibly as a result of the collapse of the late Paleozoic marginal sea (Churkin and McKee, 1974) and the resulting accumulation of a thick pile of sedimentary and volcanic rocks of late Paleozoic and Early Triassic age in southern Idaho.

Support for the presence of the rift and possibly also for the eastward displacement of marginal basin material and subjacent oceanic floor may be available in Hill and Pakiser (1967, Fig. 5). Their profile, reproduced here as

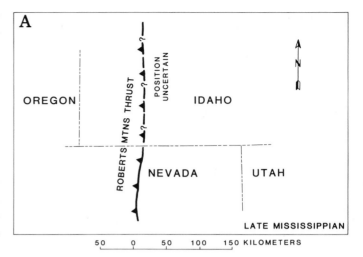

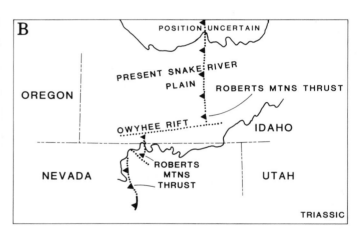

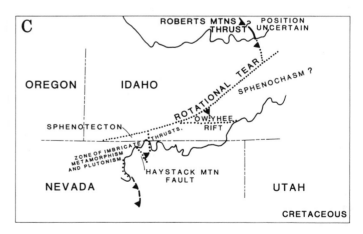

Figure 7. Sketch maps illustrating the concept that shifting on a right-lateral east-trending fault north of Elko County has displaced both lower and upper Paleozoic rocks, and the Roberts Mountains thrust.

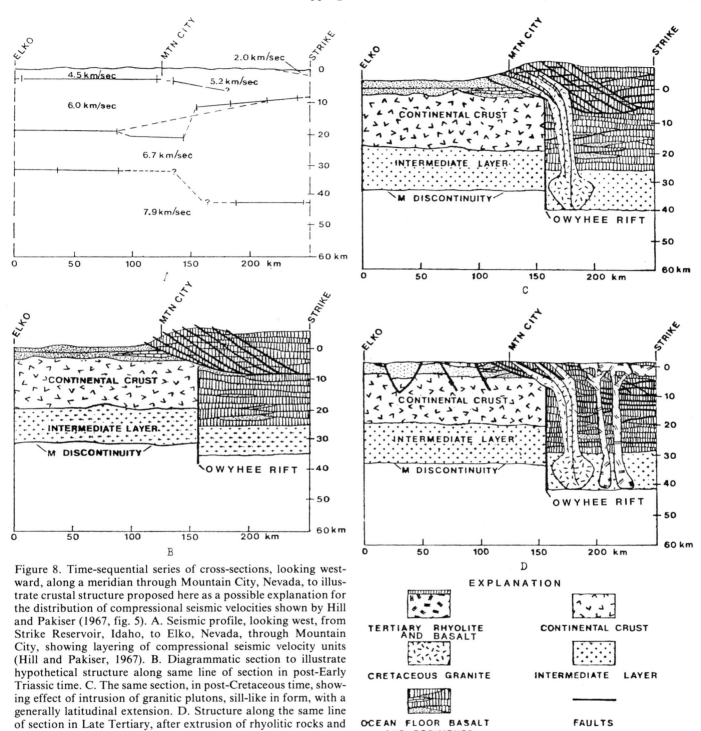

Figure 8. Time-sequential series of cross-sections, looking west-ward, along a meridian through Mountain City, Nevada, to illustrate crustal structure proposed here as a possible explanation for the distribution of compressional seismic velocities shown by Hill and Pakiser (1967, fig. 5). A. Seismic profile, looking west, from Strike Reservoir, Idaho, to Elko, Nevada, through Mountain City, showing layering of compressional seismic velocity units (Hill and Pakiser, 1967). B. Diagrammatic section to illustrate hypothetical structure along same line of section in post-Early Triassic time. C. The same section, in post-Cretaceous time, showing effect of intrusion of granitic plutons, sill-like in form, with a generally latitudinal extension. D. Structure along the same line of section in Late Tertiary, after extrusion of rhyolitic rocks and basalt of the Owyhee Plateau.

Figure 8A, is based upon seismic exploration between Eureka, Nevada, and Boise, Idaho, through Mountain City, Nevada, and indicates an abrupt northward increase in the thickness of the intermediate (mafic) layer (6.7 km/ sec) at about 30 km north of Mountain City. The thickness of this mafic layer increases there from about 13 km to

about 30 km. The thickness of the silicic (6.0 km/sec) layer coincidentally decreases from about 26 km, the normal Basin and Range province thickness, to a thickness too small for illustration. The surficial layer, here interpreted as being largely Cenozoic sedimentary and silicic volcanic rocks, increases in thickness going northward, beginning near the latitude of Mountain City. One possible interpretation is that continental granitic crust and its western margin were transported about 100 km eastward (Coats, 1980), and that beneath the Owyhee Plateau, north of Mountain City, its place was taken by a greatly thickened mafic layer, having thicknesses more appropriate to ocean floor or even to the roots of a volcanic arc. The Owyhee and Mountain City quadrangles in the extreme northwestern part of the area of Figure 3 expose possible candidates for the mafic layer: slivers and slices of upper Paleozoic volcanic rocks interbedded with marine sedimentary rocks have been seen only in the extreme northwestern part of the part of Elko County occupied by Paleozoic rocks, in the Nelson Formation of Late Mississippian age, the Reservation Hill Formation of Pennsylvanian(?) and Permian(?) age, the slate of Sikikareh Mountain of Paleozoic age, and the Mitchell Creek Formation of Early(?) Pennsylvanian age (Coats, 1969). Minor amounts of mafic volcanic rocks are found in other allochthonous upper Paleozoic and Triassic formations exposed farther south, such as the Schoonover Formation of Fagan (1962). The Owyhee rift may be Triassic in age, for Early Triassic and older rocks displaced eastward before southward displacement are intruded by Jurassic plutons. Figures 8B, 8C, and 8D are sections illustrating diagrammatically the hypothesized structure at the end of Triassic, the end of Cretaceous, and during late Tertiary times.

Roberts and Thomasson (1964) have proposed that the Roberts Mountains thrust extends into the Hailey area, Idaho (Fig. 1), although the thrust is everywhere concealed by upper Paleozoic rocks. Their suggestion was based on the presence of coarse, supposedly Antler, debris of Mississippian age in that area. They suggested a north-northwestward trend for the buried thrust in central Idaho. This trend would account for the contrast between the lower Paleozoic eastern-facies rocks of the Bayhorse region (Ross, 1937) and the eugeosynclinal rocks (Phi Kappa Formation) of the Trail Creek region (Ross, 1934; Churkin, 1963). Dover (1980) regarded the precise location and even the presence of the Roberts Mountains thrust in central Idaho as not yet documented structurally. His stratigraphic reconstructions suggest a questioned north-trending thrust of possible Early Mississippian age near the western border of Idaho (Dover, 1980, Fig. 11). He suggested that the Cordilleran hinterland in Idaho could have undergone as much as 200 km of crustal shortening, divided about equally between Antler movements and post-Antler compression of the orogenic belt in southwestern Idaho in the late Paleozoic. If Roberts and Thomasson correctly interpreted the Roberts Mountains thrust as trending north-northwesterly through central Idaho, then its trend is skewed counterclockwise from its trend in Nevada.

In order to explain this change in trend, and also north-south compression in northern Nevada, we speculate that central Idaho has rotated about 30° counterclockwise since the Antler orogeny. The southward and southeastward thrusting we have documented in northern Elko County indicates compression that is at least in part of Late Triassic or Early Jurassic age. East-west trends of the upper Paleozoic formations just south of the Idaho line are further evidence for north-south compression. These trends in turn appear to condition the east-west trend of Jurassic and Cretaceous plutons (Fig. 3). Further suggestions of compression come from the presence of dynamically metamorphosed rocks in the northwest part of the area of Figure 3. Paleozoic formations with considerable content of mafic volcanic rocks are metamorphosed to green schist grade, and evidence suggests a metamorphism predating the Cretaceous batholiths. (This metamorphism seems totally unrelated to metamorphism observed in the Ruby Range, the Pequop Mountains, Pilot Range, and East Humboldt Range.)

Rotation of a generally convexly polygonal block (the Idaho block) with respect to a once nicely mated block south of it (the Nevada block) could give rise to locally intense compressional deformation (Fig. 5C) if the rotated block were not free to move northward. In a general way, such a localized tectonic area is the counterpart of the sphenochasm of Carey (1958) and may, by analogy, be called a sphenotecton. Carey (1958, Fig. 37) termed such belts "compression wedges." As Carey suggested for the Pyrenees, compression resulting from such simultaneous rotation and translation may well have been adequate to generate the thrusting, plutonism, and regional metamorphism observed in northern Elko County. Southward gravity sliding of the compressed pile might contribute to the expanded coverage of northern Elko County by western-facies and overlap-assemblage rocks. Comparison of paleomagnetic pole positions of the eastern-assemblage rocks in central Idaho and in northern Nevada could afford a ready means of testing this hypothesis.

ACKNOWLEDGMENTS

The junior author thanks the Society of the Sigma Xi for its generosity in awarding him three grants that permitted him to complete the study of the southern H-D Range and restudy portions of the geology of nearby ranges and mountains. He acknowledges gratefully the aid, both indirect and direct, afforded to him by the Kemp Memorial Fund of Columbia University and the unstinting support and generosity of the late Professor Marshall Kay. Thanks are also due to Dr. Brian S. Oversby, now of Canberra, Australia, for his help and collaboration, and to the late Professor E. R. Larson of the University of Nevada for his continuous support and encouragement throughout this work.

Much of the fieldwork of the senior author has been

supported in part by the Nevada Bureau of Mines and Geology. The senior author wishes to acknowledge the benefit of numerous discussions with colleagues in the U.S. Geological Survey, and with other members of the Penrose Conference on the Antler orogeny, held in September 1979, in Elko, Nevada, but must absolve them of any responsibility for the far-reaching hypothesis offered here, and for any errors of interpretation of the work of others that he may have committed. This paper is somewhat modified from one originally presented orally at the 25th International Geological Congress in Sydney (Coats and Riva, 1976).

We are indebted for painstaking and useful reviews to Don Whitebread, Chester Wrucke, and R. W. Kistler of the U.S. Geological Survey, to Andrew Alden and Keith Howard of the U.S. Geological Survey, and to Elizabeth Coats for careful editorial work. Mick McCollum contributed useful comments.

REFERENCES CITED

Albers, John P., 1967, Belt of sigmoidal bending and right-lateral faulting in the western Great Basin: Geological Society of America Bulletin, v. 78, p. 143–156.

Bezzerides, T. L., 1967, Triassic stratigraphy and geology of the O'Neil Pass area, Elko County, Nevada [M.S. thesis]: Eugene, University of Oregon, 62 p.

Bushnell, Kent, 1965, Geology of the Rowland quadrangle, Elko County, Nevada [Ph.D. thesis]: New Haven, Conn., Yale University, 152 p.; Dissertation Abstracts, v. 2C, no. 4.

—— 1967, Geology of the Rowland quadrangle, Elko County, Nevada: Nevada Bureau of Mines, Bulletin 67, p. 8–10.

Carey, S. W., 1958, The tectonic approach to continental drift, *in* Continental drift; a symposium: Tasmania, University Geology Department, p. 177–355.

Churkin, Michael, Jr., 1963, Graptolite beds in thrust plates of central Idaho and their correlation with sequences in Nevada: American Association of Petroleum Geologists Bulletin, v. 47, p. 1611–1623.

Churkin, Michael, Jr., and McKee, E. H., 1974, Thin and layered subcontinental crust of the Great Basin in western North America inherited from Paleozoic marginal ocean basins?: Tectonophysics, v. 23, p. 1–15.

Coash, J. R., 1967, Geology of the Mount Velma quadrangle, Elko County, Nevada: Reno, University of Nevada, and Nevada Bureau of Mines Bulletin 63, 21 p.

Coats, R. R., 1969, Upper Paleozoic formations of the Mountain City area, Elko County, Nevada, *in* Cohee, G. V., Bates, R. G., and Wright, W. B., eds., Changes in stratigraphic nomenclature by the U.S. Geological Survey, 1967: U.S. Geological Survey Bulletin 1274–A, p. A22–A27.

——, 1971, Geologic map of the Owyhee quadrangle, Nevada-Idaho: U.S. Geological Survey Miscellaneous Geologic Investigations Map I–665, scale 1:48,000.

——, 1980, The Roberts Mountains thrust in central Twin Falls County, Idaho [abs.]: Geological Society of America Abstracts with Programs, v. 12, no. 6, p. 270.

Coats, R. R., and Gordon, Mackenzie, Jr., 1972, Tectonic implications of the presence of the Edna Mountain Formation in northern Elko County, Nevada, *in* Geological Survey research 1972: U.S. Geological Survey Professional Paper 800–C, p. C85–C94.

Coats, R. R., and McKee, E. H., 1972, Ages of plutons and types of mineralization, northwestern Elko County, Nevada, *in* Geological

Survey Research 1972: U.S. Geological Survey Professional Paper 800–C, p. C165–C168.

Coats, R. R., and Riva, John, 1976, Eastward obduction of early Paleozoic eugeosynclinal sediments, and early Mesozoic transverse thrusting resulting from southward movement of a compressed Paleozoic sedimentary pile, northeast Great Basin, Nevada, U.S.A.: International Geological Congress, 25th, Sydney, Abstracts, v. 1, p. 80.

Coats, R. R., Marvin, R. F., and Stern, T. W., 1965, Reconnaissance of mineral ages of plutons in Elko County, Nevada, and vicinity, *in* Geological Survey research 1965: U.S. Geological Survey Professional Paper 525–D, p. D11–D15.

Coats, R. R., Greene, R. C., Cress, L. D., and Marks, L. Y., 1977, Mineral resources of the Jarbidge Wilderness and adjacent areas, Elko County, Nevada, *with a section on* An aeromagnetic interpretation by W. E. Davis: U.S. Geological Survey Bulletin 1439, 74 p.

Decker, R. W., 1962, Geology of the Bull Run quadrangle, Elko County, Nevada: Nevada Bureau of Mines Bulletin 60, 65 p.

Dott, R. H., 1955, Pennsylvanian stratigraphy of Elko and northern Diamond Ranges, northeastern Nevada: American Association of Petroleum Geologists Bulletin, v. 39, p. 2211–2305.

Dover, J. H., 1980, Status of the Antler orogeny in central Idaho—clarifications and constraints from the Pioneer Mountains, *in* Fouch, T. D., and Magathan, E. R., eds., Paleozoic paleogeography of the west-central United States: Rocky Mountain Section, Society of Economic Paleontologists and Mineralogists, West-central Paleogeography Symposium 1, Denver, Colo., (preprint).

Evans, J. G., and Ketner, K. B., 1971, Geologic map of the Swales Mountain quadrangle and part of the Adobe Summit quadrangle, Elko County, Nevada: U.S. Geological Survey Miscellaneous Geological Investigations Map I–667, scale 1:24,000.

Evans, J. G., and Theodore, T. G., 1978, Deformations of the Roberts Mountains allochthon in north-central Nevada: U.S. Geological Survey Professional Paper 1060, 17 p.

Fagan, J. J., 1962, Carboniferous cherts, turbidites, and volcanic rocks in northern Independence Range, Nevada: Geological Society of America Bulletin, v. 73, p. 595–612.

Fails, T. G., 1960, Permian stratigraphy at Carlin Canyon, Nevada: American Association of Petroleum Geologists Bulletin, v. 44, p. 1692–1703.

Gardner, D. H., 1968, Structure and stratigraphy of the northern part of the Snake Mountains, Elko County, Nevada [Ph.D. thesis]: Eugene, University of Oregon, 221 p.

Hill, D. P., and Pakiser, L. C., 1967, Seismic-refraction study of crustal structure between the Nevada test site and Boise, Iaho: Geological Society of America Bulletin, v. 88, p. 685–704.

Hope, R. A., and Coats, R. R., 1976, Preliminary geologic map of Elko County, Nevada: U.S. Geological Survey Open-File Map 76–779, scale 1:100,000.

Hose, R. K., and Repenning, C. A., 1959, Stratigraphy of Pennsylvanian, Permian, and Lower Triassic rocks of Confusion Range, west-central Utah: American Association of Petroleum Geologists Bulletin, v. 43, p. 2167–2196.

Kerr, J. W., 1962, Paleozoic sequences and thrust slices of the Seetoya Mountains, Independence Range, Elko County, Nevada: Geological Society of America Bulletin, v. 73, p. 439–460.

Ketner, K. B., 1973, Preliminary geologic map of the Coal Mine Basin quadrangle, Elko County, Nevada: U.S. Geological Survey Miscellaneous Field Studies Map MF–1528, scale 1:24,000.

——, 1974, Preliminary geologic map of the Blue Basin quadrangle, Elko County, Nevada: U.S. Geological Survey Miscellaneous Field Studies Map MF–559, scale 1:24,000.

——, 1974, Late Paleozoic orogeny and sedimentation, southern California, Nevada, Idaho, and Montana, *in* Stewart, J. H., Stevens, C. H., and Fritsche, A. E., eds., Paleozoic paleogeography of the western United States: Society of Economic Paleontologists and Mineralogists, Pacific Coast Paleogeography Symposium 1: Pacific Section, Los Angeles, California.

Ketner, K. B., and Smith, J F., Jr., 1982, Mid-Paleozoic age of the Roberts thrust unsettled by new data from northern Nevada: Geology, v. 10, p. 298–303.

Lovejoy, D. W., 1959, Overthrust Ordovician and the Nannie's Peak intrusive, Lone Mountain, Elko County, Nevada: Geological Society of America Bulletin, v. 70, p. 539–564.

Miller, D. M., and Lush, A. P., 1981, Preliminary geologic map of the Pilot Peak and adjacent quadrangles, Elko County, Nevada, and Box Elder County, Utah: U.S. Geological Survey Open-File Report 81–658, scale 1:24,000.

Miller, E. L., Bateson, J., Dinter, David, Dyer, J. R., Harbaugh, Dwight, and Jones, D. L., 1981, Thrust emplacement of the Schoonover sequence, northern Independence Mountains, Nevada: Geological Society of America Bulletin pt. 1, v. 92, p. 730–737.

Mullens, T. E., 1980, Stratigraphy, petrology, and some fossil data of the Roberts Mountains Formation, north-central Nevada: U.S. Geological Survey Professional Paper 1063, 67 p.

Nilsen, T. H., and Stewart, J. H., 1980, Penrose Conference report: The Antler Orogeny-mid-Paleozoic tectonism in western North America: Geology, v. 8, no. 6, p. 298–302.

Nolan, T. B., 1943, The Basin and Range province in Utah, Nevada and California: U.S. Geological Survey Professional Paper 197–D, p. 141–196.

——, 1974, Stratigraphic evidence on the age of the Roberts Mountains thrust, Eureka and White Pine Counties, Nevada: U.S. Geological Survey Journal of Research, v. 2, no. 4, p. 411–416.

Oversby, B. S., 1969, An early Antlerian emplacement of allochthonous rocks in northeastern Nevada [Ph.D. thesis]: New York, Columbia University, 152 p.

——, 1972, Thrust sequences of the Windermere Hills, northeastern Elko County, Nevada: Geological Society of America Bulletin, v. 83, p. 2677–2688.

——, 1973, New Mississippian formation in northeastern Nevada and its possible significance: American Association of Petroleum Geologists Bulletin, v. 57, p. 1779–1783.

Peterson, B. L., 1968, Stratigraphy and structure of the Antelope Peak area, Snake Mountains, Elko County, northeastern Nevada [M.S. thesis]: Eugene, University of Oregon, 78 p.

Pilger, R. H., Jr., 1972, Structural geology of part of the northern Toana Range, Elko County, Nevada [M.S. thesis]: Lincoln, University of Nebraska, 65 p., map scale 1:24,000.

Riva, John, 1962a, Easternmost occurrence of allochthonous Ordovician-Silurian western facies in Nevada [abs.]: Geological Society of America Special Paper 68, p. 253–254.

——, 1962b, Allochthonous Ordovician-Silurian cherts, argillites, and volcanic rocks on Knoll Mountain, Elko County, Nevada [Ph.D. thesis]: New York, Columbia University, 103 p.

——, 1970, Thrusted Paleozoic rocks in the northern and central HD Range, northeastern Nevada: Geological Society of America Bulletin, v. 81, p. 2685–2716.

Roberts, R. J., 1951, Geology of the Antler Peak quadrangle, Nevada: U.S. Geological Survey Geologic Quadrangle Map GQ–10, scale 1:62,500.

——, 1964, Stratigraphy and structure of the Antler Peak quadrangle, Humboldt and Lander Counties, Nevada: U.S. Geological Survey Professional Paper 459–A, 93 p.

Roberts, R. J., and Crittenden, M. D., Jr., 1973, Orogenic mechanisms, Sevier orogenic belt, Nevada and Utah, *in* Dejong, K. A., and Scholten, R., eds., Gravity and tectonics: John Wiley, New York, p. 409–428.

Roberts, R. J., and Thomasson, M. R., 1964, Comparison of late Paleozoic depositional history of northern Nevada and central Idaho, Art.

122, *in* Short papers in geology and hydrology: U.S. Geological Survey Professional Paper 475–D, p. D1–D6.

Roberts, R. J., Crittenden, M. D., Jr., Tooker, E. W., Morris, H. T., Hose, R. K., and Cheney, T. M., 1965, Pennsylvanian and Permian basins in northwestern Utah, northeastern Nevada, and south-central Idaho: American Association of Petroleum Geologists Bulletin, v. 42, no. 11, p. 1926–1956.

Roberts, R. J., Hotz, P. E., Gilluly, James, and Ferguson, H. G., 1958, Paleozoic rocks in north-central Nevada: American Association of Petroleum Geologists Bulletin, v. 42, p. 2813–2857.

Ross, C. P., 1934, Correlation and interpretation of Paleozoic stratigraphy in south-central Idaho: Geological Society of America Bulletin, v. 43, p. 937–1000.

——, 1937, Geology and ore deposits of the Bayhorse region, Custer County, Idaho: U.S. Geological Survey Bulletin 877, 161 p.

Schaeffer, F. E., and Anderson, W. L., 1960, Geology of the central and southern Silver Island Mountains, Box Elder and Tooele Counties, Utah, and Elko County, Nevada: Utah Geological Society Guidebook to the Geology of Utah, no. 15, 185 p.

Sheehan, P. M., 1979, Silurian continental margin in northern Nevada and northwestern Utah: Laramie, University of Wyoming, Contributions to Geology, v. 17, no. 1, p. 25–35.

Silberling, N. J., 1973, Geologic events during Permian-Triassic time along the Pacific Margin of the United States, *in* Logan, A., and Hills, L. V., eds., The Permian and Triassic systems and their mutual boundary: Alberta Society of Petroleum Geologists, Calgary, Alberta, Canada, p. 345–362.

Silberling, N. J., and Roberts, R. J., 1962, Pre-Tertiary stratigraphy and structure of northwestern Nevada: Geological Society of America Special Paper 72, 53 p.

Slack, J. F., 1974, Jurassic suprastructure in the Delano Mountains, northeastern Elko County, Nevada: Geological Society of America Bulletin, v. 85, p. 260–272.

Smith, J F., Jr., and Ketner, K. B., 1977, Tectonic events since early Paleozoic in the Carlin-Pinon Range area, Nevada: U.S. Geological Survey Professional Paper 867–C, 18 p.

Steele, Grant, 1959, Basin and Range structure reflects Paleozoic tectonics and sedimentation [abs.]: Geological Society of America Bulletin, v. 43, no. 5, p. 1105.

——, 1960, Pennsylvanian-Permian stratigraphy of east-central Nevada and adjacent Utah: Intermountain Association of Petroleum Geologists Guidebook, 11th Annual Field Conference, p. 91–113.

Stevens, C. H., 1979, Lower Permian of the central Cordilleran miogeosyncline: Geological Society of America Bulletin, pt. 1, p. 140–142; pt. 2, p. 381–455.

——, 1981, Evaluation of the Wells fault, northeastern Nevada and northwestern Utah: Geology, v. 9, p. 534–537.

Stewart, J. H., 1980, Geology of Nevada: Nevada Bureau of Mines and Geology Special Publication 4, 136 p.

Stewart, J. H., and Poole, F. G., 1974, Lower Paleozoic and uppermost Precambrian, Cordilleran miogeosyncline, Great Basin, Western United States, *in* Dickinson, W. R., ed., Tectonics and sedimentation: Society of Economic Paleontologists and Mineralogists Special Publication 22.

Thorman, C. H., 1968, Mesozoic and Tertiary strike-slip faulting, northeast Nevada: Geological Society of America, Special Paper 121, p. 570–571.

——, 1970, Metamorphosed and nonmetamorphosed Paleozoic rocks in the Wood Hills and Pequop Mountains, northeast Nevada: Geological Society of America Bulletin, v. 81, p. 2417–2448.

Thorman, C. H., and Ketner, K., 1979, West-northwest strike-slip faults and other structures in allochthonous rocks in central and eastern Nevada and western Utah, *in* Newman, G. W., and Goode, H. D., eds., Basin and Range Symposium: Rocky Mountain Association of

Geologists and Utah Geological Association, p. 123–133.

Wrucke, C. T., and Theodore, T. G., 1970, Direction of movement of the Roberts Mountains thrust determined from folds, northern Shoshone Range and Battle Mountain, Nevada [abs.]: Geological Society of America Abstracts with Programs, v. 2, no. 5, p. 356.

Youngquist, Walter, and Haegele, J. R., 1956, Geological reconnaissance of the Cassia Mountain region, Twin Falls and Cassia Counties, Idaho: Idaho Bureau of Mines and Geology Pamphlet 110, 18 p.

MANUSCRIPT ACCEPTED BY THE SOCIETY JANUARY 11, 1983

Typeset by WESType Publishing Services, Inc., Boulder, Colorado
Printed in U.S.A. by Malloy Lithographing, Inc., Ann Arbor, Michigan